Microorganisms for Sustainability

Volume 2

Series Editor

Naveen Kumar Arora, Environmental Microbiology, School for Environmental Science, Babasaheb Bhimrao Ambedkar University, Lucknow, Uttar Pradesh, India

Microorganisms perform diverse roles on our planet most of which are important to make earth a habitable and sustainable ecosystem. Many properties of microorganisms are being utilized as low input biotechnology to solve various problems related to the environment, food security, nutrition, biodegradation, bioremediation, sustainable agriculture, bioenergy and biofuel, bio-based industries including microbial enzymes/ extremozymes, probiotics etc. The book series covers all the wider aspects and unravels the role of microbes towards achieving a sustainable world. It focuses on various microbial technologies related to sustenance of ecosystems and achieving targets of Sustainable Development Goals. Series brings together content on microbe based technologies for replacing harmful chemicals in agriculture, green alternatives to fossil fuels, use of microorganisms for reclamation of wastelands/ stress affected regions, bioremediation of contaminated habitats, biodegradation purposes. Volumes in the series also focus on the use of microbes for various industrial purposes including enzymes, extremophilic microbes and enzymes, effluent treatment, food products.

The book series is a peer reviewed compendium focused on bringing up contemporary themes related to microbial technology from all parts of the world, at one place for its readers, thereby ascertaining the crucial role of microbes in sustaining the ecosystems.

More information about this series at http://www.springer.com/series/14379

Naheed Mojgani • Maryam Dadar
Editors

Probiotic Bacteria and Postbiotic Metabolites: Role in Animal and Human Health

Springer

Editors
Naheed Mojgani
Agriculture Research, Education and Extension Organization (AREEO)
Razi Vaccine and Serum Research Institute (RVSRI)
Karaj, Iran

Maryam Dadar
Agriculture Research, Education and Extension Organization (AREEO)
Razi Vaccine and Serum Research Institute (RVSRI)
Karaj, Iran

ISSN 2512-1901 ISSN 2512-1898 (electronic)
Microorganisms for Sustainability
ISBN 978-981-16-0222-1 ISBN 978-981-16-0223-8 (eBook)
https://doi.org/10.1007/978-981-16-0223-8

This Springer imprint is published by the registered company Springer Nature Singapore Pte Ltd.
The registered company address is: 152 Beach Road, #21-01/04 Gateway East, Singapore 189721, Singapore

To all those who have love for science and realize the importance of microbes in healthy environment and healthy life

Preface

Since centuries, our ancestors employed microbes for preparing several trad fermented products that at the time were considered health promoting. Altho that time the presence or importance of these microbes was not much realiz with the passage of time the vital role played by these microorganisms evolution of life on earth and in maintenance of environmental health a lives was highly acknowledged.

With the knowledge of integrative medicine, we have understood the relationship existing between heath and diet. Hence, the significance of pr microbes, which exist mainly in our gut, has become the spotlight and fore vast research, gasping the attention of academics, students, industrialists, ar importantly consumers.

In this book, we have focused on the health aspects of probiotic micro their postbiotic metabolites, either added as an adjunct to food (dairy or r foods) or in animal feed or as a supplement for human use with some health Several metabolites produced by lactic acid bacteria have been known to the flavor, texture, and sensory characteristics of fermented foods with health impacts on the host when consumed in adequate amounts. The b reviews the use of these microbes having metabolic functions, for the deve of novel functional foods and functional pharmaceuticals. This book covers aspects of probiotic bacteria and their metabolites in terms of their therap technological applications and benefits. We had put our maximum efforts to current knowledge and a holistic review of the related topics considering tl use in plants, animals, and man, and it is our hope that the provided inf compiled by the expert authors who contributed in this book would prove cant contribution to the expanding knowledge of probiotics and postbiotic

In the end, I highly acknowledge and appreciate the contributing author for sharing their knowledge and expertise but also for their high co-operations, patience. I express my sincere gratitude to all those who their time and energy in preparing these chapters.

Karaj, Iran — Nahee
Karaj, Iran — Mary

Contents

To all those who have love for science and realize the importance of microbes in healthy environment and healthy life

Preface

Since centuries, our ancestors employed microbes for preparing several traditional fermented products that at the time were considered health promoting. Although at that time the presence or importance of these microbes was not much realized, but with the passage of time the vital role played by these microorganisms in the evolution of life on earth and in maintenance of environmental health and our lives was highly acknowledged.

With the knowledge of integrative medicine, we have understood the direct relationship existing between heath and diet. Hence, the significance of probiotic microbes, which exist mainly in our gut, has become the spotlight and forefront of vast research, gasping the attention of academics, students, industrialists, and most importantly consumers.

In this book, we have focused on the health aspects of probiotic microbes and their postbiotic metabolites, either added as an adjunct to food (dairy or nondairy foods) or in animal feed or as a supplement for human use with some health claims. Several metabolites produced by lactic acid bacteria have been known to improve the flavor, texture, and sensory characteristics of fermented foods with positive health impacts on the host when consumed in adequate amounts. The book also reviews the use of these microbes having metabolic functions, for the development of novel functional foods and functional pharmaceuticals. This book covers different aspects of probiotic bacteria and their metabolites in terms of their therapeutic and technological applications and benefits. We had put our maximum efforts to provide current knowledge and a holistic review of the related topics considering their wide use in plants, animals, and man, and it is our hope that the provided information compiled by the expert authors who contributed in this book would prove a significant contribution to the expanding knowledge of probiotics and postbiotics.

In the end, I highly acknowledge and appreciate the contributing authors not only for sharing their knowledge and expertise but also for their high sense of co-operations, patience. I express my sincere gratitude to all those who dedicated their time and energy in preparing these chapters.

Karaj, Iran — Naheed Mojgani
Karaj, Iran — Maryam Dadar

Contents

About the Series Editor

Naveen Kumar Arora, Ph.D. in microbiology, associate professor in the Department of Environmental Microbiology, Babasaheb Bhimrao Ambedkar University (a central university), Lucknow, Uttar Pradesh, India, is a renowned researcher in the field of environmental microbiology and biotechnology. His specific area of research is rhizosphere biology and PGPRs. He has more than 50 research papers published in premium international journals and several articles published in magazines and dailies. He is editor of three books, published by Springer. He is member of several national and international societies and reviewer of several international journals. He has delivered lectures in conferences and seminars around the globe. He has a long-standing interest in teaching at the PG level and is involved in taking courses in bacteriology, microbial physiology, environmental microbiology, agriculture microbiology, and industrial microbiology. He has been advisor to 57 postgraduate and 8 doctoral students. Recently, he was awarded for excellence in research by the honorable governor of Uttar Pradesh. Although an academician and researcher by profession, he has a huge obsession for the wildlife and its conservation and has authored a book, *Splendid Wilds*. He has a dedicated website www.naveenarora.co.in for the cause of wildlife and environment conservation.

About the Editors

Naheed Mojgani, Ph.D., professor of microbiology at Razi Vaccine and Serum Research Institute, Iran, is renowned for her research work in the field of bacterial infectious diseases, probiotics, paraprobiotics, and postbiotics, conjugate and recombinant vaccines, etc. She has formulated and commercialized several probiotic supplements for man and animals. Her continuous efforts, dedication, and hard work in the field of probiotic have led her to be recognized in the country as an eminent specialist in the field. Recently, she has been awarded by the Ministry of Education and Ministry of Agriculture for her novel approach providing efficient therapy for diarrheal diseases. Additionally, she is committee member at Iranian National Standard Organization and National Veterinary Organization for writing and setting standards for probiotic products.

Dr Mojgani has authored several books in Persian language, including Probiotics in Poultry Industry, Technology of Chicken Egg Yolk Antibodies, Peptides and Proteins of Lactic Acid Bacteria; Bacteriocins, Probiogenics, and Probiotics and Postbiotics in Dairy and Pharma Industry.

Maryam Dadar is an assistant professor of molecular biology at Razi Vaccine and Serum Research Institute, Karaj, Iran. She received her D.V.M. and Ph.D. from Shahid Chamran University, Ahvaz, Iran, and earned her spot as the supervising researcher at the Centre for Biology and Biotechnology Research at Shahid Chamran University, Ahvaz, Iran. Her research interests mainly focus on improving the understanding, design, and performance of vaccine through the application of recombinant proteins. She has also investigated the gene expression of immune system under different experimental conditions, such as the use of immunostimulants, probiotics, and specific vaccine adjuvants. She is the author of more than 120 articles in recognized peer-reviewed journals in the field of veterinary microbiology, biomedicine, and public health.

Bacillus spp. in Aquaculture - Mechanisms and Applications: An Update View

1

Hien Van Doan

Abstract

Probiotics have been widely applied in aquaculture industry as sustainable and environmentally friendly tools to sustain host's health and the well-being. Among probiotics, *Bacillus* species have great potential applications in aquaculture because they can form the spores that makes them able to survive in the harsh environmental conditions. Moreover, they are nonpathogenic and nontoxic to aquacultural environments and animals. In addition, *Bacillus* species are able to produce antimicrobial substances making them more suitable candidates compared to other probiotics. In this chapter, we discussed the role of *Bacillus* in sustainable aquaculture as alternative strategies to enhance growth performance, disease resistance, and immune response of different aquaculture farmed animals.

Keywords

Bacillus · Aquaculture · Probiotics · Disease resistance

1.1 Introduction

Aquaculture is one of the world's fastest growing food sectors (Willer and Aldridge 2019). It is necessary to meet the global seafood demand, which is being accountable for 50% of the world's seafood consumption (Gómez et al. 2019). However, sustainable development of aquaculture industry is constantly defeated by the outbreak of diseases, which is considered as main obstacles to the economical

H. Van Doan (✉)
Faculty of Agriculture, Department of Animal and Aquatic Sciences, Chiang Mai University, Chiang Mai, Thailand

N. Mojgani, M. Dadar (eds.), *Probiotic Bacteria and Postbiotic Metabolites: Role in Animal and Human Health*, Microorganisms for Sustainability 2,
https://doi.org/10.1007/978-981-16-0223-8_1

profitability of the industry (FAO 2020). The outbreak of diseases is also linked to application of antibiotics, posing a significant danger to the public health (World Health Organization 2014). Thus, new and natural alternatives that prohibit the incidence of diseases and improve human and animal health are urgently needed. The use of probiotics, "live organisms that can give a health benefit to the host when administered in the appropriate amounts," is a potential alternative to boost the global health (FAO/WHO 2001). The scientific community has been searching for the environmentally friendly solutions to prevent aquacultural disease, where probiotics emerged as crucial alternative to antibiotics due to advert effects of antibiotics, such as the modulation of microbiota in the aquaculture systems and the development of resistance bacteria (Kuebutornye et al. 2019; Resende et al. 2012; Ringø 2020; Wang et al. 2019a). Consequently, wide range of probiotics, such as *Bacillus, Enterococcus, Lactobacillus, Lactococcus, Micrococcus, Pediococcus, Enterobacter, Vibrio, Pseudomonas, Rhodopseudomonas, Roseobacter*, and *Shewanella*, have been found and applied to improve growth performance, immune response, and disease resistance of farmed fish and shellfish (Abd El-Rhman et al. 2009; Adel et al. 2017; Feng et al. 2019; Kuebutornye et al. 2019; Li et al. 2006; Li et al. 2020; Ringø 2020; Yang et al. 2019). In aquaculture, probiotics have been applied as functional feed additives to boost host's health and well-being via increasing growth, supplying nutrient, modulation gut microbiota, enhancing immunity, improving feed efficiency, increasing digestive enzyme activities and digestibility, and controlling diseases (Kuebutornye et al. 2019; Ringø 2020; Selim and Reda 2015).

Bacillus species are one of the most commonly used probiotics in the aquaculture industry because of their ability to form endospores, which is a benefit for industrial applications without losing their characteristics (Hong et al. 2005; Kuebutornye et al. 2019; Cutting 2011; Hai 2015). In addition, *Bacillus* is known to generate natural antimicrobial compounds, which are able to prohibit the proliferation of harmful bacteria in the aquaculture systems and host's intestines (Abriouel et al. 2011; Caulier et al. 2019; Sumi et al. 2015). Similarly, *Bacillus* species are known to stimulate the digestive enzymes, antioxidant enzymes, relative immune gene expression, and stress-related genes, which in turn improve disease resistance of the host against pathogenic bacteria (Elshaghabee et al. 2017; Nayak 2010; Soltani et al. 2019). *Bacillus* species also increase the use of feed in fish, contributing to better growth rates (Mukherjee et al. 2019; Nair et al. 2020; Xia et al. 2020). Therefore, these chapters gather recent data on the role of *Bacillus* species in promoting growth performance, disease resistance, and immune response in aquaculture.

1.2 Mode of Action of Probiotics in Aquaculture

Probiotics can affect the host's immune responses, as well as the interrelationship between these responses and their gastrointestinal microflora (Hemarajata and Versalovic 2013; La Fata et al. 2018; Yan and Polk 2011). Over the past decades, extensive researches on probiotics have provided insight into the significance of probiotics and their modes of action and numerous mechanisms have been suggested (Santacroce et al. 2019; Shi et al. 2016).

1. Probiotics improve feed efficiency and growth rate of farmed fish and shellfish (Ringø 2020; Romano 2021). They also enhance the host's appetite and feed digestion via decomposition of indigestible components, enhance vitamin productions, and detox diet's substances (Ashaolu 2020; Cencic and Chingwaru 2010; Hoseinifar et al. 2018).
2. Probiotics could compete the exclusion of gastrointestinal harmful bacteria via the secretion of peroxide, bacteriocin, siderophore, and lysozyme enzymes (Vieco-Saiz et al. 2019; Yang et al. 2014). The physiological and immunological effects are considered as one of the most essential modes of action of probiotics (Klaenhammer et al. 2012; Plaza-Diaz et al. 2019; Vieco-Saiz et al. 2019).
3. Probiotics could enhance aquaculture animal's disease resistance to stress caused by various environmental threats during aquaculture activities (Hlordzi et al. 2020; Mohapatra et al. 2013; Reverter et al. 2020).

These mechanisms display the favorable impacts of probiotics in farmed fish and shellfish. Future studies, however, on the relationship between probiotics and hosts, including metagenomics and proteomic studies, is important to clarify mode of action of probiotics.

1.3 *Bacillus* Applications in Aquaculture

1.3.1 Improve Growth Performance

The utmost target of aquaculture practice is to acquire the rapidest growth and lowest production cost. To achieve this goal, several means have been established to boost growth rate and feed consumption by adding functional feed additives and growth natural growth promoters (Hernández et al. 2016; Katya et al. 2014). Probiotics are potential tools to maintain the normal growth, health, and well-being of farmed fish and shellfish because they serve as nutrients source, vitamins, and digestive enzymes. These substances for their part will contribute significantly on feed consumption, nutrients uptake, and host's growth rate (Lauriano et al. 2016; Nath et al. 2019). Probiotics consumption have been speculated to improve the host's appetite or boost organisms' digestibility (Irianto and Austin 2002). Probiotics can improve feed efficiency of fish and shellfish by stimulating the excretion of digestive enzymes and maintaining the balance of intestinal microbes, which lead to the improvement of nutrients absorption and utilization, as well as the survival and growth of the host (Ibrahem 2015; Irianto and Austin 2002). Studies on diets containing probiotics revealed the possible involvement of these probiotics on the improvement of intestinal microflora balance and the production of extracellular enzymes to elevate the feed efficiency and growth of cultured species as growth promoters (Giri et al. 2013; Ringø et al. 2018). Most of the studies using *Bacillus* in aquaculture focus on growth performance and survival rate (Table 1.1).

1.3.1.1 Tilapias

During past decades, *Bacillus* spp. have been intensively applied in Nile tilapia aquaculture. Han et al. (2015) indicated that 10 weeks feeding trial with *B. licheniformis* significantly enhanced growth performance. However, there were no significant discrepancies in survival rate and feed conversion ratio (FCR) and in villi length and muscular layer thickness of anterior intestine among the treatments. In contrast, Iwashita et al. (2015) reveal that administration of the probiotic had no significant effect on the growth rates of Nile tilapias, although the fish fed probiotics had better feed conversion. Likewise, no significant difference in growth performance and FCR was observed in Nile tilapia fed *Bacillus amyloliquefaciens* (Silva et al. 2015). This can be explained due to the low temperatures during experimental period. Marcusso et al. (2015) reported that the homeostasis of Nile tilapia rearing at temperatures below 24 °C could be affected, enhancing the susceptibility to bacterial infections and impairing the growth performance. No effects were observed on the growth performance of Nile tilapia fed *Bacillus subtilis* (Aqua NZ and AP193) and *Bacillus subtilis* strains (Addo et al. 2017a, 2017b). These results are not unexpected given the short duration of this trial. This statement agrees with Apún-Molina et al. (2009) who observed a tendency toward improved growth in Nile tilapia fry (0.14 g) only after 75 d of feeding with diets composed of *Bacillus* or *Lactobacillus* probiotics. On the contrary, dietary inclusion of *Bacillus subtilis* significantly improved body weight, percent weight gain, specific growth rate, and feed conversion ratio (Liu et al. 2017). It is well documented that *Bacillus* exoenzymes are very efficient at metabolizing a large variety of carbohydrate, lipids, and proteins (Liu et al. 2009). The exoenzymatic activity of *Bacillus* spp. is one of the main reasons for its ability to improve digestive enzyme activities (Han et al. 2015). Higher enzyme activities in the digestive tract enhance digestive capability and growth performance of the host. It is widely accepted that the level of digestive enzyme activity is a useful comparative indicator of food utilization rate, digestive capacity, and growth performance of the host (Suzer et al. 2008; Ueberschär 1995). Liu et al. (2017) also reported that 4-week *B. subtilis* HAINUP40 diet supplementation significantly increased protease and amylase activities of tilapia. This is because *B. subtilis* HAINUP40 could secrete exoenzymes; the improvement of indigestive tract enzyme activities may be partially due to enzymes synthesized by the bacteria. However, the proportion of enzymes contributed by bacteria cannot be assessed since the probiotic may also stimulate the production of endogenous enzymes in the fish (Dawood et al. 2016; Suzer et al. 2008; Wu et al. 2012; Ziaei-Nejad et al. 2006). In the same trend, supplementation of *B. subtilis* and *B. licheniformis* or *B. subtilis* and *Bacillus licheniformis* (BS) combined with traditional Chinese medicine (TCM) significantly enhanced weight gain and specific growth rate of Nile tilapia and Mozambique tilapia (Abarike et al. 2018b; Abarike et al. 2018a; Gobi et al. 2018). It is known that an increase in the body weight gain in fish fed with probiotic supplemented diets, could contribute to the increase in digestive enzyme activity, increase in

Table 1.1 Weight gain (WG), specific growth rate (SGR), food conversion efficiency (FCE), food conversion ratio (FCR), protein efficiency ratio (PER), survival rate (SR), digestive enzyme, and disease resistance of fish and shellfish fed *Bacillus* probiotics. → no change, ↑ increase, ↓ decrease

Species	Source	Doses and duration	Fish species	Parameters investigated	References
Bacillus licheniformis	Commercial probiotic	0%, 0.02%, 0.04%, 0.06%, 0.08% and 0.1% containing live germ 2×10^{10} CFU/g) 10 weeks	Juvenile Nile tilapia (*O. niloticus*) 3.83 ± 0.03 g	WG, FBW, SGR, and SR ↑ FCR → Villi length → Muscular layer thickness of anterior intestinal → Resistance against to *S. iniae* ↑	Han et al. (2015)
Bacillus subtilis (combined *with S. cerevisiae* and *A. oryzae)*	Commercial probiotic	0; 5 g kg^{-1} probiotic mixture (*B. subtilis* 1.5×10^9, *S. cerevisiae* 10^9 and *A. oryzae* 2×10^9); and 10 g kg^{-1} probiotic mixture (*B. subtilis* 3.0×10^9, *S. cerevisiae* 2.0×10^9 and *A. oryzae* 4.0×10^9) (CFU g^{-1}) 6 weeks	Juvenile Nile tilapia (*O. niloticus*) 25 ± 0.05 g	Growth rates → Resistance against to *A. hydrophila* and *S. iniae* ↑	Iwashita et al. (2015)
Bacillus amyloliquefaciens	Commercial probiotic	0; 1×10^6; 5×10^6 and 1×10^7 CFU g^{-1} 90 days	Nile tilapia (*O. niloticus*) 35 ± 5 g	Growth performance → Proximal composition → Blood glucose and hemoglobin ↓ Villi height and number of goblet cells ↑	Silva et al. (2015)

(continued)

Table 1.1 (continued)

Species	Source	Doses and duration	Fish species	Parameters investigated	References
Bacillus NP5	Commercial probiotic	10^{10} CFU g^{-1} in feed with dose of 0.5, 1 and 2% 28 days	Nile tilapia (*O. niloticus*) 6.38 ± 0.05 g	Growth performance ↑	Utami and Suprayudi (2015)
Bacillus subtilis (Aqua NZ and AP193)	Commercial probiotic	4.2×10^7 CFU g^{-1} of feed 8 weeks	Nile tilapia (*O. niloticus*) 7.47 ± 0.11 g	WG and FCR → Thermal growth coefficient → Resistance against *A. hydrophila* ↑	Addo et al. (2017a)
Bacillus subtilis strains SB3086, SB3295, SB3615, and AP193	Commercial probiotic	4×10^7 CFU/g of feed in 21 days	Nile tilapia (*O. niloticus*) 16.5 ± 0.2 g	Growth performance → Resistance against *Streptococcus agalactiae* ↑	Addo et al. (2017b)
Bacillus subtilis HAINUP40	Isolated from the aquatic environment	10^8 CFU/g 8 weeks	Nile tilapia (*O. niloticus*) 95 ± 8 g	FW, WG, and SGR ↑ FCR ↓ Protease and amylase activity ↑ Total antioxidant capacity (T-AOC) ↑ Serum superoxide dismutase (SOD) ↑ Resistance against *S. agalactiae* ↑	Liu et al. (2017)
B. subtilis and *B. licheniformis*	Commercial probiotic	0, 3, 5, 7 and 10 g kg^{-1} 4 weeks	Nile tilapia (*O. niloticus*) 53.01 ± 1.0 g	WG and SGR ↑ FCR ↓ Resistance against *S. agalactiae* ↑	Abarike et al. (2018a)

Bacillus subtilis and *Bacillus licheniformis* (BS) combined with traditional Chinese medicine (TCM)	Commercial probiotic	0; TCM at 3 and BS at 7 (g/kg); TCM at 5 and BS at 5 (g/kg); TCM at 7 and BS at 3 (g/kg). 4 weeks	Nile tilapia (*O. niloticus*) 57 ± 2 g	WG and SGR ↑ FCR ↓ Resistance against *S. agalactiae* ↑	Abarike et al. (2018a)
Bacillus licheniformis Dahb1	Commercial probiotic	0, 10^5 and 10^7 CFU/g 4 weeks	Mozambique tilapia (*Oreochromis mossambicus*)	FW and SGR ↑ FCR ↓ Resistance against *A. hydrophila* ↑	Gobi et al. (2018)
Bacillus licheniformis and *B. subtilis*	Commercial probiotic	1×10^6 CFU mL^{-1} was feed for *Artemia urmiana* nauplii and *Brachionus plicatilis* 8 h	Pacific white shrimp larvae (*Litopenaeus vannamei*)	Growth performance ↑ Survival rate ↑	Jamali et al. (2015)
Bacillus licheniformis and *Lactobacillus rhamnosus*	Commercial probiotic	*B. licheniformis* 10^9 CFU/kg, *L. rhamnosus* 8×10^8 CFU/kg 120 days	Pacific white shrimp (*L. vannamei*) PL14	WG, SGR, and WG ↑	Swapna et al. (2015)
Bacillus spp.	Isolated from pustulose ark	1×10^6, 2×10^6, 4×10^6, and 6×10^6 CFU g $feed^{-1}$ 32 days	Pacific white shrimp (*L. vannamei*) 1 ± 0.1 g	Growth performance ↑	Sánchez-Ortiz et al. (2016)
Bacillus coagulans ATCC 7050	Commercial probiotic	(0 (BO), 1×10^6 (BC1), 1×10^7 (BC2) and 1×10^8 (BC3) CFU g^{-1} feed) 56 days	Pacific white shrimp larvae (*L. vannamei*) 0.57 ± 0.001	FW, WG, and SGR ↑ FCR ↓ Condition factor ↑ Lipase, amylase, and trypsin ↑ Villus height ↑ Villus width ↑ Muscle thickness ↑	Amoah et al. (2019)

(continued)

Table 1.1 (continued)

Species	Source	Doses and duration	Fish species	Parameters investigated	References
Bacillus subtilis and *Bacillus licheniformis*	Commercial probiotic	1×10^4 and 1×10^8 CFU/g 60 days	Pacific white shrimp (*Litopenaeus vannamei*)	WG, total length, SGR, and SR ↑ Dry matter and crude protein ↑	Sadat Hoseini Madani et al. (2018)
Bacillus spp.	Isolate from shrimp gut	Added to water: 1×10^2, 1×10^3, 1×10^4 and 1×10^5 CFU/mL			Kewcharoen and Srisapoome (2019)
B. licheniformis	Commercial probiotic	Control diet (CON), 0.2% MOS (MOS), 0.1% *B. licheniformis* (BL), 0.2% MOS plus 0.1% BL (SYN) 8 weeks	Pacific white shrimp (*Litopenaeus vannamei*)	WG, SGR, PER ↑ Villus number (VN) ↑ Villus height (VH) ↑ Thicker submucosa ↑ Propionic acid content ↑ Resistance against ammonia ↑	Chen et al. (2020a)
B. licheniformis		Basal diet (Control); 0.5% hydrolyzed yeast (HY); 0.1% *B. licheniformis* (BL) and 0.5% hydrolyzed yeast +0.1% *B. licheniformis* (SYN) 8 weeks	Pacific white shrimp (*Litopenaeus vannamei*	Growth and body composition → PER ↑ Intestinal villus height ↑ Villus number (VN) ↑ Villus height (VH) ↑	Chen et al. (2020b)
Bacillus cereus and *Pediococcus acidilactici*	Commercial probiotic	*P. acidilactici* (106 CFU/mL) and *B. cereus* (106 CFU/mL) to the water pond 110 days	Pacific white shrimp (*Litopenaeus vannamei*) 0.002 ± 0.001	WG and SR ↑	Khademzade et al. (2020)

Bacillus subtilis WB60, Pediococcus pentosaceus, and *Lactococcus lactis*	Commercial probiotic	*B. subtilis* at 10^7 CFU/g diet, *B. subtilis*, *P. pentosaceus*, and *L. lactis* at 10^8 CFU/g diet, and oxytetracycline at 4 g/kg 8 weeks	Pacific white shrimp (*Litopenaeus vannamei*) 1.41 ± 0.05	Growth performance ↑ Resistance against *Vibrio parahaemolyticus* ↑	Won et al. (2020a, 2020b)
Bacillus subtilis WB60 and *Lactobacillus plantarum* KCTC3928	Commercial probiotic	0; *B. subtilis* at 10^6, 10^7, 10^8 and *L. plantarum* at 10^6, 10^7, 10^8 CFU/g diet	Japanese eel (*Anguilla japonica*) 8.29 ± 0.06 g	WG, FE, and PER ↑ Resistance against *V. anguillarum* ↑	Lee et al. (2017)
Bacillus subtilis WB60 and mannanoligosaccharide (MOS)	Commercial probiotic	BS: 0.0, 0.5, and 1.0 × 107 CFU/g diet and MOS: 0 and 5 g/kg diet 8 weeks	Japanese eel (*Anguilla japonica*) 9.00 ± 0.11 g	WG, FW, SGR and PER ↑ Resistance against *Vibrio anguillarum* ↑	Lee et al. (2018)
Bacillus subtilis or *licheniformis*) and (mannan or fructo oligosaccharide)		0, Probiotics (1.0 × 10^8 CFU/g diet) and prebiotics (5 g/kg diet) 12 weeks	Japanese eel (*Anguilla japonica*) 12.8 ± 0.47	WG and SGR ↑ Intestinal villi length ↑ Resistance against *A. hydrophila* ↑	Park et al. (2020)
B. megaterium PTB 1.4	Commercial probiotic	0 and 1% 30 days	Catfish (*Clarias* sp.) 11.41 ± 0.23 g	Growth performance ↑ Protease and amylase enzymes ↑ Total amount of probiotic bacteria ↑	Afrilasari and Meryandini (2016)
B. subtilis, B. amyloliquefaciens, B. cereus and a commercial B. amyloliquefaciens	Isolated from the intestine of African catfish	10^{10} CFU/ml 60 days	African catfish (*Clarias gariepinus*) 75.23 ± 1.6	BW, WG, and SGR ↑ FCR ↓ Hemogram blood parameters ↑ Serum antioxidant and digestive enzymes ↑ Resistance against *Aeromonas sobria* ↑	Reda et al. (2018)

(continued)

Table 1.1 (continued)

Species	Source	Doses and duration	Fish species	Parameters investigated	References
Bacillus cereus EN25	Isolated from mud of sea cucumber culturing water bodies.	0, 10^5, 10^7 and 10^9 CFU/g 30 days	Juvenile sea cucumber (*Apostichopus japonicus*) 0.375 ± 0.024 g	Growth performance → Resistance against *V. splendidus* ↑	Zhao et al. (2016)
Bacillus baekryungensis MS1	Isolated from a sea cucumber pond in winter	0 and 10^7 CFU/ml 60 days	Sea cucumber (*Apostichopus japonicus*) 4.17 g ± 0.22 g	Growth performance ↑ Resistance to *Vibrio splendidus* ↑	Liu et al. (2020)
Bacillus subtilis and *Saccharomyces cerevisiae*	Commercial probiotic	*Bacillus subtilis* 10^9 UFC/g and *Saccharomyces cerevisiae* 10^9 UFC/g 90 days	Tambaqui (*Colossoma macropomum*) 2.13 ± 0.75 g	Growth performance → Body composition → Hematological parameters ↑ Resistance against *S. agalactiae* ↑	da Paixão et al. (2017)
Bacillus cereus	Commercial probiotic	0, 4.2×10^4, 3.9×10^6 and 3.3×10^8 CFU/g 120 days	Tambaqui (*Colossoma macropomum*) 0.94 ± 0.02 g	Weight and length gains ↑ Neutrophils and thrombocyte count ↑ Resistance against *Aeromonas hydrophila* ↑	Dias et al. (2018)
B. licheniformis and *B. subtilis*	Commercial probiotic	1.6×10^9 CFU/g dry pellet in 60 days	Kutum (*Rutilus frisii*) 0.4 ± 0.1 g	FW, WG, and SGR ↑ Red blood cells ↑ White blood cells ↑ Neutrophils ↑ Lymphocytes ↑ Mean cell volume ↑ Mean cell hemoglobin ↑ Mean cell hemoglobin concentration ↑	Azarin et al. (2015)

Bacillus sp. PP9	Isolated from mrigal gut	2×10^4, 2×10^5 and 2×10^6 CFU 60 days	Mrigal (*Cirrhinus mrigala*) 2.5 ± 0.20	Growth performance ↑ Maximum RNA DNA ratio ↑ FCR ↓ Intestinal protease and α-amylase activity ↑ Hepatic glutamic oxaloacetic transaminase ↑ Glutamate pyruvate transaminase levels ↑	Bandyopadhyay et al. (2015)
B. subtilis and *B. circulans*	Commercial probiotic	1×10^4, 2×10^4, 3×10^4, and 4×10^4 CFU/g) 30 days	Three spot gourami (*Trichopodus trichopterus*)	Larval growth → Larval resistance against the challenge ↑	Jafariyan et al. (2015)
Bacillus coagulant	Commercial probiotic	0, 10^5, 10^7 and 10^9 cfu g^{-1} 60 days	Freshwater prawn (*Macrobrachium rosenbergii*) 2.4 ± 0.35 g	Growth performance ↑ Feed utilization ↑ Protease, amylase, and lipase digestive enzymes ↑	Gupta et al. (2016)
Virgibacillus proomii and *Bacillus mojavensis*		5.8×10^4, 9.6×10^4, and 9.8×10^4 CFU/ml 60 days	Sea bass (*Dicentrarchus labrax*) larvae	Length, weight, and the survival rate ↑ FCR ↓ Phosphatase alkaline and amylase activities ↑	Hamza et al. (2016)
Bacillus spp.	Commercial probiotic	The live rotifers were enriched with: Algamac 3050; Algamac 3050 and a commercial mix of *Bacillus* spp. 0.5 g L^{-1}; additional probiotics in water (5 g m^{-3})	Florida pompano (*Trachinotus carolinus*) Larvae	Growth performance ↑ Survival rate → Trypsin-specific activity ↑ Alkaline phosphatase activity ↑	Hauville et al. (2016)

(continued)

Table 1.1 (continued)

Species	Source	Doses and duration	Fish species	Parameters investigated	References
Bacillus pumilus SE5		0 and 1.0 × 10^8 CFU g^{-1} 60 days	Grouper (*Epinephelus coioides*) 14.6 ± 0.2 g	FW, WG, and SGR ↑ FCR ↓	Yan et al. (2016)
Bacillus mycoides (BS) and organic selenium (OS)		BS 10^8 CFU g^{-1}, OS 0.2 g kg^{-1} and combination BS, OS	Marron (*Cherax cainii*) 10.83 ± 0.28 g	Growth performance ↑ The glutathione peroxidase ↑ Total hemocyte counts ↑ Intestinal bacterial population ↑	Ambas et al. (2017)
Bacillus amyloliquefaciens-JFP2		0 and 1.4 × 10^6 (CFU/g) of feed 90 days	Rock Bream (*Oplegnathus fasciatus*) 25.4 ± 0.13 g	BW, WG, and SGR ↑ FCR ↓ Serum protein and glucose level ↑ Resistance against *Streptococcus iniae* ↑	Kim et al. (2017)
B. siamensis B44v	Isolated from Thai pickled vegetables (Phak-dong)	10^7 CFU/g feed	Hybrid catfish (*C. macrocephalus* × *C. gariepinus*)	Protease and cellulase enzymes ↑ Gastrointestinal conditions ↑ Improve growth ↑ Resistance against to *A. hydrophila* and *S. iniae* ↑	Meidong et al. (2017)
Bacillus aerophilus KADR3	Commercial probiotic	0, 10^7, 10^8 and 10^9 CFU g^{-1} 6 weeks	Rohita labeo (*Labeo rohita*) 35–40 g	Serum lysozyme activity ↑ Phagocytic activity ↑ Serum total protein, Respiratory burst activity ↑	Ramesh et al. (2015)

				Serum IgM levels ↑ Superoxide dismutase activity ↑ Alternative complement pathway activity ↑ Resistance against *A. hydrophila* ↑	
Bacillus subtilis and *Bacillus cereus toyoi*	Commercial probiotic	0, 6×10^3 and 1.5×10^6 CFU g^{-1} of diet. 9 and 20 weeks	Rainbow trout (*Oncorhynchus mykiss*) and brown trout (*Salmo trutta*) 15.6 g	Growth performance → Body composition → Intestinal *lamina propria* ↑ Submucosa ↑	Ramos et al. (2017)
B. amyloliquefaciens 54A and B. *pumilus* 47B	Isolated from gut of striped catfish	1×10^8, 3×10^8, and 5×10^8 CFU g^{-1} feed 90 days	Striped catfish (*Pangasianodon hypophthalmus*)	WG ↑ SGR and FCR → Resistance against to *E. ictaluri* ↑	Truong Thy et al. (2017)
Bacillus sp. DDKRC1	Isolated from the gut of Asian seabass (*Lates calcarifer*)	0, 2.94×10^7 CFU/100 g feed and diet fermented with *Bacillus* sp. DDKRC1 42 days	Tiger shrimp (*Penaeus monodon*) 2.73 ± 0.01	PER ↑ FCR ↓ Dry matter and cellulose digestibility ↑ Hemicellulose and lipid digestibility ↑ Cellulase, amylase, and protease activities ↑	De et al. (2018)
Bacillus amyloliquefaciens (GB) and Yarrowia lipolytica lipase 2 (YLL2)	Commercial probiotic	0, 5.0 g/kg GB-9, 4.0 g/kg YLL2, and 5.0 g/kg GB-9 + 4.0 g/kg YLL2 12 weeks	Hybrid sturgeon (*Acipenser schrenckii* ♂and *Acipenser baerii* ♀) 5.0 g	Final weight ↑ Docosahexaenoic acid (DHA) ↑ Eicosapentaenoic acid (EPA) concentration ↑	Fei et al. (2018)

(continued)

Table 1.1 (continued)

Species	Source	Doses and duration	Fish species	Parameters investigated	References
Bacillus licheniformis	Commercial probiotic	10^3, 10^5, and 10^7 CFU/mL 8 weeks	Abalone (*Haliotis discus* hannai Ino.) 4.17 ± 0.32 g	SGR and FI ↑ FCR ↓ Resistance to *V. parahaemolyticus* ↑	Gao et al. (2018)
B. subtilis E20	Commercial probiotic	0, 10^8, 10^9, and 10^{10} CFU kg^{-1} 56 days	Parrotfish (*Oplegnathus fasciatus*)	Growth performance ↑ Resistance against *V. alginolyticus* ↑	Liu et al. (2018)
Bacillus aerius B81e	Isolated from healthy hybrid catfish	0 and 10^7 CFU g^{-1} feed 60 days	Basa fish (*Pangasius bocourti*) 69 g	WG and SGR ↑ FCR ↓ Resistance against *A. hydrophila* ↑	Meidong et al. (2018)
Bacillus subtilis	Commercial probiotic	0, 1×10^4, 1×10^6, 1×10^8 and 1×10^{10} CFU kg^{-1} diet 60 days	Red sea bream (*Pagrus major*) 3.99 ± 0.01	FW, WG, and SGR ↑ Feed utilization (FI, FCE, PER and PG) ↑ Amylase, protease, and lipase enzymes ↑	Zaineldin et al. (2018)
BetaPlus® (*B. subtilis* (DSM 5750) *B. licheniformis* (DSM 5749)) and Isomaltooligosaccharides	Commercial probiotic	0 and 2 g kg^{-1} IMOS + 1 g kg^{-1} BetaPlus® in 7 weeks	Caspian Brown Trout (*Salmo trutta caspius*) 9 g	SR, BW, FW, and SGR ↑ FCR ↓ White blood cells ↑ Monocytes, neutrophils, and hematocrit ↑ Mean corpuscular volume and lymphocytes ↑ Serum triglycerides ↑ Cholesterol, total protein, and albumin ↑ Albumin/globulin ratio ↑	Aftabgard et al. (2019)

Bacillus subtilis and β-glucan	Commercial probiotic	1 g kg^{-1} β-glucan and 1×10^9 CFU kg^{-1} *B. subtilis* 70 days	Pengze crucian carp (*Carassius auratus* var. Pengze) 12.89 ± 0.04 g	Growth performance → Textures of muscle ↑ Cholesterol activity ↑ High-density lipoprotein ↑ Low-density lipoprotein ↑ Acid phosphatase activity ↑ Alkaline phosphatase activity ↑ Catalase activity ↑ Fold height and microvillus height ↑ Amylase, lipase, and trypsase activities ↑	Cao et al. (2019)
B. licheniformis and *B. amyloliquefaciens*	Commercial probiotic	(1) probiotics supplemented to the water and live feed, (2) probiotics supplemented to the water only, and (3) no probiotic controls with 1×10^{10} CFU g^{-1} in	Larval common snook (*Centropomus undecimalis*)	Growth performance ↑ Innate enzyme activities ↑ Inhibition of opportunistic bacteria ↑ Water quality parameters ↑	Tarnecki et al. (2019)
Bacillus subtilis	Isolated from the shrimp gut	2×10^2, 4×10^4, 6×10^6, 8×10^8 and 10×10^{10} CFU/100 g of feed 40 days	Indian prawn (*Penaeus indicus*) 16.8 ± 0.11 g	Bacterial growth ↑ Bacteriocin production ↑	Ock Kim et al. (2020)

(continued)

Table 1.1 (continued)

Species	Source	Doses and duration	Fish species	Parameters investigated	References
Bacillus licheniformis	Isolated from grass carp	1×10^5 cfu/g and 1×10^6 cfu/g 56 day	Grass carp (*Ctenopharyngodon idella*) 16.5 g	WG and SGR ↑ Resistance against *A. hydrophila* ↑	Qin et al. (2020)

appetite, increase in the production of vitamin, breakdown of indigestible components, as well as possible improvement of intestine morphology (Irianto and Austin 2002). In addition, *Bacillus* spp. could secret several digestive enzymes like protease, amylase, and lipase (Cai et al. 2019; Caulier et al. 2019).

1.3.1.2 Shrimps

Shrimp is a commercially important aquatic species with high economic value and good flavor, which has been widely farmed in the world, particularly in some Asia countries (Chen et al. 2020a). However, the shrimp industry has suffered severe economic losses because of the frequent outbreaks of diseases such as early mortality syndrome (EMS) and white spot syndrome virus (WSSV) (Alavandi et al. 2019; Castex et al. 2009; Chang et al. 2012). Chemotherapeutant and antibiotics are usually applied to settle this problem. Unfortunately, prolonged use of chemotherapeutant and antibiotics could lead to severe outcomes such as resistant bacteria, drug residues, and toxins, which pose a substantial threat to human beings and environment (Dash et al. 2015). Therefore, to seek an alternative way to solve this threat has caused increasing concern (Huynh et al. 2018). Probiotics have been widely applied in shrimp aquaculture. Jamali et al. (2015) revealed that dietary enrichment with *B. licheniformis* and *B. subtilis* significantly enhanced growth performance and survival rate of Pacific white shrimp, *Litopenaeus vannamei*. Elevation of growth performance has been demonstrated as the *Bacillus* could colonize shrimp digestive tract. In *P. monodon*, *Bacillus*, when used as a probiotic, was able to colonize both the culture water and the shrimp digestive tract; the *Bacillus* also was able to replace *Vibrio* spp. in the gut of the shrimp, thereby increasing shrimp survival (Rengpipat et al. 1998), via out-competing other bacteria for nutrients and space by producing antibiotics (Moriarty 1998; Verschuere et al. 2000). Similarly, significant improvement in growth performance of *L. vannamei supplemented with Bacillus* spp. *has been reported in previous studies* (Sadat Hoseini Madani et al. 2018; Sánchez-Ortiz et al. 2016; Swapna et al. 2015). Also, Amoah et al. (2019) indicated that dietary inclusion of *B. coagulans* significantly improved growth performance and feed utilization of Pacific white shrimp. The nutritive values as reported by Vijayavel and Balasubramanian (2006) is highly dependent on their biochemical constituents such as crude protein, crude lipid, ash content, and moisture, which also is noted to be an indication of improved meat quality. In addition, higher inclusion levels of probiotic BC at 1×10^8 CFU g^{-1} feed in diets could modulate gut microflora of *L. vannamei* (Amoah et al. 2019), which play an ardent role in the digestive enzyme activities and the intestinal health. It has been well documented that digestive enzymes are known to break down food and absorb nutrients (Gobi et al. 2018). The digestive enzymes including amylase, lipase, and trypsin (Rawlings and Barrett 1994; Svendsen 2000) in this study significantly increased in the treated group compared to the untreated. Similar results of improved digestive enzyme activities in *Litopenaeus vannamei* (Zokaeifar et al. 2012) and *Fenneropenaeus indicus* (Ziaei-Nejad et al. 2006) have been established. Verschuere et al. (2000) in their work also noted that, *Bacillus* genus secrets a wide range of exoenzymes which aid in the nutritional enhancement of the host. More recently, significant increase in growth performance, villus number, villus height, thicker submucosa, and propionic acid content has been reported in *L. vannamei* fed different *Bacillus* species (*B. licheniformis*, *B. cereus*, and *B. subtilis*) singularly or combined with other probiotics (*Pediococcus acidilactici*, *P. pentosaceus,* and *Lactococcus lactis*) (Chen et al. 2020a; Chen et al. 2020b; Khademzade et al. 2020; Won et al. 2020a).

Dietary supplementation of *Bacillus coagulans* on growth and feed utilization of freshwater prawn *Macrobrachium rosenbergii* showed that growth performance and feed utilization were found to be significantly higher ($P < 0.05$) in prawn fed 109 cfu g^{-1} diet. In addition, the specific activities of protease, amylase, and lipase digestive enzymes were significantly higher ($P < 0.05$) for 109 cfu g^{-1} diet (Gupta et al. 2016).

In Marron (*Cherax cainii*), Ambas et al. (2017) found that synbiotic use of *B. mycoides* and organic selenium (OS) significantly improved some immune parameters of marron, particularly the glutathione peroxidase, and to some extent total hemocyte counts. However, the synbiotic feed did not synergistically improve marron growth; in fact, the use of *B. mycoides*-supplemented diet alone demonstrated significantly higher growth in marron compared with the growth of marron fed on other test diets. A study conducted by Ock Kim et al. (2020), it was indicated that strain *Bacillus subtilis* isolated from the gut of *Penaeus indicus* and added at 2×10^2 CFU 100 g^{-1} as probiotics in feed, resulted in weight gain of the juvenile shrimp (16.8 $\pm$ 0.11 g) after 40 days. The weight gain was 16.8 $\pm$ 0.11 CFU 100 g^{-1} at 10×10^2 CFU 100 g^{-1} probiotic concentration.

1.3.1.3 Catfish

To the best of our knowledge, there were few studies regarding the use of *Bacillus* spp. on this fish. Afrilasari and Meryandini (2016) reported that *Bacillus megaterium* PTB 1.4 increased the activity of digestive enzymes and the growth of catfish. It is known that isolate PTB 1.4 is *B. megaterium,* where *Bacillus* spp. group is known to have ability to produce extracellular enzymes (Moriarty 1998). Probiotic bacteria are capable of producing digestive enzymes that help fish use feed nutrients and digest (Bairagi et al. 2002). Generally, endogenous enzyme can be produced by fish, but the presence of probiotics can improve digestive enzyme. Probiotics improve digestive enzyme activity by stimulating the synthesis of endogenous enzyme in the digestive tract (Mohapatra et al. 2012). Similarly, combination of *B. subtilis, B. amyloliquefaciens*, *B. cereus,* and a commercial *B. amyloliquefaciens* significantly improved growth performance of *C. gariepinus* (Reda et al. 2018). This improvement could be attributed to the production of amylase and protease by the same strain (Selim et al. 2019). In addition, *Bacillus* sp. are capable to detoxify the harmful substance in feed, produce essential vitamins such as vitamin B12 and biotin, and increase the intestinal villus heights (Ramirez and Dixon 2003; Reda and Selim 2015; Sugita et al. 1992).

In hybrid catfish (*C. macrocephalus* $\times$ *C. gariepinus*), Meidong et al. (2017) indicated that *Bacillus siamensis* strain B44v, selectively isolated from Thai pickled vegetables (Phak-dong), displayed a high potential as a probiotic in catfish culture.

Fish fed diet containing strain B44v (10^7 CFU g^{-1} feed) displayed not only no mortality but also growth improvement. The potential probiotic *B. siamensis* strain B44v could produce cellulase and protease, whereas the *Bacillus* sp. strain B51f produced protease and amylase enzymes. Ability to produce some hydrolytic enzymes is beneficial to the host. Enzymes increase the digestion of macromolecules in animal feed and improve feed intake by reducing digesta viscosity and increasing nutrient absorption in host animals (Ray et al. 2012).

For striped catfish, *Pangasianodon hypophthalmus*, the mixture of probiotics (*B. amyloliquefaciens* 54A and *B. pumilus* 47B) isolated from striped catfish at concentrations of 1×10^8, 3×10^8, and 5×10^8 CFU g^{-1} was added to the fish feed and conducted for 90 days. Truong Thy et al. (2017) reported that AWG (476.6 ± 7.81 g $fish^{-1}$) of fish fed probiotics at 5×10^8 CFU g^{-1} was significantly higher than the control (390 ± 25.7 g $fish^{-1}$) after 90 days of feeding, but there was no significant ($P > 0.05$) effect of probiotics on FCR and SGR. However, in basa fish, *Pangasius bocourti* (Meidong et al. 2018) reported that the administration of strain B81e isolated from the fish's gut (1×10^7 CFU g^{-1}) for 60 days had significant effects ($p < 0.05$) on weight gain, specific growth rate, and feed utilization efficiency of *P. bocourti*. This growth improvement might be related to the capability of the putative probiotics in producing extracellular protease and lipase within fish gut and thus exert beneficial effects to the digestive processes of the host fish as bacterial enzymes can help degrade the proteinaceous and lipid substrates (Ramesh et al. 2015; Ray et al. 2012). The significant reduction in FCR indicated that the fish utilized dietary nutrients more efficiently when feed was supplemented with strain B81e.

1.3.1.4 Japanese eel (*Anguilla japonica*)

Bacillus spp. supplementations have been recently applied in Japanese eel. Lee et al. (2017) indicated that dietary supplementation of *Bacillus subtilis* WB60 at 10^8 CFU g^{-1} in diet of Japanese eel (*Anguilla japonica*) resulted in better weight gain, feed efficiency, and protein efficiency ratio compared to the control and *Lactobacillus plantarum* diets. Similar results were observed in Japanese eel fed *Bacillus subtilis* WB60 and mannanoligosaccharide (MOS), as well as (*Bacillus subtilis* or licheniformis) and (mannan or fructooligosaccharide) (Lee et al. 2018; Park et al. 2020). There is growing evidence that gastrointestinal bacteria facilitate the decomposition of nutrients in the host organism and provide physiologically active materials, such as enzymes, amino acids, and vitamins (Cencic and Chingwaru 2010; Morowitz et al. 2011; Wang et al. 2020a). These materials can positively influence the digestive tract and improve feed digestion and utilization (Bairagi et al. 2004; Dawood et al. 2019; Ramirez and Dixon 2003; Wang et al. 2020b).

1.3.1.5 Sea Cucumber (*Apostichopus japonicus*)

Supplementation of *Bacillus cereus* EN25 at 0 (control), 10^5, 10^7, and 10^9 CFU g^{-1} for 30 days showed no significant effects on growth of sea cucumbers *A. japonicus* (Zhao et al. 2016). Growth performance of sea cucumbers was one of the important indices to evaluate the effects of potential *Bacillus* spp. on culturing of sea cucumbers. Previous studies had proved that dietary *Bacillus* spp., such as indigenous *B. subtilis* T13 (Zhao et al. 2012), indigenous *B. cereus* (Yang et al. 2015), and commercial *B. subtilis* (Zhang et al. 2010), could improve the growth performance of sea cucumbers at suitable doses. This difference could be attributed to the differences in *Bacillus* strains, sizes of sea cucumbers, sources of sea cucumbers, experimental period, and experimental conditions. The present study was conducted with the same source of sea cucumbers at the same experimental period and conditions with Zhao et al. (2012), except *Bacillus* strain and initial sizes of sea cucumbers. Recently, Liu et al. (2020) indicated that dietary supplementation of *B. baekryungensis* MS1 at 10^7 cfu g^{-1} for a total of 60 days significantly improved the growth performance of the sea cucumber cultured under low temperature. This is related to the mode of action of probiotics, including the production of digestive enzymes, the production of antibacterial substances, immune stimulation, and interference of quorum sensing, all of which depend on the long-term growth and reproduction of probiotics. Studies have also shown that probiotics work by managing community assembly of the water and gut microbiota (Selim and Reda 2015; Wang et al. 2017a).

1.3.1.6 Tambaqui (*Colossoma macropomum*)

Dietary inclusion of *Bacillus subtilis* (10^9 UFC g^{-1}) and *Saccharomyces cerevisiae* (10^9 UFC g^{-1}) showed that no differences were found for the growth parameters between the treatments with probiotics (da Paixão et al. 2017). Although probiotics are supposed to be beneficial, the literature mentions possible synergistic effects. The total replacement of indigenous populations with probiotics may not be desirable to improve growth performance (Merrifield et al. 2010). The control of the endogenous balance between pathogenic and beneficial bacteria is still the target of many studies. According to Merrifield et al. (2010), the lack of improvements regarding growth and feed use may be explained by the level of gastrointestinal colonization that could be too high and any possible synergistic effect with the normal gut microbiota was negated. Thus, it is expected that the beneficial effects of probiotics for tambaquis are not on its performance but on its health and welfare. However, in another study with tambaqui, Dias et al. (2018) indicated that the use of the autochthonous bacteria *B. cereus* improves the growth performance, productivity, hematological profile, and survival of tambaqui juveniles. This enhanced growth performance of fish supplemented with probiotics is probably due to an improvement in digestion as well as an increase in the synthesis and absorption of nutrients (Hoseinifar et al. 2017). Similar results were obtained by El-Haroun et al. (2006) reporting increased growth performance and feed efficiency in tilapia fed the probiotics *Bacillus licheniformis* and *Bacillus subtilis*. According to these authors, the added probiotics improved digestibility, dietary protein, and energy utilization. These positive effects can be attributed to the capacity of the probiotics to promote an increase in the gut absorbent surface area, and stimulate and/or produce several enzymes on the intestinal tract, which improve digestibility and nutrient retention, leading to higher growth rates (El-Haroun et al. 2006; Ibrahem 2015).

1.3.1.7 Carp Species

Dietary administration of BioPlus 2B, a probiotic containing *Bacillus licheniformis* and *B. subtilis*, and Ferroin solution indicate that the combination of probiotic and Ferroin solution represents an effective dietary supplement for improving carcass quality, growth performance, and hematological parameters in kutum fry (Azarin et al. 2015). In mrigal fingerlings, *Cirrhinus mrigala* (avg.wt. 2.5 ± 0.20 g) were fed with three different doses (2×10^4, 2×10^5, and 2×10^6 CFU) of *Bacillus* sp. PP9 admixed with 100 g feed for a period of 60 days. It was found that the feed with Bacillus concentration of 2×10^4 CFU exhibited significantly higher growth and lower food conversion ratio compared to the control and other supplemented diets (Bandyopadhyay et al. 2015). More recently, Qin et al. (2020) found that dietary inclusion of *B. licheniformis* at the low-dose 1×10^5 cfu g^{-1} and the high-dose (HD) group with 1×10^6 cfu g^{-1} led to significantly ($p < 0.05$) improved percent weight gain (PWG) and specific growth rate (SGR) parameters. The improvement of growth performance parameters such as PWG and SGR with increasing concentrations of supplemented *B. licheniformis* FA6 observed in this study is in agreement with Han et al. (2015) observed a significant increase in the growth performance of tilapia fed with *B. licheniformis*. The increase in the growth performance of grass carp may due to the secretion of digestive enzymes by *B. licheniformis*, which improves feed digestibility (Kuebutornye et al. 2019).

In Pengze crucian carp, *Carassius auratus*, dietary supplementation with prebiotics β-glucan (BG group) and probiotics *Bacillus subtilis* (BS group) resulted in better growth performance than other groups whereas feed efficiency was unaffected by dietary treatments. The textures of muscle in terms of hardness, springiness, cohesiveness, gumminess, chewiness, and resilience were higher in BG and BS groups than the control group. Supplementation of β-glucan and *B. subtilis* acted as a hypolipidemic in terms of decreasing the total cholesterol, high-density lipoprotein, and low-density lipoprotein, whereas increased the immune responses in serum measured by acid phosphatase, alkaline phosphatase, and catalase activities. Dietary supplementation of β-glucan and *B. subtilis* significantly improved the fold height and microvillus height in contrast to basal diet. Moreover, β-glucan could significantly increase digestive capacity observed in terms of an increase in amylase and trypsase activities, and *B. subtilis* significantly increased amylase and lipase activities in intestine (Cao et al. 2019).

1.3.1.8 Trout

A commercial probiotic (4.2×10^9 CFU g^{-1} of additive) was supplemented to the experimental diets at 0% (control), 0.03% (P_1; 6×10^3 CFU $g^{-1}$of diet), or 0.06% (P_2; 1.5×10^6 CFU g^{-1} of diet) and fed to brown trout (*Salmo trutta*) and rainbow trout (*Oncorhynchus mykiss*) for 9 and 20 weeks, respectively. Rainbow trout showed significantly better growth performance than brown trout, regardless of the dietary treatment. No effect of dietary probiotic supplementation was detected on growth performance and body composition (Ramos et al. 2017). However, in Caspian Brown Trout (*Salmo trutta caspius*)Aftabgard et al. (2019) found that the combined effects of IMOS, a prebiotic, and BetaPlus®, a probiotic containing

B. subtilis and *B. licheniformis*, demonstrated a better performance of select growth indices, including BWI and FCR, than fish that were fed the control diet; these results were probably due to improved nutrition and digestive processes (Cerezuela et al. 2011).

1.3.1.9 Other Aquacultured Species

Two probiotics (*Virgibacillus proomii* and *Bacillus mojavensis*) were used to study their effects on the digestive enzyme activity, survival, and growth of sea bass, *Dicentrarchus labrax* at various ontogenetic stages in three separate experiments (Hamza et al. 2016). The results indicated that the two probiotics *V. proomii* and *B. mojavensis* were adequate for improved growth performance and survival and for healthy gut microenvironment of the host (Hamza et al. 2016).

In the study of Hauville et al. (2016) Florida pompano (*Trachinotus carolinus*) larvae were fed either live feed enriched with Algamac 3050 (Control), Algamac 3050, and probiotics (PB), or the previous diet combined with a daily addition of probiotics to the tank water (PB+). The results indicated that a mix of *Bacillus* sp. can promote growth through an early maturation of the digestive system during the early larval stages of pompano and snook.

In grouper *Epinephelus coioides* (Yan et al. 2016, juveniles (14.6 ± 0.2 g) were fed either a basal control diet (without probiotic) or the basal diet supplemented with 1.0×10^8 CFU g^{-1} live (T1) and heat-inactivated *B. pumilus* SE5 (T2). The results indicated that the heat-inactivated probiotic significantly improved the final weight, weight gain (WG), and specific growth rate (SGR) at day 60 and significantly decreased the feed conversion ratio (FCR) at day 30 and 60, while the viable probiotic significantly decreased the FCR at day 60 ($P < 0.05$). This suggested that live and heat-inactivated *B. pumilus* could promote the efficient utilization of dietary nutrients. Interestingly, significant increased growth was only observed in fish fed the heat-inactivated *B. pumilus* containing diet for 60 days, but not in fish fed the live *B. pumilus* containing diet. Likewise, Hoseinifar et al. (2011) observed that dietary supplementation of 20 g kg^{-1} inactive brewer's yeast *Saccharomyces cerevisiae var. ellipsoideus* significantly improved the growth performance in juvenile beluga sturgeon (*Huso huso*). In rock bream, *Oplegnathus fasciatus*, Kim et al. (2017), revealed that supplementation of *B. amyloliquefaciens* spores at a concentration of 1.4×10^6 colony-forming units per gram (CFU g^{-1}) of feed for 90 days resulted in significant improvements in body weight (BW), weight gain (WG), specific growth rate (SGR), and food conversion ratio (FCR) when compared with control group fish.

In hybrid sturgeon, *Acipenser schrenckii* ♂ and *Acipenser baerii* ♀, fish were fed with *Bacillus amyloliquefaciens* (GB-9) and *Yarrowia lipolytica* lipase2 (YLL2): Diet 1 (0-control), Diet 2 (5.0 g kg^{-1} GB-9), Diet 3 (4.0 g kg^{-1} YLL2), and Diet 4 (5.0 g kg^{-1} GB-9 + 4.0 g kg^{-1} YLL2), respectively (Fei et al. 2018). The results indicated that supplementations of GB-9 + YLL2 resulted in a significant increase in final weight, Docosahexaenoic acid (DHA) and Eicosapentaenoic acid (EPA) concentration, compared with that of control ($p < 0.05$). This might be because the DHA and EPA hydrolyzed by YYL2 improved the poor establishment of the GB-9 in the gastrointestinal tract of hybrid sturgeon and might have promoted the growth of GB-9 (Menni et al. 2017). Similarly, combination of *B. licheniformis* and *B. amyloliquefaciens* indicated up to 2.5 times higher survival with probiotic addition, as well as 20% higher survival 7 days following a transport event. These benefits could not be explained by faster larval growth. In fact, CONT larvae were significantly longer than probiotic-treated larvae, likely due to decreased competition for food in CONT tanks which exhibited significantly lower survival. The other differing morphometric in this study was oil globule volume which was lowest in CONT larvae, suggesting that CONT larvae were consuming their endogenous reserves more quickly than probiotic-treated larvae. Retention of oil globules allows for a longer transition time to exogenous feeding, and studies indicate larvae that retain their endogenous reserves longer demonstrate increased survival (Avila and Juario 1987; Berkeley et al. 2004). The probiotic may alter development of the digestive tract and thus the start of exogenous feeding, as has been demonstrated in previous studies involving *Bacillus* probiotics and common snook (Hauville et al. 2016).

Dietary supplementation of *B. subtilis* has been reported to improve the growth performance, feed utilization, amylase, protease, and lipase enzymes of parrotfish (*Oplegnathus fasciatus*) and red sea bream (*Pagrus major*) (Liu et al. 2018; Zaineldin et al. 2018). The observed improvement in growth performance might be ascribed to the enhanced intestinal digestive enzyme activity and beneficial intestinal microbiota (Dawood et al. 2014; Liu et al. 2009; Sun et al. 2010). *Bacillus* sp. can produce certain essential micronutrients to promote better growth and feed utilization of hosts (Sanders et al. 2003). Further, *Bacillus* species may participate in digestion processes to break down nutrients such as carbohydrates, proteins, and lipids by producing extracellular enzymes (Liu et al. 2009; Sun et al. 2010). In abalone, *Haliotis discus hannai*, Gao et al. (2018) indicated that the food containing 10^5 cfu mL^{-1} *Bacillus licheniformis* promoted food intake and growth of abalones. *Bacillus licheniformis* is an aerobic nonpathogenic bacterium that inhabits the intestinal microbial community in the form of spores, which can reduce intestinal pH, reduce ammonia concentration, and promote decomposition of starch and cellulose. Thus, it is generally considered to be a relatively stable probiotic (Hong et al. 2005; Vine et al. 2006).

1.3.2 Increase Disease Resistance

Probiotics have been proven as an effective tool for disease prevention in aquaculture (Hoseinifar et al. 2018). Probiotics can interact with or antagonize other enteric bacteria by resisting colonization or by directly inhibiting and reducing the incidence of opportunistic pathogens (Chiu et al. 2017). They can also improve host's health and well-being via physiological or immune modulation (Butt and Volkoff 2019). Probiotics can produce effective molecules that have bactericidal activity on intestinal pathogenic bacteria of the host, providing a barrier against the proliferation of opportunistic pathogens (Martínez Cruz et al. 2012; Seghouani et al. 2017). The functional molecules produced during the bactericidal activity are antibiotics, bacteriocins, siderophores, enzymes and/or hydrogen peroxide as well as the alteration of the intestinal pH due to the generation of organic acids (Verschuere et al. 2000). The inhibition of intestinal related diseases has been reported in several cultured species by probiotic incorporation in aquafeeds (Ringø et al. 2018; Serra et al. 2019; Wanka et al. 2018). Thus, it can be confirmed that the ability of aquatic animals to avoid the infectious diseases mainly depends on the immunomodulatory effect that happened due to the administration of beneficial bacterial cells.

1.3.2.1 Tilapias

Dietary inclusion of *B. licheniformis* at 0%, 0.02%, 0.04%, 0.06%, 0.08%, and 0.1% containing live germ 2×10^{10} CFU/g for 10 weeks significantly increased disease resistance of Nile tilapia, *Oreochromis niloticus* against *Streptococcus iniae* (Han et al. 2015). *Bacillus* strains supplementation in diet could increase disease resistance in fish through the stimulation of both the cellular and humoral immune function, such as phagocytic activity, lysozyme activity, and complement activity (Arena et al. 2006; Queiroz and Boyd 1998; Sookchaiyaporn et al. 2020; Zhou et al. 2010). It was reported that *Bacillus* bacteria are able to outcompete other bacteria for nutrients and space and can exclude other bacteria through the production of antibiotics, and as usually lead to the enhanced immunity of fish (Cha et al. 2013). Similarly, dietary inclusion of *B. licheniformis* Dahb1 at 10^7 cfu g^{-1} could improve disease resistance of Mozambique tilapia (*Oreochromis mossambicus*) against *A. hydrophila* (Gobi et al. 2018). In terms of *Bacillus subtilis* HAINUP40, H. Liu et al. (2017) reported that dietary supplement of *B. subtilis* HAINUP40 at 10^8 cfu g^{-1} can effectively enhance disease resistance of Nile tilapia against *Streptococcus agalactiae*. In addition, combination of *B. subtilis* with *S. cerevisiae* and *A. oryzae*; *Bacillus subtilis* with Aqua NZ and AP193; *Bacillus subtilis* strains SB3086, SB3295, SB3615 with AP193; *B. subtilis* and *B. licheniformis*, and *Bacillus subtilis* and *Bacillus licheniformis* (BS) combined with traditional Chinese medicine (TCM) *A. hydrophila* and *S. iniae*. Higher intestinal *Bacillus* spp. counts can regulate the gut microbiota of fish, selectively stimulate other beneficial probiotic bacteria, and depress some potential harmful bacteria (Yang et al. 2012).

1.3.2.2 Shrimps

The efficiency of these isolates in controlling pathogens, which is a key factor in selecting appropriate bacteria as probiotics, was evaluated (Kesarcodi-Watson et al. 2008). Based on in vitro laboratory results, B4, B6, and B12 inhibited *V. parahaemolyticus*; however, only *B. subtilis* AQHPS001 (B12) showed the highest antagonistic property against VP_{AHPND} strains. However, among the VP_{AHPND} strains, there were different sizes of the inhibitory clear zone, and $VP_{AHPNDAQH3.2}$ was the only strain that resisted B12. This suggests that there are varieties of VP_{AHPND} and that each strain may employ different mechanisms in response to the target B12 (Kewcharoen and Srisapoome 2019). Previous reports found that *Bacillus* spp. could produce many kinds of bacteriocins, such as subtilin, subtilosin, coagulin, megacin, bacillin, bacillomycin, mycosubtilin, toximycin, and xanthobacidin, which could reduce pathogen colonization by directly inhibiting pathogens while having no resulting effects on the virulence resistance genes of pathogenic bacteria (Desriac et al. 2010; Hammami et al. 2012; Joseph et al. 2013). Zhao et al. (2015) also reported that *Bacillus* spp. could secrete quorum-quenching enzymes, which are expected to be quorum-sensing blockers to reduce disease infection. These results suggest that *B. subtilis* AQAHBS001 possesses more effective characteristics that are important for controlling the various harmful VP_{AHPND} strains than other candidates. For these reasons, it was further chosen to study its application on a laboratory scale. Similarly, dietary inclusion of *Bacillus subtilis* WB60, *Pediococcus pentosaceus*, and *Lactococcus lactis* at 10^8 CFU g^{-1} could improve disease resistance of whiteleg shrimp *Litopenaeus vannamei* against *Vibrio parahaemolyticus* (Won et al. 2020b). Generally, administration of probiotics in the shrimp diet was shown to decrease mortality rates compared to the CON diet (Balcázar et al. 2007; Sapcharoen and Rengpipat 2013; Zhang et al. 2009). Previous studies demonstrated that probiotic supplementation can be used for modulating fish health and disease resistance (Wang et al. 2018; Zuo et al. 2019). Indeed, probiotics can beneficially influence the disease resistance of fish to pathogen bacteria by producing antimicrobial substances and competing with pathogens for physical occupation of space (Lim et al. 2020). As a result, the enhanced survival and cumulative survival rates could be due to probiotic supplementation. Chen et al. (2020a) recently indicated that dietary MOS and/or *B. licheniformis* supplementation could positively increase ammonia resistance of *Litopenaeus vannamei*. According to Chen et al. (2012), immune parameters decrease after ammonia stress, yet these parameters recover faster when they were initially stimulated by a probiotic. Faster recovery of immune parameters might have contributed to the increased survival after ammonia stress for the *Rps. palustris* fed shrimp.

1.3.2.3 Catfish

Meidong et al. (2017) revealed that *Bacillus siamensis* strain B44v and *Bacillus* sp. strain B51f, derived from indigenous fermented foods, displayed strongly antagonistic activity against the bacterial fish pathogens, *A. hydrophila* and *S. agalactiae*. Both strains effectively inhibited Gram-positive and Gram-negative bacteria, indicating their broad spectrum as a useful antagonistic property as the two most striking bacterial fish pathogens in aquaculture in Thailand belong to the genera the *Aeromonas* and *Streptococcus* (Maisak et al. 2013). Besides fish pathogens, the bacteriocin-like substance from *B. siamensis* strain B44v inhibited several foodborne pathogens suggesting potential applications in human foods (Sivamaruthi et al. 2018). Likewise, Reda et al. (2018) showed that supplementation of three autochthonous *Bacillus strains* (*B. subtilis*, *B. amyloliquefaciens*, and *B. cereus*) and a commercial *B. amyloliquefaciens* at a dose of 1×1010 CFU kg^{-1} significantly increased disease resistance of African catfish against *Aeromonas sobria*. This may be returned to the ability of *Bacillus* spore to resist gastrointestinal conditions, survive and transit cross gastrointestinal tract, germinate and vegetate with heterologous antigen expression before being excreted (Duc et al. 2003). In striped catfish, Truong Thy et al. (2017) indicated that the mixed probiotics of *Bacillus amyloliquefaciens* 54A and *B. pumilus* 47B isolated from striped catfish (*Pangasianodon hypophthalmus*) intestine significantly enhanced disease resistance of the fish against *Edwardsiella ictaluri* and ammonia tolerance. Antimicrobial activity of probiotics has been demonstrated on many in vitro and in vivo studies in animals. The study of Corr et al. (2007) reported trial mice received protection from *Lactobacillus salivarius* against *Listeria monocytogenes* involved bacteriocin produced by *L. salivarius*UCC118. Additionally, antimicrobial activities of probiotics against pathogens include secretion of hydrogen peroxide (Pridmore et al. 2008), lactic acid (Fayol-Messaoudi et al. 2005), competitive exclusion (Lee et al. 2003), and stimulation of immune system (Ryan et al. 2009). The positive effect on barrier function of probiotics is to protect the host intestine by prevention of pathogen attachment to epithelial cells on gut surface (Mennigen et al. 2009). In basa fish, *Pangasius bocourti*, Meidong et al. (2018) found that *B. aerius* B81e has beneficial effects on growth performance, innate immunity, and disease resistance of *P. bocourti* against *Aeromonas hydrophila* and *Streptococcus agalactiae*. Bacterial co-aggregation has considerable significance in the host gut as co-aggregation ability of bacterial probiotics might interfere with the ability of pathogenic bacteria to infect the host and can prevent colonization of the pathogens (Spencer and Chesson 1994). In addition, *B. aerius* B81e has an absence of hemolysin and is susceptible to most of the common antibiotics tested which demonstrated that it is likely a nonpathogen and has an inability to transfer antibiotic-resistant genes to recipient bacteria in the host gut, thus preventing the development of antibiotic-resistant pathogens (Meidong et al. 2018).

1.3.2.4 Japanese eel (*Anguilla japonica*)

The combination of *Bacillus subtilis* WB60 and *Lactobacillus plantarum* KCTC3928 or *Bacillus subtilis* WB60 and mannanoligosaccharide (MOS) significantly improved disease resistance of Japanese eel against *V. anguillarum* (Lee et al. 2017, 2018). Similarly, Park et al. (2020) reported that dietary inclusion of *B. subtilis* with FOS (BSF) and *B. licheniformis* significantly increased disease resistance against *Aeromonas hydrophila*. Significant increase in disease resistance in these works may be attributable to the stimulation of cellular and humoral immune function.

1.3.3 Sea Cucumber (*Apostichopus japonicus*)

Zhao et al. (2016) indicated that the cumulative mortality after *V. splendidus* challenge decreased significantly in sea cucumbers fed with EN25 at 10^7 CFU g^{-1} ($P < 0.05$). The present study confirmed dietary *B. cereus* EN25 at 10^7 CFU g^{-1} could significantly improve disease resistance in juvenile *A. japonicus*. Recently, Liu et al. (2020) showed that *B. baekryungensis* MS1 significantly reduced the mortality of sea cucumbers infected with *Vibrio splendidus*. By regulating the expression of immune-related genes and signaling pathways, B. *baekryungensis* MS1 improved the immunity of sea cucumber in winter and effectively controlled the infection of pathogenic bacteria such as *V. splendidus*.

1.3.4 Tambaqui (*Colossoma macropomum*)

da Paixão et al. (2017) indicated that supplementation of two probiotics *Bacillus subtilis* and *Saccharomyces cerevisiae* at 10^9 UFC g^{-1} significantly increased disease resistance of tambaqui, *Colossoma macropomum*, against *Streptococcus agalactiae*. Similarly, Dias et al. (2018) reported that *B. cereus* (4.2×10^4, 3.9×10^6 and 3.3×10^8 CFU g^{-1}) supplemented as probiotics to *C. macropomum* for 120 days significantly increased disease resistance against *Aeromonas hydrophila*. The probiotic promoted a nonspecific response against bacterial infection, increasing fish survival after challenge with *A. hydrophila*.

1.3.5 Other Species

In rock bream, *Oplegnathus fasciatus*, Kim et al. (2017) demonstrated the benefit of incorporation of *B. amyloliquefaciens* as a feed supplement to improve the health status of *Oplegnathus fasciatus* challenged with *Streptococcus iniae*. The enhancement of the innate immune response with a *B. amyloliquefaciens* enriched probiotic diet and decreased mortality rate, thereby protecting the fish against *S. iniae*. Similarly, dietary inclusion of *B. subtilis* at 10^8 CFU kg^{-1} significantly increased disease resistance of parrotfish, *Oplegnathus fasciatus*, against *Vibrio alginolyticus* (Liu et al. 2018). The growth performance and health status improvement of aquatic animal might be involved with the gut microbiota change after probiotic administration. The previous study has also demonstrated the positive effects of *B. subtilis* E20 in terms of intestinal presence and subsequent health benefits for *L. vannamei* (Liu et al. 2009; Tseng et al. 2009) and *E. coioides* (Liu et al. 2010). In the same trend, dietary inclusion of *Bacillus licheniformis* significantly improved disease resistance of abalone, *Haliotis discus hannai Ino.*, against *V. parahaemolyticus* and grass carp, *Ctenopharyngodon idella*, against *A. hydrophila* (Gao et al. 2018; Qin et al. 2020).

1.4 Immune Effects of *Bacillus*

Enhancement of host immunity is one important benefit of probiotic diet supplementation (Kuebutornye et al. 2019). As stated by Verschuere et al. (2000), probiotics can modulate innate immunity through the modulation of humoral immune responses and expression of immune-related genes. Effects of Bacillus on immune response of different fish and shellfish are displayed in Table 1.2.

Table 1.2 Immune responses of fish and shellfish fed different *Bacillus* probiotics. → no change, ↑ increase, ↓ decrease

Species	Source	Doses and duration	Fish species	Parameters investigated	References
B. amyloliquefaciens	Commercial probiotic	0; 1×10^4 and 1×10^6 CFU/g 30 days	Nile tilapia (*O. niloticus*) 27.7 ± 0.22 g	Serum killing percentages ↑ Phagocytic activities ↑ Lysozyme activities ↑ Nitric oxide assay ↑ The percentage of serum killing ↑ Serum nitric oxide activity ↑ Serum lysozyme activity ↑	(Selim & Reda 2015)
Bacillus subtilis (combined with *S. cerevisiae* and *A. oryzae*)	Commercial probiotic	0; 5 g kg^{-1} probiotic mixture (*B. subtilis* 1.5×10^9, *S. cerevisiae* 10^9 and *A. oryzae* 2×10^9); and 10 g kg^{-1} probiotic mixture (*B. subtilis* 3.0×10^9, *S. cerevisiae* 2.0×10^9 and *A. oryzae* 4.0×10^9) (CFU g^{-1}) 6 weeks	Juvenile Nile tilapia (*O. niloticus*) 25 ± 0.05 g	Respiratory burst activity ↑ Erythrocyte fragility ↑ Levels of white blood cells ↑	(Iwashita et al. 2015)
Bacillus licheniformis	Commercial probiotic	0%, 0.02%, 0.04%, 0.06%, 0.08% and 0.1% containing live germ 2×10^{10} (CFU/g) 10 weeks	Juvenile Nile tilapia (*O. niloticus*) 3.83 ± 0.03 g	Lysozyme activity ↑ Content of complement C3 ↑	(Han et al. 2015)
B. subtilis and *B. licheniformis*	Commercial probiotic	0, 3, 5, 7 and 10 g kg^{-1} 4 weeks	Nile tilapia (*O. niloticus*) 53.01 ± 1.0 g	Lysozyme and protease activity ↑ Anti-protease activity ↑ Superoxide dismutase activity ↑	(Abarike et al. 2018a)

				Immunoglobulin M level ↑ Myeloperoxidase activity ↑ Expression of C-lysozyme ↑ Heat shock protein 70 ↑ β-defensin ↑ Transforming growth factor beta ↑ Small body size decapentaplegic homolog 3 ↑	
Bacillus subtilis and *Bacillus licheniformis* (BS) combined with traditional Chinese medicine (TCM)	Commercial probiotic	0; TCM at 3 and BS at 7 (g/kg); TCM at 5 and BS at 5 (g/kg); TCM at 7 and BS at 3 (g/kg) 4 weeks	Nile tilapia (*O. niloticus*) 57 ± 2 g	Lysozyme activity ↑ Superoxide dismutase activity ↑ Catalase, protease, and antiprotease ↑ Expression of C-lysozyme ↑ Heat shock protein 70 ↑ β-defensin ↑ Transforming growth factor beta ↑	(Abarike et al. 2018b)
Bacillus licheniformis Dahb1	Commercial probiotic	0, 10^5 and 10^7 CFU/g 4 weeks	Mozambique tilapia (*Oreochromis mossambicus*)	Alkaline phosphatase ↑ Myeloperoxidase ↑ Lysozyme ↑ Reactive oxygen species ↑ Reactive nitrogen species ↑ Superoxide dismutase ↑ Glutathione peroxidase ↑	(Gobi et al. 2018)
Bacillus subtilis HAINUP40	Isolated from the aquatic environment	10^8 CFU/g 8 weeks	Nile tilapia (*Oreochromis niloticus*) 95 ± 8 g	Respiratory burst activity ↑ Serum lysozyme activity ↑	(Liu et al. 2017)

(continued)

Table 1.2 (continued)

Species	Source	Doses and duration	Fish species	Parameters investigated	References
Bacillus cereus	Commercial probiotic	Exp 1: 0, 1.0×10^4 and 1.0×10^5 CFU mL^{-1} in 42 days on tanks Exp 2: 0, 1.0×10^7 and 1.0×10^8 CFU g^{-1} 42 days on cages	Nile tilapia (*Oreochromis niloticus*) 20.20 ± 0.17 g	Serum lysozyme ↑ Peroxidase activity ↑ Alkaline phosphatase activity ↑ Total superoxide dismutase activity ↑ Autochthonous gut bacteria community ↑ Potentially beneficial bacteria ↑	(Wang et al. 2017b)
Bacillus pumilus AQAHBS01	Isolated from farmed fish	10^6, 10^7, 10^8 and 10^9 CFU/g 30 days	Nile tilapia (*Oreochromis niloticus*) 50 g	Phagocytic activity ↑ Superoxide anion levels ↑	(Srisapoome & Areechon 2017)
Bacillus subtilis strains SB3086, SB3295, SB3615, and AP193	Commercial probiotic	4×10^7 CFU/g of feed in 21 days	Nile tilapia (*Oreochromis niloticus*) 16.5 ± 0.2 g	Serum bactericidal activity ↑ Lysozyme activity ↑	(Addo et al. 2017b)
Bacillus subtilis (Aqua NZ and AP193)	Commercial probiotic	4.2×10^7 CFU g^{-1} of feed 8 weeks	Nile tilapia (*Oreochromis niloticus*) 59.5 ± 0.99 g	Lysozyme activity → Respiratory burst activity →	(Addo et al. 2017a)
*Bacillus cereus*NY5 and *Alcaligenes faecalis* Y311	Nile tilapia intestine	1.0×10^4 CFU mL^{-1} 3 months	Nile tilapia (*Oreochromis niloticus*) 5.20 ± 0.17 g	The total superoxide dismutase activities ↑ Alkaline phosphatase activities ↑	(Wang et al. 2020a, 2020b)

Bacillus licheniformis HGA8B		1×10^6 and 1×10^8 CFU g^{-1} 60 days	Pacific white shrimp (*Litopenaeus vannamei*) 0.57 ± 0.001 g	Lysozyme activity ↑ Phosphatase activity ↑ Superoxide dismutase (SOD) ↓ Total protein (TP) ↑ Albumin (ALB) in serum ↑ Glutathione peroxidase (GSH-Px) ↑	(Amoah et al. 2019)
Bacillus subtilis and *Bacillus licheniformis*	Commercial probiotic	1×10^4 and 1×10^8 CFU/g 60 days	Pacific white shrimp (*Litopenaeus vannamei*)	Lysozyme and hemocyte cell count ↑	(Sadat Hoseini Madani et al. 2018)
B. licheniformis	Commercial probiotic	Control diet (CON), 0.2% MOS (MOS), 0.1% *B. lincheniformis* (BL), 0.2% MOS plus 0.1% BL (SYN) 8 weeks	Pacific white shrimp (*Litopenaeus vannamei*)	Expression levels of catalase ↑ Expression levels of glutothion peroxidase ↑ Expression of superoxide dismutase (SOD) ↑ Expression of penaeidin -3a (Pen-3a) ↑ Expression of heat shock protein (Hsp-70 ↑	(Chen et al. 2020a)
B. licheniformis		Basal diet (Control); 0.5% hydrolyzed yeast (HY); 0.1% *B. lincheniformis* (BL) and 0.5% hydrolyzed yeast + 0.1% *B. licheniformis* (SYN) 8 weeks	Pacific white shrimp (*Litopenaeus vannamei*) 2.0 ± 0.01 g	GPX, SOD, and ACP ↑ MDA ↓ Expression of CAT, GPX, and SOD ↑ Expression of Pen-3a and PPO ↑ Pen-3a and SOD ↓	(Chen et al. 2020b)

(continued)

Table 1.2 (continued)

Species	Source	Doses and duration	Fish species	Parameters investigated	References
Bacillus licheniformis MAt32, *B. subtilis* MAt43 and *B. subtilis* subsp. *subtilis* GAtB1	Isolated from pustulose ark *Anadara tuberculosa*	0, 1×10^6, 2×10^6, 4×10^6, and 6×10^6 CFU g^{-1} of feed. 32 days	Pacific white shrimp (*Litopenaeus vannamei*) 1 ± 0.1 g	Expression of proPO gene ↑ Expression of LvToll1 and SOD genes ↑ expression of the Hsp70 gene ↓ Expression of TGase gene →	(Sánchez-Ortiz et al. 2016)
Bacillus spp	Isolated from pustulose ark	1×10^6, 2×10^6, 4×10^6, and 6×10^6 CFU g feed^{-1} 32 days	Pacific white shrimp (*L. vannamei*) 1 ± 0.1 g	proPO gene ↑ LvToll1 gene ↑ SOD gene ↑ TGase gene →	(Sánchez-Ortiz *et al.* 2016)
Bacillus subtilis WB60, Pediococcus pentosaceus, and *Lactococcus lactis*	Commercial probiotic	*B. subtilis* at 10^7 CFU/g diet, *B. subtilis*, *P. pentosaceus*, and *L. lactis* at 10^8 CFU/g diet, and oxytetracycline at 4 g/kg 8 weeks	Pacific white shrimp (*Litopenaeus vannamei*) 1.41 ± 0.05 g	Superoxide dismutase activity ↑ Lysozyme activity ↑ Immune-related gene expression ↑	(Won et al. 2020)
Bacillus cereus and *Pediococcus acidilactici*	Commercial probiotic	*P. acidilactici* (10^6 CFU/mL) and *B. cereus* (10^6 CFU/mL) to the water pond 110 days	Pacific white shrimp (*Litopenaeus vannamei*) 0.002 ± 0.001 g	Total hemocyte count ↑ Total protein ↑ Lysozyme activity ↑	(Khademzade et al. 2020)
Bacillus aryabhattai TBRC8450	Commercial probiotic	1×10^8 CFU/g diet 6 weeks	Pacific white shrimp (*Litopenaeus vannamei*) 0.9 ± 0.1 g	C-type lectin ↑ Penaeidin-3 ↑ Heat shock protein 60 ↑ Thioredoxin, and ferritin ↑ phenoloxidase activity ↑ Total antioxidant activity ↑ Total hemocyte count → Superoxide dismutase →	(Tepaamorndech et al. 2019)

Bacillus subtilis E20		10^9 cfu (kg diet)$^{-1}$ 8 weeks	Pacific white shrimp (*Litopenaeus vannamei*) 3.78 ± 0.21 g	Antioxidant enzymes gene ↑ Pattern recognition protein genes ↑ Antimicrobial molecule ↑ Hexosamine biosynthesis pathway ↑ UDP-N-acetylglucosamine-peptide N-acetylglucosaminyltransferase ↑	(Chien et al. 2020)
Bacillus sp. DDKRC1	Isolated from the gut of Asian seabass (*Lates calcarifer*)	0, 2.94 × 10^7 CFU/100 g feed and diet fermented with *Bacillus sp.* DDKRC1 42 days	Tiger shrimp (*Penaeus monodon*) 2.73 ± 0.01 g	Total heterotrophic count ↑ Amylolytic ↑ Cellulolytic and proteolytic bacterial counts ↑ Phagocytic activity ↑	(De et al. 2018)
Bacillus coagulant	Commercial probiotic	0, 10^5, 10^7 and 10^9 cfu g^{-1} 60 days	Freshwater prawn (*Macrobrachium rosenbergii*) 2.4 ± 0.35 g	Lysozyme activity ↑ Respiratory burst activity ↑	(Gupta et al. 2016)
Bacillus sp. PP9	Isolated from mrigal gut	2 × 10^4, 2 × 10^5 and 2 × 10^6 CFU 60 days	Mrigal (*Cirrhinus mrigala*) 2.5 ± 0.20 g	Hemoglobin percentage ↑ Total erythrocyte count ↑ Total leukocyte count ↑ Corpuscular hemoglobin ↑ Total serum protein ↑ Albumin globulin ratio ↑ Serum bactericidal activity ↑	(Bandyopadhyay et al. 2015)

(continued)

Table 1.2 (continued)

Species	Source	Doses and duration	Fish species	Parameters investigated	References
Bacillus subtilis KADR1	Commercial probiotic	10^6; 10^8 and 10^{10} CFU/g 4 weeks	*Labeo rohita*	Serum lysozyme ↑ Phagocytosis ↑ Serum total protein ↑ Respiratory burst ↑ Serum IgM levels ↑ Superoxide dismutase ↑ Alternative complement pathway ↑	(Ramesh & Souissi 2018)
Bacillus subtilis FPTB13 and chitin	Isolated from an indigenous fermented fish product "Shidal"	*B. subtilis* 10^9 cells g^{-1}, chitin 2% and the combition 2 weeks	Labeo catla (*Catla catla*) 40.0 ± 1.9 g	Oxygen radical production ↑ Myeloperoxidase content ↑ Lysozyme activity ↑ Total protein content and alkaline ↑ Phosphatase activity ↑	(Sangma & Kamilya 2015)
B. subtilis, B. licheniformis, and *B. cereus*		0; 1 × 10^5 cfu/g of *B. subtilis;* 1 × 10^5 cfu/g of *B. subtilis* and *B. licheniformis and* 1 × 10^5 cfu/g of *B. subtilis*, *B. licheniformis*, and *B. cereus* 45 days	Common carp (*Cyprinus carpio*) 57.40 ± 0.43 g	Phagocytic percentage ↑ Phagocytic index ↑ Serum immunoglobulin M ↑ Serum lysozyme activity ↑ Intestinal mucosal secretory immunoglobulin A ↑ Peripheral blood lymphocyte proliferation ratio ↑ Superoxide dismutase activity ↑ Glutathione peroxidase activity ↑ Catalase activity ↑ Total antioxidant maleic dialdehyde activity ↑ Glutathione activity ↑	(Wang et al. 2017a)

Bacillus subtilis		10^9 CFU/g 60 days	Gibel carp (*Carassius auratus gibelio*) 60.51 ± 0.51 g	Protective effects against lead toxicity ↑ Superoxide dismutase ↑ Catalase and glutathione ↑ Lysozyme and IgM levels ↑ immune-related genes ↑	(Yin et al. 2018)
Bacillus amyloliquefaciens		0, 10^5, 10^7 and 10^9 CFU/g 70 days	Roho labeo (*Labeo rohita*) 20.23 g	Serum protein and globulin ↑ Albumin, lysozyme, and IgM ↑ Malondialdehyde ↑ Catalase, and superoxide dismutase ↑ Serum aspartate transaminase ↑ Serum alanine transaminase activity ↑ Liver malondialdehyde level ↑	(Nandi et al. 2018)
Bacillus subtilis and β-glucan	Commercial probiotic	1 g kg^{-1} β-glucan and 1×10^9 CFU kg^{-1} *B. subtilis* 70 days	Pengze crucian carp (*Carassius auratus* var. Pengze) 12.89 ± 0.04 g	Acid phosphatase activity ↑ Alkaline phosphatase activity ↑ Glutathione peroxidase activity ↑ Glutathione activity ↓ Catalase activity ↓ Total superoxide dismutase activity ↓	(Cao et al. 2019)

(continued)

Table 1.2 (continued)

Species	Source	Doses and duration	Fish species	Parameters investigated	References
Bacillus amyloliquefaciens FPTB16		10^7, 10^8 and 10^9 cells g^{-1} diet 4 weeks	Indian major carp (*Catla catla*) 25.98 ± 2.57 g	Oxygen radical production ↑ Serum lysozyme activity ↑ Total serum protein content ↑ Myeloperoxidase activity ↑ Alkaline phosphatase activity ↑ Expression of IL-1β, TNF-α, C3 and iNOS ↑ IFN-γ expression ↓	(Singh et al. 2017)
Bacillus subtilis	Isolated from the gut of grass carp	0, high-fat diet, and high-fat diet + *B. subtilis* (1×10^7 CFU g^{-1}) for 8 weeks	Grass carp (*Ctenopharyngodon idellus*) 50.24 ± 1.38 g	Serum low-density lipoprotein cholesterol ↑ Aspartate aminotransferase ↑ Hepatic mRNA expression of fatty acid synthase ↓ Carnitine palmitoyl transferases ↑ Glutathione ↑ Hydrogen peroxide (H_2O_2) ↓ Malondialdehyde (MDA) contents ↓	(Zhao et al. 2020)
Bacillus licheniformis	Isolated from grass carp	1×10^5 cfu/g and 1×10^6 cfu/g 56 days	Grass carp (*Ctenopharyngodon idella*) 16.5 g	Superoxide dismutase (SOD) activity ↑ Malondialdehyde (MDA) levels ↓ Antioxidant enzymes *MnSOD* ↑ Catalase (CAT) in the intestine ↑ Proinflammatory cytokines ↓ Anti-inflammatory cytokine ↑ *ZO-1*, *occludin*, and *claudin-c* ↑	(Qin et al. 2020)

B. subtilis YB-1 and *B. cereus* YB-2	Commercial probiotic	0, 10^7 and 10^{10} cfu/g diet 32 days	Sea cucumber (*Apostichopus japonicus*) 50 ± 0.5 g	Phagocytic activity ↑ Superoxide anion production ↑ Lysozyme activity ↑ Catalase activity ↑ Phenoloxidase activity ↑	(Li et al. 2015)
Bacillus baekryungensis MS1	Isolated from a sea cucumber pond in winter	0 and 10^7 CFU/ml 60 days	Sea cucumber (*Apostichopus japonicus*) 4.17 g ± 0.22 g	Superoxide dismutase activity ↑ Catalase activity ↑ Alkaline phosphatase activity ↑ Acid phosphatase activity ↑ Nitric oxide synthetase activity ↑ Phagocytosis and respiratory burst ↑ Ubiquitin-mediated proteolysis pathway ↑	(Liu et al. 2020)
Bacillus cereus EN25	Isolated from mud of sea cucumber culturing water bodies	0, 10^5, 10^7 and 10^9 CFU/g for 30 days	Juvenile sea cucumber (*Apostichopus japonicus*) 0.375 ± 0.024 g	Total coelomocytes count → Acid phosphatase activity → Phagocytosis activity ↑ Respiratory burst activity ↑ Total nitric oxide synthase activity ↑ Superoxide dismutase activity →	(Zhao et al. 2016)
B. amyloliquefaciens 54A and B. *pumilus* 47B	Isolated from gut of striped catfish	1×10^8, 3×10^8, and 5×10^8 CFU g^{-1} feed 90 days	Striped catfish (*Pangasianodon hypophthalmus*)	Phagocytic activity ↑ Respiratory bursts ↑ Lysozyme activity ↑	(Truong Thy et al. 2017)

(continued)

Table 1.2 (continued)

Species	Source	Doses and duration	Fish species	Parameters investigated	References
Bacillus aerius B81e	Isolated from healthy hybrid catfish	0 and 10^7 CFU g^{-1} feed 60 days	Basa fish (*Pangasius bocourti*) 69 g	Serum lysozyme activity ↑ Bactericidal activity ↑ Alternative complement activity ↑ Phagocytic activity ↑ Respiratory burst activity ↑	(Meidong et al. 2018)
B. subtilis, B. amyloliquefacien, B. cereus and a commercial B. amyloliquefaciens	Isolated from the intestine of African catfish	10^{10} CFU/ml 60 days	African catfish (*Clarias gariepinus*) 75.23 ± 1.6 g	Lysozyme activity ↑ Nitric oxide and IgM ↑ Myostatin cDNA levels ↑ PACAP expression ↑	(Reda et al. 2018)
Bacillus subtilis WB60 and *Lactobacillus plantarum* KCTC3928	Commercial probiotic	0; *B. subtilis* at 10^6, 10^7, 10^8 and *L. plantarum* at 10^6, 10^7, 10^8 CFU/g diet	Japanese eel (*Anguilla japonica*) 8.29 ± 0.06 g	Lysozyme activity ↑ Superoxide dismutase (SOD) ↑ Myeloperoxidase (MPO) ↑ Level of intestine glyceraldehyde-3-phosphate dehydrogenase (GAPDH) ↑ Heat shock protein 70, 90 ↑ Immunoglobulin (IgM) ↑	(Lee et al. 2017)
Bacillus subtilis WB60 and mannanoligosaccharide (MOS)	Commercial probiotic	BS: 0.0, 0.5, and 1.0×10^7 CFU/g diet and MOS: 0 and 5 g/kg diet. 8 weeks	Japanese eel (*Anguilla japonica*) 9.00 ± 0.11 g	Nonspecific enzymatic activities ↑ Heat shock protein 70 mRNA levels ↑ Immunoglobulin M expressions ↑	(Lee et al. 2018)
Bacillus subtilis or *licheniformis*) and (mannan or fructo oligosaccharide)		0, Probiotics (1.0×10^8 CFU/g diet) and prebiotics (5 g/kg diet) 12 weeks	Japanese eel (*Anguilla japonica*) 12.8 ± 0.47 g	Expression of heat shock protein 70 ↑ Expression of immunoglobulin M ↑	(Park et al. 2020)

B. subtilis E20	Commercial probiotic	0, 10^8, 10^9, and 10^{10} CFU kg^{-1} 56 days	Parrotfish (*Oplegnathus fasciatus*)	Lysozyme activity ↑ Respiratory burst ↑ Phagocytic activity ↑	(Liu et al. 2018)
B. velezensis V4 and *Rhodotorula mucilaginosa* compound	Isolated from the water of RAS rearing salmonid	0; *B. velezensis* V4 5×10^6, *R. mucilaginosa* 5×10^7 (CFU g^{-1}), (*B. velezensis* V4 1.5×10^7, *R. mucilaginosa* 1.5×10^8 (CFU g^{-1}), and *B. velezensis* V4 2.5×10^7, *R. mucilaginosa* 2.5×10^8 (CFU g^{-1}) 62 days	Juvenile Atlantic salmon (*Salmo salar L.*) 180.18 ± 3.64 g	Acid phosphatase ↑ IgM ↑ Nitric oxide ↑ Glutamic pyruvic transaminase ↑ Glutamic oxalacetic transaminase ↑ Lysozyme ↑ Total superoxide dismutase malondialdehyde ↑ Glutathione ↑ Glutathione peroxide ↑ Total antiodant capacity ↑ Malondiadehyde ↑	(Wang et al. 2019a, 2019b)
Bacillus subtilis	Commercial probiotic	0, 1×10^4, 1×10^6, 1×10^8 and 1×10^{10} CFU kg^{-1} diet 60 days	Red sea bream (*Pagrus major*) 3.99 ± 0.01 g	Hematocrit and hemoglobin values ↑ Nitro blue tetrazolium value ↑ Serum bactericidal activity ↑ Serum lysozyme activity ↑ Serum peroxidase activity ↑ Catalase activity ↑	(Zaineldin et al. 2018)
B. subtilis and *B. licheniformis*	Commercial probiotic	0.6 g/kg 84 days	Turbots (*Scophthalmus maximus*) 95.8 ± 17.7 g	Plasma lysozyme activity → Neutrophil reactive oxygen species (ROS) → Production, and total plasma protein levels → Plasma glucose and triglyceride ↑ Glucose levels ↑ Cortisol levels ↓	(Fuchs et al. 2017)

(continued)

Table 1.2 (continued)

Species	Source	Doses and duration	Fish species	Parameters investigated	References
Bacillus sp. SJ-10 plus β-glucooligosaccharides	Identified from traditional Korean fermented fish	0; 1×10^8 CFU g^{-1} BSJ-10; 0.1% BGO, and 1×10^8 CFU g^{-1} BSJ-10 + 0.1% BGO. 8 weeks	Olive flounder (*Paralichthys olivaceus*) 10 ± 0.25 g	Respiratory burst activity ↑ Superoxide dismutase ↑ Lysozyme activity ↑ Expression of interleukin (IL)-1β ↑ Tumor necrosis factor (TNF)-α ↑	(Hasan et al. 2018)
Bacillus amyloliquefaciens R8		0 and 92 x 10^6 CFU g^{-1} 30 days	Zebrafish (*Danio rerio*) 0.48 g	Xylanase activity ↑ mRNA expressions of glycolysis-related genes ↑ Enzyme activities ↑ Expression of innate immune-related genes ↑ Expressions of oxidative stress-related genes ↑	(Lin et al. 2019)
Bacillus pumilus SE5		0 and 1.0×10^8 CFU g^{-1} 60 days	Grouper (*Epinephelus coioides*) 14.6 ± 0.2 g	Phagocytic activity ↑ Serum complement C3 and IgM levels ↑ SOD activity ↑ Expression of TLR2 and pro-inflammatory cytokines ↑	(Yan et al. 2016)
Bacillus amyloliquefaciens (GB) and Yarrowia lipolytica lipase 2 (YLL2)	Commercial probiotic	0, 5.0 g/kg GB-9, 4.0 g/kg YLL2, and 5.0 g/kg GB-9 + 4.0 g/kg YLL2 12 weeks	Hybrid sturgeon (*Acipenser schrenkii* ♂and *Acipenser baeri* ♀) 5.0 g	Skin mucus lysozyme activity ↑ Leukocytes phagocytosis activity ↑ Reactive oxygen species level ↑ Alternative complement pathway activity ↑ Peroxidase and lysozyme activity ↑	(Fei et al. 2018)

Bacillus subtilis and *Bacillus cereus toyoi*	Commercial probiotic	0, 6 × 10^3 and 1.5 × 10^6 CFU g^{-1} of diet. 9 and 20 weeks	Rainbow trout (*Oncorhynchys mykiss*) and brown trout (*Salmo trutta*) 15.6 g	Plasma lysozyme activity ↑ Alternative complement activity ↑ Peroxidase activity ↑	(Ramos et al. 2017)
Bacillus amyloliquefaciens-JFP2		0 and 1.4 × 10^6 (CFU/g) of feed 90 days	Rock Bream (*Oplegnathus fasciatus*) 25.4 ± 0.13 g	Serum antioxidant and lysozyme activity ↑ Triglyceride and total cholesterol ↓ Alanine aminotransferase ↑ Aspartate aminotransferase ↑	(Kim et al. 2017)
BetaPlus® (*B. subtilis*, *B. licheniformis* and Isomaltooligosaccharides	Commercial probiotic	0 and 2 g kg^{-1} IMOS + 1 g kg^{-1} BetaPlus® in 7 weeks	Caspian Brown Trout (*Salmo trutta caspius*) 9 g	Immunoglobulin M levels ↑ Alanine aminotransferase activity ↑ Lactate dehydrogenase activity ↑	(Aftabgard et al. 2019)
Bacillus licheniformis	Commercial probiotic	10^3, 10^5, and 10^7 CFU/mL 8 weeks	Abalone (*Haliotis discus* hannai Ino.) 4.17 ± 0.32 g	Blood lymphocytes ↑ Activity of acid phosphatase ↑ Expression level of heat shock protein 70 ↑ Phagocytic activity ↑ Myeloperoxidase and catalase ↑ Expression levels of CAT ↑ Expression of thioredoxin ↑ Superoxide dismutase (SOD) ↑ Respiratory burst of blood lymphocytes ↑ Expression levels of Mn-SOD ↑	(Gao et al. 2018)

1.4.1 Tilapias

Selim and Reda (2015) found that *Bacillus amyloliquefaciens* spores supplementation at concentrations of 1×10^6 (G3) and 1×10^4 (G2) colony-forming units per gram (CFU g^{-1}) of feed significantly enhanced serum killing, serum nitric oxide, serum lysozyme activities, as well as IL-1 and TNF α mRNA levels in the kidneys of Nile tilapia, *O. niloticus*. The cell wall components of both Gram-positive and Gram-negative bacteria are able to stimulate cytokine production (Henderson et al. 1999). Probiotic bacteria colonize in the gut and are involved with the gut-associated lymphoid tissue to stimulate systemic signals that end with cytokine production (Kesarcodi-Watson et al. 2008; Rangavajhyala et al. 1997; Rescigno et al. 2001; Ringø 2011). Similarly, dietary inclusion of *B. subtilis* singularly or *B. subtilis* combined with *S. cerevisiae* and *A. oryzae*; *B. subtilis* with *B. licheniformis*; *B. subtilis* and *Bacillus licheniformis* (BS) combined with traditional Chinese medicine (TCM), and *B. subtilis* with Aqua NZ and AP193 significantly enhanced innate immune response, growth, relative immune, and antioxidant gene expressions of Nile tilapia (Abarike et al. 2018a; Abarike et al. 2018b; Addo et al. 2017a, 2017b; Iwashita et al. 2015; Liu et al. 2017; Wang et al. 2020a). Dietary inclusion of *B. licheniformis* has been found to increase alkaline phosphatase, myeloperoxidase, lysozyme, reactive oxygen species, reactive nitrogen species, superoxide dismutase, and glutathione peroxidase of Mozambique tilapia (*Oreochromis mossambicus*) (Gobi et al. 2018). Also, supplementation of *Bacillus licheniformis* HGA8B significantly improved lysozyme activity and content of complement C3 (Han et al. 2015). It is well documented that, the immune system can be nonspecifically modulated by probiotics (Hoseinifar et al. 2015; Lazado and Caipang 2014; Nayak 2010). Moreover, colony formation and adhesion of probiotics in the intestine of fish are necessary to enhance the immune responses (Ausubel 2005). Interaction between probiotic cells and immune system are through microbe associated molecular patterns (MAMPs) consisting of specific cell wall polysaccharides, peptidoglycan, lipoprotein anchors, and lipoteichoic acids (Hosoi et al. 2003). Cells or components of immune system can interact with MAMPs by pattern recognition receptor such as toll-like receptors, C-type receptor, and nucleotide oligomerization domain-like receptors (Bron et al. 2012; Kleerebezem et al. 2010). This fact may indicate that, addition of fresh culture of *B. licheniformis* to the diet maintains a high level of probiotics in the diet and improve the immune responses in fish. Similar results have been reported in Nile tilapia fed *B. cereus* and *B. pumilus* (Srisapoome and Areechon 2017; Wang et al. 2017b).

1.4.2 Shrimps

In shrimp, *B. licheniformis* has been intensively applied in Pacific white shrimp (*Litopenaeus vannamei*). Amoah et al. (2019) indicated that dietary inclusion of 1×10^8 CFU g^{-1} feed significantly enhanced activity of lysozyme (LYZ), acid phosphatase (ACP), superoxide dismutase (SOD), total protein (TP), albumin (ALB) in serum, glutathione peroxidase (GSH-Px) in serum and liver of Nile tilapia. Similarly, dietary administration of *B. licheniformis* significantly upregulated the expression of catalase, glutathione peroxidase, superoxide dismutase (SOD), penaeidin-3a (Pen-3a), and heat shock protein (Hsp-70) genes of Pacific white shrimp, *Litopenaeus vannamei* (Chen et al. 2020a, 2020b). In addition, the combination of *B. licheniformis* with *B. subtilis* significantly enhanced lysozyme and hemocyte cell count and upregulated the expression of proPO, LvToll1 and SOD, Hsp70, and TGase genes (Sadat Hoseini Madani et al. 2018; Sánchez-Ortiz et al. 2016). Likewise, dietary inclusion of *B. subtilis* E20 singularly or combined with other probiotics significantly innate immune response and related immune gene expression of Pacific white shrimp, *Litopenaeus vannamei* (Chien et al. 2020; Won et al. 2020a). Also, Khademzade et al. (2020) reported that dietary inclusion of *Bacillus cereus* and *Pediococcus acidilactici* significantly enhanced total hemocyte count, total protein, and lysozyme activities of *L. vannamei*. Similar results were found in tiger shrimp and freshwater pawn fed *Bacillus* sp. and *Bacillus coagulant* where significant increase in total heterotrophic count, amylolytic, cellulolytic, and proteolytic bacterial counts, phagocytic, lysozyme, and respiratory burst activities was recorded (De et al. 2018; Gupta et al. 2016). At molecular levels, Sánchez-Ortiz et al. (2016) indicated that dietary supplementation of *Bacillus* spp. resulted in upregulation of proPO, LvToll1, SOD genes, except the TGase gene expression. Similarly, Tepaamorndech et al. (2019) revealed that dietary inclusion of *Bacillus aryabhattai* TBRC8450 significantly upregulated C-type lectin, penaeidin-3, and heat shock protein 60 genes, as well as enhanced thioredoxin, ferritin, phenoloxidase, and total antioxidant activities of Pacific white shrimp, *Litopenaeus vannamei*. However, no significant increase in total hemocyte count, and superoxide dismutase were observed (Tepaamorndech et al. 2019).

1.4.3 Carps

In mrigal, *Cirrhinus mrigala*, Bandyopadhyay et al. (2015) indicated that dietary inclusion of *Bacillus* sp. PP9 significantly improved hemoglobin percentage, total erythrocyte count, total leukocyte count, corpuscular hemoglobin, total serum protein, albumin globulin ratio, and serum bactericidal activity. Similarly, dietary supplementation of *B. subtilis* singularly or combined with other *Bacillus* sp. and prebiotics significantly stimulated hematological, antioxidant, and immunological parameters of *Labeo rohita* (Ramesh and Souissi 2018); Labeo catla, *Catla catla* (Sangma and Kamilya 2015); common carp, *Cyprinus carpio* (Wang et al. 2017a); grass carp, *Ctenopharyngodon idellus* (Zhao et al. 2020), and Pengze crucian carp, *Carassius auratus* var. Pengze (Cao et al. 2019). At gene level, Yin et al. (2018) found that supplementation of *B. subtilis* resulted in higher protective effects against lead toxicity, superoxide dismutase, catalase and glutathione, lysozyme and IgM levels, as well as immune-related genes of gibel carp, *Carassius auratus gibelio*. Likewise, dietary inclusion of *B. amyloliquefaciens* significantly stimulated innate immune response, antioxidant, and relative immune gene expressions of roho labeo,

Labeo rohita (Nandi et al. 2018); Indian major carp, *Catla catla* Singh et al. (2017), and grass carp, *Ctenopharyngodon idella* (Qin et al. 2020).

1.4.4 Sea Cucumber (*Apostichopus japonicus*)

Supplementation of *B. cereus* singularly or combined with *B. subtilis* significantly enhanced total coelomocytes count, acid phosphatase, phagocytosis, respiratory burst, total nitric oxide synthase, catalase, phenoloxidase, and superoxide dismutase activities (Li et al. 2015). Recently, Liu et al. (2020) indicated that dietary administration of *B. baekryungensis* significantly enhanced superoxide dismutase, catalase, alkaline phosphatase, acid phosphatase, nitric oxide synthetase, phagocytosis, respiratory burst activities, and ubiquitin-mediated proteolysis pathway. Ubiquitin-mediated proteolysis plays an important role in the dynamic regulation of host defense against pathogen infection. It has been reported that a number of key joint molecules in the natural immune and antiviral signaling pathways can be modified by ubiquitination to regulate the antiviral immune response of the body (Chuang and Ulevitch 2004; Liu and Chen 2011). Ubiquitination plays an important role in the Toll-like receptor (TLR) signaling pathway. The activation of this pathway leads to the upregulated expression of Toll-like receptors and enhances nonspecific immunity (Bhoj and Chen 2009). The upregulation of *TLR* in this study is consistent with the above theory. In the immune system, mTOR signaling plays an important role in maintaining immune homeostasis, for example, the survival and migration of natural immune cells and the secretion of inflammatory factors (Katholnig et al. 2013; Weichhart et al. 2008). Studies have found that the mTOR signaling pathway negatively regulates nonspecific immune responses (Weichhart et al. 2008). Therefore, the downregulation of the mTOR pathway in sea cucumber is beneficial to improve sea cucumber immunity.

1.4.5 Catfish

In striped catfish, *Pangasianodon hypophthalmus*, Truong Thy et al. (2017) reported that dietary inclusion of *B. amyloliquefaciens* and *B. pumilus* significantly enhanced phagocytic, respiratory bursts, and lysozyme activities. Similar results were observed in basa fish, *Pangasius bocourti* fed *B. aerius* (Meidong et al. 2018). Likewise, combination of *B. subtilis*, *B. amyloliquefaciens*, *B. cereus*, and *B. amyloliquefaciens* (Reda et al. 2018).

1.4.6 Japanese eel

Dietary inclusion of *B. subtilis* and *Lactobacillus plantarum* significantly enhanced lysozyme, superoxide dismutase (SOD), myeloperoxidase (MPO), level of intestine glyceraldehyde-3-phosphate dehydrogenase (GAPDH), heat shock protein 70, 90, and immunoglobulin (IgM). Similarly, dietary inclusion of *B. subtilis* and mannanoligosaccharide (MOS) significantly improved nonspecific enzymatic activities, heat shock protein 70 mRNA levels, and immunoglobulin M expressions (Lee et al. 2018). More recently, Park et al. (2020) indicated that dietary inclusion of *B. subtilis* or *B. licheniformis* and mannan or fructo oligosaccharide upregulated heat shock protein 70 and immunoglobulin M genes.

1.4.7 Other Species

Dietary inclusion of *B. subtilis* singularly or combined with *B. licheniformis*, *Bacillus cereus toyoi*, and isomaltooligosaccharides significantly stimulated hematological, innate immune response, antioxidant, and gene expression of parrotfish, *Oplegnathus fasciatus* (Liu et al. 2018); red sea bream, *Pagrus major* (Zaineldin et al. 2018); turbots, *Scophthalmus maximus* (Fuchs et al. 2017); rainbow trout, *Oncorhynchus mykiss* and brown trout, *Salmo trutta* (Ramos et al. 2017), and Caspian brown trout, *Salmo trutta caspicus* (Aftabgard et al. 2019). Regarding *B. amyloliquefaciens*, dietary inclusion of *B. amyloliquefaciens* singularly or combined with Yarrowia lipolytica lipase 2 (YLL2), *B. licheniformis* significantly enhanced innate immune response, antioxidant, and gene expression of rock bream, *Oplegnathus fasciatus* (Kim et al. 2017); hybrid sturgeon, *Acipenser schrenckii* ♂and *Acipenser baerii* ♀ (Fei et al. 2018), and zebrafish, *Danio rerio* (Lin et al. 2019).

In juvenile Atlantic salmon (*Salmo salar* L.), Wang et al. (2019a) reported that *B. velezensis* V4 and *Rhodotorula mucilaginosa* compound led to an increase in acid phosphatase, IgM, nitric oxide, glutamic pyruvic transaminase, glutamic oxalacetic transaminase, lysozyme, total superoxide dismutase malondialdehyde, glutathione, glutathione peroxide, total antioxidant capacity, and malondialdehyde. Similarly, dietary inclusion of *Bacillus licheniformis* significantly enhanced hematological, innate immune response, and Mn-SOD gene expression (Gao et al. 2018). Also, significant increase in innate immune response and relative immune gene expressions were observed in grouper, *Epinephelus coioides,* fed *Bacillus pumilus* (Yan et al. 2016) and in olive flounder, *Paralichthys olivaceus,* fed *Bacillus* sp. SJ-10 plus β-glucooligosaccharides (Hasan et al. 2018).

1.5 Conclusion

This chapter addressed the role of *Bacillus* probiotics in sustainable aquaculture. Although a wide range of researches have indicated beneficial effects of *Bacillus* species on grow rate, immunity, and disease resistance of farmed fish and shellfish, the investigated effects were species specific. In order to evaluate in vivo adherence and colonization of *Bacillus* bacteria within the complex microbial ecosystem of the intestine, detection of green fluorescence protein (GFP) tagged strains or fluorescence in situ hybridization (FISH) targeting 16S rRNA to identify the probiotics on the mucus surface must be carried out. Furthermore, mucus-associated (autochthonous) microbiome must be investigated by next-generation sequencing (NGS), transcriptomic, metagenomics or proteomic profiling, and not the allochthonous microbiome; mostly investigated *per sc*. In addition, we recommend that gnotobiotic approaches are used in future studies, as the gnotobiotic approaches have been reported to have important roles to understand the function of gut microbiota on numerous biological processes of the host. Moreover, data is needed to understand the mechanisms by which the immune system of the intestinal mucosa discriminates between pathogenic, probiotics, and commensal microorganisms.

References

Abarike ED, Cai J, Lu Y, Yu H, Chen L, Jian J et al (2018a) Effects of a commercial probiotic BS containing *Bacillus subtilis* and *Bacillus licheniformis* on growth, immune response and disease resistance in Nile tilapia, *Oreochromis niloticus*. Fish Shellfish Immunol 82:229–238

Abarike ED, Jian J, Tang J, Cai J, Yu H, Lihua C, Jun L (2018b) Influence of traditional Chinese medicine and *Bacillus* species (TCMBS) on growth, immune response and disease resistance in Nile tilapia, *Oreochromis niloticus*. Aquac Res 49(7):2366–2375

Abd El-Rhman AM, Khattab YAE, Shalaby AME (2009) *Micrococcus luteus* and *Pseudomonas* species as probiotics for promoting the growth performance and health of Nile tilapia, *Oreochromis niloticus*. Fish Shellfish Immunol 27(2):175–180. https://doi.org/10.1016/j.fsi.2009.03.020

Abriouel H, Franz CM, Omar NB, Gálvez A (2011) Diversity and applications of Bacillus bacteriocins. FEMS Microbiol Rev 35(1):201–232

Addo S, Carrias AA, Williams MA, Liles MR, Terhune JS, Davis DA (2017a) Effects of *Bacillus subtilis* strains and the prebiotic Previda® on growth, immune parameters and susceptibility to *Aeromonas hydrophila* infection in Nile tilapia, *Oreochromis niloticus*. Aquac Res 48(9):4798–4810

Addo S, Carrias AA, Williams MA, Liles MR, Terhune JS, Davis DA (2017b) Effects of *Bacillus subtilis* strains on growth, immune parameters, and *Streptococcus iniae* susceptibility in Nile tilapia, *Oreochromis niloticus*. J World Aquac Soc 48(2):257–267

Adel M, Yeganeh S, Dawood MAO, Safari R, Radhakrishnan S (2017) Effects of *Pediococcus pentosaceus* supplementation on growth performance, intestinal microflora and disease resistance of white shrimp, *Litopenaeus vannamei*. Aquac Nutr 23(6):1401–1409. https://doi.org/10.1111/anu.12515

Afrilasari W, Meryandini A (2016) Effect of probiotic Bacillus megaterium PTB 1.4 on the population of intestinal microflora, digestive enzyme activity and the growth of catfish (Clarias sp.). HAYATI J Biosci 23(4):168–172. https://doi.org/10.1016/j.hjb.2016.12.005

Aftabgard M, Salarzadeh A, Mohseni M, Bahri Shabanipour AH, Zorriehzahra MEJ (2019) The combined efficiency of dietary isomaltooligosaccharides and Bacillus spp. on the growth, hemato-serological, and intestinal microbiota indices of caspian brown Trout (Salmo trutta caspius Kessler, 1877). Probiotics Antimicrob Proteins 11(1):198–206. https://doi.org/10.1007/s12602-017-9361-z

Alavandi SV, Muralidhar M, Syama Dayal J, Rajan JS, Ezhil Praveena P, Bhuvaneswari T et al (2019) Investigation on the infectious nature of running mortality syndrome (RMS) of farmed Pacific white leg shrimp, *Penaeus vannamei* in shrimp farms of India. Aquaculture 500:278–289. https://doi.org/10.1016/j.aquaculture.2018.10.027

Ambas I, Fotedar R, Buller N (2017) Synbiotic effect of Bacillus mycoides and organic selenium on immunity and growth of marron, Cherax cainii (Austin, 2002). Aquac Res 48(6):2729–2740. https://doi.org/10.1111/are.13105

Amoah K, Huang Q-C, Tan B-P, Zhang S, Chi S-Y, Yang Q-H et al (2019) Dietary supplementation of probiotic Bacillus coagulans ATCC 7050, improves the growth performance, intestinal morphology, microflora, immune response, and disease confrontation of Pacific white shrimp, Litopenaeus vannamei. Fish Shellfish Immunol 87:796–808. https://doi.org/10.1016/j.fsi.2019.02.029

Apún-Molina JP, Santamaría-Miranda A, Luna-González A, Martínez-Díaz SF, Rojas-Contreras M (2009) Effect of potential probiotic bacteria on growth and survival of tilapia Oreochromis niloticus L., cultured in the laboratory under high density and suboptimum temperature. Aquac Res 40(8):887–894

Arena A, Maugeri TL, Pavone B, Iannello D, Gugliandolo C, Bisignano G (2006) Antiviral and immunoregulatory effect of a novel exopolysaccharide from a marine thermotolerant *Bacillus licheniformis*. Int Immunopharmacol 6(1):8–13

Ashaolu TJ (2020) Immune boosting functional foods and their mechanisms: a critical evaluation of probiotics and prebiotics. Biomed Pharmacother 130:110625. https://doi.org/10.1016/j.biopha.2020.110625

Ausubel FM (2005) Are innate immune signaling pathways in plants and animals conserved? Nat Immunol 6(10):973–979. https://doi.org/10.1038/ni1253

Avila EM, Juario JV (1987) Yolk and oil globule utilization and developmental morphology of the digestive tract epithelium in larval rabbitfish, *Siganus guttatus* (Bloch). Aquaculture 65 (3–4):319–331

Azarin H, Aramli MS, Imanpour MR, Rajabpour M (2015) Effect of a probiotic containing *Bacillus licheniformis* and *Bacillus subtilis* and ferroin solution on growth performance, body composition and haematological parameters in kutum (*Rutilus frisii kutum*) fry. Probioticsand AntimicrobProteins 7(1):31–37. https://doi.org/10.1007/s12602-014-9180-4

Bairagi A, Ghosh KS, Sen SK, Ray AK (2002) Enzyme producing bacterial flora isolated from fish digestive tracts. Aquac Int 10(2):109–121. https://doi.org/10.1023/a:1021355406412

Bairagi A, Sarkar Ghosh K, Sen S, Ray A (2004) Evaluation of the nutritive value of *Leucaena leucocephala* leaf meal, inoculated with fish intestinal bacteria *Bacillus subtilis* and *Bacillus circulans* in formulated diets for rohu, *Labeo rohita* (Hamilton) fingerlings. Aquac Res 35 (5):436–446

Balcázar JL, Rojas-Luna T, Cunningham DP (2007) Effect of the addition of four potential probiotic strains on the survival of pacific white shrimp (*Litopenaeus vannamei*) following immersion challenge with *Vibrio parahaemolyticus*. J Invertebr Pathol 96(2):147–150

Bandyopadhyay P, Sarkar B, Mahanty A, Rathore RM, Patra BC (2015) Dietary administered Bacillus sp. PP9 enhances growth, nutrition and immunity in Cirrhinus mrigala (Hamilton). Proc Natl Acad Sci India Sect B 85(3):759–766. https://doi.org/10.1007/s40011-015-0561-6

Berkeley SA, Chapman C, Sogard SM (2004) Maternal age as a determinant of larval growth and survival in a marine fish, *Sebastes melanops*. Ecology 85(5):1258–1264

Bhoj VG, Chen ZJ (2009) Ubiquitylation in innate and adaptive immunity. Nature 458 (7237):430–437

Bron PA, van Baarlen P, Kleerebezem M (2012) Emerging molecular insights into the interaction between probiotics and the host intestinal mucosa. Nat Rev Microbiol 10(1):66–78

Butt RL, Volkoff H (2019) Gut microbiota and energy homeostasis in fish. Front Endocrinol 10(9). https://doi.org/10.3389/fendo.2019.00009

Cai D, Rao Y, Zhan Y, Wang Q, Chen S (2019) Engineering Bacillus for efficient production of heterologous protein: current progress, challenge and prospect. J Appl Microbiol 126 (6):1632–1642. https://doi.org/10.1111/jam.14192

Cao H, Yu R, Zhang Y, Hu B, Jian S, Wen C et al (2019) Effects of dietary supplementation with β-glucan and Bacillus subtilis on growth, fillet quality, immune capacity, and antioxidant status of Pengze crucian carp (Carassius auratus var. Pengze). Aquaculture 508:106–112. https://doi.org/10.1016/j.aquaculture.2019.04.064

Castex M, Lemaire P, Wabete N, Chim L (2009) Effect of dietary probiotic *Pediococcus acidilactici* on antioxidant defences and oxidative stress status of shrimp *Litopenaeus stylirostris*. Aquaculture 294(3–4):306–313

Caulier S, Nannan C, Gillis A, Licciardi F, Bragard C, Mahillon J (2019) Overview of the antimicrobial compounds produced by members of the *Bacillus subtilis* group. Front Microbiol 10:302–302. https://doi.org/10.3389/fmicb.2019.00302

Cencic A, Chingwaru W (2010) The role of functional foods, nutraceuticals, and food supplements in intestinal health. Nutrients 2(6):611–625. https://doi.org/10.3390/nu2060611

Cerezuela R, Meseguer J, Esteban MA (2011) Current knowledge in synbiotic use for fish aquaculture: a review. Res Aquac Res Dev S1(008). https://doi.org/10.4172/2155-9546.S1-008

Cha J-H, Rahimnejad S, Yang S-Y, Kim K-W, Lee K-J (2013) Evaluations of *Bacillus* spp. as dietary additives on growth performance, innate immunity and disease resistance of olive flounder (*Paralichthys olivaceus*) against Streptococcus iniae and as water additives. Aquaculture 402–403:50–57. https://doi.org/10.1016/j.aquaculture.2013.03.030

Chang Y-P, Liu C-H, Wu C-C, Chiang C-M, Lian J-L, Hsieh S-L (2012) Dietary administration of zingerone to enhance growth, non-specific immune response, and resistance to *Vibrio alginolyticus* in Pacific white shrimp (*Litopenaeus vannamei*) juveniles. Fish Shellfish Immunol 32(2):284–290

Chen Y-Y, Sim SS, Chiew SL, Yeh S-T, Liou C-H, Chen J-C (2012) Dietary administration of a *Gracilaria tenuistipitata* extract produces protective immunity of white shrimp L*itopenaeus vannamei* in response to ammonia stress. Aquaculture 370:26–31

Chen M, Chen X-Q, Tian L-X, Liu Y-J, Niu J (2020a) Beneficial impacts on growth, intestinal health, immune responses and ammonia resistance of pacific white shrimp (*Litopenaeus vannamei*) fed dietary synbiotic (mannan oligosaccharide and *Bacillus licheniformis*). Aquac Rep 17:100408. https://doi.org/10.1016/j.aqrep.2020.100408

Chen M, Chen X-Q, Tian L-X, Liu Y-J, Niu J (2020b) Enhanced intestinal health, immune responses and ammonia resistance in Pacific white shrimp (*Litopenaeus vannamei*) fed dietary hydrolyzed yeast (*Rhodotorula mucilaginosa*) and *Bacillus licheniformis*. Aquac Rep 17:100385. https://doi.org/10.1016/j.aqrep.2020.100385

Chien C-C, Lin T-Y, Chi C-C, Liu C-H (2020) Probiotic, *Bacillus subtilis* E20 alters the immunity of white shrimp, *Litopenaeus vannamei* via glutamine metabolism and hexosamine biosynthetic pathway. Fish Shellfish Immunol 98:176–185. https://doi.org/10.1016/j.fsi.2020.01.014

Chiu L, Bazin T, Truchetet M-E, Schaeverbeke T, Delhaes L, Pradeu T (2017) Protective microbiota: from localized to long-reaching co-immunity. Front Immunol 8:1678. https://doi.org/10.3389/fimmu.2017.01678

Chuang T-H, Ulevitch RJ (2004) Triad3A, an E3 ubiquitin-protein ligase regulating Toll-like receptors. Nat Immunol 5(5):495–502

Corr SC, Li Y, Riedel CU, O'Toole PW, Hill C, Gahan CG (2007) Bacteriocin production as a mechanism for the antiinfective activity of *Lactobacillus salivarius* UCC118. Proc Natl Acad Sci 104(18):7617–7621

Cutting SM (2011) Bacillus probiotics. Food Microbiol 28(2):214–220

da Paixão AEM, dos Santos JC, Pinto MS, Pereira DSP, de Oliveira Ramos CEC, Cerqueira RB et al (2017) Effect of commercial probiotics (Bacillus subtilis and Saccharomyces cerevisiae) on growth performance, body composition, hematology parameters, and disease resistance against Streptococcus agalactiae in tambaqui (Colossoma macropomum). Aquac Int 25(6):2035–2045. https://doi.org/10.1007/s10499-017-0173-7

Dash G, Raman RP, Prasad KP, Makesh M, Pradeep M, Sen S (2015) Evaluation of paraprobiotic applicability of *Lactobacillus plantarum* in improving the immune response and disease protection in giant freshwater prawn, *Macrobrachium rosenbergii* (de Man, 1879). Fish Shellfish Immunol 43(1):167–174

Dawood MA, El-Dakar A, Mohsen M, Abdelraouf E, Koshio S, Ishikawa M, Yokoyama S (2014) Effects of using exogenous digestive enzymes or natural enhancer mixture on growth, feed utilization, and body composition of rabbitfish, *Siganus rivulatus*. J Agric Sci Technol B 4(3B)

Dawood MAO, Koshio S, Ishikawa M, El-Sabagh M, Esteban MA, Zaineldin AI (2016) Probiotics as an environment-friendly approach to enhance red sea bream, *Pagrus major* growth, immune response and oxidative status. Fish Shellfish Immunol 57:170–178. https://doi.org/10.1016/j.fsi.2016.08.038

Dawood MAO, Koshio S, Abdel-Daim MM, Hien DV (2019) Probiotic application for sustainable aquaculture. Rev Aquac 11(3):907–924. https://doi.org/10.1111/raq.12272

De D, Ananda Raja R, Ghoshal TK, Mukherjee S, Vijayan KK (2018) Evaluation of growth, feed utilization efficiency and immune parameters in tiger shrimp (Penaeus monodon) fed diets supplemented with or diet fermented with gut bacterium Bacillus sp. DDKRC1. isolated from gut of Asian seabass (Lates calcarifer). Aquac Res 49(6):2147–2155. https://doi.org/10.1111/are.13669

Desriac F, Defer D, Bourgougnon N, Brillet B, Le Chevalier P, Fleury Y (2010) Bacteriocin as weapons in the marine animal-associated bacteria warfare: inventory and potential applications as an aquaculture probiotic. Mar Drugs 8(4):1153–1177

Dias JAR, Abe HA, Sousa NC, Couto MVS, Cordeiro CAM, Meneses JO et al (2018) Dietary supplementation with autochthonous Bacillus cereus improves growth performance and survival in tambaqui Colossoma macropomum. Aquac Res 49(9):3063–3070. https://doi.org/10.1111/are.13767

Duc LH, Hong HA, Cutting SM (2003) Germination of the spore in the gastrointestinal tract provides a novel route for heterologous antigen delivery. Vaccine 21(27–30):4215–4224

El-Haroun E, Goda AS, Kabir Chowdhury M (2006) Effect of dietary probiotic Biogen® supplementation as a growth promoter on growth performance and feed utilization of Nile tilapia *Oreochromis niloticus* (L.). Aquac Res 37(14):1473–1480

Elshaghabee FM, Rokana N, Gulhane RD, Sharma C, Panwar H (2017) *Bacillus* as potential probiotics: status, concerns, and future perspectives. Front Microbiol 8:1490

FAO (2020) The State of World Fisheries and Aquaculture (SOFIA). FAO, Rome

FAO/WHO (2001) Health and nutritional properties of probiotics in food including powder milk with liver lactic acid bacteria

Fayol-Messaoudi D, Berger CN, Coconnier-Polter M-H, Lievin-Le Moal V, Servin AL (2005) pH-, Lactic acid-, and non-lactic acid-dependent activities of probiotic Lactobacilli against *Salmonella enteric*a Serovar Typhimurium. Appl Environ Microbiol 71(10):6008–6013

Fei H, Lin G-D, Zheng C-C, Huang M-M, Qian S-C, Wu Z-J et al (2018) Effects of Bacillus amyloliquefaciens and Yarrowia lipolytica lipase 2 on immunology and growth performance of Hybrid sturgeon. Fish Shellfish Immunol 82:250–257. https://doi.org/10.1016/j.fsi.2018.08.031

Feng J, Chang X, Zhang Y, Yan X, Zhang J, Nie G (2019) Effects of *Lactococcus lactis* from *Cyprinus carpio* L. as probiotics on growth performance, innate immune response and disease resistance against *Aeromonas hydrophila*. Fish Shellfish Immunol 93:73–81. https://doi.org/10.1016/j.fsi.2019.07.028

Fuchs VI, Schmidt J, Slater MJ, Buck BH, Steinhagen D (2017) Influence of immunostimulant polysaccharides, nucleic acids, and Bacillus strains on the innate immune and acute stress response in turbots (Scophthalmus maximus) fed soy bean- and wheat-based diets. Fish Physiol Biochem 43(6):1501–1515. https://doi.org/10.1007/s10695-017-0388-6

Gao X, Zhang M, Li X, Han Y, Wu F, Liu Y (2018) Effects of a probiotic (Bacillus licheniformis) on the growth, immunity, and disease resistance of Haliotis discus hannai Ino. Fish Shellfish Immunol 76:143–152. https://doi.org/10.1016/j.fsi.2018.02.028

Giri SS, Sukumaran V, Oviya M (2013) Potential probiotic *Lactobacillus plantarum* VSG3 improves the growth, immunity, and disease resistance of tropical freshwater fish, *Labeo rohita*. Fish Shellfish Immunol 34(2):660–666. https://doi.org/10.1016/j.fsi.2012.12.008

Gobi N, Vaseeharan B, Chen J-C, Rekha R, Vijayakumar S, Anjugam M, Iswarya A (2018) Dietary supplementation of probiotic *Bacillus licheniformis* Dahb1 improves growth performance, mucus and serum immune parameters, antioxidant enzyme activity as well as resistance against *Aeromonas hydrophila* in tilapia *Oreochromis mossambicus*. Fish Shellfish Immunol 74:501–508. https://doi.org/10.1016/j.fsi.2017.12.066

Gómez B, Munekata PES, Zhu Z, Barba FJ, Toldrá F, Putnik P et al (2019) Chapter 7: Challenges and opportunities regarding the use of alternative protein sources: aquaculture and insects. In: Toldrá F (ed) Advances in food and nutrition research, vol 89. Academic Press, San Diego, CA, pp 259–295

Gupta A, Verma G, Gupta P (2016) Growth performance, feed utilization, digestive enzyme activity, innate immunity and protection against Vibrio harveyi of freshwater prawn, Macrobrachium rosenbergii fed diets supplemented with Bacillus coagulans. Aquac Int 24 (5):1379–1392

Hai NV (2015) The use of probiotics in aquaculture. J Appl Microbiol 119(4):917–935

Hammami I, Jaouadi B, Bacha AB, Rebai A, Bejar S, Nesme X, Rhouma A (2012) *Bacillus subtilis* bacteriocin Bac 14B with a broad inhibitory spectrum: purification, amino acid sequence analysis, and physicochemical characterization. Biotechnol Bioprocess Eng 17(1):41–49

Hamza A, Fdhila K, Zouiten D, Masmoudi AS (2016) Virgibacillus proomii and Bacillus mojavensis as probiotics in sea bass (Dicentrarchus labrax) larvae: effects on growth performance and digestive enzyme activities. Fish Physiol Biochem 42(2):495–507. https://doi.org/10.1007/s10695-015-0154-6

Han B, Long W-Q, He J-Y, Liu Y-J, Si Y-Q, Tian L-X (2015) Effects of dietary *Bacillus licheniformis* on growth performance, immunological parameters, intestinal morphology and resistance of juvenile Nile tilapia (*Oreochromis niloticus*) to challenge infections. Fish Shellfish Immunol 46(2):225–231

Hasan MT, Jang WJ, Kim H, Lee B-J, Kim KW, Hur SW et al (2018) Synergistic effects of dietary *Bacillus* sp. SJ-10 plus β-glucooligosaccharides as a synbiotic on growth performance, innate immunity and streptococcosis resistance in olive flounder (*Paralichthys olivaceus*). Fish Shellfish Immunol 82:544–553. https://doi.org/10.1016/j.fsi.2018.09.002

Hauville MR, Zambonino-Infante JL, Gordon Bell J, Migaud H, Main KL (2016) Effects of a mix of Bacillus sp. as a potential probiotic for Florida pompano, common snook and red drum larvae performances and digestive enzyme activities. Aquac Nutr 22(1):51–60. https://doi.org/10.1111/anu.12226

Hemarajata P, Versalovic J (2013) Effects of probiotics on gut microbiota: mechanisms of intestinal immunomodulation and neuromodulation. Therap Adv Gastroenterol 6(1):39–51. https://doi.org/10.1177/1756283X12459294

Henderson B, Wilson M, McNab R, Lax A (1999) The innate immune response. *B. Henderson, M. Wilson, M*, 311–353

Hernández AJ, Romero A, Gonzalez-Stegmaier R, Dantagnan P (2016) The effects of supplemented diets with a phytopharmaceutical preparation from herbal and macroalgal origin on disease resistance in rainbow trout against Piscirickettsia salmonis. Aquaculture 454(Supplement C):109–117. https://doi.org/10.1016/j.aquaculture.2015.12.016

Hlordzi V, Kuebutornye FKA, Afriyie G, Abarike ED, Lu Y, Chi S, Anokyewaa MA (2020) The use of Bacillus species in maintenance of water quality in aquaculture: a review. Aquac Rep 18:100503. https://doi.org/10.1016/j.aqrep.2020.100503

Hong HA, Duc LH, Cutting SM (2005) The use of bacterial spore formers as probiotics. FEMS Microbiol Rev 29(4):813–835

Hoseinifar SH, Mirvaghefi A, Merrifield DL (2011) The effects of dietary inactive brewer's yeast *Saccharomyces cerevisiae* var. ellipsoideus on the growth, physiological responses and gut microbiota of juvenile beluga (*Huso huso*). Aquaculture 318(1–2):90–94

Hoseinifar SH, Esteban MÁ, Cuesta A, Sun Y-Z (2015) Prebiotics and fish immune response: a review of current knowledge and future perspectives. Rev Fisher Sci Aquac 23(4):315–328

Hoseinifar SH, Dadar M, Ringø E (2017) Modulation of nutrient digestibility and digestive enzyme activities in aquatic animals: the functional feed additives scenario. Aquac Res 48 (8):3987–4000

Hoseinifar SH, Sun Y-Z, Wang A, Zhou Z (2018) Probiotics as means of diseases control in aquaculture, a review of current knowledge and future perspectives. Front Microbiol 9:2429. https://doi.org/10.3389/fmicb.2018.02429

Hosoi T, Hirose R, Saegusa S, Ametani A, Kiuchi K, Kaminogawa S (2003) Cytokine responses of human intestinal epithelial-like Caco-2 cells to the nonpathogenic bacterium Bacillus subtilis (natto). Int J Food Microbiol 82(3):255–264. https://doi.org/10.1016/S0168-1605(02)00311-2

Huynh T-G, Cheng A-C, Chi C-C, Chiu K-H, Liu C-H (2018) A synbiotic improves the immunity of white shrimp, *Litopenaeus vannamei*: Metabolomic analysis reveal compelling evidence. Fish Shellfish Immunol 79:284–293

Ibrahem MD (2015) Evolution of probiotics in aquatic world: Potential effects, the current status in Egypt and recent prospectives. J Adv Res 6(6):765–791. https://doi.org/10.1016/j.jare.2013.12.004

Irianto A, Austin B (2002) Use of probiotics to control furunculosis in rainbow trout, *Oncorhynchus mykiss* (Walbaum). J Fish Dis 25(6):333–342. https://doi.org/10.1046/j.1365-2761.2002.00375.x

Iwashita MKP, Nakandakare IB, Terhune JS, Wood T, Ranzani-Paiva MJT (2015) Dietary supplementation with Bacillus subtilis, Saccharomyces cerevisiae and Aspergillus oryzae enhance immunity and disease resistance against Aeromonas hydrophila and Streptococcus iniae infection in juvenile tilapia Oreochromis niloticus. Fish Shellfish Immunol 43(1):60–66. https://doi.org/10.1016/j.fsi.2014.12.008

Jafariyan H, Sahandi J, Taati M, Eslamloo K (2015) The use of Bacillus probiotics in-feed improved stress resistance of Trichopodus trichopterus (Pallas, 1770) larvae. J Coast Life Med 3(10):757–760

Jamali H, Imani A, Abdollahi D, Roozbehfar R, Isari A (2015) Use of probiotic Bacillus spp. in rotifer (Brachionus plicatilis) and artemia (Artemia urmiana) enrichment: effects on growth and survival of Pacific White Shrimp, Litopenaeus vannamei, Larvae. Probiotics Antimicrob Proteins 7(2):118–125. https://doi.org/10.1007/s12602-015-9189-3

Joseph B, Dhas B, Hena V, Raj J (2013) Bacteriocin from *Bacillus subtilis* as a novel drug against diabetic foot ulcer bacterial pathogens. Asian Pac J Trop Biomed 3(12):942–946

Katholnig K, Linke M, Pham H, Hengstschläger M, Weichhart T (2013) Immune responses of macrophages and dendritic cells regulated by mTOR signalling. Biochem Soc Trans 41(4):927–933. https://doi.org/10.1042/BST20130032

Katya K, Yun Y-h, Park G, Lee J-Y, Yoo G, Bai SC (2014) Evaluation of the efficacy of fermented by-product of mushroom, pleurotus ostreatus, as a fish meal replacer in Juvenile Amur Catfish, Silurus asotus: effects on growth, serological characteristics and immune responses. Asian Australas J Anim Sci 27(10):1478–1486. https://doi.org/10.5713/ajas.2014.14038

Kesarcodi-Watson A, Kaspar H, Lategan MJ, Gibson L (2008) Probiotics in aquaculture: the need, principles and mechanisms of action and screening processes. Aquaculture 274(1):1–14. https://doi.org/10.1016/j.aquaculture.2007.11.019

Kewcharoen W, Srisapoome P (2019) Probiotic effects of *Bacillus* spp. from Pacific white shrimp (*Litopenaeus vannamei*) on water quality and shrimp growth, immune responses, and resistance to *Vibrio parahaemolyticus* (AHPND strains). Fish Shellfish Immunol 94:175–189. https://doi.org/10.1016/j.fsi.2019.09.013

Khademzade O, Zakeri M, Haghi M, Mousavi SM (2020) The effects of water additive *Bacillus cereus* and *Pediococcus acidilactici* on water quality, growth performances, economic benefits, immunohematology and bacterial flora of whiteleg shrimp (*Penaeus vannamei* Boone, 1931) reared in earthen ponds. Aquac Res 51(5):1759–1770

Kim D-H, Subramanian D, Heo M-S (2017) Dietary effect of probiotic bacteria, Bacillus amyloliquefaciens-JFP2 on growth and innate immune response in rock bream Oplegnathus fasciatus, challenged with Streptococcus iniae

Klaenhammer TR, Kleerebezem M, Kopp MV, Rescigno M (2012) The impact of probiotics and prebiotics on the immune system. Nat Rev Immunol 12(10):728–734. https://doi.org/10.1038/nri3312

Kleerebezem M, Hols P, Bernard E, Rolain T, Zhou M, Siezen R, Bron P (2010) The extracellular biology of the lactobacilli. FEMS Microbiol Rev 34(2):199–230

Kuebutornye FK, Abarike ED, Lu Y (2019) A review on the application of *Bacillus* as probiotics in aquaculture. Fish Shellfish Immunol 87:820–828

La Fata G, Weber P, Mohajeri MH (2018) Probiotics and the gut immune system: indirect regulation. Probiotics Antimicrob Proteins 10(1):11–21. https://doi.org/10.1007/s12602-017-9322-6

Lauriano ER, Pergolizzi S, Capillo G, Kuciel M, Alesci A, Faggio C (2016) Immunohistochemical characterization of Toll-like receptor 2 in gut epithelial cells and macrophages of goldfish *Carassius auratus* fed with a high-cholesterol diet. Fish Shellfish Immunol 59:250–255. https://doi.org/10.1016/j.fsi.2016.11.003

Lazado CC, Caipang CMA (2014) Mucosal immunity and probiotics in fish. Fish Shellfish Immunol 39(1):78–89. https://doi.org/10.1016/j.fsi.2014.04.015

Lee Y-K, Puong K-Y, Ouwehand AC, Salminen S (2003) Displacement of bacterial pathogens from mucus and Caco-2 cell surface by lactobacilli. J Med Microbiol 52(10):925–930

Lee S, Katya K, Park Y, Won S, Seong M, Hamidoghli A, Bai SC (2017) Comparative evaluation of dietary probiotics Bacillus subtilis WB60 and Lactobacillus plantarum KCTC3928 on the growth performance, immunological parameters, gut morphology and disease resistance in Japanese eel, Anguilla japonica. Fish Shellfish Immunol 61:201–210. https://doi.org/10.1016/j.fsi.2016.12.035

Lee S, Katya K, Hamidoghli A, Hong J, Kim D-J, Bai SC (2018) Synergistic effects of dietary supplementation of *Bacillus subtilis* WB60 and mannanoligosaccharide (MOS) on growth performance, immunity and disease resistance in Japanese eel, *Anguilla japonica*. Fish Shellfish Immunol 83:283–291. https://doi.org/10.1016/j.fsi.2018.09.031

Li J, Tan B, Mai K, Ai Q, Zhang W, Xu W et al (2006) Comparative study between probiotic bacterium *Arthrobacter* XE-7 and chloramphenicol on protection of *Penaeus chinensis* post-larvae from pathogenic vibrios. Aquaculture 253(1):140–147. https://doi.org/10.1016/j.aquaculture.2005.07.040

Li J, Xu Y, Jin L, Li X (2015) Effects of a probiotic mixture (Bacillus subtilis YB-1 and Bacillus cereus YB-2) on disease resistance and non-specific immunity of sea cucumber, Apostichopus japonicus (Selenka). Aquac Res 46(12):3008–3019. https://doi.org/10.1111/are.12453

Li C, Zhang B, Liu C, Zhou H, Wang X, Mai K, He G (2020) Effects of dietary raw or *Enterococcus faecium* fermented soybean meal on growth, antioxidant status, intestinal microbiota, morphology, and inflammatory responses in turbot (*Scophthalmus maximus* L.). Fish Shellfish Immunol 100:261–271. https://doi.org/10.1016/j.fsi.2020.02.070

Lim S-Y, Loo KW, Wong W-L (2020) Synergistic antimicrobial effect of a seaweed-probiotic blend against acute hepatopancreatic necrosis disease (AHPND)-causing *Vibrio parahaemolyticus*. Probiotics Antimicrob Proteins 12(3):906–917

Lin Y-S, Saputra F, Chen Y-C, Hu S-Y (2019) Dietary administration of Bacillus amyloliquefaciens R8 reduces hepatic oxidative stress and enhances nutrient metabolism and immunity against Aeromonas hydrophila and Streptococcus agalactiae in zebrafish (Danio rerio). Fish Shellfish Immunol 86:410–419. https://doi.org/10.1016/j.fsi.2018.11.047

Liu S, Chen ZJ (2011) Expanding role of ubiquitination in NF-κB signaling. Cell Res 21(1):6–21

Liu CH, Chiu CS, Ho PL, Wang SW (2009) Improvement in the growth performance of white shrimp, *Litopenaeus vannamei*, by a protease-producing probiotic, *Bacillus subtilis* E20, from natto. J Appl Microbiol 107(3):1031–1041. https://doi.org/10.1111/j.1365-2672.2009.04284.x

Liu K-F, Chiu C-H, Shiu Y-L, Cheng W, Liu C-H (2010) Effects of the probiotic, *Bacillus subtilis* E20, on the survival, development, stress tolerance, and immune status of white shrimp, *Litopenaeus vannamei* larvae. Fish Shellfish Immunol 28(5–6):837–844

Liu H, Wang S, Cai Y, Guo X, Cao Z, Zhang Y et al (2017) Dietary administration of *Bacillus subtilis* HAINUP40 enhances growth, digestive enzyme activities, innate immune responses and disease resistance of tilapia, *Oreochromis niloticus*. Fish Shellfish Immunol 60:326–333. https://doi.org/10.1016/j.fsi.2016.12.003

Liu C-H, Wu K, Chu T-W, Wu T-M (2018) Dietary supplementation of probiotic, Bacillus subtilis E20, enhances the growth performance and disease resistance against Vibrio alginolyticus in parrot fish (Oplegnathus fasciatus). Aquac Int 26(1):63–74. https://doi.org/10.1007/s10499-017-0189-z

Liu B, Zhou W, Wang H, Li C, Wang L, Li Y, Wang J (2020) Bacillus baekryungensis MS1 regulates the growth, non-specific immune parameters and gut microbiota of the sea cucumber Apostichopus japonicus. Fish Shellfish Immunol 102:133–139. https://doi.org/10.1016/j.fsi.2020.04.023

Maisak H, Jantrakajorn S, Lukkana M, Wongtavatchai J (2013) Antibacterial activity of tannin from sweet chestnut wood against aeromonas and streptococcal pathogens of tilapia (*Oreochromis niloticus*). Thai J Vet Med 43(1):105

Marcusso PF, Aguinaga JY, da Silva Claudiano G, Eto SF, Fernandes DC, Mello H et al (2015) Influence of temperature on *Streptococcus agalactiae* infection in Nile tilapia. Braz J Vet Res Anim Sci 52(1):57–62

Martínez Cruz P, Ibáñez AL, Monroy Hermosillo OA, Ramírez Saad HC (2012) Use of probiotics in aquaculture. ISRN Microbiol 2012:916845. https://doi.org/10.5402/2012/916845

Meidong R, Doolgindachbaporn S, Jamjan W, Sakai K, Tashiro Y, Okugawa Y et al (2017) A novel probiotic Bacillus siamensis B44v isolated from Thai pickled vegetables (Phak-dong) for potential use as a feed supplement in aquaculture. J Gen Appl Microbiol 63(4):246–253

Meidong R, Khotchanalekha K, Doolgindachbaporn S, Nagasawa T, Nakao M, Sakai K, Tongpim S (2018) Evaluation of probiotic *Bacillus aerius* B81e isolated from healthy hybrid catfish on growth, disease resistance and innate immunity of Pla-mong *Pangasius bocourti*. Fish Shellfish Immunol 73:1–10. https://doi.org/10.1016/j.fsi.2017.11.032

Menni C, Zierer J, Pallister T, Jackson MA, Long T, Mohney RP et al (2017) Omega-3 fatty acids correlate with gut microbiome diversity and production of N-carbamylglutamate in middle aged and elderly women. Sci Rep 7(1):1–11

Mennigen R, Nolte K, Rijcken E, Utech M, Loeffler B, Senninger N, Bruewer M (2009) Probiotic mixture VSL# 3 protects the epithelial barrier by maintaining tight junction protein expression and preventing apoptosis in a murine model of colitis. Am J Physiol Gastrointest Liver Physiol

Merrifield D, Bradley G, Baker R, Davies S (2010) Probiotic applications for rainbow trout (*Oncorhynchus mykiss* Walbaum) II. Effects on growth performance, feed utilization, intestinal microbiota and related health criteria postantibiotic treatment. Aquac Nutr 16(5):496–503

Mohapatra S, Chakraborty T, Prusty A, Das P, Paniprasad K, Mohanta K (2012) Use of different microbial probiotics in the diet of rohu, *Labeo rohita* fingerlings: effects on growth, nutrient digestibility and retention, digestive enzyme activities and intestinal microflora. Aquac Nutr 18 (1):1–11

Mohapatra S, Chakraborty T, Kumar V, DeBoeck G, Mohanta KN (2013) Aquaculture and stress management: a review of probiotic intervention. J Anim Physiol Anim Nutr (Berl) 97 (3):405–430. https://doi.org/10.1111/j.1439-0396.2012.01301.x

Moriarty DJW (1998) Control of luminous *Vibrio* species in penaeid aquaculture ponds. Aquaculture 164(1–4):351–358. https://doi.org/10.1016/S0044-8486(98)00199-9

Morowitz MJ, Carlisle EM, Alverdy JC (2011) Contributions of intestinal bacteria to nutrition and metabolism in the critically ill. Surg Clin North Am 91(4):771–778. https://doi.org/10.1016/j.suc.2011.05.001

Mukherjee A, Chandra G, Ghosh K (2019) Single or conjoint application of autochthonous *Bacillus* strains as potential probiotics: effects on growth, feed utilization, immunity and disease resistance in Rohu, *Labeo rohita* (Hamilton). Aquaculture 512:734302. https://doi.org/10.1016/j.aquaculture.2019.734302

Nair AV, Leo Antony M, Praveen NK, Sayooj P, Raja Swaminathan T, Vijayan KK (2020) Evaluation of in vitro and in vivo potential of Bacillus subtilis MBTDCMFRI Ba37 as a candidate probiont in fish health management. Microb Pathog:104610. https://doi.org/10.1016/j.micpath.2020.104610

Nandi A, Banerjee G, Dan SK, Ghosh K, Ray AK (2018) Evaluation of in vivo probiotic efficiency of *Bacillus amyloliquefaciens* in *Labeo rohita* challenged by pathogenic strain of *Aeromonas hydrophila* MTCC 1739. Probiotics Antimicrob Proteins 10(2):391–398

Nath S, Matozzo V, Bhandari D, Faggio C (2019) Growth and liver histology of *Channa punctatus* exposed to a common biofertilizer. Nat Prod Res 33(11):1591–1598. https://doi.org/10.1080/14786419.2018.1428586

Nayak SK (2010) Probiotics and immunity: a fish perspective. *Fish Shellfish Immunol 29*(1):2–14. https://doi.org/10.1016/j.fsi.2010.02.017

Ock Kim Y, Mahboob S, Viayaraghavan P, Biji D, Abdullah Al-Ghanim K, Al-Misned F et al (2020) Growth promoting activity of Penaeus indicus by secondary metabolite producing probiotic bacterium Bacillus subtilis isolated from the shrimp gut. J King Saud Univ Sci 32 (2):1641–1646. https://doi.org/10.1016/j.jksus.2019.12.023

Park Y, Kim H, Won S, Hamidoghli A, Hasan MT, Kong I-S, Bai SC (2020) Effects of two dietary probiotics (*Bacillus subtilis* or licheniformis) with two prebiotics (mannan or fructo oligosaccharide) in Japanese eel, *Anguilla japonica*. Aquac Nutr 26(2):316–327. https://doi.org/10.1111/anu.12993

Plaza-Diaz J, Ruiz-Ojeda FJ, Gil-Campos M, Gil A (2019) Mechanisms of action of probiotics. Adv Nutr (Bethesda, Md) 10(Suppl_1):S49–S66. https://doi.org/10.1093/advances/nmy063

Pridmore RD, Pittet A-C, Praplan F, Cavadini C (2008) Hydrogen peroxide production by *Lactobacillus johnsonii* NCC 533 and its role in anti-Salmonella activity. FEMS Microbiol Lett 283 (2):210–215

Qin L, Xiang J, Xiong F, Wang G, Zou H, Li W et al (2020) Effects of *Bacillus licheniformis* on the growth, antioxidant capacity, intestinal barrier and disease resistance of grass carp (*Ctenopharyngodon idella*). Fish Shellfish Immunol 97:344–350. https://doi.org/10.1016/j.fsi.2019.12.040

Queiroz JF, Boyd CE (1998) Effects of a bacterial inoculum in channel catfish ponds. J World Aquac Soc 29(1):67–73

Ramesh D, Souissi S (2018) Effects of potential probiotic Bacillus subtilis KADR1 and its subcellular components on immune responses and disease resistance in Labeo rohita. Aquac Res 49(1):367–377. https://doi.org/10.1111/are.13467

Ramesh D, Vinothkanna A, Rai AK, Vignesh VS (2015) Isolation of potential probiotic *Bacillus* spp. and assessment of their subcellular components to induce immune responses in *Labeo rohita* against *Aeromonas hydrophila*. Fish Shellfish Immunol 45(2):268–276. https://doi.org/10.1016/j.fsi.2015.04.018

Ramirez RF, Dixon BA (2003) Enzyme production by obligate intestinal anaerobic bacteria isolated from oscars (*Astronotus ocellatus*), angelfish (*Pterophyllum scalare*) and southern flounder (*Paralichthys lethostigma*). Aquaculture 227(1–4):417–426

Ramos MA, Gonçalves JFM, Costas B, Batista S, Lochmann R, Pires MA et al (2017) Commercial Bacillus probiotic supplementation of rainbow trout (Oncorhynchys mykiss) and brown trout (Salmo trutta): growth, immune responses and intestinal morphology. Aquac Res 48 (5):2538–2549. https://doi.org/10.1111/are.13090

Rangavajhyala N, Shahani K, Sridevi G, Srikumaran S (1997) Nonlipopolysaccharide components) of Lactobacillus addophilus stimulate (s) the production of interleukin-1α and tumor necrosis factor-α by murine macrophages

Rawlings ND, Barrett AJ (1994) [2] Families of serine peptidases. Methods Enzymol:244, 19–261

Ray A, Ghosh K, Ringø E (2012) Enzyme-producing bacteria isolated from fish gut: a review. Aquac Nutr 18(5):465–492

Reda RM, Selim KM (2015) Evaluation of *Bacillus amyloliquefaciens* on the growth performance, intestinal morphology, hematology and body composition of Nile tilapia, *Oreochromis niloticus*. Aquac Int 23(1):203–217

Reda RM, El-Hady MA, Selim KM, El-Sayed HM (2018) Comparative study of three predominant gut Bacillus strains and a commercial *B. amyloliquefaciens* as probiotics on the performance of *Clarias gariepinus*. Fish Shellfish Immunol 80:416–425. https://doi.org/10.1016/j.fsi.2018.06.031

Rengpipat S, Phianphak W, Piyatiratitivorakul S, Menasveta P (1998) Effects of a probiotic bacterium on black tiger shrimp *Penaeus monodon* survival and growth. Aquaculture 167 (3):301–313. https://doi.org/10.1016/S0044-8486(98)00305-6

Rescigno M, Urbano M, Valzasina B, Francolini M, Rotta G, Bonasio R et al (2001) Dendritic cells express tight junction proteins and penetrate gut epithelial monolayers to sample bacteria. Nat Immunol 2(4):361–367

Resende JA, Silva VL, Fontes CO, Souza-Filho JA, de Oliveira TLR, Coelho CM et al (2012) Multidrug-resistance and toxic metal tolerance of medically important bacteria isolated from an aquaculture system. Microbes Environ:ME12049

Reverter M, Sarter S, Caruso D, Avarre J-C, Combe M, Pepey E et al (2020) Aquaculture at the crossroads of global warming and antimicrobial resistance. Nat Commun 11(1):1870. https://doi.org/10.1038/s41467-020-15735-6

Ringø E (2011) Evaluation of probiotic strain *Bacillus subtilis* C-3102 as a feed supplement for koi carp (*Cyprinus carpio*)

Ringø E (2020) Probiotics in shellfish aquaculture. Aquac Fisher 5(1):1–27. https://doi.org/10.1016/j.aaf.2019.12.001

Ringø E, Hoseinifar SH, Ghosh K, Doan HV, Beck BR, Song SK (2018) Lactic acid Bacteria in finfish—An update. Front Microbiol 9(1818). https://doi.org/10.3389/fmicb.2018.01818

Romano N (2021) Chapter 5: Probiotics, prebiotics, biofloc systems, and other biocontrol regimens in fish and shellfish aquaculture. In: Kibenge FSB, Baldisserotto B, Chong RS-M (eds) Aquaculture pharmacology. Academic Press, San Diego, CA, pp 219–242

Ryan KA, O'Hara AM, van Pijkeren J-P, Douillard FP, O'Toole PW (2009) *Lactobacillus salivarius* modulates cytokine induction and virulence factor gene expression in Helicobacter pylori. J Med Microbiol 58(8):996–1005

Sadat Hoseini Madani N, Adorian TJ, Ghafari Farsani H, Hoseinifar SH (2018) The effects of dietary probiotic Bacilli (*Bacillus subtilis* and *Bacillus licheniformis*) on growth performance, feed efficiency, body composition and immune parameters of whiteleg shrimp (*Litopenaeus vannamei*) postlarvae. Aquac Res 49(5):1926–1933

Sánchez-Ortiz AC, Angulo C, Luna-González A, Álvarez-Ruiz P, Mazón-Suástegui JM, Campa-Córdova ÁI (2016) Effect of mixed-*Bacillus* spp isolated from pustulose ark *Anadara tuberculosa* on growth, survival, viral prevalence and immune-related gene expression in shrimp *Litopenaeus vannamei*. Fish Shellfish Immunol 59:95–102. https://doi.org/10.1016/j.fsi.2016.10.022

Sanders ME, Morelli L, Tompkins T (2003) Sporeformers as human probiotics: Bacillus, *Sporolactobacillus*, and *Brevibacillus*. Compr Rev Food Sci Food Saf 2(3):101–110

Sangma T, Kamilya D (2015) Dietary Bacillus subtilis FPTB13 and chitin, single or combined, modulate systemic and cutaneous mucosal immunity and resistance of catla, Catla catla (Hamilton) against edwardsiellosis. Comp Immunol Microbiol Infect Dis 43:8–15. https://doi.org/10.1016/j.cimid.2015.09.003

Santacroce L, Charitos IA, Bottalico L (2019) A successful history: probiotics and their potential as antimicrobials. Expert Rev Anti Infect Ther 17(8):635–645. https://doi.org/10.1080/14787210.2019.1645597

Sapcharoen P, Rengpipat S (2013) Effects of the probiotic *Bacillus subtilis* (BP 11 and BS 11) on the growth and survival of Pacific white shrimp, *Litopenaeus vannamei*. Aquac Nutr 19 (6):946–954

Seghouani H, Garcia-Rangel C-E, Füller J, Gauthier J, Derome N (2017) Walleye autochthonous Bacteria as promising probiotic candidates against *Flavobacterium columnare*. Front Microbiol 8:1349. https://doi.org/10.3389/fmicb.2017.01349

Selim KM, Reda RM (2015) Improvement of immunity and disease resistance in the Nile tilapia, Oreochromis niloticus, by dietary supplementation with Bacillus amyloliquefaciens. Fish Shellfish Immunol 44(2):496–503. https://doi.org/10.1016/j.fsi.2015.03.004

Selim KM, El-Sayed HM, El-Hady M, Reda RM (2019) *In vitro* evaluation of the probiotic candidates isolated from the gut of *Clarias gariepinus* with special reference to the in vivo assessment of live and heat-inactivated *Leuconostoc mesenteroides* and *Edwardsiella* sp. Aquac Int 27(1):33–51

Serra CR, Almeida EM, Guerreiro I, Santos R, Merrifield DL, Tavares F et al (2019) Selection of carbohydrate-active probiotics from the gut of carnivorous fish fed plant-based diets. Sci Rep 9 (1):6384. https://doi.org/10.1038/s41598-019-42716-7

Shi LH, Balakrishnan K, Thiagarajah K, Mohd Ismail NI, Yin OS (2016) Beneficial properties of probiotics. Tropical Life Sci Res 27(2):73–90. https://doi.org/10.21315/tlsr2016.27.2.6

Silva TFA, Petrillo TR, Yunis-Aguinaga J, Marcusso PF, da Silva Claudiano G, de Moraes FR, de Moraes JR (2015) Effects of the probiotic Bacillus amyloliquefaciens on growth performance, hematology and intestinal morphometry in cage-reared Nile tilapia. Lat Am J Aquat Res 43 (5):963–971

Singh ST, Kamilya D, Kheti B, Bordoloi B, Parhi J (2017) Paraprobiotic preparation from Bacillus amyloliquefaciens FPTB16 modulates immune response and immune relevant gene expression in Catla catla (Hamilton, 1822). Fish Shellfish Immunol 66:35–42. https://doi.org/10.1016/j.fsi.2017.05.005

Sivamaruthi BS, Kesika P, Chaiyasut C (2018) Thai fermented foods as a versatile source of bioactive microorganisms—A comprehensive review. Sci Pharm 86(3):37

Soltani M, Ghosh K, Hoseinifar SH, Kumar V, Lymbery AJ, Roy S, Ringø E (2019) Genus *bacillus*, promising probiotics in aquaculture: aquatic animal origin, bio-active components, bioremediation and efficacy in fish and shellfish. Rev Fisher Sci Aquac 27(3):331–379. https://doi.org/10.1080/23308249.2019.1597010

Sookchaiyaporn N, Srisapoome P, Unajak S, Areechon N (2020) Efficacy of *Bacillus* spp. isolated from Nile tilapia *Oreochromis niloticus* Linn. on its growth and immunity, and control of pathogenic bacteria. Fisher Sci:1–13

Spencer R, Chesson A (1994) The effect of Lactobacillus spp. on the attachment of enterotoxigenic Escherichia coli to isolated porcine enterocytes. J Appl Bacteriol 77(2):215–220

Srisapoome P, Areechon N (2017) Efficacy of viable *Bacillus pumilus* isolated from farmed fish on immune responses and increased disease resistance in Nile tilapia (*Oreochromis niloticus*): Laboratory and on-farm trials. Fish Shellfish Immunol 67:199–210. https://doi.org/10.1016/j.fsi.2017.06.018

Sugita H, Takahashi J, Deguchi Y (1992) Production and consumption of biotin by the intestinal microflora of cultured freshwater fishes. Biosci Biotechnol Biochem 56(10):1678–1679

Sumi CD, Yang BW, Yeo I-C, Hahm YT (2015) Antimicrobial peptides of the genus Bacillus: a new era for antibiotics. Can J Microbiol 61(2):93–103

Sun Y-Z, Yang H-L, Ma R-L, Lin W-Y (2010) Probiotic applications of two dominant gut *Bacillus* strains with antagonistic activity improved the growth performance and immune responses of grouper *Epinephelus coioides*. Fish Shellfish Immunol 29(5):803–809. https://doi.org/10.1016/j.fsi.2010.07.018

Suzer C, Çoban D, Kamaci HO, Saka Ş, Firat K, Otgucuoğlu Ö, Küçüksari H (2008) *Lactobacillus* spp. bacteria as probiotics in gilthead sea bream (*Sparus aurata*, L.) larvae: effects on growth performance and digestive enzyme activities. Aquaculture 280(1–4):140–145

Svendsen A (2000) Lipase protein engineering. Biochim Biophys Acta/Protein Struct Mol Enzymol 1543(2):223–238

Swapna B, Venkatrayulu C, Swathi AJEJOEB (2015) Effect of probiotic bacteria Bacillus licheniformis and Lactobacillus rhamnosus on growth of the Pacific white shrimp Litopenaeus vannamei (Boone, 1931). Eur J Exp Biol 5(11):31–36

Tarnecki AM, Wafapoor M, Phillips RN, Rhody NR (2019) Benefits of a Bacillus probiotic to larval fish survival and transport stress resistance. Sci Rep 9(1):1–11

Tepaamorndech S, Chantarasakha K, Kingcha Y, Chaiyapechara S, Phromson M, Sriariyanun M et al (2019) Effects of Bacillus aryabhattai TBRC8450 on vibriosis resistance and immune enhancement in Pacific white shrimp, Litopenaeus vannamei. Fish Shellfish Immunol 86:4–13. https://doi.org/10.1016/j.fsi.2018.11.010

Truong Thy HT, Tri NN, Quy OM, Fotedar R, Kannika K, Unajak S, Areechon N (2017) Effects of the dietary supplementation of mixed probiotic spores of *Bacillus amyloliquefaciens* 54A, and *Bacillus pumilus* 47B on growth, innate immunity and stress responses of striped catfish (*Pangasianodon hypophthalmus*). Fish Shellfish Immunol 60:391–399. https://doi.org/10.1016/j.fsi.2016.11.016

Tseng D-Y, Ho P-L, Huang S-Y, Cheng S-C, Shiu Y-L, Chiu C-S, Liu C-H (2009) Enhancement of immunity and disease resistance in the white shrimp, Litopenaeus vannamei, by the probiotic, *Bacillus subtilis* E20. Fish Shellfish Immunol 26(2):339–344

Ueberschär B (1995) The use of tryptic enzyme activity measurement as a nutritional condition index: laboratory calibration data and field application. Paper presented at the ICES Marine Science Symposia

Utami D, Suprayudi MA (2015) Quality of dried Bacillus NP5 and its effect on growth performance of Tilapia (Oreochromis niloticus). Pak J Biol Sci: PJBS 18(2):88–93

Verschuere L, Rombaut G, Sorgeloos P, Verstraete W (2000) Probiotic bacteria as biological control agents in aquaculture. Microbiol Mol Biol Rev 64(4):655–671

Vieco-Saiz N, Belguesmia Y, Raspoet R, Auclair E, Gancel F, Kempf I, Drider D (2019) Benefits and inputs from lactic acid bacteria and their bacteriocins as alternatives to antibiotic growth promoters during food-animal production. Front Microbiol 10(57). https://doi.org/10.3389/fmicb.2019.00057

Vijayavel K, Balasubramanian MP (2006) Fluctuations of biochemical constituents and marker enzymes as a consequence of naphthalene toxicity in the edible estuarine crab *Scylla serrata*. Ecotoxicol Environ Saf 63(1):141–147

Vine NG, Leukes WD, Kaiser H (2006) Probiotics in marine larviculture. FEMS Microbiol Rev 30 (3):404–427. https://doi.org/10.1111/j.1574-6976.2006.00017.x

Wang L, Ge C, Wang J, Dai J, Zhang P, Li Y (2017a) Effects of different combinations of Bacillus on immunity and antioxidant activities in common carp. Aquac Int 25(6):2091–2099. https://doi.org/10.1007/s10499-017-0175-5

Wang M, Liu G, Lu M, Ke X, Liu Z, Gao F et al (2017b) Effect of *Bacillus cereus* as a water or feed additive on the gut microbiota and immunological parameters of Nile tilapia. Aquac Res 48 (6):3163–3173

Wang H, Wang C, Tang Y, Sun B, Huang J, Song X (2018) *Pseudoalteromonas* probiotics as potential biocontrol agents improve the survival of *Penaeus vannamei* challenged with acute hepatopancreatic necrosis disease (AHPND)-causing *Vibrio parahaemolyticus*. Aquaculture 494:30–36

Wang C, Liu Y, Sun G, Li X, Liu Z (2019a) Growth, immune response, antioxidant capability, and disease resistance of juvenile Atlantic salmon (Salmo salar L.) fed Bacillus velezensis V4 and Rhodotorula mucilaginosa compound. Aquaculture 500:65–74. https://doi.org/10.1016/j.aquaculture.2018.09.052

Wang A, Ran C, Wang Y, Zhang Z, Ding Q, Yang Y et al (2019b) Use of probiotics in aquaculture of China—a review of the past decade. Fish Shellfish Immunol 86:734–755. https://doi.org/10.1016/j.fsi.2018.12.026

Wang C, Chuprom J, Wang Y, Fu L (2020a) Beneficial bacteria for aquaculture: nutrition, bacteriostasis and immunoregulation. J Appl Microbiol 128(1):28–40. https://doi.org/10.1111/jam.14383

Wang M, Yi M, Lu M, Gao F, Liu Z, Huang Q et al (2020b) Effects of probiotics *Bacillus cereus* NY5 and *Alcaligenes faecalis* Y311 used as water additives on the microbiota and immune enzyme activities in three mucosal tissues in Nile tilapia *Oreochromis niloticus* reared in outdoor tanks. Aquac Rep 17:100309. https://doi.org/10.1016/j.aqrep.2020.100309

Wanka KM, Damerau T, Costas B, Krueger A, Schulz C, Wuertz S (2018) Isolation and characterization of native probiotics for fish farming. BMC Microbiol 18(1):119. https://doi.org/10.1186/s12866-018-1260-2

Weichhart T, Costantino G, Poglitsch M, Rosner M, Zeyda M, Stuhlmeier KM et al (2008) The TSC-mTOR signaling pathway regulates the innate inflammatory response. Immunity 29 (4):565–577

Willer DF, Aldridge DC (2019) Microencapsulated diets to improve bivalve shellfish aquaculture for global food security. Glob Food Sec 23:64–73. https://doi.org/10.1016/j.gfs.2019.04.007

Won S, Hamidoghli A, Choi W, Bae J, Jang WJ, Lee S, Bai SC (2020a) Evaluation of potential probiotics *Bacillus subtilis* WB60, *Pediococcus pentosaceus*, and *Lactococcus lactis* on growth performance, immune response, gut histology and immune-related genes in Whiteleg Shrimp, *Litopenaeus vannamei*. Microorganisms 8(2):281

Won S, Hamidoghli A, Choi W, Bae J, Jang WJ, Lee S, Bai SCJM (2020b) Evaluation of potential probiotics Bacillus subtilis WB60, Pediococcus pentosaceus, and Lactococcus lactis on growth performance, immune response, gut histology and immune-related genes in Whiteleg Shrimp. Litopenaeus vannamei 8(2):281

World Health Organization (2014) *Antimicrobial resistance: global report on surveillance*. World Health Organization, Geneva

Wu ZX, Feng X, Xie LL, Peng XY, Yuan J, Chen XX (2012) Effect of probiotic *Bacillus subtilis* Ch9 for grass carp, *Ctenopharyngodon idella* (Valenciennes, 1844), on growth performance, digestive enzyme activities and intestinal microflora. J Appl Ichthyol 28(5):721–727. https://doi.org/10.1111/j.1439-0426.2012.01968.x

Xia Y, Wang M, Gao F, Lu M, Chen G (2020) Effects of dietary probiotic supplementation on the growth, gut health and disease resistance of juvenile Nile tilapia (Oreochromis niloticus). Anim Nutr (Zhongguo xu mu shou yi xue hui) 6(1):69–79. https://doi.org/10.1016/j.aninu.2019.07.002

Yan F, Polk DB (2011) Probiotics and immune health. Curr Opin Gastroenterol 27(6):496–501. https://doi.org/10.1097/MOG.0b013e32834baa4d

Yan YY, Xia HQ, Yang HL, Hoseinifar SH, Sun YZ (2016) Effects of dietary live or heat-inactivated autochthonous Bacillus pumilus SE5 on growth performance, immune responses and immune gene expression in grouper Epinephelus coioides. Aquac Nutr 22(3):698–707. https://doi.org/10.1111/anu.12297

Yang HL, Sun YZ, Ma RL, Ye JD (2012) PCR-DGGE analysis of the autochthonous gut microbiota of grouper *Epinephelus coioides* following probiotic *Bacillus clausii* administration. Aquac Res 43(4):489–497

Yang S-C, Lin C-H, Sung CT, Fang J-Y (2014) Antibacterial activities of bacteriocins: application in foods and pharmaceuticals. Front Microbiol 5:241–241. https://doi.org/10.3389/fmicb.2014.00241

Yang G, Tian X, Dong S, Peng M, Wang D (2015) Effects of dietary *Bacillus cereus* G19, B. cereus BC-01, and *Paracoccus marcusii* DB11 supplementation on the growth, immune response, and expression of immune-related genes in coelomocytes and intestine of the sea cucumber (*Apostichopus japonicus* Selenka). Fish Shellfish Immunol 45(2):800–807. https://doi.org/10.1016/j.fsi.2015.05.032

Yang Q, Lü Y, Zhang M, Gong Y, Li Z, Tran NT et al (2019) Lactic acid bacteria, *Enterococcus faecalis* Y17 and *Pediococcus pentosaceus* G11, improved growth performance, and immunity of mud crab (*Scylla paramamosain*). Fish Shellfish Immunol 93:135–143. https://doi.org/10.1016/j.fsi.2019.07.050

Yin Y, Zhang P, Yue X, Du X, Li W, Yin Y et al (2018) Effect of sub-chronic exposure to lead (Pb) and Bacillus subtilis on Carassius auratus gibelio: bioaccumulation, antioxidant responses and immune responses. Ecotoxicol Environ Saf 161:755–762. https://doi.org/10.1016/j.ecoenv.2018.06.056

Zaineldin AI, Hegazi S, Koshio S, Ishikawa M, Bakr A, El-Keredy AM et al (2018) *Bacillus subtilis* as probiotic candidate for red sea bream: Growth performance, oxidative status, and immune response traits. Fish Shellfish Immunol 79:303–312

Zhang L, Mai K, Tan B, Ai Q, Qi C, Xu W et al (2009) Effects of dietary administration of probiotic *Halomonas* sp. B12 on the intestinal microflora, immunological parameters, and midgut histological structure of shrimp. J World Aquac Soc 40(1):58–66

Zhang Q, Ma H, Mai K, Zhang W, Liufu Z, Xu W (2010) Interaction of dietary *Bacillus subtilis* and fructooligosaccharide on the growth performance, non-specific immunity of sea cucumber, *Apostichopus japonicus*. Fish Shellfish Immunol 29(2):204–211

Zhao Y, Zhang W, Xu W, Mai K, Zhang Y, Liufu Z (2012) Effects of potential probiotic *Bacillus subtilis* T13 on growth, immunity and disease resistance against Vibrio splendidus infection in juvenile sea cucumber *Apostichopus japonicus*. Fish Shellfish Immunol 32(5):750–755. https://doi.org/10.1016/j.fsi.2012.01.027

Zhao J, Chen M, Quan C, Fan S (2015) Mechanisms of quorum sensing and strategies for quorum sensing disruption in aquaculture pathogens. J Fish Dis 38(9):771–786

Zhao Y, Yuan L, Wan J, Sun Z, Wang Y, Sun H (2016) Effects of potential probiotic Bacillus cereus EN25 on growth, immunity and disease resistance of juvenile sea cucumber Apostichopus japonicus. Fish Shellfish Immunol 49:237–242. https://doi.org/10.1016/j.fsi.2015.12.035

Zhao H, Luo Y e, Zhang Y, Chen X, Wang H, Guo D, Wu Z (2020) Effects of Bacillus subtilis on hepatic lipid metabolism and oxidative stress response in grass carp (Ctenopharyngodon idellus) fed a high-fat diet. Marine Life Sci Technol 2(1):50–59. https://doi.org/10.1007/s42995-019-00005-2

Zhou X, Tian Z, Wang Y, Li W (2010) Effect of treatment with probiotics as water additives on tilapia (*Oreochromis niloticus*) growth performance and immune response. Fish Physiol Biochem 36(3):501–509

Ziaei-Nejad S, Rezaei MH, Takami GA, Lovett DL, Mirvaghefi A-R, Shakouri M (2006) The effect of *Bacillus* spp. bacteria used as probiotics on digestive enzyme activity, survival and growth in the Indian white shrimp *Fenneropenaeus indicus*. Aquaculture 252(2):516–524. https://doi.org/10.1016/j.aquaculture.2005.07.021

Zokaeifar H, Balcázar JL, Saad CR, Kamarudin MS, Sijam K, Arshad A, Nejat N (2012) Effects of *Bacillus subtilis* on the growth performance, digestive enzymes, immune gene expression and disease resistance of white shrimp, *Litopenaeus vannamei*. Fish Shellfish Immunol 33(4):683–689

Zuo Z-H, Shang B-J, Shao Y-C, Li W-Y, Sun J-S (2019) Screening of intestinal probiotics and the effects of feeding probiotics on the growth, immune, digestive enzyme activity and intestinal flora of *Litopenaeus vannamei*. Fish Shellfish Immunol 86:160–168. https://doi.org/10.1016/j.fsi.2018.11.003

Immunity and Gut Microbiome: Role of Probiotics and Prebiotics

2

T. R. Keerthi, Rakhie Narayanan, K. Sreelekshmi, and C. Honey Chandran

Abstract

The chapter comprises the role of gut microbiome and beneficial bacteria (probiotic) to boost the overall health. The GI tract is described as the body's largest immune organ. The intestinal microbiota has a vital role in the body's defence system. The most important factor that determines gut health is the microflora or gut flora. Our gut comprises diverse and hundreds of trillion bacteria and it comes in both good, i.e. beneficial bacteria, and bad, i.e. harmful bacteria. So it is mandatory to keep the beneficial bacteria for a better health. Microbes considered to be beneficial usually ferment carbohydrates, do not produce toxins and may have a range of potential benefits for the host. Such microbes include *Bifidobacterium*, *Eubacterium* and *Lactobacillus*. These beneficial bacteria are called probiotic. To get flourish good bacteria in the gut we need to feed it with a proper nutrient called prebiotic. A prebiotic is a special type of soluble non-digestible plant fibres that nourish the beneficial good bacteria in the gut. The synergy of the probiotic and prebiotic components in the gut provides a stable and relatively uniform gut microbiome and thereby boost the gut health and immune system.

Supplementations of prebiotics improve the establishment of microbial community which benefits the overall health. By metabolizing these fibres gastrointestinal tract community produces short-chain fatty acids which elicit many immune pathways and recruit immune cells to the gut. The pattern recognition receptors of immune cells recognize the pathogen-associated molecular pattern and initiate a cascade of immune pathways that ends in production of cytokines or helps in recruiting more immune cells. However, the specialized macrophages

T. R. Keerthi (✉) · R. Narayanan · K. Sreelekshmi · C. Honey Chandran
School of Biosciences, Mahatma Gandhi University, Kerala, India

N. Mojgani, M. Dadar (eds.), *Probiotic Bacteria and Postbiotic Metabolites: Role in Animal and Human Health*, Microorganisms for Sustainability 2,
https://doi.org/10.1007/978-981-16-0223-8_2

and neutrophils of gut regulate the overexpression of inflammatory cytokines and help to protect the gut lumen from inflammation.

Keywords

Gut microbiota · Immunomodulation · Probiotics · Prebiotics · Synbiotics · Postbiotics

2.1 Introduction

In human body, the gastrointestinal (GI) tract represents the largest contact area between the body and the external environment. GI tract is the natural habitat of trillions of microflora collectively known as gut microbiota or microflora. Microflora coevolved in a symbiotic relationship with the human intestinal mucosa in such a way that the indigenous microbiota is essential for gut homeostasis. So, the microbiota is considered as 'Super organisms' and is an integral part of the GI tract.

The GI tract harbours millions of bacteria which continuously stimulate the immune system both by its own structural component and by the metabolic products. The gut flora elicits not only a local immune response but also systemic response, thereby affecting whole body. Use of microorganisms that can contribute to a healthy gut, the probiotics, is common nowadays. The micro organisms inhabiting the gut environment act as immune eliciting agents and also their metabolites. Both cell-mediated and humoral immune systems get upregulated. But the gut immune system has a capacity to regulate the immune response to commensal bacteria.

Nutrients also affect gut immunity, and strategies that restore a healthy gut microbial community by affecting the microbial composition are being developed as new therapeutic approaches to treat several inflammatory diseases. The use of probiotics and prebiotics is a promising strategy for the reduction and prevention of GI infections. One of the main differences between probiotics and prebiotics is that probiotics are viable food supplements whereas prebiotics are nonviable food component. Prebiotics are non-digestible oligosaccharides, remain intact through the digestive system, and act as nutrients for already established microflora. These are oligosaccharide that overcomes several limitations of introducing probiotic bacteria into the GI tract. Therefore, using prebiotics is possibly a more practical and efficient way to manipulate the gut microflora. Prebiotics are defined as functional components of food which are metabolized by particular commensal bacteria in the gut conferring various health benefits to the host. Prebiotics reach the large intestine without changing their chemical and structural properties. Prebiotics are capable to escape the digestive processes in the upper part of the gut due to their molecular and structural composition, which makes them essentially resistant to our digestive enzymes.

Most of the studies about prebiotics have been focused on fructans, such as inulin, fructo- oligosaccharides (FOS) and galacto-oligosaccharides (GOS). They are also valuable functional ingredients for the food industry with the potential to improve the sensory properties of food. Other important prebiotics include lactulose, Xylooligosaccharides (XOS) and Mannan-oligosaccharides (MOS). These prebiotics stimulate the growth of bacteria in the colon, where lactobacilli and bifidobacteria are respectively preferred. As a product of the beneficial fermentation, short-chain fatty acid (SCFA) are produced, mainly the butyrate. The most studied SCFA is butyrate. Butyrate is the major energy source of colonic epithelial cells that affect the proliferation and barrier function of the colonic epithelium and reduce DNA damage. Other roles of prebiotics include reducing the level of cholesterol, reducing constipation, stimulating immune system, reduction of antibiotic-associated diarrhoea, reduction in inflammation and symptoms associated with inflammatory bowel disease, protective effects for prevention of colon cancer, and increasing the uptake of minerals, including calcium, magnesium iron, etc.

Numerous studies prove that a number of mechanisms mediating the health benefits of beneficial bacterial cells require viability. However, recent evidence suggests that bacterial viability is not necessary to attain the health-promoting activity. The newest member of biotic family (including probiotic, prebiotic, synbiotic, postbiotic,), the postbiotics also known as metabiotics, or simply metabolites, can confer health promotion in the host. The cell-free supernatants (CFS) refer to soluble factors (products or metabolic by products) secreted by live bacteria or released after bacterial lysis. These by-products offer physiological benefits to the host by providing additional bioactivity. Such soluble factors have been collected from different bacterial strains; examples include SCFAs, enzymes, peptides, teichoic acids, peptidoglycan-derived muropeptides, endo and exopolysaccharides, cell surface proteins, vitamins, plasmalogens, and organic acids.

Synbiotics refer to nutritional supplements combining probiotics and prebiotic food ingredients and in a form of synergism that improves the survival and implantation of live microbial dietary supplements in the GI tract, either by stimulating growth or by metabolically activating the health-promoting bacteria. Synbiotics products offer the potential to develop prebiotics targeted at specific probiotic strains to optimize health benefits.

This chapter mainly focused on the effects of dietary components, commensal bacteria and their metabolites in host immune system.

2.2 An Introduction to Gut Microbiome

The human GI tract harbours very complex population of microorganisms collectively called the gut microbiome that influences the host metabolism, homeostasis and pathogenesis. The gut microbiota has evolved with host and forms a mutually beneficial relationship and their number has been estimated to exceed 10^{14} (Cani et al. 2008). The colonization of gut is generally believed to begin from birth onwards and the members colonized immediately depend on the mode of delivery. Vaginally delivered infants harbour abundant Lactobacilli and Bifidobacteria during the first few days which is an indication of its abundance in vagina (Aagaard et al.

2012; Dominguez-Bello et al. 2010) while infants of caesarean session harbour *Clostridium* sp. The gut of a healthy adult is predominantly constituted by phylum Firmicutes and Bacteroidetes followed by Actinobacteria and Verrucomicrobia. Gut microbiota exhibits great variation in their distribution throughout the GI tract (Ramakrishna 2007). Streptococcus is the dominant genus in the oesophagus, duodenum and jejunum. In addition to Streptococcus, Prevotella, Veillonella and Rothia inhabit the stomach. A few number of *Helicobacter pylori* is also seen as a commensal in stomach. Besides Firmicutes and Bacteroidetes, human colon also inhabits pathogens like Salmonella, Vibrio, *E. coli* and camphylobacter but constitutes less than 0.1% of the entire gut microbiome. The intestinal microbiota exhibits an axial difference from intestinal lumen to mucosal surface. Bifidobacterium, Enterobacteriaceae, Enterococcus, Clostridium, Lactobacillus, Ruminococcus, Streptococcus and Bacteroides predominate in the lumen while Lactobacillus, Enterococcus, Clostridium and Akkermansia predominate in mucosal layer (Jandhyala et al. 2015). Age, diet, antibiotic consumption, host genetics, and life events are some of the factors which alter the normal gut microbiota.

2.3 Gut Immune System

Gut is the primary interface between the environment and immune system. It is important that the gut immune system must eliminate invading pathogens and simultaneously maintain self-tolerance to avoid autoimmunity. The homeostasis between the two is essential to maintain host health (Chassaing et al. 2014).

2.3.1 Immune Barriers in the Gut

Gut interacts with all other organs and it is the connection link between external environment and internal organs. Many microorganisms including pathogens and food antigens are ingested along with food. It is the gut which determines what needed and what not. The gut plays a major role in preventing the pathogens. At the same time the immune systems get boosted. The lining of the gut, mucus itself is a barrier and it is the first line of defence against pathogens and antigens throughout the gut. But in *Helicobacter pylori* infection, gastric neoplasia, colorectal polyps and cancer, the composition of mucin has altered (Jass and Walsh 2001). The next line of physical barrier is the gastrointestinal epithelium. It is primarily composed of enterocytes which selectively transport nutrients, electrolytes and water to underlying cells by various pumps and thereby eliminating the antigens entering into immune system (Chassaing et al. 2014). Beyond a physical barrier, intestinal epithelial cells (IEC) secrete cytokines and chemokines which regulate chemotaxis of both innate and adaptive immune cells. Neutrophils are recruited to the gut by epithelia derived chemokine IL-8 and epithelial neutrophil attractant 78. The monocyte chemotactic protein (MCP 1), RANTES/CC L5 and macrophage inflammatory protein (MIP1 α) regulate the chemotaxis of monocytes. T cells are recruited by interferon inducible protein (IP-10) and interferon γ. IEC also secretes proinflammatory cytokines TNF-α and IL-6, nitric oxide synthase, cyclooxygenase and reactive oxygen species (ROS). Another structural component is Paneth cells which contain defensin-rich granules (defensin 5 and 6) and can regulate composition and number of microbes. They also secrete antibacterial peptides like lysozyme and secretary phospholipase A2. The microfold cells or M cells in the epithelium are the main site of invasion of pathogens and normal microbiota. The main function of M cells is antigen sampling that is the uptake of antigen and microorganisms and presentation to lymphoid follicle. Hence they are also the site of immunological functions. Intestinal macrophages play an important role in gut immune homeostasis. They do not respond to TLR ligand and secrete proinflammatory cytokines or generate ROS or nitric oxide. But they express high levels of CD 36 which facilitates phagocytosis and apoptosis. Intestinal macrophages differ from other tissue macrophages in its property of "inflammation anergy" in which macrophage maintains the overexpression of proinflammatory cytokines to normal flora (Smythies et al. 2005). Intestinal dendritic cells are also distinguishable from other tissues. They also maintain a tolerogenic immune response by decreased expression of pattern recognition receptors, increased levels of anti-inflammatory cytokine IL-10, low level of antigen presentation by reducing co-stimulatory molecules, and favouring differentiation of Treg and IgA secreting B cells (Coombes and Powrie 2009). Secretary IgA gives protection against *Vibrio cholera*, Salmonella, rotavirus, and *Escherichia coli*. Another obstacle to activation of gut immune system is the controlled activation of pattern recognition receptors (PRR). TLR-2 (Toll like receptor 2) and TLR-4, specific for bacterial peptidoglycan and lipopolysaccharide (LPS), respectively, are abundant in IEC when they are migrating to the surface epithelium but are barely expressed once they reach the surface. Moreover, the cofactors for TLR 4 activation, LPS binding protein, CD 14, and myeloid differentiating factor are also limited in the intestine. Likewise, TLR 5, specific for flagellin is expressed only on the basolateral side to respond only if invaded by flagellated microorganisms (Vamadevan et al. 2010; Carvalho et al. 2012). In addition to all the microbial metabolites especially short-chain fatty acids produced by gut microbes play a vital role in intestinal immunity by regulating Treg cells (Smith et al. 2013).

2.4 Gut Microbiota and Immunity

The gut microbial community can regulate local as well as systemic immune responses. As the gut contains commensal and beneficial bacteria as well as opportunistic pathogens, the immune system always maintains a balance that will not disturb the beneficial bacteria, but when there is any increase in pathogens or antigens it will act immediately to eliminate the effect. The normal flora of the gut has a profound role in shaping the immune system. Its effect not only confines to gut but also to other organs.

2.4.1 Gut-Associated Lymphoid Tissue (GALT) Development

The role of microbiota in GALT development was evident from studies in germ-free (GF) mice. GALT includes Peyer's patches, crypt patches and isolated lymphoid follicle (ILF). Microbial stimulation is required for the development of these tissues. It was observed that the maturation of ILF was incomplete in mice deficient of PRRs like TLR2, TLR3, NOD2 (nucleotide binding oligomerization domain 2) and MyD88 (myeloid differentiation primary response protein 88) suggesting the role of microbial stimulation for proper development of an immune system (Hendricks et al. 2014).

2.4.2 Modulation of Innate Immune Cells

As mentioned earlier, antigen-presenting cells (APCs) of gut have co-evolved with gut microbiota and develop the ability to protect from invading pathogens but maintaining tolerance to normal flora. The dendritic cells (DCs) of Peyer's patches produce high levels of anti-inflammatory cytokine IL-10 when compared to splenic dendritic cells (Iwasaki and Kelsall 1999). Likewise, the intestinal macrophages developed the inflammation synergy, the non-inflammatory profile. However, experiments in GF animal showed a reduction in the number of intestinal dendritic cells but colonization with *E. coli* was sufficient for recruiting DCs to intestine. Similarly, macrophage activity was reduced in GF mice and major histocompatibility complex class II was also absent (Mikkelsen et al. 2004). The gut flora influences the neutrophil activity also. The peripheral blood neutrophils of GF rats exhibited decreased phagocytic activity and impaired generation of free radicles (Ohkubo et al. 1999). The role of gut microbes in systemic immune system is evident from the enhanced activity of bone marrow neutrophils when cytosolic receptor–nucleotide oligomerization domain 1 (NOD 1) gets activated by peptidoglycans of gut microbiota (Clarke et al. 2010). Overall for the complete maturation and activation of phagocytic cells microbial stimulation is required either the whole organism or the structural components (antigens). Natural killer (NK) cells produce IFN γ and perforins to eliminate damaged and infected cells. But the specialized NK cells, NKp46$^+$ of intestine is limited in its production; instead they express IL-22 and the nuclear hormone receptor retinoic acid receptor-related orphan receptor gamma t (RORγt). The absence of this NKp46$^+$ in GF mice explains the role of microbes in gut for its development (Yan and Polk 2002). Mice lacking IL-22 producing NKp46$^+$ cells were susceptible to pathogenic infection. An important immune barrier in gut is IECs. They produce various antimicrobials among which defensins and cathelicidins are important. There are two types of defensins, alpha and beta. Human β-defensin 1, 2, and 3 and mice β-defensin 2/3 can regulate the chemotaxis of immature DCs and memory T cells. Human β-defensin 3 enhances the expression of co-stimulatory factors CD 40, CD 80 and CD 86 on monocyte and myeloid DCs. The major function of cathelicidins is antibacterial activity. It shows broad activity towards gram positive and gram negative bacteria. Other antimicrobials secreted by

IECs include antimicrobial C-type lectins, angiogenin 4, phospholipase A2 type IIA and lysozyme C (Muniz et al. 2012). But in GF mice a lower cell proliferation and expression of genes for these antimicrobial was observed (Reikvam et al. 2011).

2.4.3 Modulation of Adaptive Immune System

$CD4^+$ cells are the most important component of adaptive immune system. Upon stimulation by microbiota $CD4^+$ cells in Lamina propria (LP) differentiates into its subtype Th1, Th2, Th17 and Treg cells. In GF mice there is a decrease in $CD4^+$ cells of LP and defects in spleen and mesenteric lymph nodes were observed. The polysaccharide A of *Bacteroides fragilis* induces a Th1 systemic response and also suppresses Th17 response by signalling through TLR 2 on Treg cells. Conversely, segmented filamentous bacteria induce LP Th17 cell response (Macpherson et al. 2002). Similarly, $CD8^+$ cells, commonly seen in intraepithelial compartment of gut, are also minimal in GF mice indicating the critical role of microbial stimulation for maintaining $CD8^+$ population. Gut microbiota also stimulates the cytolytic activity of γδT cells, the connecting link between innate and adaptive immunity and their number is very high in intestine compared to lymph node and spleen. Peyer's patches are rich in IgA secreting plasma cells but are considerably low in GF mice. To induce IgA production in GF mice a large dose of bacteria approximately 10^9 CFU/ml was required (Hapfelmeier et al. 2010).

2.5 Communication of Gut Microbiota to Other Organs

Gut microbiota communicates to other organs mainly through metabolic, endocrine, autonomous nervous system and immune pathways. Bacterial fermentation in intestine produces many metabolites. Fermentation of dietary fibres produces many SCFA like acetate, propionate, butyrate, etc. SCFAs have profound role in many signalling pathways. The GALT comprises 70% of the body's immune system and can be considered as the largest immune organ of the body. Enterohormones, metabolites, immune cells and cytokines derived from this complex mucosal and submucosal network have systemic impacts on other organs such as the kidney, cardiovascular system, bone marrow and brain via the circulation (Yang et al. 2018). Here we mainly discuss on the immune axis between gut microbiota and other organs (Fig. 2.1).

The gut lung axis is bidirectional. The metabolites produced by gut microbes enter the blood stream and reach the lungs, and the immune factors from lungs also elicit an immune response in gut. The immune cells induced by antigens move through the lymphatic system between gut and lungs and thereby elicit immune response in both organs. Kalliomäki et al. (2001) showed that reduction in *Bifidobacteria* and increase in *Clostridia* in gut are associated with asthma. Respiratory tract infection by influenza virus reduces *Lactobacilli* and *Lactococci* and increases *Enterobacteriaceae.* A study in Canadian children whom at the risk of

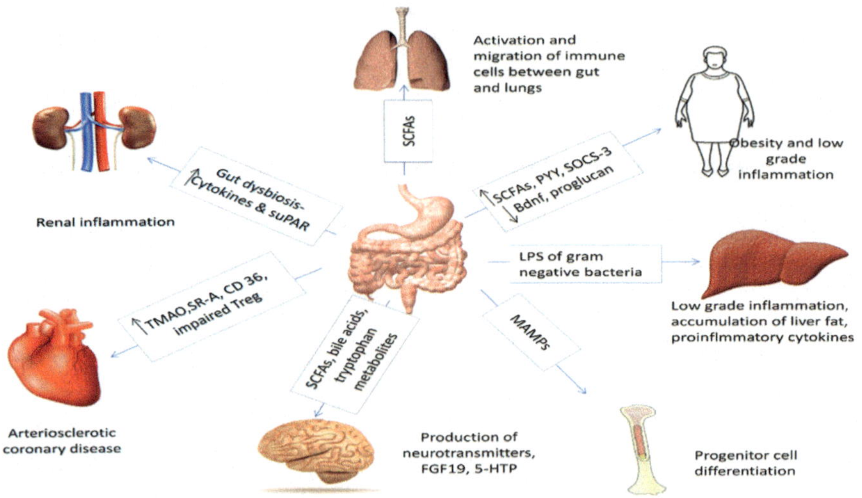

Fig. 2.1 Communication of gut to other systems. Microbial products or microbially derived host metabolites exerts effect on different organs in different ways

asthma showed decreased number of *Veillonella, Rothia, Lachnospira* and *Faecalibacteria* (Arrieta et al. 2015). The gut microbiota not only modulates gut immune system but also immune progenitor cells in bone marrow. Renal dysfunctions have been seen in patients receiving bone marrow transplants which indicate the role of bone marrow in kidney inflammation. In the immune pathway, immune cells originating from the bone marrow encounter dysbiotic microbiota and become overactivated within the intestine. Inflammatory cells, cytokines and soluble urokinase plasminogen activator surface receptor (suPAR) generated in the gut contribute to renal inflammation via the circulation (Hingorani et al. 2007; Hahm et al. 2017).

Nutrients and bacterial components reach liver by portal circulation. LPS of gram negative bacteria circulating through blood triggers low-grade inflammation by TLR signalling. Besides, they also enhance accumulation of liver fat but this is not evident in mice lacking LPS receptor CD14/TLR4 (Cani et al. 2008). Duparc et al. (2017) demonstrated that a deletion of Myd88 in hepatic cells affects gut microbiota. Administration of *Bifidobacterium longum* plus inulin type fructans significantly reduces inflammatory markers (TNF- α and C-reactive proteins), Steatosis and non-alcoholic steatohepatitis index (Malaguarnera et al. 2012). In alcoholic liver disease, gut permeability increased due to disruption of tight junction by alcohol and aldehyde causes the entry of LPS, endotoxin and bacterial DNA into circulation and thereby enters liver. As a result, the Kupffer cells in liver produces proinflammatory cytokines through TLR4 or TLR9 (Yoon-Seok and Ekihiro 2013).

Gut microbiome also improves cardiovascular diseases. The accumulation of foam cells in subendothelium constitutes the first step in arteriosclerosis. The gut microbiome derived trimethylamine N-oxide (TMAO) increases the expression of the receptor SR-A and CD 36 which can ultimately lead to the formation of foam cells. TMAO can also activate NLRP3 inflammasome (Wright et al. 2000; Duewell et al. 2012). Also impaired Treg cell increases the incidence of arteriosclerosis (Duewell et al. 2012). The gut invasion by probiotic can prevent arteriosclerotic coronary diseases.

A comparative study in GF and conventionally raised mice showed that conventionally raised mice with gut colonization were prone to become obese than GF mice when consumed with high carbohydrate high fat diet. The proposed mechanism is that the SCFAs produced by gut bacteria binds with GPR41 (G-protein coupled receptor) and GPR43 (FFAR3 and FFAR2) and promotes nutrient uptake and adipose tissue development. GPR41 also induces the secretion of pancreatic peptide YY (PYY) which increases the transit time. As the transit time increases, more nutrients will be absorbed, mainly glucose, which also contributes to obesity (Samuel et al. 2008). A similar study by Schéle et al. (2013) also showed weight gain in conventionally raised mice. They showed a reduced expression of two genes for antiobesity peptide, brain derived neurotrophic factor (Bdnf) and proglucagon (precursor of glucagon like peptide 1, GLP-1). In the presence of gut microbiota, leptin signalling inhibitor SOCS-3 was upregulated, thereby reducing the sensitivity to leptin. Since the leptin signal was not received by hypothalamus, mice became obese. Leptin is a neurohormone which is secreted by adipocytes. Its level is proportional to fat mass. As the size of adipocytes increases, more leptin is secreted. Leptin has proinflammatory properties and upregulates TNF-α, IL-16 and IL-12. This might be a reason of low-grade inflammation in obese persons. Leptin is also involved in innate and adaptive immunity. It significantly increases $CD4^+$ and $CD8^+$ cells and is also involved in DC maturation, proliferation of monocytes, neutrophil chemotaxis, reactive oxygen species generation, NK cell proliferation and activation of various pathways (IRS-1, PI3k/Akt, NF-kB and STAT-3) for production of interleukins (Naylor and Petri Jr 2016).

Gut microbiota communicates to the brain directly or indirectly through microbe-derived products. The gut brain axis involves neural, immune and endocrine pathways. The metabolic products of gut bacteria interact with ganglionated plexus of enteric nervous system and aids in peristalsis. Inflammation of GI tract increases anxiety like response and anorexia. It is clear that dysbiosis of the gut causes such behavioural changes. The gut microbe modulates CNS by neuroimmune and neuro-endocrine pathways through metabolites like SCFA, bile acids and tryptophan metabolites. In addition, the microbiota produce or can stimulate the production of γ-aminobutyric acid, norepinephrine, dopamine and serotonine. The bile acids induce the production of fibroblast growth factor (FGF19), enter the circulation and cross blood brain barrier, and activate arcuate nucleus of hypothalamus which regulates glucose and energy metabolism (Tomlinson et al. 2002). Gut microbes contribute to the development and function of microglial cells in CNS. Defective and compromised glial cells found in GF mice can be normalized by SCFA supplementation or colonization with microbes (Erny et al. 2015). SCFAs also aid in the release of norepinephrine by activation of GPR 41 (Kimura et al. 2011). Serotonin (5-hydroxytryptamine, 5-HTP) is an important neurotransmitter which has immune functions also. Above 90% of body's 5-HTP is present in enterochromaffin cells of intestine. The indigenous spore-forming bacteria of gut, *Clostridium* sp. promotes the biosynthesis. The SCFA produced by fermentation induces TPH1 gene expression and leads to the utilization of tryptophan for 5-HTP. It regulates cytokine secretion in macrophages and monocytes and reduces the level of TNF-α, IL-1β and also neutrophil recruitment to the inflammation site and T cell activation (Yano et al. 2015).

2.6 Probiotics in Immunomodulation

The composition of gut microbiota may vary with age, clinical status, diet, mental stress, antibiotic consumption, etc. Thus, a dysbiosis of gut leads to impaired immune system which ultimately leads to inflammatory bowel disease (IBD) and polyps in gut. The scope of probiotics lies here. Probiotics are live microorganisms which when administrated in adequate amounts confer a health benefit to the host (FAO/WHO 2001). The use of microorganisms for improving health was first appeared in the book of Ilya Ilyich Metchnikof in early twentieth century. Now probiotics are available in many fermented dairy and non-dairy products as well as capsules and probiotic drinks. Health benefits of probiotics are numerous. It can be exploited for all gut-related problems as evident from recent studies. Antipathogenicity, immunomodulation and anticancer effects of probiotics are the major research areas. Probiotics exert their beneficial effects through their metabolites, competitive exclusion of pathogens and by boosting innate and adaptive immunity of the host.

Most extensively studied organisms in immunomodulation are Lactobacillus and Bifidobacteria, the classic probiotics. The commercial strain *Lactobacillus rhamnosus* GG can be used to treat inflammatory bowel disease (IBD) since it showed equal effect as that of mesalazine, the drug for IBD (Zocco et al. 2006), and also exhibits anti-apoptotic property by activating Akt/protein kinase B. *Lactobacillus casei* can induce IL-12 production by TLR-2, TLR-4 or TLR-9 deficient mcrophage but not by MyD88 deficient macrophage (Ichikawa et al. 2007). Another Strain, *L. casei* DN114001, downregulates TNF-α production by inflamed mucosa in Crohn's disease (CD) patients (Borruel et al. 2001). Dietary supplementation of *L. rhamnosus* HN001 and *L. casei* Shirota enhances the number and cytolytic activity of NK cells in the peripheral blood in adults (Dong et al. 2010; Gill et al. 2001). The mutant strain of *L. acidophilus* NCK2025 lacking normal lipoteichoic acid lowers the level of IL- 12 and TNF α but enhanced IL-10 in DCs than its wild type (Gill et al. 2001). Administration of *B. lactis, L. rhamnosus* and *B. breve* can upregulate Treg cells, thereby reducing allergic response (Sagar et al. 2014).

An improvement in lung cancer was observed when *Enterococcus hirae* and *Barnesiella intestinihominis* were given along with chemotherapeutics (Daillère et al. 2016). The colon cancer cell lines CaCO-2 and HT-29 secrete IL-8 when stimulated by TNF–α. But pretreatment of CaCO-2 by *L. rhamnosus* GG reduces the level of IL-8 and inhibited the secretion of IL-8 in HT-29 when treated with Bifidobacterium genomic DNA. Clinical trials in humans proved the efficiency of various probiotics in IBD treatment. Administration of probiotic tablet VSL#3 (consisting of *L. plantarum, L. delbrueckii* subsp. *bulgaricus, L. casei, L. acidophilus, Bifidobacterium breve, B. longum, B. infantis*, and *Streptococcus salivarius* subsp. *thermophilus*) for 9 months reduced the relapse rate of pouchitis. Similarly, fermented milk containing Bifidobacterium improves ulcerative colitis (Gionchetti et al. 2000). Oral administration of a combination of *Lactobacillus helveticus* R0052 and *Bifidobacterium longum* R0175 for a period of one month has been reported to improve depression, anxiety, and lower the level of the stress hormone cortisol in humans (Messaoudi et al. 2011). A three-week consumption of a probiotic-containing milk drink that contained *Lactobacillus casei* Shirota showed improved mood in healthy volunteers (Benton et al. 2007).

The commensals as well as probiotics have profound effect on the overall homeostasis of the body. The survivability of these beneficial organisms in gut can be increased by administration of additional substances called prebiotics.

2.7 Prebiotics

The human gastrointestinal microbiota is one of the most densely populated microbial communities that provide metabolic, immunological and protective functions that play an important role in human health (Jumpertz et al. 2011; Goldsmith and Sartor 2014; Wang et al. 2011). Genetics, host physiology (Age of host, diseases, stress, etc.), and environmental factors including living conditions and use of medications are the number of factors influenced by the gastrointestinal microbiota (Wang et al. 2011; Greenblum et al. 2012; Wang et al. 2012a; Goodrich et al. 2014). However, the key environmental factor is a diet that mediates the composition and metabolic function of the gastrointestinal microbiota. Actually, the consumption of specific dietary ingredients, fibre and prebiotics is an important strategy that stimulates the functions of gut microbiota. Some dietary fibres can also be classified as prebiotics. Prebiotics are widely defined as a food ingredient that is composed of oligosaccharides, that are non-digestible by the host and has a beneficial effect on host health through selective stimulation of the growth and activity of specific members of the gut microbiota (Vieira et al. 2013). These types of food supplements have innumerable and composite effects on the intestinal microbiota and gut immune system. The US recommended daily fibre intake is 25–38 g but only an average of 15 g is consumed (American Dietetic Association), recommending dietary prebiotics could positively influence total fibre intake.

The major source of prebiotics is dietary fibre. They occur naturally in fruits and vegetables including chicory root, Jerusalem Artichokes, Raw dandelion Greens, Garlic, Leeks, Onion, Asparagus, Wheat Bran, Banana, Barley, Oats, Apple, Konjac root, Cocoa, Burdock root, Flaxseeds, Yacon root, Jicama root, sea weeds, etc. But they are present in the form of nutritional supplements for maximum health benefits. Prebiotics are also found in human milk, cow's milk, and yoghurt in the form of galactooligosaccharides. The most important prebiotic, inulin, is generally found in plants, bacteria and some fungi. It is known in more than 36,000 fruits and vegetables (Chicory, Banana, Onion, etc.). Soybean oligosaccharides (SOS) are another type of naturally occurring oligosaccharides present in soybean, which consist of raffinose and stachyose.

Another major naturally occurring prebiotics is acacia gum. More than 20 studies have been performed since the late 1970s to understand the relationships between acacia gum and the colonic microflora. It is a soluble dietary fibre obtained from the stems and branches of *Acacia Senegal* and *Acacia seyal.* It is composed mainly of complex polysaccharides (95%) that consist of highly branched galactan polymers, with galactose and/or arabinose side chains, possibly terminated by rhamnose or glucuronic acid residues (Cherbut et al. 2003b). It is present in different names including gum Arabic, Gum Hashab, Kordofanian gum and Acacia gum. 80% above production is used by the food industry for various applications such as food additives, emulsification, encapsulation, coating, gum cadies, thickener, demulcent, suspension agent, and foam stabilizer in cosmetics, bath and body products, and other skincare applications, etc. Gum is traditionally consumed by African and Indian population to improve digestive comfort and intestinal transit. Acacia gum induces bifidogenic effect, specific stimulation of SCFAs production and high gut tolerance. Guar gum, in its intact state, is a gel-forming galactomannan made from the endosperm of the plant *Cyamopsis tetragonolobus*, and is composed primarily of high molecular weight polysaccharides ([1,4]-linked β-D-mannopyranosyl units with [1,6]-linked α-D-galactopyranosyl side-chain residues (Kolida et al. 2000). Guar is commonly used in dairy, bakery, cereal and meat products.

Besides a range of naturally occurring prebiotics, there are synthetic prebiotics including Lactosucrose (LS) produced by combining lactose and sucrose using β-fructofuranosidase, Lactulose produced from lactose (it is not hydrolysed and absorbed in the small intestine), and Isomaltooligosaccharide (IMO) produced from starch (it can be digested in the small intestine) (Mudgil et al. 2014). Glucooligosaccharides are synthesized with glucosyl transferase, which is produced by *Leuconostoc mesenteroides* or may be extracted from β-glucan of oak tree. XOS can be hydrolysed by Bifidobacteria and Lactobacilli are found to be more effective than FOS in increasing the population of the probiotics and in decreasing the number in harmful bacteria.

Prebiotics provide nutrition to the host, inhibit the growth of potential pathogens and promote beneficial microbiota. The latter causes fermentation of non-digestible fibres, saves energy, synthesizes vitamin B and K, produces SCFA and polyamines, leads to improvement in gastrointestinal motility and function, reduces the level of cholesterol and stimulates the immune system. Other benefits of prebiotic consumption include reduction in the prevalence and duration of infectious and antibiotic-associated diarrhoea, reduction in inflammation and symptoms associated with inflammatory bowel disease, protective effects for prevention of colon cancer, enhancement of the bioavailability and uptake of minerals, including calcium, magnesium and possibly iron, lowering some risk factors for cardiovascular diseases, promotion of satiety and weight loss and prevention of obesity, reduced constipation and gas formation, etc. (Tomar et al. 2015).

2.7.1 Role of Prebiotics in Gut Immunity

Non-digestible carbohydrates of plant origin are the main substrates of gut microflora and include resistant starch as well as non-starch polysaccharides such as cellulose, hemi cellulose, pectin and inulin which are referred to as dietary fibre. But breakdown of dietary fibres is different, based on the matrix and the type of polysaccharides present. Mucus, slough epithelial cells, lysed bacteria, etc. are the other principal substrates of the gut microflora. These substrates provide carbon and energy for growth of the gut microorganisms.

The bacterial metabolism in the human colon is primarily anaerobic, because more than 99% of the bacteria encountered in an adult's faecal flora are strict anaerobes (Moore and Holdeman 1974). The available substrates are broken down to the SCFAs acetate, propionate, butyrate and the gases hydrogen (H_2) and carbon dioxide (CO_2). Formate, Valerate and Carporate are formed in small amounts only. Lactate, ethanol and succinate are intermediate which are also converted to SCFAs. The most studied SCFAs is butyrate. The major energy source of colonic epithelial cells is butyrate that affects the proliferation and barrier function of the colonic epithelium and reduces oxidative DNA damage (Gibson et al. 1999; Wang et al. 2012b). This energy source is transported into cells via monocarboxylated transporters, such as MCT–1107 (Ritzhaupt et al. 1998). Butyrates modulate the immune system in different ways. Initial studies using primary human leukocytes found that butyrate inhibits IL-12 production by *S. aureus* stimulated human monocytes (Säemann et al. 2000). The same study also found that in anti-CD3 stimulated monocytes, butyrate enhanced IL-10 and IL-4 secretion, but inhibited IL-2 and IFN-gamma release, presenting an anti-inflammatory profile for butyrate. Other in vitro studies have found that butyrate inhibits vascular cell adhesion molecule (VCAM-1)-mediated leukocyte adhesion to endothelial cells (Menzel et al. 2004). Ex vivo studies in mice found that butyrate suppresses colonic immune activation through Fas-mediated apoptosis of T cells through histone deacetylase (HDAC) 1-dependent Fas upregulation. This work also provided evidence that butyrate inhibits IFN-γ-mediated inflammatory signalling, particularly through STAT1 and iNOS, and that loss of butyrate signalling induces increased expression of inflammatory genes in mice (Zimmerman et al. 2012). Other in vitro findings demonstrate that butyrate inhibits the IFN-α/STAT1 axis (Klampfer et al. 2003), which is important because enhanced activation of STAT1 occurs in CD patients (Schreiber et al. 2002). Human ex vivo studies found that butyrate was able to decrease pro-inflammatory cytokine (TNF-β, IL-1b, IL-6) mRNA expression as well as TNF secretion in intestinal biopsies and peripheral blood mononuclear cells of CD patients, through inhibition of NFκB (Segain et al. 2000).

Acetate is the second most abundant short-chain fatty acids in the colon. Many researchers reported the anti-inflammatory effects of acetate on the inflammatory response (Maslowski et al. 2009; Kim et al. 2013; Smith et al. 2013), but most studies mainly focused on butyrate. The receptors GPR41 (Ffar1), GPR109A and GPR43 (Ffar2) were identified as receptors of butyrate, propionate and acetate, respectively (Brown et al. 2003). GPR41 is primarily expressed by adipose tissue and is also present at very low levels in peripheral blood mononuclear cells (PBMCs). GPR43 expression is entirely related to the immune system and is particularly high on polymorph nuclear cells (eosinophils and neutrophils). Maslowski et al. (2009) showed that mice that lack the Gpr43 gene have increased inflammation and a poor ability to resolve inflammation because their immune cells cannot bind to SCFAs. Hence they were more susceptible to IBD. However, the effect of activation of acetate/Gpr43 helps to increase the clinical and inflammatory response in experimental mice. Intestinal bacteria are useful in the elevation of human health, but certain components of microflora in genetically susceptible individuals contribute to various pathological disorders, including inflammatory bowel disease (IBD). A change in gut microbiota composition is considered as one of many factors involved in the pathogenesis of either inflammatory bowel disease or irritable bowel syndrome. For these reasons, the use of prebiotics in IBD such as Crohn's disease, ulcerative colitis and pouchitis is very important because they restore the balance of GI microflora, reducing and preventing intestinal inflammation (Cherbut et al. 2003a; Schultz et al. 2004; Furrie et al. 2005; Kelly et al. 2005). These diseases are characterized by persistent mucosal inflammation at different levels of the GIT (Guarner 2007). In the GIT, the inflammatory capacity of commensal bacteria varies because some bacteria are pro-inflammatory, whereas others attenuate inflammatory responses.

Prebiotics fermentation in large intestine also produces propionate that shows anti-inflammatory effects with respect to colon cancer (Makivuokko et al. 2009). Acetate is largely produced in the colon but reaches a high concentration in the blood, so we could observe systemic anti-inflammatory effects of this SCFA in other diseases, such as asthma and arthritis and also decrease in the luminal pH. A low pH can stimulate the growth of Lactobacilli and Bifidobacteria which are adapted to low pH. While a low pH suppresses growth of harmful bacteria. SCFAs may play an important role for the optimal functioning of the colonic epithelium and the absorption of various cations including Ca^{2+}, Mg^{2+} and Fe^{2+}.

Recent study investigated the effects of prebiotic oligosaccharide on microbiota composition and immune function (NK cells, phagocytosis and cytokines) in healthy elderly volunteers. The study also found significant positive effect on immune response, evidenced by an improvement in NK cell activity and phagocytosis, increased secretion of the anti-inflammatory cytokines, IL-10, and decreased secretion of proinflammatory cytokines (IL-6, IL-1β and TNF-α) (Vulevic et al. 2008). Oral lactulose increases stool water content (Hebden et al. 1999) and increases stool frequency in constipation (Bass and Dennis 1981; Freedman et al. 1997). This beneficial effect arises out of a combination of increased bacterial mass, increased stool water as well as increased colonic tone resulting in accelerated transit (Jouet et al. 2006). Numerous other poorly absorbed storage carbohydrates are used for their laxative effect with important effects on gut microbiota although their specific mechanism of action is less well worked out than for lactulose.

Prebiotics can administer along with live bacteria (probiotics) that are most able to utilize that energy source to improve the health benefits to the host. The synergistic combinations of probiotics and prebiotics are called synbiotics. Probiotics, prebiotics and synbiotics can influence the intestinal microbiota and modulate the immune response. In the study on the effect of the synbiotic product containing a blend of probiotics (*Lactobacillus casei, Lactobacillus rhamnosus, Streptococcus thermophilus, Bifidobacterium breve, Lactobacillus acidophilus, Bifidobacterium longum, Lactobacillus bulgaricus*) and fructooligosaccharides, 52 adults participated for 28 weeks. It was found that supplementation with the synbiotic resulted in the inhibition of NFκB and reduced production of TNF-*a* (Eslamparast et al. 2014). The use of a synbiotic containing five probiotics (*Lactobacillus plantarum, Lactobacillus delbrueckii* spp. *bulgaricus, Lactobacillus acidophilus, Lactobacillus rhamnosus, Bifidobacterium bifidum*) and inulin as a prebiotic in adult subjects with NASH (non-alcoholic steatohepatitis) confirmed a significant reduction of IHTG (intrahepatic triacylglycerol) within six months (Wong et al. 2013). It is also known that lipopolysaccharides (LPSs) induce proinflammatory cytokines, such as the tumour necrosis factor alpha (TNF-*a*), playing a crucial role in insulin resistance and inflammatory cell uptake in NAFLD (non-alcoholic fatty liver disease). Other beneficial effects of synbiotics include improved hepatic function in patients suffering from cirrhosis, prevention of bacterial translocation and reduced incidence of nosocomial infections in patient post-surgical procedures and similar intervention.

2.8 Postbiotic Modulation of Immune System

2.8.1 Postbiotics and Host Microbiota

Mechanism of actions of postbiotics is not fully elucidated. One of the possible immunomodulatory effects by postbiotics in humans could be derived from an in vitro experiment showing the innate response of macrophages to non-viable *Lactobacillus casei* cells. Heat killed bacterial cell suspension increases in expression of proinflammatory cytokines and increases the transcription of Toll-like receptors (TLR-2, TLR-3, TLR-4 and TLR-9) (Wang et al. 2013). But the heat treated *Bifidobacterium* cells induce cellular immune and anti-inflammatory responses by inhibiting IL-8 secretion in intestinal epithelial cells obtained from patients with ulcerative colitis (Imaoka et al. 2008).

The postbiotic compounds from *Lactobacilli* spp. can exert immunomodulation activity by increasing levels of Th1-associated cytokines and reducing Th2-associated cytokines (Ou et al. 2011). Likewise, retinoic acid produced by *L. reuteri* 17,938 influences the phenotype and function of mucosal like dendritic cells and also increases the level of IL-10, CD103 and CD1d and downregulates inflammation-associated genes like NFκB and TNF (Haileselassie et al. 2016).

Similar findings were reported in another research conducted by Sokol et al. (2008), who reported that increased IL-8 levels in Caco-2 cells when exposed to intracellular extracts and the supernatant fraction of *Faecalibacterium prausnitzii*. Cultural supernatants of *Lactobacillus rhamnosus* GG collected at different stages of growth (middle and late exponential, stationary, and overnight) were able to protect human colonic smooth muscle cells (HSMCs) against lipopolysaccharide (LPS)-induced myogenic damage. Increased level of protective effect was observed with supernatants of the late stationary phase, which reverted 84.1% of LPS-induced cell shortening, and inhibited 85.5% of acetylcholine-induced contraction and 92.7% LPS-induced IL-6 secretion (Cicenia et al. 2016).

Exopolysaccharides and extracellular vesicles (EV) are two important fermentation products that are associated with health benefits. Exopolysaccharide from *Lactobacillus plantarum* 70,810 was found to function as antitumor agents in vitro by inhibiting the proliferation of HepG-2, BGC-823 and HT-29 tumour cells (Wang et al. 2014). Extracellular vesicles are spherical lipid bilayer structures that can be secreted by both gram negative and gram positive bacteria. Extracellular vesicles have ability to carry a different type of compounds such as nucleic acids, proteins, phospholipids, polysaccharides and glycolipids. EVs can be differentiated into two; they are outer membrane vesicles (OMVs) for gram negative bacteria and membrane vesicles (MVs) for gram positive bacteria. EVs derived from *Akkermansia muciniphila* and commensal *Escherichia coli* have shown respectively to decrease gut permeability and activate signalling through the intestinal epithelial barrier in vitro (Chelakkot et al. 2018; Fábrega et al. 2016). However, human clinical trials are needed to establish safety and potential for the use of EVs as therapeutic agents in humans.

The potential of innate and adaptive immunity to trigger inflammation in response to abundant microbial compounds including lipoteichoic acids and S-layer proteins was elucidated by Konstantinov et al. (2013). The major metabolite of gut bacteria, the SCFA, has numerous health-promoting activities such as butyrate enhances the intestinal barrier function and mucosal immunity and butyrate and small amount of propionate act as histone deacetylase (HDAC) inhibitors, etc. As a result, they promote anti-inflammatory and immune effects through suppression of lamina propria macrophages and cause differentiation of dendritic cells from bone marrow stem cells (Koh et al. 2016; Johnstone 2002; Singh et al. 2010; Lukovac et al. 2014). SCFAs can also activate some SCFAs specific G-protein-coupled receptors (GPRs) present on gut epithelial cells and others. It helps to modulate cellular activity (Gill et al. 2018). SCFAs have antitumor effects, anti-inflammatory effects on the colonic epithelium, protection from development of immune disorders and control of obesity. Table 2.1 illustrates the important gut bacterial products/metabolites which elicit an immunomodulatory effect. All these studies suggest that postbiotics have ability to increase health by providing better and specific physiological effects, although the exact mechanisms remain to be elucidated.

Table 2.1 Gut microbial metabolites or postbiotics and their role in immunity

Postbiotic metabolites/ compounds	Immunomodulatory functions	Reference
Butyrate	Boost extra thymic Treg cell generation in mice	Arpaia et al. (2013)
	Differentiation of colonic Treg cells	Furusawa et al. (2013)
	Downregulation of LPS induced proinflammatory mediators like NO, IL-6 and IL-12 in macrophages in vitro	Chang et al. (2014)
Butyrate and propionate	Downregulation of proinflammatory cytokines IL-6, IL-12p40, CCL3, CCL4, CCL5, CCL-9, CCL10, CCL11 in human monocyte derived DC	Nastasi et al. (2015)
Butyrate and acetate	Increased GPR43 expression and decreased proinflammatory monocyte chemoattractant protein MCP-1, IL-β and inhibit oxidative stress in high glucose treated glomerular mesangial cells	Huang et al. (2017)
Aryl hydrocarbon receptor ligand	Necessary for the postnatal expansion of RORγt	Kiss et al. (2011)
Polyamines	Increased production of sIgA in rats	Buts et al. (1993)
	Enhances the integrity of IECs	Chen et al. (2007)
	Modulates adaptive immunity by accelerating the maturation of $CD4^+$ and $CD8^+$ T cells	Pérez-Cano et al. (2010)
Polysaccharide A	Anti-inflammatory effect by increasing the levels of IL-10	Round et al. (2011)
	Maintains balance between T_H1 and T_H2 cells in GF mice	Mazmanian et al. (2005)
Formyl peptides	Helps in recruiting leukocytes and production of proinflammatory cytokines	Liu et al. (2014)
D-glycero β-D-Manno-heptose-1,7-bisphosphate	Stimulates innate immune response by activating NFκB pathway	Gaudet et al. (2015)

LPS lipopolysaccharide, *NO* nitric oxide, *IL* interleukin, *CCL* chemokines, *DC* dendritic cells, *RORγt* retinoic acid receptor related orphan nuclear receptor gamma, *sIgA* secretary immunoglobulin A, *IEC* intestinal epithelial cells

2.9 Conclusion

Gut is the largest immune organ of the body and it is considered as the second brain due to the complex enteric, endocrine, neuron and immune networks. The food and microbiota in the gut influences all organs through these networks. If the microorganisms inhabiting the gut are beneficial, body homeostasis will be maintained; otherwise it gets disrupted. Administration of probiotic is an alternative to a dysbiotic gut. Prebiotics are given along with probiotics for their establishment and survival. The synergistic effects of both improve gut health. Fermentation of prebiotics in gut produces metabolites which circulates through the body and activates different systems. Usage of synbiotics is an approach to prevent aberrations in the gut. The latest trend in the biological modulation of immunity is the administration of microbial by-products, metabolites or inactivated cells itself. Upon further studies and clinical trials, the postbiotics can be used as non-specific immune boosting vaccines which can activate the low-immune children as well as adult.

It is important to maintain proper health and immunity through natural means, since we are facing new challenges day by day. In current pandemic COVID 19, several positive cases were asymptomatic or they may be immune. Boosting immunity through functional food helps in preventing infection and staying healthy. As this chapter indicates, consuming healthy food always improves the health through the nutrients and metabolites that are present in it or by the by-products of gut microflora. Age and health status of an individual are always a critical factor that increases the severity of infection, but a proper diet and hygiene does help improving the condition.

References

Aagaard K, Kevin R, Jun M et al (2012) A metagenomic approach to characterization of the vaginal microbiome signature in pregnancy. PLoS One 7:e36466

Arpaia N, Clarissa C, Xiying F et al (2013) Metabolites produced by commensal bacteria promote peripheral regulatory T-cell generation. Nature 504:451–455

Arrieta M, Leah T, Pedro AD et al (2015) Early infancy microbial and metabolic alterations affect risk of childhood asthma. Sci Transl Med 7:1–14

Bass P, Dennis S (1981) The laxative effects of lactulose in normal and constipated subjects. J Clin Gastroenterol 3:23–28

Benton D, William C, Brown A (2007) Impact of consuming a milk drink containing a probiotic on mood and cognition. Eur J Clin Nutr 61:355–361

Borruel N, Francesc C, Maria A et al (2001) Increased mucosal TNF-α production in Crohn's disease can be modulated locally by probiotics. Gastroenterology 120:278–279

Brown AJ, Susan MG, Ashley AB et al (2003) The orphan G protein-coupled receptors GPR41 and GPR43 are activated by propionate and other short chain carboxylic acids. J Biol Chem 278:11312–11319

Buts JP, Nadine DK, Jaroslaw K et al (1993) Maturation of villus and crypt cell functions in rat small intestine - role of dietary polyamines. Digest Dis Sci 38:1091–1098

Cani PD, Rodrigo B, Claude K et al (2008) Changes in gut microbiota control metabolic diet–induced obesity and diabetes in mice. Diabetes 57:1470–1481

Carvalho FA, Jesse DA, Matam VK et al (2012) Toll-like receptor–gut microbiota interactions: perturb at your own risk! Ann Rev Physiol 74:177–198

Chang PV, Liming H, Stefan O et al (2014) The microbial metabolite butyrate regulates intestinal macrophage function via histone deacetylase inhibition. PNAS 111:2247–2252

Chassaing B, Manish K, Mark TB et al (2014) Mammalian gut immunity. Biom J 37:246–258

Chelakkot C, Youngwoo C, Dae-Kyum K et al (2018) Akkermansia Muciniphila -derived extracellular vesicles influence gut permeability through the regulation of tight junctions. Exp Mol Med 50:1–11

Chen J, Jaladanki NR, Tongtong Z et al (2007) Polyamines are required for expression of Toll-like receptor 2 modulating intestinal epithelial barrier integrity. Am J Physiol Gastr L 293:568–576

Cherbut C, Michel C, Lecannu G (2003a) The prebiotic characteristics of fructooligosaccharides are necessary for reduction of TNBS-induced colitis in rats. J Nutr 133:21–27

Cherbut C, Michel C, Raison V et al (2003b) Acacia gum is a bifidogenic dietary fibre with high digestive tolerance in healthy humans. Microb Ecol Health Dis 15:43–50

Cicenia A, Floriana S, Lucrezia G et al (2016) Protective role of postbiotic mediators secreted by *Lactobacillus rhamnosus GG* versus lipopolysaccharide-induced damage in human colonic smooth muscle cells. J Clin Gastroentrol 50:140–144

Clarke TB, Kimberly MD, Elena SL et al (2010) Recognition of peptidoglycan from the microbiota by nod1 enhances systemic innate immunity. Nat Med 16:228–231

Coombes JL, Powrie F (2009) Dendritic cells in intestinal immune regulation. Nat Rev Immunol 8:435–446

Daillère R, Marie V, Nadine W et al (2016) *Enterococcus hirae* and *Barnesiella intestinihominis* facilitate cyclophosphamide-induced therapeutic immunomodulatory effects. Immunity 45:931–943

Dominguez-Bello MG, Elizabeth KC, Monica C et al (2010) Delivery mode shapes the acquisition and structure of the initial mirobiota across multiple body habitats in newborns. PNAS 107:11971–11975

Dong H, Rowland I, Tuohy KM, Thomas LV et al (2010) Selective effects of *Lactobacillus Casei* Shirota on T cell activation, natural killer cell activity and cytokine production. Clin Exp Immunol 161:378–388

Duewell P, Hajime K, Katey JR et al (2012) NLRP3 inflamasomes are required for atherogenesis and activated by cholesterol crystals that form early in disease. Pan Afr Med J 464:1357–1361

Duparc T, Hubert P, Vannina GM et al (2017) Hepatocyte MyD88 affects bile acids, gut microbiota and metabolome contributing to regulate glucose and lipid metabolism. Gut 66:620–632

Erny D, Anna LH, Diego J et al (2015) Host microbiota constantly control maturation and function of microglia in the CNS. Nat Neurosci 18(7):965–977

Eslamparast T, Hossein P, Farhad Z et al (2014) Synbiotic supplementation in nonalcoholic fatty liver disease: A randomized double blind placebo controlled pilot study. Am J Clin Nutr 99:535–542

Fábrega MJ, Laura A, Rosa G et al (2016) Activation of immune and defense responses in the intestinal mucosa by outer membrane vesicles of commensal and probiotic *Escherichia coli* strains. Front Microbiol 7:1–14

FAO/WHO (2001) Probiotics in food. Health and Nutritional Properties and Guidelines for Evaluation

Freedman MD, Jeffrey S, Robert R et al (1997) Tolerance and efficacy of polyethylene glycol 3350/ electrolyte solution versus lactulose in relieving opiate induced constipation: a double-blinded placebo-controlled trial. J Clin Pharmacol 37:904–907

Furrie E, Macfarlane S, Kennedy A et al (2005) Synbiotic therapy (*Bifidobacterium longum*/ synergy 1) initiates resolution of inflammation in patients with active ulcerative colitis: a randomised controlled pilot trial. Gut 54:242–249

Furusawa Y, Obata Y, Fukuda S et al (2013) Commensal microbe-derived butyrate induces the differentiation of colonic regulatory T cells. Nature 504:446–450

Gaudet RG, Anna S, Carolyn MB et al (2015) Cytosolic detection of the bacterial metabolite HBP activates TIFA-dependent innate immunity. Science 348:1251–1255

Gibson PR, Ourania R, Andrew JW et al (1999) Colonic epithelial cell activation and the paradoxical effects of butyrate. Carcinogenesis 20:539–544

Gill HS, Rutherfurd KJ, Cross ML (2001) Dietary probiotic supplementation enhances natural killer cell activity in the elderly: an investigation of age-related immunological changes. J Clin Immunol 21:264–271

Gill PA, Zelm MC, Muir JG et al (2018) Short chain fatty acids as potential therapeutic agents in human gastrointestinal and inflammatory disorders. Alimnent Pharmacol Ther 48:15–34

Gionchetti P, Fernando R, Alessandro V et al (2000) Oral bacteriotherapy as maintenance treatment in patients with chronic pouchitis: a double-blind, placebo-controlled trial. Gastroenterology 119:305–309

Goldsmith JR, Sartor RB (2014) The role of diet on intestinal microbiota metabolism: downstream impacts on host immune function and health, and therapeutic implications. J Gastroenterol 23:1–7

Goodrich JK, Jillian LW, Angela CP et al (2014) Human genetics shape the gut microbiome. Cell 159:789–799

Greenblum S, Turnbaugh PJ, Borenstein E (2012) Metagenomic systems biology of the human gut microbiome reveals topological shifts associated with obesity and inflammatory bowel disease. PNAS 109:594–599

Guarner F (2007) Prebiotics in inflammatory bowel diseases. Br J Nutr 98:85–89

Hahm E, Changil W, Isabel F et al (2017) Bone marrow-derived immature cells are a main source of circulating SuPAR contributing to proteinuric kidney disease. Nat Med 23:100–106

Haileselassie Y, Navis M, Vu N et al (2016) Postbiotic modulation of retinoic acid imprinted mucosal-like dendritic cells by probiotic *Lactobacillus reuteri* 17938 in vitro. Front Immunol 7:96

Hapfelmeier S, Melissaa EL, Emma S et al (2010) Reversible microbial colonization of germ-free mice reveals the dynamics of IgA immune responses. Science 328:1705–1709

Hebden JM, Peter JG, Alan CP et al (1999) Stool water content and colonic drug absorption: contrasting effects of lactulose and codeine. Pharm Res 16:1254–1259

Hendricks JM, Lowe DC, Hardy ME (2014) Differential induction of isolated lymphoid follicles in the gut by 18β-glycyrrhetinic acid. PLoS One 9:1–8

Hingorani S, Guthrie KA, Schoch G et al (2007) Chronic kidney disease in long-term survivors of hematopoietic cell transplant. Bone Marrow Transplant 39:223–229

Huang W, Heng LG, Xian D et al (2017) Short-chain fatty acids inhibit oxidative stress and inflammation in mesangial cells induced by high glucose and lipopolysaccharide. Exp Clin Endocrinol Diabetes 125:98–105

Ichikawa S, Rei F, Daisuke F et al (2007) MyD88 but not TLR2, 4 or 9 is essential for IL-12 induction by lactic acid bacteria. Biosci Biotechnol Biochem 71:3026–3032

Imaoka A, Tatsuichiro S, Kimitoshi K et al (2008) Anti-inflammatory activity of probiotic Bifidobacterium: enhancement of IL-10 production in peripheral blood mononuclear cells from ulcerative colitis patients and inhibition of IL-8 secretion in HT-29 cells. World J Gastroenterol 14:2511–2516

Iwasaki A, Kelsall BL (1999) Freshly isolated Peyer's patch, but not spleen, dendritic cells produce interleukin 10 and induce the differentiation of T helper type 2 cells. J Exp Med 190:229–240

Jandhyala SM, Rupjyoti T, Chivkula S et al (2015) Role of the normal gut microbiota. World J Gastroenterol 21:8836–8847

Jass JR, Walsh MD (2001) Altered mucin expression in the gastrointestinal tract: a review. J Cell Mol Med 5:327–351

Johnstone RW (2002) Histone-deacetylase inhibitors: novel drugs for the treatment of cancer. Nat Rev 1:287–299

Jouet P, Sabaté JM, Cuillerier E et al (2006) Low-dose lactulose produces a tonic contraction in the human colon. J Neurogastroenterol Motil 18:45–52

Jumpertz R, Duc SL, Peter JT et al (2011) Energy-balance studies reveal associations between gut microbes, caloric load, and nutrient absorption in humans. Am J Clin Nutr 94:58–65

Kalliomäki M, Pirkka K, Erkki E et al (2001) Distinct patterns of neonatal gut microflora in infants in whom atopy was and was not developing. J Allergy Clin Immunol 107:129–134

Kelly D, Conway S, Aminov R (2005) Commensal gut bacteria: mechanisms of immune modulation. Trends Immunol 26:326–333

Kim MH, Seung GK, Jeong HP et al (2013) Short-chain fatty acids activate GPR41 and GPR43 on intestinal epithelial cells to promote inflammatory responses in mice. Gastroenterology 145:396–406

Kimura I, Daisuke I, Takeshi M et al (2011) Short-chain fatty acids and ketones directly regulate sympathetic nervous system via G protein-coupled receptor 41 (GPR41). PNAS 108:8030–8035

Kiss EA, Cedric V, Stefanie K et al (2011) Natural aryl hydrocarbon receptor ligands control organogenesis of intestinal lymphoid follicles. Science 334:1561–1565

Klampfer L, Jie H, Takehiko S et al (2003) Inhibition of interferon γ signaling by the short chain fatty acid butyrate. Mol Cancer Res 1:855–862

Koh A, Filipe DV, Petia KD et al (2016) From dietary fiber to host physiology: short-chain fatty acids as key bacterial metabolites. Cell 165:1332–1345

Kolida S, Tuohy K, Gibson GR (2000) The human gut flora in nutrition and approaches for its dietary modulation. Nutr Bull 25:223–231

Konstantinov SR, Kuipers EJ, Peppelenbosch MP (2013) Functional genomic analyses of the gut microbiota for CRC screening. Nat Rev Gastroenterol Hepatol 10:741–745

Liu M, Keqiang C, Teizo Y et al (2014) Formylpeptide receptors mediate rapid neutrophil mobilization to accelerate wound healing. PLoS One 9:1–7

Lukovac S, Clara B, Linette P et al (2014) Differential modulation by *Akkermansia Muciniphila* and *Faecalibacterium Prausnitzii* of host peripheral lipid metabolism and histone acetylation in mouse gut organoids. mBio 5:1–10

Macpherson AJ, Martinic MM, Harris N (2002) The functions of mucosal T cells in containing the indigenous commensal flora of the intestine. Cell Mol Life Sci 59:2088–2096

Makivuokko H, Jussi N, Paivi N et al (2009) In vitro effects on polydextrose by colonic bacteria and Caco-2 cell cyclooxygenase gene expression. Nut Cancer 52:94–104

Malaguarnera M, Marco V, Tijana A (2012) *Bifidobacterium Longum* with fructo-oligosaccharides in patients with non alcoholic steatohepatitis. Digest Dis Sci 57:545–553

Maslowski KM, Angelica TV, Aylwin NG et al (2009) Regulation of inflammatory responses by gut microbiota and chemoattractant receptor GPR43. Nature 461:1282–1286

Mazmanian SK, Hua LC, Arthur OT et al (2005) An immunomodulatory molecule of symbiotic bacteria directs maturation of the host immune system. Cell 122:107–118

Menzel T, Hardi L, Sabine Z (2004) Butyrate inhibits leukocyte adhesion to endothelial cells via modulation of VCAM-1. Inflamm Bowel Dis 10:122–128

Messaoudi M, Robert L, Nicolas V et al (2011) Assessment of psychotropic-like properties of a probiotic formulation (*Lactobacillus Helveticus* R0052 and *Bifidobacterium Longum* R0175) in rats and human subjects. Br J Nutr 105:755–764

Mikkelsen HB, Charly G, Jørgen T et al (2004) Macrophages in the small intestinal muscularis externa of embryos, newborn and adult germ- free mice. J Mol His 35:377–387

Moore WE, Holdeman LV (1974) Human fecal flora: the normal flora of 20 japanese-hawaiians. Appl Microbiol 27:961–979

Mudgil D, Barak S, Khatkar BS (2014) Guar gum: processing, properties and food applications - a review. J Food Sci Tech 51:409–418

Muniz LR, Knosp C, Yeretssian G (2012) Intestinal antimicrobial peptides during homeostasis, infection, and disease. Front Immunol 3:1–13

Nastasi C, Marco C, Charlotte MB et al (2015) The effect of short-chain fatty acids on human monocyte-derived dendritic cells. Sci Rep 5:1–10

Naylor C, Petri WA Jr (2016) Leptin regulation of immune responses. Trends Mol Med 22:88–98

Ohkubo T, Tsuda M, Suzuki S et al (1999) Peripheral blood neutrophils of germ-free rats modified by in vivo granulocyte-colony-stimulating factor and exposure to natural environment. Scand J Immunol 49:73–77

Ou CC, Shiao LL, Jaw JT, Meei YL (2011) Heat-killed lactic acid bacteria enhance immunomodulatory potential by skewing the immune response toward Th1 polarization. J Food Sci 76(5): M260–M267

Pérez-Cano FJ, Ana GC, Cristina C et al (2010) Influence of breast milk polyamines on suckling rat immune system maturation. Dev Com Immunol 34:210–218

Ramakrishna BS (2007) The normal bacterial flora of the human intestine and its regulation. J Clin Gastroenterol 41:2–6

Reikvam DH, Alexander E, Anders S et al (2011) Depletion of murine intestinal microbiota: effects on gut mucosa and epithelial gene expression. PLoS One 6:1–13

Ritzhaupt A, Stuart WI, Antony E et al (1998) Identification and characterization of a monocarboxylate transporter (MCT1) in pig and human colon: its potential to transport L-lactate as well as butyrate. J Physiol 513:719–732

Round JL, Melanie L, Jennifer L et al (2011) The toll-like receptor pathway establishes commensal gut colonization. Science 332:974–977

Säemann MD, Georg AB, Christoph HO et al (2000) Anti-inflammatory effects of sodium butyrate on human monocytes: potent inhibition of IL-12 and up-regulation of IL-10 production. FASEB J 14:2380–2382

Sagar S, Mary EM, Si C et al (2014) *Bifidobacterium breve* and *Lactobacillus rhamnosus* treatment is as effective as budesonide at reducing inflammation in a murine model for chronic asthma. Respir Res 15:1–17

Samuel BS, Shaito A, Motoike T et al (2008) Effects of the gut microbiota on host adiposity are modulated by the short-chain fatty-acid binding G protein-coupled receptor, Gpr41. PNAS 105:16767–16772

Schéle E, Louise G, Fredrik A et al (2013) The gut microbiota reduces leptin sensitivity and the expression of the obesity-suppressing neuropeptides proglucagon (Gcg) and brain-derived neurotrophic factor (Bdnf) in the central nervous system. Endocrinology 154:3643–3651

Schreiber S, Rosenstiel P, Hampe J (2002) Activation of signal transducer and activator of transcription (STAT) 1 in human chronic inflammatory bowel disease. Gut 51:379–385

Schultz M, Munro K, Tannock GW et al (2004) Effects of feeding a probiotic preparation (SIM) containing inulin on the severity of colitis and on the composition of the intestinal microflora in HLA-B27 transgenic rats. Clin Diagnostic Lab Immunol 11:581–587

Segain JP, Raingeard D, Bourreille A et al (2000) Butyrate inhibits inflammatory responses through NFkappaB inhibition: implications for Crohn's disease. Gut 47:397–403

Singh N, Muthusamy T, Puttur DP et al (2010) Blockade of dendritic cell development by bacterial fermentation products butyrate and propionate through a transporter (Slc5a8)-dependent inhibition of histone. J Biol Chem 285:27601–276018

Smith PM, Michael RH, Nicolai P et al (2013) The microbial metabolites, short chain fatty acids, regulate colonic Treg cell homeostasis. Science 341:569–573

Smythies LE, Marty S, Ronald HC et al (2005) Human intestinal macrophages display profound inflammatory anergy despite avid phagocytic and bacteriocidal activity. J Clin Invest 115:66–75

Sokol H, Bénédicte P, Laurie W et al (2008) *Faecalibacterium Prausnitzii* is an anti-inflammatory commensal bacterium identified by gut microbiota analysis of Crohn disease patients. PNAS 105:16731–16736

Tomar SK, Anand S, Sharma P et al (2015) Role of probiotic, prebiotics, synbiotics and postbiotics in inhibition of pathogens. The Battle Against Microbial Pathogens: Basic Science, Technological Advances and Educational Programs (October 2016):717–732

Tomlinson E, Ling F, Linu J et al (2002) Transgenic mice expressing human fibroblast growth factor-19 display increased metabolic rate and decreased adiposity. Endocrinology 143:1741–1747

Vamadevan AS, Masayuki F, Elizabeth TA et al (2010) Regulation of TLR4-associated MD-2 in intestinal epithelial cells: a comprehensive analysis. Innat Immunol 16:93–103

Vieira AT, Teixeira MM, Martins FD (2013) The role of probiotics and prebiotics in inducing gut immunity. Front Immonol 4:1–12

Vulevic J, Alexandra D, Parveen Y et al (2008) Modulation of the fecal microflora profile and immune function by a novel trans-galactooligosaccharide mixture (B-GOS) in healthy elderly volunteers. Am J Clin Nutr 88:1438–1446

Wang Z, Elizabeth K, Brian JB et al (2011) Gut flora metabolism of phosphatidylcholine promotes cardiovascular disease. Nature 472:57–65

Wang J, Junjie Q, Yingrui L et al (2012a) A metagenome-wide association study of gut microbiota in type 2 diabetes. Nature 490:55–60

Wang HB, Peng YW, Xin W et al (2012b) Butyrate enhances intestinal epithelial barrier function via up-regulation of tight junction protein claudin-1 transcription. Digest Dis Sci 57:3126–3135

Wang Y, Jiming X, Na W et al (2013) *Lactobacillus Casei* Zhang modulate cytokine and toll-like receptor expression and beneficially regulate Poly I:C-induced immune responses in RAW264.7 macrophages. Microbiol Immunol 57:54–62

Wang K, Wei L, Xin R et al (2014) Characterization of a novel exopolysaccharide with antitumor activity from *Lactobacillus Plantarum* 70810. Int J Bio Macromol 63:133–139

Wong VWS, Grace LHW, Angel MLC (2013) Treatment of nonalcoholic steatohepatitis with probiotics. A proof-of-concept study. Annl Hepatol 12:256–262

Wright SD, Charlotte B, Melba H (2000) Infectious agents are not necessary for murine atherogenesis. J Exp Med 191:1437–1442

Yan F, Polk DB (2002) Probiotic bacterium prevents cytokine-induced apoptosis in intestinal epithelial cells. J Biol Chem 277:50959–50965

Yang T, Elaine MR, Carl JP et al (2018) The gut microbiota and the brain-gut-kidney axis in hypertension and chronic kidney disease. Nat Rev Nephrol 14:442–456

Yano JM, Kristie Y, Gregory PD et al (2015) Indigenous bacteria from the gut microbiota regulate host serotonin biosynthesis. Cell 161:264–276

Yoon-Seok R, Ekihiro S (2013) Toll-like receptors in alcoholic liver disease, non-alcoholic steatohepatitis and carcinogenesis. Gastroenterol Hepatol 28:38–42

Zimmerman MA, Nagendra S, Pamela MM et al (2012) Butyrate suppresses colonic inflammation through HDAC1-dependent Fas upregulation and Fas-mediated apoptosis of T cells. Am J Physiol Gastrointest Liver Physiol 302(12):G1405–G1415

Zocco MA, Zileri DV, Cremonini F et al (2006) Efficacy of Lactobacillus GG in maintaining remission of ulcerative colitis. Aliment Pharmacol Ther 23:1567–1574

Preventive Effects of Probiotics and Prebiotics in Food Allergy: Potentials and Promise

3

Youcef Shahali and Maryam Dadar

Abstract

The significant increase in food allergy severity and prevalence stresses the need for efficient preventive strategies to reduce life-threatening allergic reactions, particularly among children. At present, there is no cure for food allergy and the eviction of triggers remains the main preventive strategy. The gut microbiome was found to play a key role in the development and pathogenesis of food allergy, opening new therapeutic possibilities. Differences in gut microbiomes were reported between allergic and healthy individuals, suggesting that imbalances in the gut microbial environment likely precede the development of food allergy. The administration of probiotics and prebiotics has been proposed as a safe non-allergen specific therapy with promising outcomes for food allergy treatment. Although numerous studies support the effective role of the probiotics and prebiotics against different allergy conditions, these beneficial impacts appeared to be highly strain specific and particularly observed in pediatric studies. This chapter tries to address the potentials of prebiotics and probiotics in the prevention or treatment of food allergy in the light of preclinical and clinical investigations.

Keywords

Probiotics · Prebiotics · Gut microbiota · Food Allergy

Y. Shahali (✉) · M. Dadar
Agriculture Research, Education and Extension Organization (AREEO), Razi Vaccine and Serum Research Institute (RVSRI), Karaj, Iran

N. Mojgani, M. Dadar (eds.), *Probiotic Bacteria and Postbiotic Metabolites: Role in Animal and Human Health*, Microorganisms for Sustainability 2,
https://doi.org/10.1007/978-981-16-0223-8_3

3.1 Introduction

According to the National Institute of Allergy and Infectious Diseases and Guidelines for the Diagnosis and Management of Food Allergy promoted by the National Institutes of Health (NIAID/NIH) in the United States, food allergy is described as an unfavorable immune response to food properties that affects approximately 5% of adults and 8% of children (Sicherer and Sampson 2006; Boyce et al. 2011). It impacts negatively the life quality of millions of people worldwide, significantly contributes to morbidity, and is related to important medical costs (Rona et al. 2007; Boyce et al. 2011). The significant increase in food allergy severity and prevalence stresses the need for efficient preventive strategies to reduce life-threatening allergic reaction, chiefly among children with 0–14 years old (Sicherer and Sampson 2014). The immediate symptoms affecting the respiratory, skin, cardiovascular, or gastrointestinal systems are mainly the result of immunoglobulin (Ig) E-mediated food allergic reactions (Boyce et al. 2011; Sicherer and Sampson 2014). Numerous epidemiological investigations reported an important increment in hospital admissions for severe food-allergic reactions in children in the USA, the UK, Australia, Iran, and Italy over the last 10 years (Ahanchian et al. 2016; Canani et al. 2013; Mullins et al. 2015; Nocerino et al. 2015; Paparo et al. 2019; Turner et al. 2015). Although more than 170 food allergens have already been described, the most common and serious allergic reactions are caused by a restricted list of food including egg, tree nuts, peanuts, milk, shellfish, fish, soy, wheat, and seeds, with national and geographical variations (Boyce et al. 2011; Sicherer and Sampson 2014; Chafen et al. 2010). At present, there is no specific cure for food allergy and the diagnosis of the offending allergen (s) allowing the eviction of triggers remains the main preventive strategy. However, the accidental consumption of food allergens is not uncommon and the treatment of symptoms with glucocorticoids, antihistamines, or epinephrine in case of systemic reactions is often recommended (Burks et al. 2018; Oyoshi et al. 2014; Sicherer and Sampson 2014). On the other hand, in the past decade, potential therapeutic alternatives for food allergy have been advocated. These emerging therapies are concentrated on suppressing Th2 effector cells, increasing levels of allergen-specific IgA or IgG, decreasing levels of allergen-specific IgE, or increasing regulatory T cells by different allergen nonspecific and allergen-specific approaches (Berin 2014; Paparo et al. 2019). Considering the variable efficacy and safety of allergen-specific therapies (subcutaneous, epicutaneous, sublingual, oral immunotherapy and heat treatment of food), strict risk evaluation and mitigation strategies are required when using these methods. Among numerous uncertainties reported for these approaches, we can mention the observation of adverse events, the onset of eosinophilic esophagitis, desensitization without proper immunological tolerance, the lack of long-term efficacy and difficulties to determine optimal dose and duration (Rachid and Keet 2018).

Besides, the gut microbiome was found to play a key role in the development and pathogenesis of food allergy, opening new therapeutic possibilities. Differences in gut microbiomes were reported between allergic and healthy individuals, suggesting that imbalances in the gut microbial environment likely precede the development of food allergy (Bunyavanich 2019). In this respect, pioneer studies revealed that the commensal gut microbiota function and composition significantly affect the immune tolerance mechanisms following antigen exposure and its dysregulation may lead to the development of food allergy (Paparo et al. 2019). Recently, the administration of probiotics and prebiotics has been proposed as a novel non-allergen specific therapy with promising outcomes for food allergy treatment. Probiotics are described as live microorganisms that when consumed in adequate amounts as oral supplements or as a food component induce a health benefit on the host by regulating its microbiota (Hill et al. 2014). The functions of probiotics are commonly regulated through the toll-like receptors or TLRs (innate immune system), inducing the production of regulatory cytokines (TGF-beta and IL-10), improvement of T helper1 differentiation, increased intestinal release of IgA (Rautava et al. 2012), although these impacts could be different according to the food matrix, probiotic strain, timing, and dose (Heine 2018). This chapter tries to highlight the potentials of prebiotics and probiotics in the prevention or treatment of food allergy in the light of preclinical and clinical investigations.

3.2 Safety of Probiotics and Prebiotics

The safety of probiotics is an important asset leading to its widespread consumption in various forms (Castellazzi et al. 2013). A wide range of people around the world ingest probiotics daily because of their purported health benefits and the global market of probiotic products is steadily expanding (Novik and Savich 2020). *Lactobacilli, lactococci*, and *bifidobacteria* have been recognized as safe (GRAS) by the United States Food and Drug Administration (FDA). This designation indicates that these bacteria are considered safe to be added to food, thereby exempting them from usual tolerance requirements for food additives. Several safety properties have been taken into account such as the absence of associated disease, including endocarditis or bacteremia, the absence of antibiotic resistance gene transformation in the gastrointestinal flora, as well as the absence of metabolic or toxic impacts on the gastrointestinal system (Aureli et al. 2011). Principally, the safe properties of a probiotic strain are associated with the absence of virulence factors and the absence of clinical or veterinary resistance to antibiotics (Daliri et al. 2019). Besides, the Italian Ministry of Health along with the Scientific Committee on Animal Nutrition of the EU (SCAN) and the EFSA panel on additives, products and substances used in animal feed (FEEDAP) proposed to add "the absence of evidence regarding the possible transfer of genes related to antibiotic resistant" as an essential parameter for microorganism safety confirmation (Ahanchian et al. 2016; Snydman 2008). Furthermore, the confirmation of safety and efficacy of probiotic strains is important for different bacterial strains of the same species which may reveal variable effects on the host immune system (Aureli et al. 2011). In recent years, the administration of probiotic products has increased in different medical conditions, because of their efficacy and safety in clinical practice. However, although limited, there are also some risks related to the consumption of probiotics such as an inappropriate immune response in vulnerable populations, the capability of some strains to transfer antibiotic resistance genes to pathogens and deleterious metabolic activities and/or production of host-deleterious metabolites (Daliri et al. 2019). These risks are considerably reduced by the use of prebiotics which represent the substrate that is selectively utilized by beneficial host microorganisms (Gibson et al. 2017). Thus, prebiotics are used as nutrients for favorable microorganisms found in the host such as resident microorganisms and specific probiotics (Monteagudo-Mera et al. 2019). It is supposed that prebiotics could stimulate modifications in the gut environment toward a different host microbial ecosystem through their selective employment by host microorganisms. The effects of prebiotics on various functional pathways including immunomodulatory effects, inhibition of pathogenic bacteria, induction of metabolic function and barrier function have been repeatedly emphasized (Quigley 2019). The fructo-oligosaccharides, galactooligosaccharides and inulin are considered as safe prebiotics because of their long history of safe use in many countries (Cremon et al. 2018; Quigley 2019). However, new prebiotic and probiotic components with different effective impacts on the host immune system have emerged. Some of these components have been designated as 'novel foods' in the EU. These novel foods are evaluated on a case-to-case basis and different production methods or sources might be reported as novel (Kumar et al. 2015).

3.3 Probiotics/Prebiotics and Immunity

The terms probiotics and prebiotics are comprehensive, and various genera, species, and strains show differential effects on the immune system (Bron et al. 2012). Several important cofounding factors such as commensal bacteria and diet could influence the gut immune system. Although specific metabolites/food components (prebiotics) and live microorganisms (probiotics) may regulate and restore the gut microbial composition, an accurate knowledge of the associated molecular pathways behind their impacts on the immune system may provide insights into therapeutic potential for many diseases, such as allergy (Vieira et al. 2013). Several investigations showed that TLRs, Nod-like receptors (NLRs), and pattern recognition receptors (PRRs) play a key role in the development of immune tolerance mediated by probiotics (Abreu 2010; Kamada et al. 2013). The outcomes of TLRs stimulation result into the overexpression of pro-inflammatory regulators that facilitate the responses of host's immune systems. Furthermore, some cytoplasmic proteins, NLRs, could modulate the activation of PRRs and inflammatory responses through the commensal microbiota that is directly involved in the gut homeostasis (Yeretssian 2012). However, disorders in the interactions of PRR-microbiota, in gut mucosal compartment and various cell types, often lead to the development of diseases and intensified inflammation (Lavelle et al. 2010; Maynard et al. 2012). The beneficial effects of probiotics in the host have been supported by a plethora of in vivo studies based on clinical experiments or animal models, suggesting their effectiveness in the necrotizing enterocolitis; post-antibiotic-related diarrhea; certain pediatric allergic disorders; prevention or treatment of acute viral gastroenteritis; and inflammatory bowel disease (IBD) (Cruchet et al. 2015). The efficient role of probiotics in the alleviation of numerous dysfunctions of the gastrointestinal system has also been reported (Vieira et al. 2013). One of the major mechanisms by which probiotics may decrease gut disorder symptoms is through the elevation of short-chain fatty acids (SCFAs) production in the colon that decrease the intestinal permeability and the invasion of pathogenic microorganisms (Morais and Jacob 2006; Szajewska and Kołodziej 2015). Several probiotic effector molecules such as the cell wall components of bacteria including lipoteichoic acid and peptidoglycan along with specific proteins effectively contribute to the immune response (Klaenhammer et al. 2012). Moreover, the regulation of several receptor-mediated signaling cascades playing a key role in the modulation of the human immune system is considered as one of the most important functions of these probiotic effector molecules (Bron et al. 2012). Probiotics also regulate the function of epithelial cells, natural killers and dendritic cells (Yahfoufi et al. 2018). These beneficial bacteria could stimulate the Treg cells and polarize the immune pathways toward Th1.

Prebiotics as not digestible food ingredients are composed of oligosaccharides that show several profitable impacts on host health by selective induction of the growth and/or function of particular microbes present in the gut microbiota (Gibson 1998). Fiber carbohydrates such as gums, pectin, cellulose, lignin, and beta-glucan are not digested in the upper gastrointestinal tract. However, residential gut bacteria selectively ferment these components into SCFAs, especially propionate, acetate, lactate, and butyrate which are fermented when reaching the colon (Horrocks and De Dombal 1978). The majority of the bacteria in the colon are severe anaerobes getting energy from fermentation. The other profitable role of prebiotics is the stimulation of the immune system through the regulation of beneficial microbes' population in the gut, particularly bifidobacteria and lactic acid bacteria. Also the expression of cytokines is another important pathway influenced by the consumption of specific probiotics and prebiotics (Shokryazdan et al. 2017). The mechanism for the beneficial impact of prebiotics on immune system is still largely unknown. Interestingly, the prebiotic metabolites are involved in the modulation of Treg cells, cytokines, and chemokines (Yahfoufi et al. 2018), while prebiotic fibers, by promoting short-chain fatty acids (SCFA) like propionate, lead to the regulation of hepatic lipogenic enzymes. In addition, it was found that inulin supplementation led to increased SCFA levels in the caecum of treated animals (Vieira et al. 2013; Artiss et al. 2006). Other possible effects of prebiotics are the modulation of mucin production, an increase in the number of lymphocyte and/or leucocyte in gut-associated lymphoid tissues (GALT) and peripheral blood, as well as elevated IgA secretion by the GALT.

3.4 Food Allergy and Microbiota

Numerous microorganisms localized in the gastrointestinal tract influence the function and shaping of host adaptive and innate immune responses. Several studies based on both clinical and animal experimentation showed that abnormalities in the microbiota composition (dysbiosis) can lead to allergic disorders through their effects on immune system. Food allergies are reported frequently in preschool children of developing and developed countries with a prevalence reaching 7% and 10%, respectively (Prescott et al. 2013). However, the etiology of food allergy is complex. The environmental/developmental/genetic combined effects involved in the food allergy may explain the global rising trends in recent decades. The key role of microbiome in the development of food allergy is now well documented (Bunyavanich 2019). A changed susceptibility to allergic disorders could thus be linked to the microbial exposure in early childhood (Cahenzli et al. 2013). Furthermore, the comparison of genetically similar populations in Finland and Russia provided more insights into the close interactions between the host microbiome, food allergy, and environment (Haahtela et al. 2015). Experiments on animals revealed that mice with food allergy have a particular gut microbiota signature that could be responsible for increased allergic susceptibility (Rivas et al. 2013). It has thus been suggested that a particular microbiota composition related to food allergy could lead to allergic sensitization and life-threatening anaphylaxis reaction. Several investigation data proposed that dysregulations in the composition of intestinal microbiota in infants are involved in the food allergy pathogenesis, although the precise composition and structure of the intestinal microbiota in human with food allergy still need to be clarified (Matsui et al. 2019). A study using high-throughput 454 sequencing to target hypervariable V1-V3 regions of the 16S rRNA gene in the feces investigated the microbial composition and diversity of 34 infants with food allergy. The results of this study confirmed remarkable modifications in the fecal microbiota of infants suffering from food allergy, showing a significant association with the development of food allergy (Ling et al. 2014). In the food allergy population, the abundance of *Firmicutes* phylum dramatically increased, while the concentration of *Proteobacteria, Actinobacteria,* and *Bacteroidetes* phyla significantly decreased. Furthermore, the phyla of *Clostridiaceae* organisms were commonly found in infants suffering from food allergy. Detailed analysis of microbiota community suggested that the dysbiosis of fecal microbiota is associated with several food allergy-related key phylotypes and may play an effective role in the development of food allergy. Another study also revealed that mice and infants with food allergy had increased IgE and decreased IgA binding to fecal bacteria (Abdel-Gadir et al. 2019). Interestingly, bacteriotherapy stimulated the expression of the transcription factor ROR-γt by Treg cells in a MyD88-dependent manner. These results were of importance as the transcription factor ROR-γt was found to be deficient and ineffectively produced by the microbiota of infants and mice affected by food allergy. Protection by bacteriotherapy is, however, abrogated following the deletion of Myd88 or Rorc in Treg cells. Thus, by activating a MyD88/ROR-γt pathway in nascent Treg cells, commensal microbiota is able to protect against food allergy, while, inversely, microbial imbalance and dysbiosis may promote disease (Abdel-Gadir et al. 2019; Aitoro et al. 2017). Comparison of fecal microbiota in a mouse model of food allergy showed that the development of the disease could be related to a specific microbiota composition (Diesner et al. 2016; Hussain et al. 2019). Different animal studies investigated the possible association of food allergy with intestinal microbiota. For example, a study reported that Germ-free (GF) mice are more likely susceptible to oral sensitization with cow's milk protein and ovalbumin compared to wild-type control mice (Cahenzli et al. 2013). Moreover, mice with antibiotic-related modifications in their microbiota were more severely affected by food allergy when compared to untreated mice (Bashir et al. 2004). Interestingly, the regulation of the microbiota of GF mice with commensals such as *Bacteroides fragilis* and *Clostridia* or short-chain fatty acids and prebiotics promoted the induction of Treg cells and reduced allergic sensitization (Geuking et al. 2011; Smith et al. 2013; Lathrop et al. 2011). Surprisingly, significant reduction in allergic diarrhea and increased levels of Treg cells were also reported among mice exposed to the human microbiota, thereby suggesting that protection or susceptibility to food allergy could be transmitted (Atarashi et al. 2013). In human, the pathogenesis of food allergy, atopic dermatitis, and asthma has been associated with altered microbiota composition (Marrs et al. 2013). However, further investigations are needed to determine implicated microbial species and their influence on the development of allergies. Preclinical and clinical studies on the efficacy of different probiotics and prebiotics on food allergy are addressed in the rest of the chapter.

3.5 Preclinical Studies on the Efficacy of Probiotics and Prebiotics in Food Allergy

As discussed above, gut microbiota and its metabolites such as short-chain fatty acids play a key role in immune tolerance (Paparo et al. 2019). In vivo benefits of probiotics are difficult to assess through in vitro studies as the extrapolation of the results is not possible (Berni Canani et al. 2012). The first in vivo studies on the effect of probiotics on food allergy were performed in Finland and revealed that hydrolysis of caseins with *L. casei* GG-derived enzymes resulted in molecules with suppressive effects on lymphocyte proliferation. These preliminary results suggested that intestinal bacteria can promote the downregulation of hypersensitivity reactions to ingested proteins in patients with food allergy (Sütas et al. 1996). It was found that the probiotic effects on immune tolerance to food allergens could be due to the regulation of gut microbiota function and composition through the elevation of butyrate production (Canani et al. 2016) and immune tolerogenic pathways through the induction of beta-defensins, sIgA production (Hardy et al. 2013), cytokines regulation and through improving the mucus thickness and gut permeability (Kim et al. 2008; Niers et al. 2005; Turner et al. 2015). Furthermore, the evaluation of probiotic and prebiotic impacts on immune cell responses has been performed by in vitro induction of mononuclear cells in human peripheral blood under selected strains of probiotics and prebiotics. The treatment of mononuclear cells of human peripheral blood with lactic acid bacteria (LAB) strains such as *Bifidobacterium adolescentis* and *Lactobacillus plantarum* induced the production of IFN-γ by T cells and the regulatory cytokine IL-10 by dendritic cells and monocytes (Cross and Gill 2001; Karlsson et al. 2004; Mohamadzadeh et al. 2005). The incubation of mononuclear cells of human peripheral blood with a mixture of probiotics including *L. acidophilus* W55, *B. infantis* W52, *L. casei* W56, *L. salivarius* W57, *L. lactis* W58, *B. lactis* W18, and *B. longum* W51 in children with food allergy induces the production of Th1 and regulatory cytokines and proliferation of T cell (Flinterman et al. 2007). Furthermore, after 3 months of incubation with these mixture, the increase in B and T cell proliferation and a decrease in production of IgE were also reported in children with food allergy (Flinterman et al. 2007). In addition, a mixture of probiotics *L. rhamnosus* GG and B. breve regulated the function of IL-23 and IL-17 inflammatory cytokines, resulting in a decrease in histone acetylation and an increase in DNA methylation in a 3D coculture model of mononuclear cells from human peripheral blood and intestinal epithelial cells used as an in vitro model of the intestinal mucosal immune system (Ghadimi et al. 2008). Animal models for food allergy were also regularly applied as experimental approaches to assess probiotics and prebiotics effects. Differential effects of oral ingestion of three LAB strains including *B. infantis* 11.322, *L. plantarum* 08.923 (Lp), and *B. coagulans* 09.712 (Bc) in a murine model induced by shrimp allergen were reported on the reduction of Th2-driven intestinal inflammation and other symptoms related to food-induced anaphylaxis (Fu et al. 2017). Oral supplementation of these probiotics remarkably increased the population of CD4+ FoxP3+ T cells and alleviated anaphylaxis symptoms in sensitized mice by FoxP3 upregulation, GATA-3 downregulation, and mTORC inhibition (Fu et al. 2017). The therapeutic and preventive effects of oral *Clostridium butyricum CGMCC0313-1* on anaphylactic symptoms in sensitized mice by a β-lactoglobulin (BLG) showed that this bacteria could increase CD4+ CD25+FoxP3Treg cell and sIgA and alleviate anaphylaxis symptoms in the spleen of sensitized mice (Zhang et al. 2017). Another study showed that casein immunogenicity after oral sensitization to cow's milk could be induced in neonatal monocolonization of germ-free mice by *L. casei* BL23 (Maiga et al. 2017). It was also revealed that oral ingestion of *B. infantis* improved the allergic conditions through the reduction of the Th2 cytokines release in the spleen and ovalbumin-specific IgG1 and IgE contents in the sera of ovalbumin-sensitized mice. Furthermore, the analysis of gut microbiota showed that the probiotics-regulated protection was induced by overexpression of Rikenella and Coprococcus at genus level (Yang et al. 2017). A decrease in IgE, IL-4, and IL-13 levels was reported following the administration of *B. infantis* CGMCC313-2 in BLG-sensitized mice (Liu et al. 2017). Another study reported that oral ingestion of VSL# 3 probiotic-mixture remarkably ameliorate the anaphylactic reactions through the decrease of the Th2 immune responses in sensitized mice (Sicherer and Sampson 2018). Also, the treatment of probiotic mixture with mouse spleen cells in sensitized mice increased the production of IL-10 and IFN-γ, while decreasing the allergen-induced IL-5 and IL-13 production (Schiavi et al. 2011). An oral supplementation of *Lactobacillus rhamnosus* GG with cholera toxin B-subunit as adjuvant decreased the cow's milk allergy in the sensitized Balb/C mice through the regulation of immune responses by shifting Th2-dominated trends toward Th1-dominated responses (Thang et al. 2011). Similar studies have been showed that the oral administration of *Lactobacillus rhamnosus* GG induced a remarkable decrease of allergic reaction and of specific production of IgE and IL-4, IL-5, IL-13, in a BLG-sensitized mouse model (Aitoro et al. 2017). Probiotics also ferment prebiotics or fiber-rich diets to SCFAs, including acetate, propionate, and butyrate. Evidence data proposed that SCFAs, especially butyrate, are involved in the homeostasis of mucosal system by the modulation of epithelial barrier integrity and stimulation of Tregs (Canani et al. 2015). The deficiency of butyrate has been reported in patients with food allergy symptom (Canani et al. 2016). Therefore, it is possible that various kinds of dysbiosis led to similar impacts in SCFAs or other production of microbiota-derived metabolites resulting in allergy occurrence. Clostridia species are known as the main source of SCFAs in the colon that has been involved in the modulation of proportions and activation of Tregs functions in the colon (Arpaia et al. 2013; Smith et al. 2013). SCFAs also stimulate G-protein-coupled receptors involved in the induction of colonic macrophages and dendritic cells, the secretion of IL-10 and increase Tregs in the mesenteric lymph nodes. Tregs are a prominent source of tolerogenic cytokines, like TGF-β and IL-10 that control inflammatory and allergic responses (Paparo et al. 2019). Another study also reported that dietary vitamin A together with fiber/SCFAs in a healthy gut microbiota could protect the food allergy development through the conservation of a tolerogenic mucosal environment and increase function of tolerogenic CD103+ dendritic cells, resulting to heighten differentiation of Tregs. In addition, mice lacking GPR109A or GPR43 receptors for SCFAs showed fewer CD103 + dendritic cells and increased food allergy symptom (Tan et al. 2016). These researches suggest that the effective role of different prebiotics on food allergy could be related to their direct effects on the gut microbiota.

3.6 Clinical Data on the Probiotics Efficacy in Food Allergy

The pioneer studies suggesting the potential of probiotic bacteria to prevent allergic diseases and regulate the immune response originate from Finland (Majamaa and Isolauri 1997). There are several investigations on the importance of the gut microbiota composition in the food allergy pathogenesis that have been supported by clinical research on the effective role of the probiotics against allergy conditions. These protective impacts appeared to be strain specific and particularly reported in the pediatric age (Paparo et al. 2019). On the other hand, tolerance and safety of prebiotic-containing starter infant formula supplemented with *Lactobacillus paracasei* and *Bifidobacterium animalis* have been reported to be an effective approach to improve the beneficial bacteria in the intestine to develop a gut flora (Vieira et al. 2013). However, in a randomized double-blind placebo-controlled trial, it has been reported that 12 months administration of hydrolyzed formula of *B. lactis* BB12 and *L. casei* CRL431 could not affect the immune tolerance responses to cow's milk proteins in infants with cow's milk allergy (Hol et al. 2008). Conversely another study showed that the supplementation of hydrolyzed casein formula (EHCF) with the *L. rhamnosus* strain GG is capable to increase the immune tolerance acquisition in infants with cow's milk allergy (Berni Canani et al. 2012). After 12-month treatment period, the group receiving EHCF+ *L. rhamnosus* strain GG (78.9%) showed higher proportion of children acquiring tolerance to cow's milk proteins when compared to other groups (Canani et al. 2013). More confirmation of a lower incidence of other atopic manifestations as well as a better resolution of IgE-regulated cow's milk allergy also was reported after treatment with EHCF+ *L. rhamnosus* strain GG after 3-year follow-up in pediatric cohort study (Canani et al. 2017). These beneficial effects could be because of regulation of *L. rhamnosus* strain GG-related immune functions by different pathways such as mast cells, enterocytes, monocytes, Tregs cell, and DCs (Canani et al. 2013; Ghadimi et al. 2008; Mileti et al. 2009) and by an expansion of butyrate-producing gut microbiota (Canani et al. 2016). Accordingly, supplementation of EHCF with *L. rhamnosus* strain GG in infants with eczema and/or CMA resulted in beneficial effects on the reducing of inflammation and gastrointestinal symptoms (Isolauri et al. 2000; Kalliomäki et al. 2010). Furthermore, it has been showed that the administration of *L. rhamnosus* strain GG for 4–12 weeks could significantly reduce in atopic dermatitis score in children aged 4–48 months. These children expressed less Scoring of Atopic Dermatitis (SCORAD) in the three components, including area of affected skin, intensity of atopic dermatitis, and patient symptoms, with a significant decrease in the mean change of intensity from baseline compared with placebo (Wu et al. 2017). Moreover, the combination of a prebiotic (galactooligosaccharides) with four probiotics (*L. rhamnosus* strain GG, *L. rhamnosus* LC705, *B. breve* Bb99, and *Propionibacterium freudenreichii* ssp. shermanii) decreased the incidence of atopic eczema and eczema and tended to decrease IgE-related diseases by modulating the infant's gut microbiota (Kukkonen et al. 2007). The same combination of probiotics without prebiotic did not show any significant impacts (Viljanen et al. 2005). The addition of prebiotic may have been the critical difference, although clear evidence of its bifidogenic effect is still lacking. Probiotics have been also proposed to reinforce the effectiveness of immunotherapy (Rachid and Keet 2018). Another randomized double-blind placebo-controlled trial study in 62 children (1–10 years) with peanut allergy showed that the probiotic *L. rhamnosus* CGMCC 1.3724 and peanut oral immunotherapy (PPOIT) could show effective impacts on the regulation of the peanut-specific immune response (Tang et al. 2015). For a total of 18 months, children received a fixed dose of probiotic (or placebo) along with peanut oral immunotherapy (or placebo) once daily. PPOIT induced high rates of desensitization (90%) and was related to decreased peanut-specific IgE levels and peanut skin prick test responses and increased levels of peanut-specific IgG4. Further investigations comparing food allergy with probiotic and prebiotics in different situation will be hotly demanded to evaluate associated mechanisms and relative contributions of probiotics versus prebiotics.

3.7 Conclusion

The key role of microbiome in the development of food allergy is now well documented and numerous investigations highlighted the importance of the gut microbiota composition in the food allergy pathogenesis. Microbial exposure was found to influence the development of oral tolerance, particularly in childhood. Dysbiosis and low diversity of gut microbiota have been linked to the pathogenesis of food allergy. The therapeutic strategy of the prebiotics and probiotics administration is to restore the gut microbiota in order to improve the microbiome immune support leading to a better tolerance to allergens. Numerous studies assessed allergy-preventive capacity of probiotics and prebiotics in food allergy and some very promising results were obtained following the administration of specific probiotic strains. Prebiotics also represent an interesting and safe alternative to some probiotics in allergy prevention, but more studies on various types and combinations of prebiotics are needed in the context of rigorous clinical investigations. Although, the use of probiotics in the prevention of food allergy and eczema among children led to new therapeutic perspectives, many variables such as the duration and mode of delivery, feeding type, optimal dose and strain combination merit to be addressed in future studies to confirm their effectiveness in the primary prevention of allergic disease.

References

Abdel-Gadir A, Stephen-Victor E, Gerber GK, Rivas MN, Wang S, Harb H, Wang L, Li N, Crestani E, Spielman S (2019) Microbiota therapy acts via a regulatory T cell MyD88/RORγt pathway to suppress food allergy. Nat Med 25:1164–1174

Abreu MT (2010) Toll-like receptor signalling in the intestinal epithelium: how bacterial recognition shapes intestinal function. Nat Rev Immunol 10:131–144

Ahanchian H, Jafari S, Behmanesh F, Haghi NM, Nakhaei AA, Kiani MA, Radbin MH, Kianifar H (2016) Epidemiological survey of pediatric food allergy in Mashhad in Northeast Iran. Electron Physician 8:1727

Aitoro R, Paparo L, Amoroso A, Di Costanzo M, Cosenza L, Granata V, Di Scala C, Nocerino R, Trinchese G, Montella M (2017) Gut microbiota as a target for preventive and therapeutic intervention against food allergy. Nutrients 9:672

Arpaia N, Campbell C, Fan X, Dikiy S, van der Veeken J, Deroos P, Liu H, Cross JR, Pfeffer K, Coffer PJ (2013) Metabolites produced by commensal bacteria promote peripheral regulatory T-cell generation. Nature 504:451–455

Artiss JD, Brogan K, Brucal M, Moghaddam M, Jen K-LC (2006) The effects of a new soluble dietary fiber on weight gain and selected blood parameters in rats. Metabolism 55:195–202

Atarashi K, Tanoue T, Oshima K, Suda W, Nagano Y, Nishikawa H, Fukuda S, Saito T, Narushima S, Hase K (2013) T reg induction by a rationally selected mixture of Clostridia strains from the human microbiota. Nature 500:232–236

Aureli P, Capurso L, Castellazzi AM, Clerici M, Giovannini M, Morelli L, Poli A, Pregliasco F, Salvini F, Zuccotti GV (2011) Probiotics and health: an evidence-based review. Pharmacol Res 63:366–376

Bashir MEH, Louie S, Shi HN, Nagler-Anderson C (2004) Toll-like receptor 4 signaling by intestinal microbes influences susceptibility to food allergy. J Immunol 172:6978–6987

Berin MC (2014) Future therapies for IgE-mediated food allergy. Current Pediatr Rep 2:119–126

Berni Canani R, Nocerino R, Terrin G, Coruzzo A, Cosenza L (2012) Effect of Lactobacillus GG on tolerance acquisition in infants with cow's milk allergy: a randomized trial. J Allergy Clin Immunol 129:580–582

Boyce JA, Assa'ad A, Burks AW, Jones SM, Sampson HA, Wood RA, Plaut M, Cooper SF, Fenton MJ, Arshad SH (2011) Guidelines for the diagnosis and management of food allergy in the United States: summary of the NIAID-sponsored expert panel report. J Am Acad Dermatol 64:175–192

Bron PA, Van Baarlen P, Kleerebezem M (2012) Emerging molecular insights into the interaction between probiotics and the host intestinal mucosa. Nat Rev Microbiol 10:66–78

Bunyavanich S (2019) Food allergy: could the gut microbiota hold the key? Nat Rev Gastroenterol Hepatol 16:201–202

Burks AW, Sampson HA, Plaut M, Lack G, Akdis CA (2018) Treatment for food allergy. J Allergy Clin Immunol 141:1–9

Cahenzli J, Köller Y, Wyss M, Geuking MB, McCoy KD (2013) Intestinal microbial diversity during early-life colonization shapes long-term IgE levels. Cell Host Microbe 14:559–570

Canani RB, Di Costanzo M, Bedogni G, Amoroso A, Cosenza L, Di Scala C, Granata V, Nocerino R (2017) Extensively hydrolyzed casein formula containing Lactobacillus rhamnosus GG reduces the occurrence of other allergic manifestations in children with cow's milk allergy: 3-year randomized controlled trial. J Allergy Clin Immunol 139:1906–1913. e1904

Canani RB, Gilbert JA, Nagler CR (2015) The role of the commensal microbiota in the regulation of tolerance to dietary allergens. Curr Opin Allergy Clin Immunol 15:243

Canani RB, Nocerino R, Terrin G, Frediani T, Lucarelli S, Cosenza L, Passariello A, Leone L, Granata V, Di Costanzo M (2013) Formula selection for management of children with cow's milk allergy influences the rate of acquisition of tolerance: a prospective multicenter study. J Pediatr 163:771–777. e771

Canani RB, Sangwan N, Stefka AT, Nocerino R, Paparo L, Aitoro R, Calignano A, Khan AA, Gilbert JA, Nagler CR (2016) Lactobacillus rhamnosus GG-supplemented formula expands butyrate-producing bacterial strains in food allergic infants. ISME J 10:742–750

Castellazzi AM, Valsecchi C, Caimmi S, Licari A, Marseglia A, Leoni MC, Caimmi D, Del Giudice MM, Leonardi S, La Rosa M (2013) Probiotics and food allergy. Ital J Pediatr 39:1–10

Chafen JJS, Newberry SJ, Riedl MA, Bravata DM, Maglione M, Suttorp MJ, Sundaram V, Paige NM, Towfigh A, Hulley BJ (2010) Diagnosing and managing common food allergies: a systematic review. JAMA 303:1848–1856

Cremon C, Barbaro MR, Ventura M, Barbara G (2018) Pre-and probiotic overview. Curr Opin Pharmacol 43:87–92

Cross M, Gill H (2001) Can immunoregulatory lactic acid bacteria be used as dietary supplements to limit allergies? Int Arch Allergy Immunol 125:112–119

Cruchet S, Furnes R, Maruy A, Hebel E, Palacios J, Medina F, Ramirez N, Orsi M, Rondon L, Sdepanian V (2015) The use of probiotics in pediatric gastroenterology: a review of the literature and recommendations by Latin-American experts. Pediatr Drugs 17:199–216

Daliri EB-M, Lee BH, Oh DH (2019) Safety of probiotics in health and disease. In: The role of functional food security in global health. Elsevier, pp 603–622

Diesner SC, Bergmayr C, Pfitzner B, Assmann V, Krishnamurthy D, Starkl P, Endesfelder D, Rothballer M, Welzl G, Rattei T (2016) A distinct microbiota composition is associated with protection from food allergy in an oral mouse immunization model. Clin Immunol 173:10–18

Flinterman A, Knol E, van Ieperen-Van Dijk A, Timmerman H, Knulst A, Bruijnzeel-Koomen C, Pasmans S, Van Hoffen E (2007) Probiotics have a different immunomodulatory potential in vitro versus ex vivo upon oral administration in children with food allergy. Int Arch Allergy Immunol 143:237–244

Fu L, Peng J, Zhao S, Zhang Y, Su X, Wang Y (2017) Lactic acid bacteria-specific induction of CD4+ Foxp3+ T cells ameliorates shrimp tropomyosin-induced allergic response in mice via suppression of mTOR signaling. Sci Rep 7:1–14

Geuking MB, Cahenzli J, Lawson MA, Ng DC, Slack E, Hapfelmeier S, McCoy KD, Macpherson AJ (2011) Intestinal bacterial colonization induces mutualistic regulatory T cell responses. Immunity 34:794–806

Ghadimi D, Fölster-Holst R, De Vrese M, Winkler P, Heller KJ, Schrezenmeir J (2008) Effects of probiotic bacteria and their genomic DNA on TH1/TH2-cytokine production by peripheral blood mononuclear cells (PBMCs) of healthy and allergic subjects. Immunobiology 213:677–692

Gibson GR (1998) Dietary modulation of the human gut microflora using prebiotics. Br J Nutr 80: S209–S212

Gibson GR, Hutkins R, Sanders ME, Prescott SL, Reimer RA, Salminen SJ, Scott K, Stanton C, Swanson KS, Cani PD (2017) Expert consensus document: the International Scientific Association for Probiotics and Prebiotics (ISAPP) consensus statement on the definition and scope of prebiotics. Nat Rev Gastroenterol Hepatol 14:491

Haahtela T, Laatikainen T, Alenius H, Auvinen P, Fyhrquist N, Hanski I, Von Hertzen L, Jousilahti P, Kosunen T, Markelova O (2015) Hunt for the origin of allergy–comparing the Finnish and Russian Karelia. Clin Exp Allergy 45:891–901

Hardy H, Harris J, Lyon E, Beal J, Foey AD (2013) Probiotics, prebiotics and immunomodulation of gut mucosal defences: homeostasis and immunopathology. Nutrients 5:1869–1912

Heine RG (2018) Food allergy prevention and treatment by targeted nutrition. Ann Nutr Metab 72:33–45

Hill C, Guarner F, Reid G, Gibson GR, Merenstein DJ, Pot B, Morelli L, Canani RB, Flint HJ, Salminen S (2014) Expert consensus document: the International Scientific Association for Probiotics and Prebiotics consensus statement on the scope and appropriate use of the term probiotic. Nat Rev Gastroenterol Hepatol 11:506

Hol J, van Leer EH, Schuurman BEE, de Ruiter LF, Samsom JN, Hop W, Neijens HJ, de Jongste JC, Nieuwenhuis EE (2008) The acquisition of tolerance toward cow's milk through probiotic supplementation: a randomized, controlled trial. J Allergy Clin Immunol 121:1448–1454

Horrocks JC, De Dombal F (1978) Clinical presentation of patients with "dyspepsia". Detailed symptomatic study of 360 patients. Gut 19:19–26

Hussain M, Bonilla-Rosso G, Chung CKK, Bäriswyl L, Rodriguez MP, Kim BS, Engel P, Noti M (2019) High dietary fat intake induces a microbiota signature that promotes food allergy. J Allergy Clin Immunol 144:157–170. e158

Isolauri E, Arvola T, Sütas Y, Moilanen E, Salminen S (2000) Probiotics in the management of atopic eczema. Clin Exp Allergy 30:1605–1610

Kalliomäki M, Antoine J-M, Herz U, Rijkers GT, Wells JM, Mercenier A (2010) Guidance for substantiating the evidence for beneficial effects of probiotics: prevention and management of allergic diseases by probiotics. J Nutr 140:713S–721S

Kamada N, Seo S-U, Chen GY, Núñez G (2013) Role of the gut microbiota in immunity and inflammatory disease. Nat Rev Immunol 13:321–335

Karlsson H, Larsson P, Wold AE, Rudin A (2004) Pattern of cytokine responses to gram-positive and gram-negative commensal bacteria is profoundly changed when monocytes differentiate into dendritic cells. Scand J Immunol 59:628–628

Kim JY, Choi YO, Ji GE (2008) Effect of oral probiotics (Bifidobacterium lactis AD011 and Lactobacillus acidophilus AD031) administration on ovalbumin-induced food allergy mouse model. J Microbiol Biotechnol 18:1393–1400

Klaenhammer TR, Kleerebezem M, Kopp MV, Rescigno M (2012) The impact of probiotics and prebiotics on the immune system. Nat Rev Immunol 12:728–734

Kukkonen K, Savilahti E, Haahtela T, Juntunen-Backman K, Korpela R, Poussa T, Tuure T, Kuitunen M (2007) Probiotics and prebiotic galacto-oligosaccharides in the prevention of allergic diseases: a randomized, double-blind, placebo-controlled trial. J Allergy Clin Immunol 119:192–198

Kumar H, Salminen S, Verhagen H, Rowland I, Heimbach J, Bañares S, Young T, Nomoto K, Lalonde M (2015) Novel probiotics and prebiotics: road to the market. Curr Opin Biotechnol 32:99–103

Lathrop SK, Bloom SM, Rao SM, Nutsch K, Lio C-W, Santacruz N, Peterson DA, Stappenbeck TS, Hsieh C-S (2011) Peripheral education of the immune system by colonic commensal microbiota. Nature 478:250–254

Lavelle EC, Murphy C, O'Neill L, Creagh EM (2010) The role of TLRs, NLRs, and RLRs in mucosal innate immunity and homeostasis. Mucosal Immunol 3:17–28

Ling Z, Li Z, Liu X, Cheng Y, Luo Y, Tong X, Yuan L, Wang Y, Sun J, Li L (2014) Altered fecal microbiota composition associated with food allergy in infants. Appl Environ Microbiol 80:2546–2554

Liu M-Y, Yang Z-Y, Dai W-K, Huang J-Q, Li Y-H, Zhang J, Qiu C-Z, Wei C, Zhou Q, Sun X (2017) Protective effect of Bifidobacterium infantis CGMCC313-2 on ovalbumin-induced airway asthma and β-lactoglobulin-induced intestinal food allergy mouse models. World J Gastroenterol 23:2149

Maiga MA, Morin S, Bernard H, Rabot S, Adel-Patient K, Hazebrouck S (2017) Neonatal mono-colonization of germ-free mice with Lactobacillus casei enhances casein immunogenicity after oral sensitization to cow's milk. Mol Nutr Food Res 61:1600862

Majamaa H, Isolauri E (1997) Probiotics: a novel approach in the management of food allergy. J Allergy Clin Immunol 99:179–185

Marrs T, Bruce KD, Logan K, Rivett DW, Perkin MR, Lack G, Flohr C (2013) Is there an association between microbial exposure and food allergy? A systematic review. Pediatr Allergy Immunol 24:311–320. e318

Matsui S, Kataoka H, Tanaka J-I, Kikuchi M, Fukamachi H, Morisaki H, Matsushima H, Mishima K, Hironaka S, Takaki T (2019) Dysregulation of intestinal microbiota elicited by food allergy induces IgA-mediated oral dysbiosis. Infect Immun 88

Maynard CL, Elson CO, Hatton RD, Weaver CT (2012) Reciprocal interactions of the intestinal microbiota and immune system. Nature 489:231–241

Mileti E, Matteoli G, Iliev ID, Rescigno M (2009) Comparison of the immunomodulatory properties of three probiotic strains of Lactobacilli using complex culture systems: prediction for in vivo efficacy. PLoS One 4:e7056

Mohamadzadeh M, Olson S, Kalina WV, Ruthel G, Demmin GL, Warfield KL, Bavari S, Klaenhammer TR (2005) Lactobacilli activate human dendritic cells that skew T cells toward T helper 1 polarization. Proc Natl Acad Sci 102:2880–2885

Monteagudo-Mera A, Rastall RA, Gibson GR, Charalampopoulos D, Chatzifragkou A (2019) Adhesion mechanisms mediated by probiotics and prebiotics and their potential impact on human health. Appl Microbiol Biotechnol 103:6463–6472

Morais MBd, Jacob CMA (2006) The role of probiotics and prebiotics in pediatric practice. J Pediatr 82:S189–S197

Mullins RJ, Dear KB, Tang ML (2015) Time trends in Australian hospital anaphylaxis admissions in 1998-1999 to 2011-2012. J Allergy Clin Immunol 136:367–375

Niers LE, Timmerman HM, Rijkers GT, van Bleek GM, van Uden NO, Knol EF, Kapsenberg ML, Kimpen JL, Hoekstra MO (2005) Identification of strong interleukin-10 inducing lactic acid bacteria which down-regulate T helper type 2 cytokines. Clin Exp Allergy 35:1481–1489

Nocerino R, Leone L, Cosenza L, Canani RB (2015) Increasing rate of hospitalizations for food-induced anaphylaxis in Italian children: an analysis of the Italian Ministry of Health database. J Allergy Clin Immunol 135:833–835. e833

Novik G, Savich V (2020) Beneficial microbiota. Probiotics and pharmaceutical products in functional nutrition and medicine. Microbes Infect 22:8–18

Oyoshi MK, Oettgen HC, Chatila TA, Geha RS, Bryce PJ (2014) Food allergy: insights into etiology, prevention, and treatment provided by murine models. J Allergy Clin Immunol 133:309–317

Paparo L, Nocerino R, Di Scala C, Della Gatta G, Di Costanzo M, Buono A, Bruno C, Canani RB (2019) Targeting food allergy with probiotics, probiotics and child gastrointestinal health. Springer, pp 57–68

Prescott SL, Pawankar R, Allen KJ, Campbell DE, Sinn JK, Fiocchi A, Ebisawa M, Sampson HA, Beyer K, Lee B-W (2013) A global survey of changing patterns of food allergy burden in children. World Allergy Organ J 6:1–12

Quigley EM (2019) Prebiotics and probiotics in digestive health. Clin Gastroenterol Hepatol 17:333–344

Rachid R, Keet CA (2018) Current status and unanswered questions for food allergy treatments. J Allergy Clin Immunol Pract 6:377–382

Rautava S, Collado MC, Salminen S, Isolauri E (2012) Probiotics modulate host-microbe interaction in the placenta and fetal gut: a randomized, double-blind, placebo-controlled trial. Neonatology 102:178–184

Rivas MN, Burton OT, Wise P, Zhang Y-Q, Hobson SA, Lloret MG, Chehoud C, Kuczynski J, DeSantis T, Warrington J (2013) A microbiota signature associated with experimental food allergy promotes allergic sensitization and anaphylaxis. J Allergy Clin Immunol 131:201–212

Rona RJ, Keil T, Summers C, Gislason D, Zuidmeer L, Sodergren E, Sigurdardottir ST, Lindner T, Goldhahn K, Dahlstrom J (2007) The prevalence of food allergy: a meta-analysis. J Allergy Clin Immunol 120:638–646

Schiavi E, Barletta B, Butteroni C, Corinti S, Boirivant M, Di Felice G (2011) Oral therapeutic administration of a probiotic mixture suppresses established Th2 responses and systemic anaphylaxis in a murine model of food allergy. Allergy 66:499–508

Shokryazdan P, Jahromi MF, Navidshad B, Liang JB (2017) Effects of prebiotics on immune system and cytokine expression. Med Microbiol Immunol 206:1–9

Sicherer SH, Sampson HA (2006) 9. Food allergy. J Allergy Clin Immunol 117:S470–S475

Sicherer SH, Sampson HA (2014) Food allergy: epidemiology, pathogenesis, diagnosis, and treatment. J Allergy Clin Immunol 133:291–307. e295

Sicherer SH, Sampson HA (2018) Food allergy: a review and update on epidemiology, pathogenesis, diagnosis, prevention, and management. J Allergy Clin Immunol 141:41–58

Smith PM, Howitt MR, Panikov N, Michaud M, Gallini CA, Bohlooly-Y M, Glickman JN, Garrett WS (2013) The microbial metabolites, short-chain fatty acids, regulate colonic Treg cell homeostasis. Science 341:569–573

Snydman DR (2008) The safety of probiotics. Clin Infect Dis 46:S104–S111

Sütas Y, Hurme M, Isolauri E (1996) Down-regulation of anti-CD3 antibody-induced IL-4 production by bovine caseins hydrolysed with Lactobacillus GG-derived enzymes. Scand J Immunol 43:687–689

Szajewska H, Kołodziej M (2015) Systematic review with meta-analysis: Saccharomyces boulardii in the prevention of antibiotic-associated diarrhoea. Aliment Pharmacol Ther 42:793–801

Tan J, McKenzie C, Vuillermin PJ, Goverse G, Vinuesa CG, Mebius RE, Macia L, Mackay CR (2016) Dietary fiber and bacterial SCFA enhance oral tolerance and protect against food allergy through diverse cellular pathways. Cell Rep 15:2809–2824

Tang ML, Ponsonby A-L, Orsini F, Tey D, Robinson M, Su EL, Licciardi P, Burks W, Donath S (2015) Administration of a probiotic with peanut oral immunotherapy: a randomized trial. J Allergy Clin Immunol 135:737–744. e738

Thang CL, Baurhoo B, Boye JI, Simpson BK, Zhao X (2011) Effects of Lactobacillus rhamnosus GG supplementation on cow's milk allergy in a mouse model. Allergy, Asthma Clin Immunol 7:20

Turner PJ, Gowland MH, Sharma V, Ierodiakonou D, Harper N, Garcez T, Pumphrey R, Boyle RJ (2015) Increase in anaphylaxis-related hospitalizations but no increase in fatalities: an analysis of United Kingdom national anaphylaxis data, 1992-2012. J Allergy Clin Immunol 135:956–963. e951

Vieira AT, Teixeira MM, Martins FdS (2013) The role of probiotics and prebiotics in inducing gut immunity. Front Immunol 4:445

Viljanen M, Savilahti E, Haahtela T, Juntunen-Backman K, Korpela R, Poussa T, Tuure T, Kuitunen M (2005) Probiotics in the treatment of atopic eczema/dermatitis syndrome in infants: a double-blind placebo-controlled trial. Allergy 60:494–500

Wu Y-J, Wu W-F, Hung C-W, Ku M-S, Liao P-F, Sun H-L, Lu K-H, Sheu J-N, Lue K-H (2017) Evaluation of efficacy and safety of Lactobacillus rhamnosus in children aged 4–48 months with atopic dermatitis: an 8-week, double-blind, randomized, placebo-controlled study. J Microbiol Immunol Infect 50:684–692

Yahfoufi N, Mallet J, Graham E, Matar C (2018) Role of probiotics and prebiotics in immunomodulation. Curr Opin Food Sci 20:82–91

Yang B, Xiao L, Liu S, Liu X, Luo Y, Ji Q, Yang P, Liu Z (2017) Exploration of the effect of probiotics supplementation on intestinal microbiota of food allergic mice. Am J Transl Res 9:376

Yeretssian G (2012) Effector functions of NLRs in the intestine: innate sensing, cell death, and disease. Immunol Res 54:25–36

Zhang J, Su H, Li Q, Wu H, Liu M, Huang J, Zeng M, Zheng Y, Sun X (2017) Oral administration of Clostridium butyricum CGMCC0313-1 inhibits β-lactoglobulin-induced intestinal anaphylaxis in a mouse model of food allergy. Gut Pathogens 9:11

An Overview of Dairy Microflora

4

Deeba Noreen Baig and Samina Mehnaz

Abstract

An assembly of bacterial and fungal communities in the milk and dairy products presents a complete picture of dairy born microflora. Fermentation and pasteurization processes are crucial for the maintenance of microflora. Chemical composition and initial colonization of bacteria and fungi define the mutualistic pattern of microbial communities. The abundance and variety of microbial communities in milk are highly variable and depend upon many factors ranging from the health of milking animals to the milking practices, storage, and transportation methods. Probiotics are beneficial microbes, specifically lactic acid bacteria such as *Lactobacilli* and *Bifidobacteria* are generally regarded as safe (GRAS) microorganisms that benefit the host physiology upon ingestion. Lactic acid bacteria are the predominant group in all dairy microbiota that display a diverse range of strains associated with the milk from different animals. Few dairy microbes behave as pathogens as well as the cause of food spoilage. Human diseases from milk-borne pathogens are usually due to raw milk or products made from raw milk. However, the enormous medicinal and health-promoting impact of microbes and their additives overcome the limited effects of few harmful bacteria in the dairy environment. In addition to the known advantages of dairy bacteria, the phenomenon of psychobiotics is introducing a new therapeutic channel for the treatment of many psychological disorders.

Keywords

Dairy microflora · Probiotics · Lactic acid bacteria · Nutraceutical · Psychobiotics

D. N. Baig (✉) · S. Mehnaz
School of Life Sciences, Forman Christian College (A Chartered University), Lahore, Pakistan
e-mail: deebabaig@fccollege.edu.pk

N. Mojgani, M. Dadar (eds.), *Probiotic Bacteria and Postbiotic Metabolites: Role in Animal and Human Health*, Microorganisms for Sustainability 2,
https://doi.org/10.1007/978-981-16-0223-8_4

4.1 Introduction

Dairy microflora refers to the assemblage of microorganisms present in milk and its associated products. Milk is an important food for human consumption and was considered as a drink of ancient times that aided in the survival of generations. For centuries it has served as a cure for a variety of diseases and as an instant source of energy (Shori 2012). Today, it is considered to host a complex microbial community with great diversity. The microbial quality of milk products is highly dependent on their initial microflora colonization. Each kind of milk and dairy products develops a specific microflora composition. The most common dairy associated microflora includes *Lactobacillus*, *Bifidobacterium*, *Enterococcus*, *Streptococcus*, *Pediococcus*, *Propionibacterium*, and *Leuconostoc* bacterial genera, and *Saccharomyces* and *Aspergillus* yeast genera (Abushelaibi et al. 2017; Amara and Shibl 2015; Ogier and Serror 2008).

Fermentation is one of the common and primitive methods for processing and preservation of the microbial community that has been used worldwide. This method conserves the food and makes sure that the food is safe for human consumption by boosting their desired microbial composition. As a source of probiotics, raw and fermented forms of milk are well known around the world. A combination of fresh and lyophilized, one or more pure microorganisms (starter cultures) are routinely used for the fermentation of dairy products (Ahmed and Kanwal 2004; Lourens-Hattingh and Viljoen 2001; Vinderola et al. 2000). Sugars are metabolized into lactic acid, which enables food preservation by providing an acidic environment that is hostile for spoilage microorganisms (Hati et al. 2013). The diversity of microorganisms is highly varied in raw and fermented milk, as well as in dairy products like yogurt, cheese, kefir, and dahi. The quality of dairy products entirely depends on the viable count of microbiota in fresh milk, breeding area, nutritive condition, breed type, age of the animal, stage of lactation, and milking practices (Khaskheli et al. 2005). Milk microbiota exploration relies on both culture-dependent and molecular culture-independent approaches, including sequencing of 16S rRNA clone libraries and metabolomics, based on 16S rRNA gene amplicon sequencing (Gill et al. 2006; Verdier-Metz et al. 2012).

4.2 Different Sources of Milk Microbes

Various bovine and non-bovine milk sources have been reported in the account of diverse microflora. Generally, all types of milk carry a variety of bacterial and fungal strains in its raw and fermented forms. However, complete specie level identification and accurate count of viable and non-viable microorganisms in pasteurized and fermented forms are not known yet. Modern high-throughput sequencing technologies (including second- and third-generation sequencing and combinations thereof) enabled the detection and inventory of animal-specific complex microbial communities. Milk microbiota is well documented in various hosts like cows (Addis et al. 2016; Falentin et al. 2016; Oikonomou et al. 2014), goats, sheep, camel,

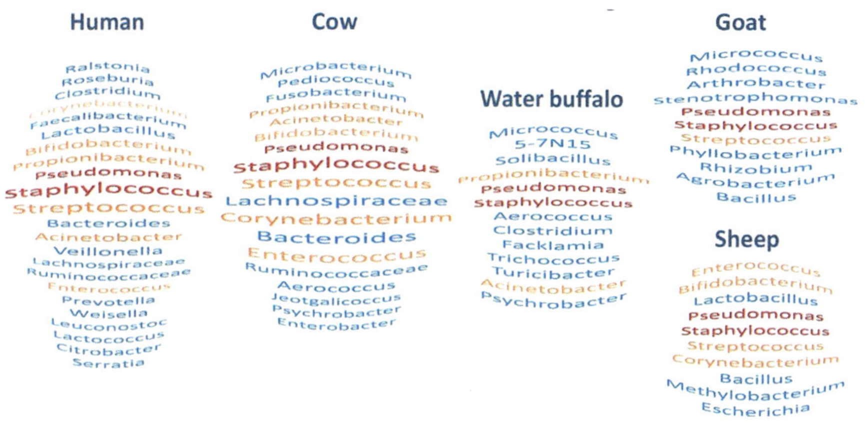

Fig. 4.1 Milk-associated microbiota in humans, cow, water buffalo, sheep, and goat. Major taxa, Red and orange taxa are shared between all human and animal species or present in three species out of five, respectively. For humans and bovines, taxa size reveals citation frequency

donkeys, buffalo, deer, reindeer, mice (Catozzi et al. 2017; De Los DoloresSoto et al. 2017; McInnis et al. 2015; Quigley et al. 2013; Treven et al. 2015), and human (Hunt et al. 2011; Jost and Lacroix 2013; Fitzstevens et al. 2017). Nevertheless, significant differences have been reported in the milk bacterial communities of different ruminants, such as water deer, reindeer, and goat, suggesting host-microbial adaptation, although the influence of environment and herd management should not be excluded. Recently, a comparison of bovine and human milk microbiota exhibited the clear metataxonomic picture and revealed the presence of common genera including *Bifidobacterium*, *Staphylococcus*, *Pseudomonas*, *Streptococcus*, *Propionibacterium*, *Corynebacterium*, *Bacteroides*, and *Enterococcus* which are among the most reported taxa in scientific reports related to bovine and human microbiota (Fig. 4.1) (Oikonomou et al. 2014).

4.2.1 Cow Milk

Culture-independent approaches described cow milk microbiota as one of the complex and diverse community comprised of 146 bacterial strains, with *Bacteroides*, *Bifidobacterium*, *Corynebacterium*, *Enterococcus*, *Propionibacterium*, *Pseudomonas*, *Staphylococcus*, and *Streptococcus* being the predominant taxa (Addis et al. 2016; Boix-Amorós et al. 2016; Cabrera-Rubio et al. 2012; Derakhshani and Naghizadeh 2018; Hoque et al. 2019; Jiménez et al. 2015; Murphy et al. 2017; Oikonomou et al. 2014; Urbaniak et al. 2016). Similar milk bacterial profiles were noticed through the shotgun metagenomic approach (Jiménez et al. 2015; Pärnänen et al. 2018) and described the presence of fungal, protozoal, and viral DNA. Colostrum microbiome depends on the lactation number and major taxonomic profile; and diversity of primiparous colostrum microbiome includes the presence of *Staphylococcus*, *Prevotella*, *Ruminococcaceae*, *Bacteroidales*, *Clostridiales,* and *Pseudomonas* (Lima et al. 2018).

4.2.2 Buffalo Milk

Differential microbial communities and diversity in buffalo milk include major taxa of *Micrococcus*, *Propionibacterium*, *Solibacillus*, *Staphylococcus*, *Aerococcus*, *Facklamia*, *Trichococcus*, *Turicibacter*, *Clostridium*, *Acinetobacter*, *Psychrobacter*, and *Pseudomonas* through Ion Torrent 16S rRNA gene sequencing (Catozzi et al. 2017).

4.2.3 Sheep Milk

Sheep milk is reported to have various genera of lactic acid bacteria. These genera are identified as *Bacillus*, *Enterococcus*, *Lactobacillus*, *Lactococcus*, and *Leuconostoc*. The species identified for these genera are *Bacillus shackletonii*, *E. casseliflavus*, *E. durans*, *E. faecium*, *Lactobacillus rhamnosus*, *Lactobacillus acidophilus*, *Lactobacillus plantarum*, *Lactobacillus casei*, *Lactobacillus delbrueckii*, *Lactococcus lactis* ssp. *cremoris*, *Lactococcus lactis* ssp. *lactis*, *Lactococcus lactis subsp. biovar diacetylactis,* and *Leuconostoc* spp. (Acurcio et al. 2014; Aziz et al. 2009; Medina et al. 2011; Patil et al. 2019).

4.2.4 Goat Milk

Lactic acid bacteria isolated from the goat milk belonged to the genera *Enterococcus*, *Lactobacillus*, *Lactococcus, Leuconostoc*, *Pediococcus,* and *Streptococcus*. Isolated species are identified as *Enterococcus faecium*, *Enterococcus durans*, *Enterococcus faecalis*, *Enterococcus hirae*, *Enterococcus avium*, *Lactobacillus acidophilus Lactobacillus delbrueckii subsp. bulgaricus*, *Lactobacillus fermentum*, *Lactobacillus lactis subsp. lactis Lactobacillus paracasei*, *Lactobacillus plantarum*, *Lactobacillus rhamnosus*, *Lactobacillus reutei*, *Lactobacillus casei*, *Lactobacillus bulgaricus*, *Lactobacillus brevis*, *Lactobacillus curvatus*, *Leuconostoc mesenteroides subsp. mesenteroides*, *Leuconostoc mesenteroides subsp. dextranicum*, *Lactococcus plantarum*, *Lactococcus lactis subsp. lactis*, *Lactococcus raffinolactis*, *Pediococcus pentosaceus, Streptococcus thermophiles*, *Streptococcus salivarius subsp. thermophillus* (Medina et al. 2011; Mittu and Girdhar 2015; Perin and Nero 2014; Pisano et al. 2019).

4.2.5 Camel Milk

Nowadays an increasing attention is being focused towards consumption of camel milk. Its composition is closer to human milk than cow's milk; therefore it is better for humans especially for infants and children. Camel milk is an enriched source of *Lactococcus*, *Lactobacillus*, *Enterococcus*, *Streptococcus*, *Weissella*, and *Pediococcus.* Isolated strains belonged to *Enterococcus durans*, *Enterococcus faecium*, *Enterococcus gallinarum*, *Lactobacillus brevis*, *Lactobacillus salivarius*, *Lactobacillus reuteri*, *Lactobacillus fermentum*, *Lactobacillus plantarum*, *Lactobacillus pentosus*, *Lactobacillus helveticus*, *Lactococcus garvieae*, *Lactococcus lactis*, *Leuconostoc pseudomesenteroides*, *Leuconostoc mesenteroides*, *Pediococcus pentosaceus*, *Pediococcus acidilactici*, *Weissella sp. t4r2c13*, *Weissella paramesenteroides*, *Weissella confusa*, *Streptococcus infantarius subsp. infantarius*, *Streptococcus equinus*, and *Str. thermophilus* (Abushelaibi et al. 2017; Amara and Shibl 2015; Bin Masalam et al. 2018; Edalati et al. 2019; Fguiri et al. 2015; Ogier and Serror 2008; Rahmeh et al. 2019).

4.3 Sources of Contaminant Microbes in Milk

The microbiological quality of dairy products reflects good hygienic practices during the milking process; raw milk contamination may occur in diseased or infected animals with environmental bacteria (Kongo et al. 2008). The detection of mesophilic aerobes and total coliforms is a clear indication of *E. coli* contamination; in addition to this the presence of *L. monocytogenes* and *Salmonella spp.* revealed poor microbiological quality of dairy products and cause interference with the native microbiota of milk. The predominant bacterial species isolated at the dairy farm comes from the water, feedstuffs, and milking equipment. In this context, *Bacillus licheniformis* and *Bacillus pallidus* act as entry points being in the form of highly heat-resistant spores in raw milk. The contamination risk of such aerobic spore-forming bacteria could lead to spoilage of milk and dairy products. The fecal material attached to the udder skin of milking animals is another source of contamination. Many species of *Lactobacillus* and *Enterococcus* are major fecal genera in the milk from rural and farm animals.

4.4 Indigenous Bacterial Community Composition

4.4.1 Raw Milk

The abundance and variety of microbial communities in raw milk varies and depends upon many factors ranging from the health of milking animal, to the milking practices, storage, and transportation methods (Kable et al. 2019; Skeie et al. 2019). The immediate cold storage of fresh milk reduces the bacterial growth and keeps milk in its native load of microflora (Li et al. 2018). The breeding practices, lactation period, and availability of feeding plants in specific geographic location of herd are important factors for the change of microbial community patterns in the milk (Kable et al. 2019; Li et al. 2018; Parente et al. 2020; Skeie et al. 2019).

Modern high-throughput metagenomic sequencing of milk is a robust tool for the identification and estimation of indigenous microbiota of milk (Ercolini 2013; Zhang et al. 2019). Recently, Li et al. (2016) reported *Proteobacteria* as predominant group in fresh buffalo milk; however the population of abundance of *Firmicutes* increased and *Proteobacteria* and *Bacteroidetes* decreased significantly during the 24 h of cold storage. Looking at the genera-level microbial population pattern, *Streptococcus, Lactococcus,* and *Pseudomonas* were found in the fresh milk, and after 24 h of refrigeration the abundance of *Lactococcus* and *Streptococcus* populations increased significantly ($P < 0.05$), with the *Lactococcus* population contributing up to 38.6% of the total microflora (Li et al. 2016). One of the noticeable aspects was the robust growth of *Pseudomonas* and *Acinetobacter* genera (62%) in 72 h of cold storage (Fig. 4.2; Li et al. 2016).

4.4.2 Pasteurized Milk

Due to risk of pathogen contamination in milk produced from healthy animals under sanitary milk conditions, pasteurization of milk prior to consumption destroys pathogens, and provides hygienic milk (Fusco et al. 2020; Melini et al. 2017). Occasionally, human illness has been linked to pasteurized milk products but these cases usually have been a result of contamination of the product after pasteurization or due to improper pasteurization.

Despite the pasteurization process, a diverse bacterial population is a key characteristic feature of milk. According to Li et al. (2016), *Paenibacillus* is a dominated taxon at genus level in the microbial population. Other predominant bacterial populations appeared after prolonged storage, were psychotropic in nature, and were mostly associated with the spoilage of dairy products (Li et al. 2016). However, pasteurization appeared sufficient for eliminating contaminants from the *Pseudomonas* and *Acinetobacter* genera. However, there is a crucial need for developing novel technologies for controlling the proliferation of *Paenibacillus* to extend the shelf life of pasteurized milk products (Doll et al. 2017; Li et al. 2016).

Pasteurized milk bacterial composition did not significantly change during a storage period of 7 days; however the population of *Lactococcus* increased, while *Streptococcus* reportedly decreased (Li et al. 2016). At phylum level, *Firmicutes* and *Proteobacteria* contributed to more than 90% of the microbial composition after 7 days of storage. However, after 14 days of storage period, there was a significant increase in the population of *Firmicutes*, with a decrease in the population of *Proteobacteria* (Li et al. 2016). The analysis of the pasteurized milk after 21 days of storage showed that the *Firmicutes* increased and contributed to 90% of the total composition, along with *Paenibacillus* which increased to 80% in the bacterial population (Fig. 4.3; Li et al. 2016).

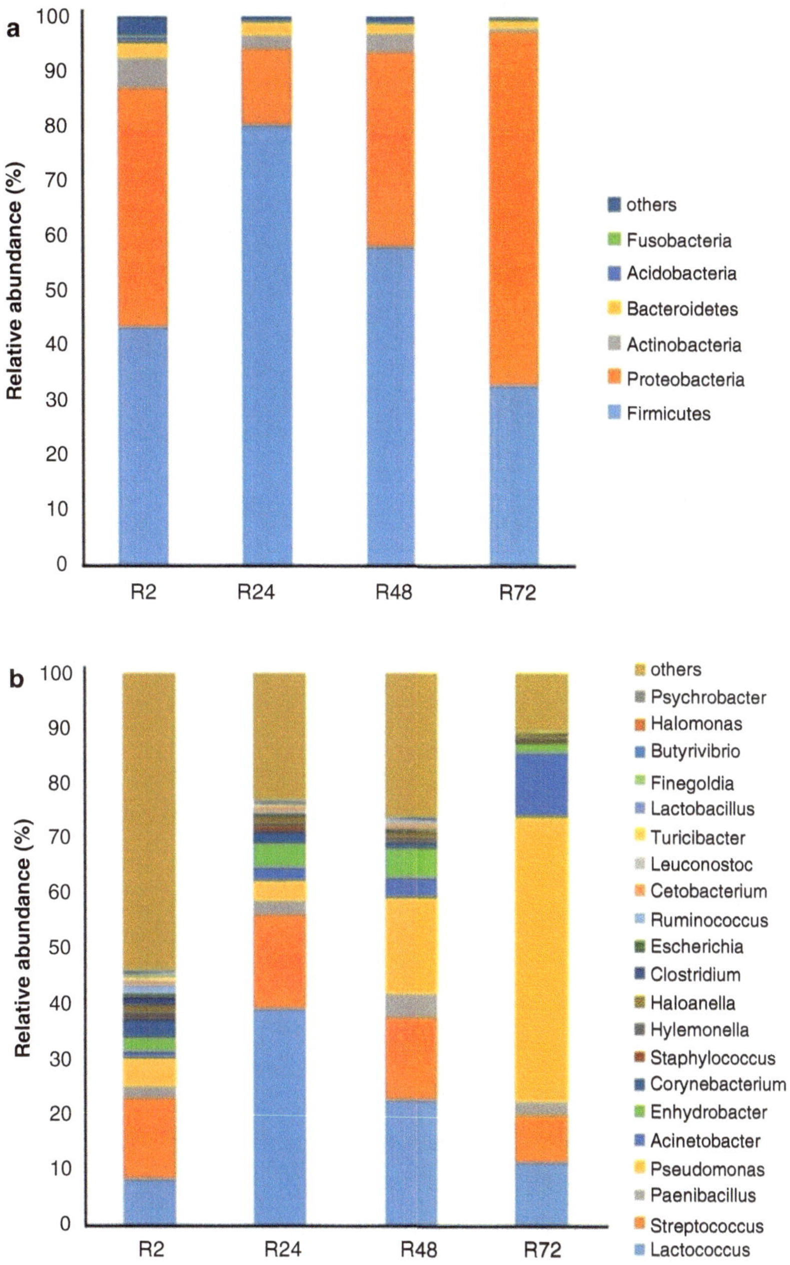

Fig. 4.2 Composition of the indigenous microflora, at the phyla (**a**) and genus (**b**) levels, in raw milk samples stored under refrigerated temperatures. Data represent the mean percentage from the metagenomics analysis of three separate raw milk samples. R_2 = raw milk samples stored for 2 h, R_{24} = raw milk samples stored for 24 h, R_{48} = raw milk samples stored for 48 h, R_{72} = raw milk samples stored for 72 h (Li et al. 2016)

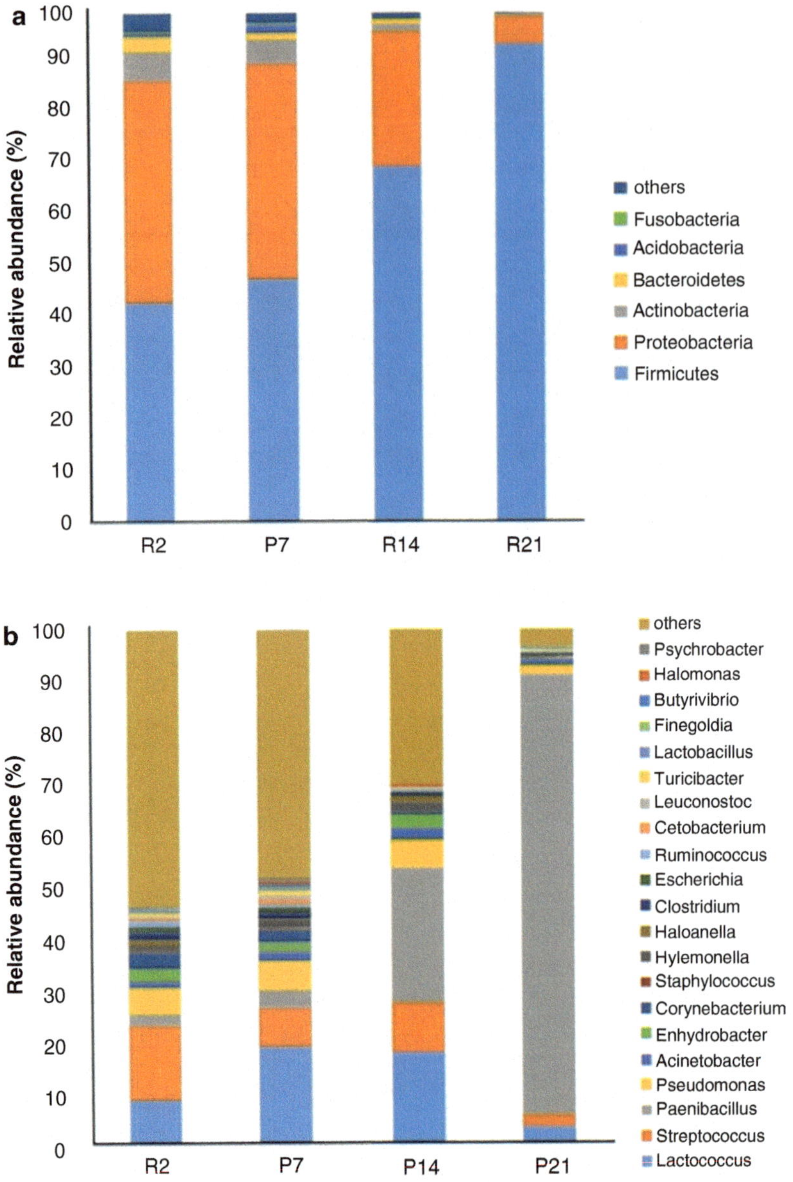

Fig. 4.3 Composition of the indigenous microflora, at both the phyla (**a**) and genus (**b**) levels, in pasteurized milk stored at refrigerated temperature. Data represent the mean percentages from the metagenomic analysis of 3 separate raw milk samples. R_2 = raw milk samples stored for 2 h, P_7 = pasteurized milk samples stored for 7 days, P_{14} = pasteurized milk samples stored for 14 days, P_{21} = pasteurized milk samples stored for 21 days (Li et al. 2016)

4.4.3 Fermented Milk

Fermented milk and its associated products are the richest and traditional source of probiotic microorganisms (Bernardeau et al. 2006). Naturally fermented milk has a variable microbial diversity in each of the resultant products, which contributes to their taste and texture (Zhong et al. 2016). Fermentation results in the functionally active microbial population to increase the bioavailability of nutrients for the consumers, while degrading toxic components to enhance the safety and bio-preservation of the final product (Tamang et al. 2016a). Low pH, fermented environment is an ideal medium to flourish beneficial microbial population (Savadogo et al. 2006; Sun et al. 2020). Fermented milk associated lactic acid bacteria (LAB) include *Enterococcus*, *Lactobacillus*, *Lactococcus*, *Leuconostoc*, *Pediococcus*, *Weissella*, *Bifidobacterium*, etc. these species of these genera are widely present in all types of milk (Axelsson et al. 2012; Tamang et al. 2016b).

Gao et al. (2017) reported *Lactococcus* as most predominant and *Lactobacillus* as subdominant genera in the milk samples collected in different times of year. Other genera found are *Leuconostoc*, *Streptococcus*, *Enterococcus*, *Chryseobacterium*, *Acetobacter*, *Weissella*, *Dysgonomonas*, *Macrococcus*, *Xenophilus*, *Pseudoclavibacter*, and *Corynebacterium* in variable proportions. Among fungal genera, *Pichia*, *Kluyveromyces*, and *Geotrichum* are found predominantly in the milk through the year. However, *Naumovozyma* and *Hanseniaspora* are subdominant genera (Fig. 4.4; Gao et al. 2017).

4.5 Types of Microbes

4.5.1 Beneficial Microbes

Beneficial bacteria are well known as "Probiotics" (usually lactic acid bacteria such as *Lactobacilli* and *Bifidobacteria*) that benefit the host physiology upon ingestion. Food and Agriculture Organization (FAO) and World Health Organization (WHO) defined probiotics as "Live microorganisms which when administered in adequate amount confers a health benefit on the host". They have become very popular over the past two decades due to their countless benefits to human health and for this reason they have been incorporated in many food-related products, mainly fermented products. Probiotic strains are marketed in the form of capsules, powder, or fermented products. The global market of probiotics is rapidly increasing annually due to consumers' interest in optimizing their health with functional foods (Di Cerbo and Palmieri 2015).

Lactic acid bacteria are generally regarded as safe (GRAS) microorganisms and are gram positive, facultative aerobes or anaerobes with bacilli, coccobacilli, or cocci morphology. These are non-respiratory, catalase-negative, acid-tolerant, and non-spore-forming bacteria, grouped on the basis of physiological, morphological, and metabolic constellation. These bacteria are normally associated with human and animal healthy mucosal surfaces and are a part of various animal and plant niches.

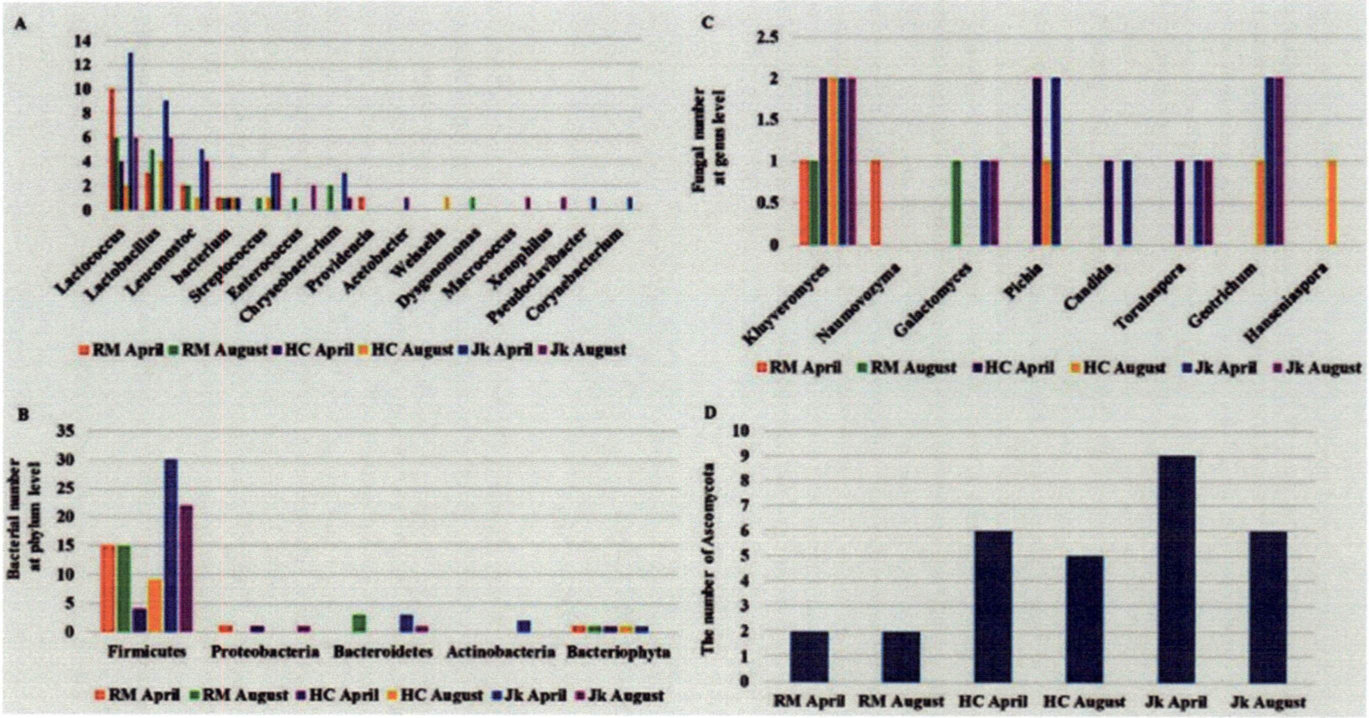

Fig. 4.4 Composition of the indigenous microflora, at the bacterial genus (**a**) and phyla (**b**) levels, and fungal genus (**c**) and Ascomycota (**d**) levels in raw and fermented milk samples. Data represents the number from the metagenomics analysis of two raw milk samples, and four fermented milk samples. RM April = raw milk samples collected in April, RM August = raw milk samples collected in August, HC April = Hurood Cheese samples collected in April, HC August = Hurood Cheese samples collected in August, Jk April = Jeuke samples collected in April, and Jk August = Jeuke samples collected in August (Gao et al. 2017)

Historically, the core genera of lactic acid bacteria include *Pediococcus*, *Lactobacillus*, *Streptococcus,* and *Leuconostoc*; however, nowadays there are 20 taxonomic revised genera. The significant LAB genera from food technology point of view are *Lactococcus*, *Leuconostoc*, *Lactobacillus*, *Enterococcus*, *Pediococcus*, *Aerococcus*, *Carnobacterium*, *Tetragenococcus*, *Vagococcus*, *Oenococcus,* and *Weissella* (Makarova et al. 2006).

Fermented milk associated LAB play a crucial role in the production of fermented beverages and other dairy products. They are strictly fermentative and produce lactic acid as a major product during the course of sugar fermentation. They are classified into two major groups based upon their fermentation potential, e.g., homofermentative or heterofermentative. Homofermentative LAB produces twice the energy from glucose fermentation as compared to heterofermentative. Homofermentation occurs through Embden Meyerhof Parna's pathway, whereas heterofermentation occurs either through hexose monophosphate or pentose phosphate pathway. The end product in the former case is mainly lactic acid, while in the latter ethanol/acetic acid and CO_2 are also significantly produced (Bassyouni et al. 2012; Çetin 2011; Rattanachaikunsopon and Phumkhachorn 2010).

A higher intake of fermented dairy products would reduce the risk of immune and metabolic disorders that will reduce the risk of obesity. Metabolizable nutrients and beneficial microorganism are incorporated due to the intestinal microbiota flourishing with the consumption of fermented dairy products. Yogurt is one of the dairy products that is well known for its numerous health benefits due to the probiotics. The intestinal health is maintained with the restoration of healthy balance between the good and bad bacteria from the probiotic intake. Moreover, it enhances the humoral and cellular immunity (Borchers et al. 2009). Despite general gut microenvironment, every individual's gut has a unique pattern of microbial community, and thus the response towards the use of probiotics is different.

Flu-like symptoms and upper respiratory infections are decreased with consumption of probiotics, as there is an immunity boost with the production of IgA antibodies, T lymphocytes, and natural killer cells. Crohn's disease, colorectal cancer, celiac disease, ulcerative colitis, and irritable bowel are some of the diseases that are improved with the use of yogurt. The severity of diarrhea is reduced with the use of probiotics, as it is among the side effects of consuming antibiotics. Therefore, doctors have suggested the use of yogurt for patients taking an antibiotic course to prevent the risk of antibiotic associated diarrhea. A study showed how certain strains of good bacteria present in the probiotics will help reduce the time of infectious diarrhea (Kechagia et al. 2013).

Another interesting research shows how the probiotics impact the mental health, as there is link between the brain and gut called the gut-brain axis (Mayer et al. 2014). Yogurt has proven to help reduce anxiety and stress which further improves the mental health of the individual. The *Bifidobacterium* and *Lactobacillus* strains for 1–2 months have been proven to positively affect the memory, obsessive compulsive disorder, autism, depression, and much more. Probiotic supplements introduced in the diet for 8 weeks decreased 40 patient's depression levels along with C-reactive protein that causes inflammation.

Probiotics have been declared to be healthy for all those suffering from chronic heart illnesses, such as angina, cardiovascular disease (CVD), heart attack, etc. due to their potential to reduce pressure and cholesterol by lowering the low-density lipoproteins (LDL). Moreover, probiotic microbes help with digestion, as the cholesterol is broken down into bile, which adds digestion. The benefit of having probiotics is the prevention of the reabsorption of the broken-down cholesterol in the blood. Studies suggest the reduction of allergies and eczema in children and infants with the consumption of probiotics in the form of milk or yogurt.

The health-promoting properties of conjugated linoleic acid (CLA) include anticarcinogenic, antiatherogenic, anti-inflammatory, and antidiabetic activity, as well as the ability to reduce body fat (Sosa-Castañeda et al. 2015). Although it is a native component of milk, the amount consumed in foods is far from that required in order to obtain desired beneficial effects. Thus, increasing the CLA content in dairy foods through milk fermentation with specific LAB offers a promising alternative. An effective way to increase CLA uptake in humans is to increase its level in dairy products by using strains with high production potential.

4.5.2 Pathogenic Microbes

Mammary glands of milking animal are natural reservoirs of microbes. Many of these bacteria are not harmful to humans, but some may be harmful to humans even though the animals are not affected and appear healthy. As listed in Table 4.1, the bacteria present in dairy products may cause disease or spoilage. Human diseases from milk-borne pathogens are usually due to the consumption of raw milk or products made from raw milk such as fresh cheeses. Till now, major dairy microorganisms are predominately associated with *Brucella* spp., *Campylobacter jejuni, Coxiella burnetii, Salmonella enterica, Listeria monocytogenes, Mycobacterium bovis, Mycobacterium paratuberculosis, Yersinia enterocolitica*, and *Escherichia coli* O157:H7 (Table 4.1).

4.5.2.1 *Brucella* spp.

Brucella species (spp.) are found in many animal species including cattle, sheep, and goats. *Brucella* spp. are destroyed by pasteurization. *Brucella* spp. cause illness with symptoms that are flu-like and include fever, sweats, headaches, back pain, and physical weakness. In some cases, long-lasting symptoms of fever, joint pain, and fatigue may occur.

4.5.2.2 *Campylobacter jejuni*

Campylobacter jejuni is found in the intestinal tract, udder, and feces of cattle, in poultry and wild birds, and in contaminated water sources. *C jejuni* is destroyed by pasteurization. *C. jejuni* is one of the most common bacterial causes of diarrheal illness. *C jejuni* generally causes illness 2–5 days after exposure, and illness typically lasts 5–10 days. Symptoms of campylobacteriosis include diarrhea, bloody diarrhea, abdominal pain, cramping, nausea, vomiting, and fever. Patients with

Table 4.1 Dairy pathogenic bacteria and associated diseases

Organism	Source of microorganism	Disease condition	Reference
Campylobacter jejuni	Intestinal tract and feces	Gastroenteritis	Facciolà et al. (2017)
Coxiella burnetii	Infected cattle, sheep, and goats	Q fever	
Escherichia coli O157:H7	Intestinal tract, and feces	Gastroenteritis, Hemolytic uremic syndrome (HUS)	
Listeria monocytogenes	Water, soil, and environment	Listeriosis	Radoshevich and Cossart (2018)
Mycobacterium bovis or tuberculosis	Infected animals	Tuberculosis	Lan et al. (2016)
Mycobacterium paratuberculosis	Infected animals	Johne's (ruminants)	Whittington et al. (2019)
Salmonella spp.	Feces, and environment	Gastroenteritis, Typhoid fever	
Yersinia enterocolitica	Environment, water, and infected animals	Gastroenteritis	Sabina et al. (2011)

Campylobacteriosis usually recover without specific treatment other than fluid and electrolyte replacement. In some persons with a compromised immune system, *C. jejuni* infection can lead to the more serious diseases like Guillan-Barré syndrome and Reiter syndrome. Guillan-Barré syndrome is a disorder that results in temporary neuromuscular paralysis, although 20% of those infected may have long-term disability and it may cause death. Reiter syndrome is a reactive arthritis that may affect multiple joints, particularly the knee joint. The prevalence of *C jejuni* is very widespread. It has been reported in bulk tank raw milk samples in Illinois, Michigan, Minnesota, Ohio, Pennsylvania, South Dakota, Tennessee, Virginia, and Wisconsin, suggesting that the organism is ubiquitous. In these studies, *C jejuni* was found in 0.4–12.3% of the bulk tank milk samples (Facciolà et al. 2017; Jayarao et al. 2006).

4.5.2.3 *Coxiella burnetii*

Coxiella burnetii is a pathogen shed in the milk, urine, and feces of cattle, goats, and sheep. *C. burnetii* is considered to be the most heat-resistant, non-spore-forming pathogen commonly found in milk, and the established conditions for milk pasteurization are specifically designed to destroy this organism. *C. burnetii* causes Q fever, an illness characterized by a sudden onset of high fever, severe headache, nausea, vomiting, diarrhea, abdominal pain, chest pain, chills, sweats, sore throat, non-productive cough, and general malaise. Fever can last for 1–2 weeks. Most patients recover without any treatment, although *C. burnetii* may result in death. The prevalence of *Coxiella burnetii* was >94% in raw milk samples from the North-eastern, Midwestern, and Western regions of the USA tested between 2001 and 2003 (Kim et al. 2005).

4.5.2.4 *Escherichia coli* O157:H7

Escherichia coli O157:H7 is one strain in a large family of bacteria. Strains of *E. coli* are considered fecal coliforms. Most strains of *E coli* do not cause illness and live in the intestinal tracts of healthy humans and animals. *E. coli* O157:H7 is found in the intestinal tract and feces of cattle and destroyed by pasteurization. *E. coli* O157:H7 produces toxins that cause illness in humans. Symptoms of illness include bloody diarrhea and abdominal cramps. In some cases, particularly in young children, *E. coli* O157:H7 infection causes hemolytic uremic syndrome, which destroys red blood cells and causes kidney damage or failure, and in some cases death. The prevalence of *E. coli* O157:H7 and Shiga-toxin producing *E. coli* have been reported for bulk tank raw milk samples in Minnesota, Pennsylvania, South Dakota, Wisconsin, and Ontario. *E. coli* O157:H7 was found in 0.87–10% of the bulk tank milk samples tested (Jayarao et al. 2001, 2006).

4.5.2.5 *Listeria monocytogenes*

Listeria monocytogenes is found in soil and water and has been isolated from a large number of environmental sources. It is destroyed by pasteurization, but if food products are contaminated after pasteurization, *it* can grow at refrigerator temperatures. Illness can occur as sporadic events or larger outbreaks. *L. monocytogenes* typically causes illness in pregnant adults, newborns, the elderly, and patients with compromised immune systems, but healthy adults and children may also become infected. Symptoms of Listeriosis include flu-like symptoms, fever, muscle aches, stiff neck, headache, septicemia, meningitis, miscarriage, still-birth, premature delivery, abortion, or death. The prevalence of *L. monocytogenes* has been reported for bulk tank raw milk samples in individual states (or grouped by region) for California, Colorado, Florida, Idaho, Illinois, Indiana, Iowa, Kentucky, Massachusetts, Michigan, Minnesota, Missouri, Nebraska, New Mexico, New York, Ohio, Pennsylvania, South Dakota, Tennessee, Texas, Washington, Wisconsin, Vermont, Virginia, and in Alberta and Ontario, Canada. *Listeria monocytogenes* was found in up to 12% of the bulk tank milk samples tested (Jayarao et al. 2001, 2006; Van Kessel et al. 2004) illustrating the widespread presence of *L. monocytogenes* in unpasteurized milk.

4.5.2.6 *Mycobacterium bovis* and *Mycobacterium tuberculosis*

Mycobacterium bovis and *Mycobacterium tuberculosis* are found in infected cattle worldwide. Both of these organisms are destroyed by pasteurization. *Mycobacterium bovis* and *Mycobacterium tuberculosis* cause tuberculosis, a lung disease. Tuberculosis in the USA is not very common today, although historically milk was a common source of tuberculosis. Tuberculosis is a concern in many parts of the world. *Mycobacterium paratuberculosis* causes Johne's disease in cattle. It has been suggested that *M. paratuberculosis* may be associated with Crohn's disease, an intestinal disorder, in humans, but this has not been confirmed (Peden 2000; Whittington et al. 2019).

4.5.2.7 *Salmonella* spp.

Salmonella species (spp.) contain several strains that cause illness in humans; the most common are the serotypes Enteriditis and Typhimurium. *Salmonella* has been found in the intestinal tracts of all warm-blooded animals including humans. *Salmonella* is destroyed by pasteurization. *Salmonella* spp. causes illness that can develop 12–72 h after exposure, and can last 4–7 days. Symptoms of Salmonellosis include diarrhea, abdominal cramps, and fever. Most people recover without treatment other than fluid and electrolyte replacement. Some cases may be severe and require hospitalization. A small number of people may develop Reiter syndrome, which is a reactive arthritis that may affect multiple joints, particularly the knee joint. The prevalence of *Salmonella* spp. has been reported for bulk tank milk samples in individual states (or grouped by region) for California, Colorado, Florida, Idaho, Illinois, Indiana, Iowa, Kentucky, Michigan, Minnesota, Missouri, New Mexico, New York, Ohio, Pennsylvania, South Dakota, Tennessee, Texas, Washington, Wisconsin, Vermont, Virginia, and Ontario, Canada. *Salmonella* spp. were found in 0.17–8.9% of the bulk tank milk samples tested (Jayarao et al. 2001, 2006; Van Kessel et al. 2004), indicating the widespread presence of *Salmonella* in unpasteurized milk.

4.5.2.8 *Yersinia enterocolitica*

Yersinia enterocolitica is found in the intestinal tract of farm animals, especially pigs, and in the environment. *Y. enterocolitica* is destroyed by pasteurization, but if food products are contaminated after pasteurization, *Y. enterocolitica* can grow at refrigerator temperature. *Yersinia enterocolitica* causes illness with symptoms of fever, abdominal pain, and diarrhea. The prevalence of *Yersinia enterocolitica* has been reported for bulk tank milk samples in Michigan, Minnesota, Pennsylvania, South Dakota, Tennessee, Wisconsin, Virginia, and Ontario, Canada. *Yersinia enterocolitica* was found in 1.2–18% of the bulk tank milk samples tested (Jayarao et al. 2001, 2006; Sabina et al. 2011).

4.5.2.9 Other Pathogens

Coliforms are a large group of bacteria that are found in the intestines of warm-blooded animals. Most coliforms are not pathogenic, but their presence indicates contamination, usually from fecal sources. Coliforms are destroyed by pasteurization. The prevalence of coliforms was detected in 62–95% of the raw bulk tank milk tested in regions that included California, Colorado, Florida, Idaho, Illinois, Indiana, Iowa, Kentucky, Michigan, Minnesota, Missouri, New Mexico, New York, Ohio, Pennsylvania, South Dakota, Tennessee, Texas, Washington, Wisconsin, Vermont, and Virginia (Jayarao et al. 2001, 2006; Van Kessel et al. 2004).

Psychotropic bacteria are capable of growing at 44.6 °F (7 °C) or less. This group of microbes is a concern in dairy products because they grow at refrigerator temperature and cause spoilage, often resulting in off-flavors. The most common psychrotrophs are in the genus *Pseudomonas*. These organisms are killed by pasteurization, but may occur in milk from contamination after pasteurization. Some bacterial pathogens are psychrotrophic, including *Listeria monocytogenes*, *Yersinia enterocolitica*, some *E. coli* strains, and some *Bacillus* strains (Radoshevich and Cossart 2018; Sabina et al. 2011).

4.6 Microbial Additives

Milk itself is a natural source for a variety of bacteria; the group of lactic acid bacteria is one of the prime sources of microbial additives. Many health-promoting effects are achieved from bioactive molecules produced by dairy fermented products. In contrast to the conventional concept of probiotic (ingestion of alive bacteria for the production of metabolites within human gut), a biologically functional food concept is based on the endogenous production of healthy metabolites in the fermented products, as a result of the metabolic response of bacterial machinery. The main biologically active molecules produced by LAB during dairy fermentation are vitamins, gamma-aminobutyric acid, bioactive peptides, bacteriocins, enzymes, conjugated linoleic acid, and exopolysaccharides.

4.6.1 Bioactive peptides

In the process of milk fermentation, lactic acid bacteria digest many proteins into short peptides through proteolytic activity. These peptides are biologically functional and exhibit antioxidative, antimicrobial, antihypertensive, immunomodulatory, and antithrombotic properties (Nongonierma and FitzGerald 2015). One of the most important bioactive peptides is Angiotensin-I-converting enzyme (ACE) inhibitory peptides. ACE inhibitory peptides display strong antihypertensive features and have been reported from a number of dairies (Fitzgerald and Murray 2006; Pritchard et al. 2010). Initially, ACE-inhibitory peptides, Ile-Pro-Pro (IPP), and Val-Pro-Pro (VPP) were extracted from milk fermented by *L. helveticus* (Slattery et al. 2010). Later on, other lactic acid bacteria including *L. rhamnosus*, *L. plantarum*, *L. delbrueckii*, *L. acidophilus*, *Lactococcus lactis*, *and S. thermophilus were reported as* dairy starter cultures in the industry as a source of inhibitory peptides of ACE (Hafeez et al. 2014). β-casein (SLVYPFPGPI) is another bioactive peptide produced by *L. delbrueckii* in fermented milk (Qian et al. 2011). Similarly, two short peptides are produced by the hydrolysis of α-S2 casein during the process of fermentation; both peptides are antimicrobial and display protective function against many human pathogens including *Saccharomyces thermophilus*, *E. coli*, *Helicobacter pylori*, *Staphylococcus aureus*, *Streptococcus pyogenes*, and *Listeria monocytogenes* (Nagpal et al. 2011). Lactoferrin is another source of antimicrobial peptides (Zivkovic et al. 2013). Peptic digestion of lactoferrin produces short peptides that displayed antimicrobial activity against a broad range of bacteria including *E. coli*, *Listeria*, *Salmonella*, *Campylobacter*, and many fungal strains, however, non-toxic to *Bifidobacterium* (Quintieri et al. 2013; Shah 2007).

4.6.2 Bacteriocins

Bacteriocins are major ribosomal antimicrobial peptides known to inhibit adhesion and invasions of pathogens through direct microbial interaction or by altering the exterior environment leading to slow to no growth of microbes (Hernández-Ledesma et al. 2014). Different types of ribosomal short peptides and their respective immunity proteins are produced by many lactic acid bacteria, which provide a broad range of antimicrobial activity against major human pathogens. Thus, bacteriocin producers are a potential alternative to pharmaceutically synthesized antibiotics and offer a means of controlling pathogen-induced inflammation (Cotter et al. 2013). Many lactic acid bacteria are generally regarded as safe (GRAS) for human consumption. These are the ideal source of bacteriocin production on a commercial scale (Nes et al. 2007). Because of the strong antimicrobial characteristics of bacteriocin, the producing strains also use as natural food-preservatives.

Nisin is the most used for food preservation due to its antimicrobial effect against spoilage and disease-associated bacteria like *Listeria* and clostridia spores. Plantaricin C is another broad-spectrum peptide produced by *L. plantarum* and documented as an immunomodulator for dendritic cells (Meijerink et al. 2010). Briefly, the use of bacteriocins directly or bacteriocin-producing bacteria as a starter culture for the generation of bacteriocins through fermentation became an efficient health-promoting strategy. Similarly, the use of lacticin-producing strain of *Lactococcus lactis* greatly inhibits the growth of *Listeria monocytogenes* in Cheddar cheese (Chen and Hoover 2003). Many other lactic acid bacteria like *L. acidophilus*, *Pediococcus acidilactici*, and *Leuconostoc mesenteroides* known for their specific bacteriocins can be added as an adjunct in many food fermentations processes as food preservatives (Anjum et al. 2014). Besides the production of antimicrobial peptides, these bacteria pose many other advantages to enhance flavor, texture, and nutritional value of the product (Gaggia et al. 2011; Jiang et al. 2012; Grosu-Tudor et al. 2013; Mitra et al. 2010; Khan et al. 2010; Tamang et al. 2009).

4.6.3 Enzymes

Many *Lactobacillus*, *Lactococcus*, and *Streptococci* species can ferment milk by producing hydrolytic enzymes. The proteolytic machinery of lactic acid bacteria (LAB) comprises membrane-bound aminopeptidases, endopeptidases, and proteinases for the production of hydrolysates. Fermentation-associated microbes depend on the degradation of milk proteins to get free amino acid residues and short peptides required for their growth. Yogurt and other conventional fermented dairy products associated with bacteria reduce lactose intolerance and improve lactose digestion by degrading lactose through the activity of microbial β-galactosidase (De Vrese et al. 2001; Patel et al. 2013).

4.6.4 Vitamins

Although milk contains many vitamins, however, in the fermented milk the vitamin producer lactic acid bacteria enhance the nutritional value of the product. Many species of *Lactobacillus* and *Bifidobacterium* genera secrete vitamin B complex (B1, B2, B7, B9, B12) during the fermentation process. Dietary depletion of vitamin B1 (thiamine) and vitamin B2 (riboflavin) can dysregulate glucose metabolism in the brain and lead to both skin and liver diseases, respectively (Russo et al. 2014). Some *Propionibacteria* and lactic acid bacteria can produce cobalamin, folic acid, and biotin, such as *L. casei* richly produce thiamine and riboflavin in fermented milk (Hugenholtz et al. 2002; Drywień et al. 2015).

Vitamin B7 (Biotin) deficiency can be genetic or dietary that affects the skin and hair health. Starter culture of lactic acid bacteria, e.g., *L. helveticus* and Propionibacteria, ferment and produce biotin-enriched milk products (Patel et al. 2013). The deficiency of vitamin B9 (Folate) is linked to neural tube impairment and cardiac issues. Limited strains of lactic acid bacteria including *Streptococcus thermophilus* CRL803/CRL415, *L. amylovorus*, and *L. bulgaricus* are designated as vital for dairy folate enrichment (Laiño et al. 2014). Among *Bifidobacteria, B. catenulatum* is known as rich folate producer.

Plants, animals, and fungi are unable to produce, thus bacteria are the exclusive source of vitamin B12 (cobalamin) (LeBlanc et al. 2011). It has been demonstrated that vitamin B12 cobalamin can be synthesized by some bacteria such as *L. reuteri, Propionibacterium freudenreichii*, and *B. animalis* (Gu et al. 2015; Moslemi et al. 2016; Patel et al. 2013; Van Wyk et al. 2011). *Propionibacterium freudenreichii* is able to secrete vitamin B12 and the pseudovitamin B12 isoforms during the milk fermentation process. Pseudovitamin B12 converts into vitamin B12 to enhance the bioavailability of cobalamin (Deptula et al. 2017).

Vitamin K is essential for arterial de-calcification to reduce the risk of cardiovascular disorders. Its deficiency can cause medical ailments such as osteoporosis and hemorrhage (LeBlanc et al. 2011). Vitamin K in nature exists in the forms of phylloquinone (vitamin K1) and menaquinone (vitamin K2). Menaquinone is microbial vitamin synthesized by *Lactococcus lactis, a common starter culture for the* industrial production of sour cream, cheese, kefir, and buttermilk (Walther et al. 2013).

4.6.5 Gamma-Aminobutyric Acid

Gamma-aminobutyric acid (GABA) is one of the exclusive inhibitory neurotransmitters (INT) of the central nervous system (CNS). Glutamate decarboxylase (GAD) catalyzes glutamate in the process of α-decarboxylation and synthesizes GABA (Tajabadi et al. 2015). Interestingly, *Bacteroides* genus is the largest GABA producer group; for example, *Bacteroides fragilis* produces GABA, polysaccharide A, and sphingolipids; the latter two are evident for the health of immune and gut systems (Tan et al. 2019; Troy and Kasper 2010). In addition to

Bacteroides, several lactic acid bacteria have been reported as the source of GABA producers including *Lactococcus lactis*, *Lactobacilli* (*L. paracasei*, *L. brevis*, *L. delbrueckii*, *L. plantarum*, *L. helveticus*, *L. buchneri*), *Streptococcus thermophilus*, and *Bifidobacterium* spp. (Barrett et al. 2012; Li and Cao 2010) which are most promising candidates.

Few strains, *S. salivarius* fmb5, *L. casei* Shirota, and *L. plantarum* NDC75017, were selected for commercial production of GABA-enriched fermented milk drink (Chen et al. 2016; Inoue et al. 2003; Shan et al. 2015). Similarly, yogurt and cheese enriched with GABA were produced by using the strain *S. thermophiles* APC151, *L. brevis* OPY-1, and *Lactococcus lactis* (Linares et al. 2016; Park and Oh 2007; Pouliot-Mathieu et al. 2013).

4.6.6 Conjugated Linoleic Acid

Polyunsaturated fatty acids (PUFA) are important metabolites of lactic acid and bifidobacteria bacteria such as conjugated linoleic acid (CLA) produced by conversion of linoleic acid. Many LAB and bifidobacterial strains like *L. casei*, *L. plantarum*, *Lactococcus lactis*, *L. rhamnosus*, *L. acidophilus*, *B. bifidum*, and *B. animalis* were reported to produce CLA in dairy products (Florence et al. 2009; Sosa-Castañeda et al. 2015; Van Nieuwenhove et al. 2007; Yang et al. 2015). These strains also used to add extra CLA contents in cheese and yogurt as adjunct cultures (Van Nieuwenhove et al. 2007).

4.6.7 Exopolysaccharides

Exopolysaccharides (EPS) are complex carbohydrates produced by a group of lactic acid bacteria, Propionibacteria, and bifidobacteria in the form of secretions during the fermentation process of dairy products and support the immune system by promoting host beneficial microflora (Salazar et al. 2016). Lactic acid bacteria including *L. delbrueckii*, *L. mucosae*, *Lactobacillus kefiranofaciens*, *Lactococcus lactis, and S. thermophilus* are predominant EPS-producing species in the yogurt and cheese and boost immune-stimulatory effects and reduce cholesterol levels (Darilmaz and Gumustekin 2012; Makino et al. 2016; Ryan et al. 2015). Specifically, *Lactobacillus kefiranofaciens* produce EPS metabolites, which dramatically inhibit the invasion of pathogens like *Listeria monocytogenes* and *Salmonella enteritidis* in the enterocytes (Jeong et al. 2017; Medrano et al. 2008). Antimicrobial effects of these metabolites may extend to other microbial species in the gut microflora.

In addition to health-promoting effects, EPS greatly enhance the quality, sensory and rheological features of dairy products. For example, *Bifidobacterium longum* and *S. thermophiles* are well known for immune-modulatory effects and high EPS production that directly reduces syneresis and improves the texture and viscosity of fermented ice-cream and yogurt (Dertli et al. 2016; Han et al. 2017; Hidalgo-Cantabrana et al. 2012; Prasanna et al. 2013).

4.6.8 Other Bio-Functional Molecules

Carbohydrate-fermenting microbes also secrete many neuroactive molecules including Clostridia metabolites, short-chain fatty acids, histamine, and diacylglycerol kinase (Karl et al. 2018; Shaw 2017). Mycelial fungi *Aspergillus, Actinomucor, Monascus, Amylomyces, Mucor, Rhizopus,* and *Neurospora* also produce various carbohydrate enzymes including ß-galactosidase, α-amylase, pectinase, maltase, cellulase, amyloglucosidase, hemi-cellulase as well as lipase and proteases.

4.7 Industrial Importance of Dairy Microbes

The dairy starter culture is used on a large scale in the food industries for the manufacturing of butter, cheese, yogurt, kefir, sour cream, and other fermented milk products. The principle purpose of the starter culture is to convert lactose and other sugars present in milk to lactic acid. The industrially important lactic acid bacteria are used as a starter culture for the preparation of many important food products and they impart various sensory characteristics to them, i.e., aroma, texture, viscosity, and flavor; henceforth, an increase in the use of LAB probiotics has been observed in the recent years. Dairy industry has become an integral part of food industries worldwide. Henceforth, the demand for starter culture is growing by leaps and bounds over the past few years. Lactic acid bacteria have also been reported to play a crucial role in the cheese ripening and giving it perfect consistency, flavor, and aroma (Hannon et al. 2003). Apart from this, many antimicrobial short peptides, exopolysaccharides, and enzymes are associated with dairy microbes to enhance nutritional value and shelf life of product.

The growth of the dairy starter culture market is driven by the growth of dairy industry. The overall increase in the production of dairy products and growing demand for dairy-based products is expected to boost the demand for the dairy starter culture globally.

4.8 Nutraceutical Properties of Milk Microbiota

Milk proteins exhibit a wide range of nutraceuticals and biological properties. Most of the dairy proteins are specific in biological functions and display many health-promoting effects. These short peptides are inactive within the endogenously secreting proteins and can be cleaved by proteolytic activity of gastrointestinal enzymes upon ingestion of milk or fermentation process. Proteins are the essential components of dairy products that have a variety of applications in several food industries.

4.8.1 Antihypertensive

Although many fermented food products exhibit medicinal characteristics, however, fermented dairy products are exceptional in the nutraceutical contents. Regular consumption of fermented dairy products displayed anticholesterol and antihypertensive properties, thus reducing the risk of cardiovascular diseases. In addition to milk proteins, fermented milk-associated probiotic bacteria secrete some proteins and metabolites and exert an overall positive impact on the health of the consumer. Kefir and Calpis contain many short peptides that are responsible for hypotensive effects. Some lactic acid bacteria functionally antihypertensive such as *L. rhamnosus*, *L. plantarum*, *L. delbrueckii* ssp. *bulgaricus*, *Lactococcus Lactis*, *L. acidophilus, and S. thermophilus in fermented milk are the commercial source of* ACE inhibitory peptides (Hafeez et al. 2014), and thus greatly reduces elevated blood pressure (Shah 2015).

4.8.2 Anticarcinogenic

Many dairy raw and fermentation-associated bacteria like *L. acidophilus* inhibit the conversion of paracarcinogenic molecules into carcinogenic forms by reducing specific enzymes including azoreductase ß-glucuronidase, and nitroreductase in human, hence, trigger and boost body immunity. In this context, South Asian fermented milk product *dahi* (yogurt) is the most known anticarcinogenic dairy product. Daily use of yogurt can reduce the risk of cervical, bladder, and colon cancer (Mohania et al. 2014).

4.8.3 Gastrointestinal Support

Many fermented dairy lactic acid bacteria significantly reduce a load of gastrointestinal diseases (Verna and Lucak 2010). Intake of *Lactobacillus* species in the food improves the symptoms of ulcerative colitis, paucities, and inflammatory bowel disease (Orel and Trop 2014). Similarly, *L. rhamnosus* specifically treat severe diarrheal issue (Szajewska et al. 2007). Moreover, probiotics in fermented dairy products manifest immunomodulatory effects and thus inhibit the growth of pathogens in the gastrointestinal tract (Balamurugan et al. 2003).

4.8.4 Anti-allergic Effects

Lactobacillus kefiranofaciens has an anti-allergic effect. In the process of fermentation, cleavage, and degradation of casein proteins of allergenic reactivity thus increases tolerance (Alessandri et al. 2012). Several species of *Lactobacillus* captured attention because of their ability to produce interleukins and interferons, and thus significantly reduce allergic reactions due to food or dermatitis. Yogurt is a rich probiotic supplement that increases glucose tolerance and reduces oxidative stress, hyperglycemia, dyslipidemia, hyperinsulinemia, indicating a lower risk of diabetes (Yadav et al. 2007).

4.8.5 Alleviation of Lactose Intolerance

Lactose intolerance both in children and in adults arises because of the unavailability of ß-D-galactosidase (Shah 2015). Lactic acid bacteria including *L. delbrueckii* and *S. thermophilus* strains are capable to secrete high contents of ß-D-galactosidase which improve the symptoms of lactose malabsorption in lactose intolerant people. Consumption of fresh yogurt (with live yogurt cultures) has demonstrated better lactose digestion and absorption than with the consumption of a pasteurized product. *Kefir* can minimize the symptoms of lactose intolerance by providing an extra source of β-galactosidase (Hertzler and Clancy 2003).

4.8.6 Brain Gut Axis Aid

Many mental conditions including psychiatric, neurodevelopmental and neurodegenerative disorders can be potentially treated with the psychobiotic microbes. These bacteria include many species of *Lactobacillus*, *Lactococcus*, *Bifidobacterium*, *Streptococcus* and few species of *Bacillus* and *Clostridium* genera. Appropriate dose management of these microbes display psychotropic potential by the production of neuroactive molecules, such as neurotransmitters (GABA, serotonin, norepinephrine, acetylcholine, glutamate), neuropeptides (neuropeptide Y, glucagon-like peptide-1 and 2, Tyr-Tyr peptide), and other molecules like cholecystokinin and substance P for the regulation of brain-associated protein like brain-derived neurotrophic factor (BDNF). The regulation of neuronal proteins is important to modulate specific behavior types. Psychobiotics employ antidepressant, antianxiety, and antidepressant properties, and improve sleep quality and energy metabolism of brain through enteric neural network, systemic, humoral, and metabolic mechanisms in the body and establish a brain gut axis. The bacteria-brain communication is important in the modulation of behaviors related to the central nervous system (Hao et al. 2019).

4.9 Dairy Psychobiotics

Lactobacillus and *Bifidobacterium* have reportedly shown potential psychobiotic activity when present in higher numbers in the human gut microbiome. Dairy products which undergo fermentation have proven to be a chief reliable source of *Lactobacillus* species. Species of *Lactobacillus* are reported to produce a variety of neurotransmitters, and their precursors *in vitro*. The gut microflora plays an important role in the regulation of bioavailability of the precursor molecules for neurotransmitters. In the brain, dopamine is converted to norepinephrine through an enzyme known as dopamine-β-hydroxylase. The inhibitors of this enzyme 4-hydroxyphenylacetate, and 4-cresol, are metabolites produced by *Clostridia*, a class of *Firmicutes*. Similarly, the microbes that ferment carbohydrates produce a short-chain fatty acid, known as butyrate that has been reported to impact the intestinal entero-chromaffin cells by stimulating them to synthesize serotonin (5-HT). As shown in Table 4.2, these precursor molecules for neurotransmitters and other metabolites produced by the probiotic microbes are neuroactive molecules, and have an influence on the modulation of enteric nervous system signaling, which in turn impacts the gut-brain axis (Yong et al. 2020).

4.9.1 *Lactobacillus rhamnosus*

Lactobacillus rhamnosus has been a commercially available probiotic for quite some time. It has been reported that *L. rhamnosus* is able to metabolize glutamate and gamma amino-butyric acid (GABA), which are the excitatory and inhibitory neurotransmitters, respectively. *L. rhamnosus* in vitro has reportedly utilized microbial enzymes glutamate decarboxylase to produce GABA and glutaminase to produce glutamate. Studies on mice models have shown that an intervention of *L. rhamnosus* in the diet resulted in alleviation of anxious and depressive behaviors. The alteration was brought about in the expression of mRNA of the receptors of GABA. However, the reduced anxious and depressive behavior of the mice was also dependent on the neural signaling from the intact vagus nerve. GABA produced by the gut microbiota is reported to utilize the H^+/GABA symporter to cross the intestinal barrier *in vitro.* The enteric neurons and the vagus afferents have a large number of GABA receptors and transporters, since it is a chief inhibitory neurotransmitter. These GABA receptors and transporters are possibly utilized by GABA molecules which are produced by microbes, such as *L. rhamnosus* (Bravo et al. 2011; Nielsen et al. 2012; Lin 2013; Yong et al. 2020).

4.9.2 *Lactobacillus casei*

Lactobacillus casei has a potential for maintaining gut health, and is known for its industrial value as a starter culture for fermentation. A dietary intervention of milk containing *L. casei* resulted in a reportedly uplifted mood in individuals. In the analysis of saliva collected from individuals who reported to be stressed, it was found that cortisol levels were high. Consequently, the high cortisol levels resulted in abdominal disturbances and flu symptoms. However, in the clinical trials, an intervention with *L. casei* reportedly alleviated the abdominal and flu symptoms, and reduced the stress frequency by lowering the cortisol levels. Similar to *L. rhamnosus*, *L. casei* was also able to produce GABA, which is involved in inhibition mechanisms. The presence of *L. casei* in a probiotic comprising a mixture of similar species resulted in a reduction in the depression levels of individuals diagnosed with

Table 4.2 The neurotransmitters produced by probiotics and their regulatory functions

Neurotransmitter	Regulatory functions	Probiotics	References
Gamma-aminobutyric acid (GABA)	• Hippocampal neurogenesis • HPA axis regulation • Mood	*L. brevis* *L. rhamnosus* *L. reuteri* *L. paracasei* *L. plantarum* *L. bulgaricus* *L. helveticus* *L. casei*	Barrett et al. (2012), Oleskin et al. (2014)
Serotonin (5-HT)	• Impulsivity • Aggression • Appetite • Circadian rhythm • Learning • HPA axis regulation • Mood	*L. plantarum* *L. helveticus*	Oleskin et al. (2014)
Dopamine (DA)	• Motivation • Concentration • Psychomotor speed • Ability to experience pleasure • Mood	*L. plantarum* *L. helveticus* *L. casei* *L. bulgaricus*	Oleskin et al. (2014)
Norepinephrine (NE)	• Aggression • Cognitive function • Sleep • Sympathetic activity • HPA axis regulation • Mood	*L. helveticus* *L. casei* *L. bulgaricus*	Oleskin et al. (2014)
Glutamate (Glu)	• Gastrointestinal reflexes • Intestinal motility • HPA axis regulation • Mood	*L. rhamnosus* *L. reuteri* *L. plantarum* *L. paracasei* *L. helveticus* *L. casei* *L. bulgaricus*	Oleskin et al. (2014)
Histamine	• Motivation • Learning • Memory • Appetite • Sleep • Sympathetic activity • Mood	*L. plantarum* *L. reuteri*	
Acetylcholine (ACh)	• Cognition • Synaptic plasticity • Analgesia • Sleep • HPA axis regulation • Mood	*L. plantarum*	

clinical depression, and those exhibiting depressive symptoms. The production of microbial GABA by *L. casei* shows that there is a possibility to have similar mechanisms, and the resultant antidepressant effect like *L. rhamnosus* (Kato-Kataoka et al. 2016; Oleskin et al. 2014; Takada et al. 2016; Yong et al. 2020).

4.9.3 *Lactobacillus brevis*

Lactobacillus brevis has a possible overlap in the underlying mechanisms for GABA production, with *L. rhamnosus* and *L. casei*, though reportedly the central GABAergic system remains uninfluenced by its presence. *L. brevis* utilizes the microbial glutamate decarboxylase to produce GABA. An increase in the total GABA content was observed in a quantitative analysis of milk fermented with a starter culture of *L. brevis*. A study on rat models for depression found that *L. brevis* exhibited antidepressive potential, much like fluoxetine, after a dietary intervention of milk fermented with *L. brevis*. Since GABA is the primary inhibitory neurotransmitter, it plays an important role in sleep quality and REM cycle, and hence its imbalance may result in sleep disorders. Sleep disorders such as insomnia are mostly treated by an increased dosage of GABA through diet, or by treatment with pharmacological benzodiazepine which targets GABA receptors. In mice models, the presence of *L. brevis* in the diet has reportedly improved the quality of sleep; therefore it shows great potential to be a therapeutic intervention for treatment of insomnia in people suffering from major depressive disorder (Ko et al. 2013; Miyazaki et al. 2014; Yamatsu et al. 2015; Yong et al. 2020).

4.9.4 *Lactobacillus reuteri*

Lactobacillus reuteri is a probiotic that enhances the immune system. *L. reuteri* is reported to have anti-inflammatory effects on the human body. Hydrogen peroxide is a chief metabolite produced by *L. reuteri*, that inhibits the activity of indoleamine 2,3 dioxygenase through peroxidase-mediated catalyzed reactions. Indoleamine 2,3 dioxygenase is reported to impact levels of kynurenine, and the microbial hydrogen peroxide can possibly cross the intestinal epithelial lining, and reduce the activity of indoleamine 2,3 dioxygenase. Hence, the suppressed activity of this key enzyme lowers the kynurenine levels. *L. reuteri* utilizes microbial histidine decarboxylase to produce histamine from the metabolism of dietary L-histidine. Diacylglycerol kinase is also a microbial enzyme produced by *L. reuteri*, which metabolizes diacylglycerol to phosphatidic acid which plays a role in the microbial histamine anti-inflammatory activity. Both the microbial histamine and the enzyme diacylglycerol kinase produced by *L. reuteri* have been reported to interact with the histamine receptors and enhance the immune response by reducing the inflammatory cytokines in the gastrointestinal tract (Jang et al. 2019; Réus et al. 2015; Yong et al. 2020).

4.9.5 *Lactobacillus plantarum*

Lactobacillus plantarum has been reported to utilize fatty acid synthase II-thioesterase to synthesize butyrate following a butyrogenic pathway mediated by glutamine. Studies on mammals have reported that a dietary intervention of *L. plantarum* has antidepressive effects. It has also been reported that there was an overall increase in levels of butyrate, as *L. plantarum* not only produces butyrate as a metabolite, it also favors the colonization of *Bacteroidetes, Lactobacillus*, and *Roseburia* which are also butyrate-producing bacteria. Supplements containing *L. plantarum* have exhibited the enhancement of hippocampal brain-derived neurotrophic factor. Similarly, analysis of butyrate levels from the cecum showed an elevation after the administration of *L. plantarum* (Botta et al. 2017; Dhaliwal et al. 2018; Yong et al. 2020).

4.9.6 *Lactobacillus gasseri*

Lactobacillus gasseri is known for its anti-inflammatory effect on the immune system. Heat-killed or live form of *L. gasseri*, both have the ability to alter the levels of gut microbiome by favoring the colonization of few microbes over others in the gastrointestinal tract. A study reported that consumption of milk containing probiotics including *L. gasseri* showed an altered gut microflora composition in stressed individuals. *L. gasseri* is reported to produce gassericins which have antibacterial properties against possible pathogens present in the gastrointestinal tract. An introduction of live *L. gasseri* resulted in reduced growth of inflammatory bacterial populations such as *Enterobacteriaceae, Clostridium cluster IV group*, and *Veillonella*, along with altered levels of short-chain fatty acids. The heat-killed form of *L. gasseri* reportedly increased the population of *Dorea longicatena*, while decreasing *Bacteroides vulgatus*. *L. gasseri* when administered in heat-killed form across multiple studies showed that it does not have a unique microbial target, but alters the gut microflora composition towards a favorable anti-inflammatory environment (Nishida et al. 2017; Sawada et al. 2017; Yong et al. 2020).

4.9.7 *Lactobacillus helveticus*

Lactobacillus helveticus is a probiotic that imparts multiple health benefits to the human body. *L. helveticus* has been reported to increase immunity by protection against pathogenic bacterial colonization, along with prevention of diseases of the gastrointestinal tract. In patients diagnosed with clinical depression and symptoms related to depression, a probiotic intervention was introduced which included *L. helveticus* and *Bifidobacterium longum*, and a positive result was observed as depressive symptoms were reduced. In a study involving cognitively impaired rodent models, it was reported that an intervention of *L. helveticus* enhanced cognitive performance and memory. Similarly, *L. helveticus* introduced as a dietary intervention improved cognition abilities such as attention, memory, and learning as reported by studies on animal models and human participants (Liang et al. 2015; Oleskin et al. 2014; Yong et al. 2020).

4.9.8 *Lactobacillus paracasei*

Lactobacillus paracasei belongs to the *Lactobacillus casei* group which also includes *L. rhamnosus* and *L. casei*. The *Lactobacillus casei* group is the most used *Lactobacillus* species, and is used as a potential therapeutic agent for health, along with being of industrial and commercial use. Lactocepin is a protein that is produced by *L. paracasei*, it is a serine protease, and hence is sensitive to high temperatures. However, studies have demonstrated that whether alive or heat-killed, *L. paracasei* exhibits antidepressive and mood uplifting mechanisms. Reportedly while an intervention of heat-killed *L. paracasei* resulted in elevated levels of dopamine in the brain, introduction of live *L. paracasei* increased the levels of serotonin. In a study on mice models, where depression was induced by corticosterone, oral administration of both forms of *L. paracasei* demonstrated potential for antidepressive agents in par with fluoxetine. Similarly, in a study done on healthy individuals in stressful times, a dietary intervention of *L. paracasei* in its heat-killed form kept the mood stable and prevented it from deteriorating (Chunchai et al. 2018; Réus et al. 2015; Wei et al. 2019; Yong et al. 2020).

4.9.9 *Lactobacillus kefiranofaciens*

Lactobacillus kefiranofaciens is reported to have a variety of physiological alterations as a result of its administration. In a study on chronically stressed depressive mice models, the oral administration of *L. kefiranofaciens* showed a marked improvement in their behavior: alleviated depressive and stress-related mood. *L. kefiranofaciens* is reported to affect the Tryptophan/Kynurenine metabolic pathway by increasing the levels of tryptophan in circulation in the body, and hence reducing the Kynurenine/Tryptophan ratio. The presence of *L. kefiranofaciens* also favors the abundance of beneficial gut microbiome such as *Akkermansia*, *Bifidobacteriaceae*, and *Lachnospiraceae*, while reducing the abundance of *Proteobacteria* in the gastrointestinal tract. *L. kefiranofaciens* impacts the immune system by increasing the level of splenic IL-10, and decreasing the levels of splenic IL-6 and IFN-γ levels. The exopolysaccharide is being considered the potential focal point for future researches, as it seems to play a role in the *L. kefirofaciens*' ability to mediate the hypothalamus-pituitary-axis, the immune system, the tryptophan/kynurenine metabolic pathway, and the colonization of gut microbiome (Jeong et al. 2017; Sun et al. 2020; Yong et al. 2020).

4.9.10 *Bifidobacterium breve*

Bifidobacterium breve is a probiotic widely known for its antidepressant potential. There has been no widely reported success in understanding and clarifying the exact mechanism of action of *B. breve*. However, a metabolite produced by *B. breve*, benzoic acid, was reported in a study to play a role in the antidepressive mechanism. *B. breve* introduced to schizophrenic patients showed reduced depressive symptoms, and hence is prescribed as an antidepressive agent. It was also reported that *B. breve* uplifted mood, and enhanced cognition in cognitively impaired elderly individuals (Okubo et al. 2019; Yong et al. 2020).

4.9.11 *Clostridium butyricum*

Clostridium butyricum belongs to *Clostridia* which are a class of bacteria responsible for fermenting free sugars and carbohydrates. *C. butyricum*, as the name suggests, produces a metabolite known as butyrate as a result of carbohydrate fermentation. Similar to *L. paracasei* and *B. infantis*, *C. butyricum* has a potential to upregulate the central BDNF-5HT system through a mechanism involving its metabolite, butyrate. This microbial butyrate-mediated upregulation results in reduced depressive symptoms. Despite being a potential antidepressant agent, not all strains of *C. butyricum* are safe for consumption, as few are reportedly pathogenic and can cause gastrointestinal complications (Anderberg et al. 2016; Cassir et al. 2016; Yong et al. 2020).

4.10 Conclusions

Conclusively, total dairy microflora presents a complete profile of differential bacterial and fungal communities that predominately depends on the chemical composition of milk. Lactic acid bacteria are the most versatile group in all dairy microbiota that display a variety of strains associated with the milk of different animals. The health-promoting advantages of microbes and their additives are overwhelming the few effects of few harmful bacteria in the dairy environment. Despite many benefits of dairy associated bacteria, the emergence of psychobiotics is directing a new avenue towards personalized treatment of many psychological disorders and enhancing the need to explore new microbes with therapeutic potential (Table 4.2).

References

Abushelaibi A, Al-Mahadin S, El-Tarabily K et al (2017) Characterization of potential probiotic lactic acid bacteria isolated from camel milk. LWT Food Sci Technol 79:316–325. https://doi.org/10.1016/j.lwt.2017.01.041

Acurcio LB, Souza MR, Nunes AC et al (2014) Isolation, enumeration, molecular identification and probiotic potential evaluation of lactic acid bacteria isolated from sheep milk. Arq Bras Med Vet e Zootec 66:940–948. https://doi.org/10.1590/1678-41625796

Addis MF, Tedde V, Puggioni GMG et al (2016) Evaluation of milk cathelicidin for detection of bovine mastitis. J Dairy Sci 99:8250–8258. https://doi.org/10.3168/jds.2016-11407

Ahmed T, Kanwal R (2004) Biochemical characteristics of lactic acid producing bacteria and preparation of camel milk cheese by using starter culture. Pak Vet J 24:87–91

Alessandri C, Sforza S, Palazzo P et al (2012) Tolerability of a fully maturated cheese in cow's milk allergic children: biochemical, immunochemical, and clinical aspects. PLoS One 7. https://doi.org/10.1371/journal.pone.0040945

Amara AA, Shibl A (2015) Role of probiotics in health improvement, infection control and disease treatment and management. Saudi Pharm J 23:107–114. https://doi.org/10.1016/j.jsps.2013.07.001

Anderberg RH, Richard JE, Hansson C et al (2016) GLP-1 is both anxiogenic and antidepressant; divergent effects of acute and chronic GLP-1 on emotionality. Psychoneuroendocrinology 65:54–66. https://doi.org/10.1016/j.psyneuen.2015.11.021

Anjum N, Maqsood S, Masud T et al (2014) Lactobacillus acidophilus: characterization of the species and application in food production. Crit Rev Food Sci Nutr 54:1241–1251. https://doi.org/10.1080/10408398.2011.621169

Axelsson L, Rud I, Naterstad K et al (2012) Genome sequence of the naturally plasmid-free Lactobacillus plantarum strain NC8 (CCUG 61730). J Bacteriol 194:2391–2392. https://doi.org/10.1128/JB.00141-12

Aziz T, Khan H, Bakhtair SM, Naurin M (2009) Incidence and relative abundance of lactic acid bacteria in raw milk of buffalo, cow and sheep. J Anim Plant Sci 19:168–173

Balamurugan K, Ortiz A, Said HM (2003) Biotin uptake by human intestinal and liver epithelial cells: role of the SMVT system. Am J Physiol Gastrointest Liver Physiol 285:73–77. https://doi.org/10.1152/ajpgi.00059.2003

Barrett E, Ross RP, O'Toole PW et al (2012) γ-Aminobutyric acid production by culturable bacteria from the human intestine. J Appl Microbiol 113:411–417. https://doi.org/10.1111/j.1365-2672.2012.05344.x

Bassyouni RH, Abdel-all WS, Fadl MG et al (2012) Characterization of lactic acid bacteria isolated from dairy products in Egypt as a Probiotic. Life Sci J 9

Bernardeau M, Guguen M, Vernoux JP (2006) Beneficial lactobacilli in food and feed: long-term use, biodiversity and proposals for specific and realistic safety assessments. FEMS Microbiol Rev 30:487–513. https://doi.org/10.1111/j.1574-6976.2006.00020.x

Bin Masalam MS, Bahieldin A, Alharbi MG et al (2018) Isolation, molecular characterization and probiotic potential of lactic acid bacteria in Saudi raw and fermented milk. Evid Based Comp Altern Med 2018:7970463. https://doi.org/10.1155/2018/7970463

Boix-Amorós A, Collado MC, Mira A (2016) Relationship between milk microbiota, bacterial load, macronutrients, and human cells during lactation. Front Microbiol 7:492. https://doi.org/10.3389/fmicb.2016.00492

Borchers AT, Selmi C, Meyers FJ et al (2009) Probiotics and immunity. J Gastroenterol 44. https://doi.org/10.1007/s00535-008-2296-0

Botta C, Acquadro A, Greppi A et al (2017) Genomic assessment in Lactobacillus plantarum links the butyrogenic pathway with glutamine metabolism. Sci Rep 7:1–13. https://doi.org/10.1038/s41598-017-16186-8

Bravo JA, Forsythe P, Chew MV et al (2011) Ingestion of Lactobacillus strain regulates emotional behavior and central GABA receptor expression in a mouse via the vagus nerve. Proc Natl Acad Sci U S A 108:16050–16055. https://doi.org/10.1073/pnas.1102999108

Cabrera-Rubio R, Collado MC, Laitinen K et al (2012) The human milk microbiome changes over lactation and is shaped by maternal weight and mode of delivery. Am J Clin Nutr 96:544–551. https://doi.org/10.3945/ajcn.112.037382

Cassir N, Benamar S, La Scola B (2016) Clostridium butyricum: from beneficial to a new emerging pathogen. Clin Microbiol Infect 22:37–45. https://doi.org/10.1016/j.cmi.2015.10.014

Catozzi C, Sanchez Bonastre A, Francino O et al (2017) The microbiota of water buffalo milk during mastitis. PLoS One 12:1–20. https://doi.org/10.1371/journal.pone.0184710

Çetin B (2011) Production of probiotic mixed pickles (turşu) and microbiological properties. African J Biotechnol 10:14926–14931. https://doi.org/10.5897/AJB11.2621

Chen H, Hoover DG (2003) Bacteriocins and their food applications. Compr Rev Food Sci Food Saf 2:82–100. https://doi.org/10.1111/j.1541-4337.2003.tb00016.x

Chen L, Zhao H, Zhang C et al (2016) γ-Aminobutyric acid-rich yogurt fermented by Streptococcus salivarius subsp. thermophiles fmb5 apprars to have anti-diabetic effect on streptozotocin-induced diabetic mice. J Funct Foods 20:267–275. https://doi.org/10.1016/j.jff.2015.10.030

Chunchai T, Thunapong W, Yasom S et al (2018) Decreased microglial activation through gut-brain axis by prebiotics, probiotics, or synbiotics effectively restored cognitive function in obese-insulin resistant rats. J Neuroinflammation 15:1–15. https://doi.org/10.1186/s12974-018-1055-2

Cotter PD, Ross RP, Hill C (2013) Bacteriocins-a viable alternative to antibiotics? Nat Rev Microbiol 11:95–105. https://doi.org/10.1038/nrmicro2937

Darilmaz DO, Gumustekin Y (2012) Research on some factors influencing acid and exopolysaccharide produced by dairy propionibacterium strains isolated from traditional home-made Turkish cheeses. J Food Prot 75:918–926. https://doi.org/10.4315/0362-028X.JFP-11-510

De Los DoloresSoto del Rio M, Dalmasso A, Bottero MT (2017) Characterization of bacterial communities of donkey milk by high-throughput sequencing. Int J Food Microbiol 251:67–72. https://doi.org/10.1016/j.ijfoodmicro.2017.03.023

De Vrese M, Stegelmann A, Richter B et al (2001) Probiotics—compensation for lactase insufficiency. Am J Clin Nutr 73:421–429

Deptula P, Chamlagain B, Edelmann M et al (2017) Food-like growth conditions support production of active vitamin B12 by Propionibacterium freudenreichii 2067 without DMBI, the lower ligand base, or cobalt supplementation. Front Microbiol 8:1–11. https://doi.org/10.3389/fmicb.2017.00368

Derakhshani E, Naghizadeh A (2018) Optimization of humic acid removal by adsorption onto bentonite and montmorillonite nanoparticles. J Mol Liq 259:76–81. https://doi.org/10.1016/j.molliq.2018.03.014

Dertli E, Mercan E, Arici M et al (2016) Characterisation of lactic acid bacteria from Turkish sourdough and determination of their exopolysaccharide (EPS) production characteristics. Lwt 71:116–124. https://doi.org/10.1016/j.lwt.2016.03.030

Dhaliwal J, Singh DP, Singh S et al (2018) Lactobacillus plantarum MTCC 9510 supplementation protects from chronic unpredictable and sleep deprivation-induced behaviour, biochemical and selected gut microbial aberrations in mice. J Appl Microbiol 125:257–269. https://doi.org/10.1111/jam.13765

Di Cerbo A, Palmieri B (2015) The market of probiotics. Pak J Pharm Sci 28:2199–2206

Doll EV, Scherer S, Wenning M (2017) Spoilage of microfiltered and pasteurized extended shelf life milk is mainly induced by psychrotolerant spore-forming bacteria that often originate from recontamination. Front Microbiol 8:1–13. https://doi.org/10.3389/fmicb.2017.00135

Drywień M, Frąckiewicz J, Górnicka M et al (2015) Effect of probiotic and storage time of thiamine and riboflavin content in the milk drinks fermented by Lactobacillus casei KNE-1. Rocz Państwowego Zakładu Hig 66:373–377

Edalati E, Saneei B, Alizadeh M et al (2019) Isolation of probiotic bacteria from raw camel's milk and their antagonistic effects on two bacteria causing food poisoning. New Microbes New Infect 27:64–68. https://doi.org/10.1016/j.nmni.2018.11.008

Ercolini D (2013) High-throughput sequencing and metagenomics: moving forward in the culture-independent analysis of food microbial ecology. Appl Environ Microbiol 79:3148–3155. https://doi.org/10.1128/AEM.00256-13

Facciolà A, Riso R, Avventuroso E et al (2017) Campylobacter: from microbiology to prevention. J Prev Med Hyg 58:79–92

Falentin H, Rault L, Nicolas A et al (2016) Bovine teat microbiome analysis revealed reduced alpha diversity and significant changes in taxonomic profiles in quarters with a history of mastitis. Front Microbiol 7:480. https://doi.org/10.3389/fmicb.2016.00480

Fguiri I, Atigui M, Ziadi M et al (2015) Biochemical and molecular identification of lactic acid bacteria isolated from camel milk in Tunisia. Emirates J Food Agric 27:716–720. https://doi.org/10.9755/ejfa.2015.04.114

Fitzgerald RJ, Murray BA (2006) Bioactive peptides and lactic fermentations. Int J Dairy Technol 59:118–125. https://doi.org/10.1111/j.1471-0307.2006.00250.x

Fitzstevens JL, Smith KC, Hagadorn JI et al (2017) Systematic review of the human milk microbiota. Nutr Clin Pract 32:354–364. https://doi.org/10.1177/0884533616670150

Florence ACR, Da Silva RC, Do Espírito Santo AP et al (2009) Increased CLA content in organic milk fermented by bifidobacteria or yoghurt cultures. Dairy Sci Technol 89:541–553. https://doi.org/10.1051/dst/2009030

Fusco V, Chieffi D, Fanelli F et al (2020) Microbial quality and safety of milk and milk products in the 21st century. Compr Rev Food Sci Food Saf 19:2013–2049. https://doi.org/10.1111/1541-4337.12568

Gaggia F, Di Gioia D, Baffoni L, Biavati B (2011) The role of protective and probiotic cultures in food and feed and their impact in food safety. Trends Food Sci Technol 22:S58–S66. https://doi.org/10.1016/j.tifs.2011.03.003

Gao ML, Hou HM, Teng XX et al (2017) Microbial diversity in raw milk and traditional fermented dairy products (Hurood cheese and jueke) from Inner Mongolia. China Genet Mol Res 16:1–13. https://doi.org/10.4238/gmr16019451

Gill SR, Pop M, DeBoy RT et al (2006) Metagenomic analysis of the human distal gut microbiome. Science 312:1355–1359. https://doi.org/10.1126/science.1124234

Grosu-Tudor SS, Zamfir M, Van der Meulen R, De Vuyst L (2013) Isolation of novel homopolysaccharide-producing lactic acid bacteria from Romanian raw milk and fermented dairy products. Eur Food Res Technol 237:609–615. https://doi.org/10.1007/s00217-013-2038-2

Gu Q, Zhang C, Song D et al (2015) Enhancing vitamin B12 content in soy-yogurt by Lactobacillus reuteri. Int J Food Microbiol 206:56–59. https://doi.org/10.1016/j.ijfoodmicro.2015.04.033

Hafeez Z, Cakir-Kiefer C, Roux E et al (2014) Strategies of producing bioactive peptides from milk proteins to functionalize fermented milk products. Food Res Int 63:71–80. https://doi.org/10.1016/j.foodres.2014.06.002

Han SH, Hong KB, Suh HJ (2017) Biotransformation of monosodium glutamate to gamma-aminobutyric acid by isolated strain Lactobacillus brevis L-32 for potentiation of pentobarbital-induced sleep in mice. Food Biotechnol 31:80–93. https://doi.org/10.1080/08905436.2017.1301821

Hannon JA, Wilkinson MG, Delahunty CM et al (2003) Use of autolytic starter systems to accelerate the ripening of Cheddar cheese. Int Dairy J 13:313–323. https://doi.org/10.1016/S0958-6946(02)00178-4

Hao Z, Wang W, Guo R, Liu H (2019) Faecalibacterium prausnitzii (ATCC 27766) has preventive and therapeutic effects on chronic unpredictable mild stress-induced depression-like and anxiety-like behavior in rats. Psychoneuroendocrinology 104:132–142. https://doi.org/10.1016/j.psyneuen.2019.02.025

Hati S, Mandal S, Prajapati JB (2013) Novel starters for value added fermented dairy products. Curr Res Nutr Food Sci 1:83–91. https://doi.org/10.12944/CRNFSJ.1.1.09

Hernández-Ledesma B, García-Nebot MJ, Fernández-Tomé S et al (2014) Dairy protein hydrolysates: peptides for health benefits. Int Dairy J 38:82–100. https://doi.org/10.1016/j.idairyj.2013.11.004

Hertzler SR, Clancy SM (2003) Kefir improves lactose digestion and tolerance in adults with lactose maldigestion. J Am Diet Assoc 103:582–587. https://doi.org/10.1053/jada.2003.50111

Hidalgo-Cantabrana C, López P, Gueimonde M et al (2012) Immune modulation capability of exopolysaccharides synthesised by lactic acid bacteria and bifidobacteria. Probiotics Antimicrob Proteins 4:227–237. https://doi.org/10.1007/s12602-012-9110-2

Hoque MN, Istiaq A, Clement RA et al (2019) Metagenomic deep sequencing reveals association of microbiome signature with functional biases in bovine mastitis. Sci Rep:1–14. https://doi.org/10.1038/s41598-019-49468-4

Hugenholtz J, Sybesma W, Groot MN et al (2002) Metabolic engineering of lactic acid bacteria for the production of nutraceuticals. Lact Acid Bact Genet Metab Appl 82:217–235. https://doi.org/10.1007/978-94-017-2029-8_13

Hunt KM, Foster JA, Forney LJ et al (2011) Characterization of the diversity and temporal stability of bacterial communities in human milk. PLoS One 6:1–8. https://doi.org/10.1371/journal.pone.0021313

Inoue K, Shirai T, Ochiai H et al (2003) Blood-pressure-lowering effect of a novel fermented milk containing γ-aminobutyric acid (GABA) in mild hypertensives. Eur J Clin Nutr 57:490–495. https://doi.org/10.1038/sj.ejcn.1601555

Jang H, Lee K, Kim D (2019) The prebventive and curative effects of lactobacillus reuteri NK33 and Bifidobacterium adolescentis NK98 on immobilization stress-induced anxiety/depression and colitis in mice. Nutrients 11:819. https://doi.org/10.3390/nu11040819

Jayarao BM, Donaldson SC, Straley BA et al (2006) A survey of foodborne pathogens in bulk tank milk and raw milk consumption among farm families in Pennsylvania. J Dairy Sci 89:2451–2458. https://doi.org/10.3168/jds.S0022-0302(06)72318-9

Jayarao BM, Pillai SR, Wolfgang DR, et.al. (2001) Herd level information and bulk tank milk analysis: tools for improving milk quality and udder health. Bov Pract 35:23–35

Jeong D, Kim DH, Kang IB et al (2017) Characterization and antibacterial activity of a novel exopolysaccharide produced by Lactobacillus kefiranofaciens DN1 isolated from kefir. Food Control 78:436–442. https://doi.org/10.1016/j.foodcont.2017.02.033

Jiang J, Shi B, Zhu D et al (2012) Characterization of a novel bacteriocin produced by Lactobacillus sakei LSJ618 isolated from traditional Chinese fermented radish. Food Control 23:338–344. https://doi.org/10.1016/j.foodcont.2011.07.027

Jiménez E, De Andrés J, Manrique M et al (2015) Metagenomic analysis of milk of healthy and mastitis-suffering women. J Hum Lact. https://doi.org/10.1177/0890334415585078

Jost T, Lacroix C (2013) Assessment of bacterial diversity in breast milk using culture-dependent and culture-independent approaches. Br J Nutr 110(7):1253–1262. https://doi.org/10.1017/S0007114513000597

Kable ME, Srisengfa Y, Xue Z et al (2019) Viable and total bacterial populations undergo equipment and time-dependent shifts during milk processing. Appl Environ Microbiol 85:1–14

Karl PJ, Hatch AM, Arcidiacono SM et al (2018) Effects of psychological, environmental and physical stressors on the gut microbiota. Front Microbiol 9:1–32. https://doi.org/10.3389/fmicb.2018.02013

Kato-Kataoka A, Nishida K, Takada M et al (2016) Fermented milk containing Lactobacillus casei strain Shirota preserves the diversity of the gut microbiota and relieves abdominal dysfunction in healthy medical students exposed to academic stress. Appl Environ Microbiol 82:3649–3658. https://doi.org/10.1128/AEM.04134-15

Kechagia M, Basoulis D, Konstantopoulou S et al (2013) Health benefits of probiotics: a review. ISRN Nutr 2013

Khan H, Flint S, Yu PL (2010) Enterocins in food preservation. Int J Food Microbiol 141:1–10. https://doi.org/10.1016/j.ijfoodmicro.2010.03.005

Khaskheli M, Arain MA, Chaudhry S et al (2005) Physicochemical quality of camel milk. J Agric Soc Sci 2:164–166

Kim SG, Kim EH, Lafferty CJ, Dubovi E (2005) Coxiella burnetti in bulk tank milk samples, United States. Emerg Infect Dis 11:619–621. https://doi.org/10.3201/eid1104.041036

Ko CY, Lin HTV, Tsai GJ (2013) Gamma-aminobutyric acid production in black soybean milk by Lactobacillus brevis FPA 3709 and the antidepressant effect of the fermented product on a forced swimming rat model. Process Biochem 48:559–568. https://doi.org/10.1016/j.procbio.2013.02.021

Kongo JM, Gomes ANAP, Malcata FX (2008) Monitoring and identification of bacteria associated with safety concerns in the manufacture of Sao Jorge, a Portuguese Traditional Cheese from Raw Cow's Milk. J Food Prot 71:986–992

Laiño JE, Juarez del Valle M, Savoy de Giori G, LeBlanc JGJ (2014) Applicability of a Lactobacillus amylovorus strain as co-culture for natural folate bio-enrichment of fermented milk. Int J Food Microbiol 191:10–16. https://doi.org/10.1016/j.ijfoodmicro.2014.08.031

Lan Z, Bastos M, Menzies D (2016) Treatment of human disease due to Mycobacterium bovis: a systematic review. Eur Respir J 48:1500–1503. https://doi.org/10.1183/13993003.00629-2016

Leblanc JG, Laiño JE, del Valle MJ et al (2011) B-Group vitamin production by lactic acid bacteria—current knowledge and potential applications. J Appl Microbiol 111:1297–1309. https://doi.org/10.1111/j.1365-2672.2011.05157.x

Li H, Cao Y (2010) Lactic acid bacterial cell factories for gamma-aminobutyric acid. Amino Acids 39:1107–1116. https://doi.org/10.1007/s00726-010-0582-7

Li L, Renye JA, Feng L et al (2016) Characterization of the indigenous microflora in raw and pasteurized buffalo milk during storage at refrigeration temperature by high-throughput sequencing. J Dairy Sci 99:7016–7024. https://doi.org/10.3168/jds.2016-11041

Li N, Wang Y, You C et al (2018) Variation in raw milk microbiota throughout 12 months and the impact of weather conditions. Sci Rep 8:1–10. https://doi.org/10.1038/s41598-018-20862-8

Liang S, Wang T, Hu X et al (2015) Administration of Lactobacillus helveticus NS8 improves behavioral, cognitive, and biochemical aberrations caused by chronic restraint stress. Neuroscience 310:561–577. https://doi.org/10.1016/j.neuroscience.2015.09.033

Lima SF, Lucas M, Bicalho DS, Bicalho RC (2018) Evaluation of milk sample fractions for characterization of milk microbiota from healthy and clinical mastitis cows. PLoS One 13

Lin Q (2013) Submerged fermentation of Lactobacillus rhamnosus YS9 for γ-aminobutyric acid (GABA) production. Brazilian J Microbiol 44:183–187. https://doi.org/10.1590/S1517-83822013000100028

Linares DM, Ross P, Stanton C (2016) Beneficial Microbes: the pharmacy in the gut. Bioengineered 7:11–20. https://doi.org/10.1080/21655979.2015.1126015

Lourens-Hattingh A, Viljoen BC (2001) Growth and survival of a probiotic yeast in dairy products. Food Res Int 34:791–796. https://doi.org/10.1016/S0963-9969(01)00085-0

Makarova K, Slesarev A, Wolf Y et al (2006) Comparative genomics of the lactic acid bacteria. PNAS 103

Makino S, Sato A, Goto A et al (2016) Enhanced natural killer cell activation by exopolysaccharides derived from yogurt fermented with Lactobacillus delbrueckii ssp. bulgaricus OLL1073R-1. J Dairy Sci 99:915–923. https://doi.org/10.3168/jds.2015-10376

Mayer EA, Knight R, Mazmanian SK et al (2014) Gut microbes and the brain: paradigm shift in neuroscience. J Neurosci 34:15490–15496. https://doi.org/10.1523/JNEUROSCI.3299-14.2014

Mcinnis EA, Kalanetra KM, Mills DA, Maga EA (2015) Analysis of raw goat milk microbiota: impact of stage of lactation and lysozyme on microbial diversity. Food Microbiol 46:121–131. https://doi.org/10.1016/j.fm.2014.07.021

Medina RB, Oliszewski R, Abeijón Mukdsi MC et al (2011) Sheep and goat's dairy products from South America: microbiota and its metabolic activity. Small Rumin Res 101:84–91. https://doi.org/10.1016/j.smallrumres.2011.09.028

Medrano M, Pérez PF, Abraham AG (2008) Kefiran antagonizes cytopathic effects of Bacillus cereus extracellular factors. Int J Food Microbiol 122:1–7. https://doi.org/10.1016/j.ijfoodmicro.2007.11.046

Meijerink M, van Hemert S, Taverne N et al (2010) Identification of genetic loci in Lactobacillus plantarum that modulate the immune response of dendritic cells using comparative genome hybridization. PLoS One 5:e10632. https://doi.org/10.1371/journal.pone.0010632

Melini F, Melini V, Luziatelli F, Ruzzi M (2017) Raw and heat-treated milk: from public health risks to nutritional quality. Beverages 3:54. https://doi.org/10.3390/beverages3040054

Mitra S, Chakrabartty PK, Biswas SR (2010) Potential production and preservation of dahi by Lactococcus lactis W8, a nisin-producing strain. LWT Food Sci Technol 43:337–342. https://doi.org/10.1016/j.lwt.2009.08.013

Mittu B, Girdhar Y (2015) Role of lactic acid bacteria isolated from goat milk in cancer prevention. Autoimmune Infect Dis 1(2). https://doi.org/10.16966/2470-1025.108

Miyazaki K, Itoh N, Yamamoto S et al (2014) Dietary heat-killed Lactobacillus brevis SBC8803 promotes voluntary wheel-running and affects sleep rhythms in mice. Life Sci 111:47–52. https://doi.org/10.1016/j.lfs.2014.07.009

Mohania D, Kansal VK, Kruzliak P, Kumari A (2014) Probiotic dahi containing lactobacillus acidophilus and bifidobacterium bifidum modulates the formation of aberrant crypt foci, Mucin-depleted foci, and cell proliferation on 1,2-dimethylhydrazine-induced colorectal carcinogenesis in wistar rats. Rejuvenation Res 17:325–333. https://doi.org/10.1089/rej.2013.1537

Moslemi M, Mazaheri Nezhad Fard R, Hosseini SM et al (2016) Incorporation of propionibacteria in fermented milks as a probiotic. Crit Rev Food Sci Nutr 56:1290–1312. https://doi.org/10.1080/10408398.2013.766584

Murphy K, Curley D, Callaghan TFO et al (2017) The composition of human milk and infant faecal microbiota over the first three months of life: a pilot study. Nat Publ Gr:1–10. https://doi.org/10.1038/srep40597

Nagpal R, Behare P, Rana R et al (2011) Bioactive peptides derived from milk proteins and their health beneficial potentials: an update. Food Funct 2:18–27. https://doi.org/10.1039/c0fo00016g

Nes IF, Yoon S-S, Diep DB (2007) Ribosomally synthesised antimicrobial peptides (bacteriocins) in lactic acid bacteria: a review. Food Sci Biotechnol 16:675–690

Nielsen CU, Carstensen M, Brodin B (2012) Carrier-mediated γ-aminobutyric acid transport across the basolateral membrane of human intestinal Caco-2 cell monolayers. Eur J Pharm Biopharm 81:458–462. https://doi.org/10.1016/j.ejpb.2012.03.007

Nishida K, Sawada D, Kuwano Y et al (2017) Daily administration of paraprobiotic Lactobacillus gasseri CP2305 ameliorates chronic stress-associated symptoms in Japanese medical students. J Funct Foods 36:112–121. https://doi.org/10.1016/j.jff.2017.06.031

Nongonierma AB, FitzGerald RJ (2015) The scientific evidence for the role of milk protein-derived bioactive peptides in humans: a review. J Funct Foods 17:640–656. https://doi.org/10.1016/j.jff.2015.06.021

Ogier JC, Serror P (2008) Safety assessment of dairy microorganisms: the Enterococcus genus. Int J Food Microbiol 126:291–301. https://doi.org/10.1016/j.ijfoodmicro.2007.08.017

Oikonomou G, Bicalho ML, Meira E et al (2014) Microbiota of cow's milk; distinguishing healthy, sub-clinically and clinically diseased quarters. PLoS One 9:e85904. https://doi.org/10.1371/journal.pone.0085904

Okubo R, Koga M, Katsumata N et al (2019) Effect of bifidobacterium breve A-1 on anxiety and depressive symptoms in schizophrenia: a proof-of-concept study. Elsevier B:V

Oleskin AV, Zhilenkova OG, Shenderov BA et al (2014) Lactic-acid bacteria supplement fermented dairy products with human behavior-modifying neuroactive compounds. J Pharm Nutr Sci 4:199–206. https://doi.org/10.6000/1927-5951.2014.04.03.5

Orel R, Trop TK (2014) Intestinal microbiota, probiotics and prebiotics in inflammatory bowel disease. World J Gastroenterol 20:11505–11524. https://doi.org/10.3748/wjg.v20.i33.11505

Parente E, Ricciardi A, Zotta T (2020) The microbiota of dairy milk: a review. Int Dairy J 107:104714. https://doi.org/10.1016/j.idairyj.2020.104714

Park KB, Oh SH (2007) Production of yogurt with enhanced levels of gamma-aminobutyric acid and valuable nutrients using lactic acid bacteria and germinated soybean extract. Bioresour Technol 98:1675–1679. https://doi.org/10.1016/j.biortech.2006.06.006

Pärnänen K, Karkman A, Hultman J et al (2018) Maternal gut and breast milk microbiota affect infant gut antibiotic resistome and mobile genetic elements. Nat Commun 9:1–12. https://doi.org/10.1038/s41467-018-06393-w

Patel A, Shah N, Prajapati JB (2013) Biosynthesis of vitamins and enzymes in fermented foods by lactic acid bacteria and related genera—a promising approach. Croat. J Food Sci Technol 5:85–91

Patil A, Disouza J, Pawar S (2019) Shelf life stability of encapsulated lactic acid bacteria isolated from sheep milk thrived in different milk as natural media. Small Rumin Res 170:19–25. https://doi.org/10.1016/j.smallrumres.2018.09.014

Peden DB (2000) Development of atopy and asthma: candidate environmental influences and important periods of exposure. Environ Health Perspect 108:475–482. https://doi.org/10.1289/ehp.00108s3475

Perin LM, Nero LA (2014) Antagonistic lactic acid bacteria isolated from goat milk and identification of a novel nisin variant Lactococcus lactis. BMC Microbiol 14:1–9. https://doi.org/10.1186/1471-2180-14-36

Pisano MB, Deplano M, Fadda ME, Cosentino S (2019) Microbiota of Sardinian Goat's milk and preliminary characterization of prevalent LAB species for starter or adjunct cultures development. Biomed Res Int 2019:6131404. https://doi.org/10.1155/2019/6131404

Pouliot-Mathieu K, Gardner-Fortier C, Lemieux S et al (2013) Effect of cheese containing gamma-aminobutyric acid-producing lactic acid bacteria on blood pressure in men. PharmaNutrition 1:141–148. https://doi.org/10.1016/j.phanu.2013.06.003

Prasanna PHP, Grandison AS, Charalampopoulos D (2013) Microbiological, chemical and rheological properties of low fat set yoghurt produced with exopolysaccharide (EPS) producing Bifidobacterium strains. Food Res Int 51:15–22. https://doi.org/10.1016/j.foodres.2012.11.016

Pritchard SR, Phillips M, Kailasapathy K (2010) Identification of bioactive peptides in commercial Cheddar cheese. Food Res Int 43:1545–1548. https://doi.org/10.1016/j.foodres.2010.03.007

Qian B, Xing M, Cui L et al (2011) Antioxidant, antihypertensive, and immunomodulatory activities of peptide fractions from fermented skim milk with Lactobacillus delbrueckii ssp. bulgaricus LB340. J Dairy Res 78:72–79. https://doi.org/10.1017/S0022029910000889

Quigley L, O'Sullivan O, Stanton C et al (2013) The complex microbiota of raw milk. FEMS Microbiol Rev 37:664–698. https://doi.org/10.1111/1574-6976.12030

Quintieri L, Pistillo BR, Caputo L et al (2013) Bovine lactoferrin and lactoferricin on plasma-deposited coating against spoilage Pseudomonas spp. Innov Food Sci Emerg Technol 20:215–222. https://doi.org/10.1016/j.ifset.2013.04.013

Radoshevich L, Cossart P (2018) Listeria monocytogenes: towards a complete picture of its physiology and pathogenesis. Nat Rev Microbiol 16:32–46. https://doi.org/10.1038/nrmicro.2017.126

Rahmeh R, Akbar A, Kishk M et al (2019) Distribution and antimicrobial activity of lactic acid bacteria from raw camel milk. New Microbes New Infect 30:100560. https://doi.org/10.1016/j.nmni.2019.100560

Rattanachaikunsopon P, Phumkhachorn P (2010) Lactic acid bacteria: their antimicrobial compounds and their uses in food production. Ann Biol Res 1:218–228

Réus GZ, Jansen K, Titus S et al (2015) Kynurenine pathway dysfunction in the pathophysiology and treatment of depression: evidences from animal and human studies. J Psychiatr Res 68:316–328. https://doi.org/10.1016/j.jpsychires.2015.05.007

Russo P, Capozzi V, Arena MP et al (2014) Riboflavin-overproducing strains of Lactobacillus fermentum for riboflavin-enriched bread. Appl Microbiol Biotechnol 98:3691–3700. https://doi.org/10.1007/s00253-013-5484-7

Ryan PM, Guinane CM, London LEE et al (2015) Genome sequence of the heteropolysaccharide-producing strain Lactobacillus mucosae DPC 6426. Genome Announc 3:2014–2015. https://doi.org/10.1128/genomeA.01350-14

Sabina Y, Rahman A, Ray RC, Montet D (2011) Yersinia enterocolitica: mode of Transmission, Molecular Insights of Virulence, and Pathogenesis of Infection. J Pathog 2011:1–10. https://doi.org/10.4061/2011/429069

Salazar N, Gueimonde M, de los Reyes-Gavilán CG, Ruas-Madiedo P (2016) Exopolysaccharides produced by lactic acid bacteria and bifidobacteria as fermentable substrates by the intestinal microbiota. Crit Rev Food Sci Nutr 56:1440–1453. https://doi.org/10.1080/10408398.2013.770728

Savadogo A, Ouattara CAT, Bassole IHN, Traore SA (2006) Bacteriocins and lactic acid bacteria—a minireview. African. J Biotechnol 5:678–684. https://doi.org/10.5897/AJB05.388

Sawada D, Kawai T, Nishida K et al (2017) Daily intake of Lactobacillus gasseri CP2305 improves mental, physical, and sleep quality among Japanese medical students enrolled in a cadaver dissection course. J Funct Foods 31:188–197. https://doi.org/10.1016/j.jff.2017.01.042

Shah NP (2007) Functional cultures and health benefits. Int Dairy J 17:1262–1277. https://doi.org/10.1016/j.idairyj.2007.01.014

Shah NP (2015) Functional properties of fermented milks. In: Health benefits of fermented foods and beverages, pp 261–274

Shan Y, Man CX, Han X et al (2015) Evaluation of improved γ-aminobutyric acid production in yogurt using Lactobacillus plantarum NDC75017. J Dairy Sci 98:2138–2149. https://doi.org/10.3168/jds.2014-8698

Shaw W (2017) Elevated urinary glyphosate and clostridia metabolites with altered dopamine metabolism in triplets with autistic spectrum disorder or suspected seizure disorder: a case study. Integr Med 16:50–57

Shori AB (2012) Comparative study of chemical composition, isolation and identification of micro-flora in traditional fermented camel milk products: Gariss, Suusac, and Shubat. J Saudi Soc Agric Sci 11:79–88. https://doi.org/10.1016/j.jssas.2011.12.001

Skeie SB, Håland M, Thorsen IM et al (2019) Bulk tank raw milk microbiota differs within and between farms: a moving goalpost challenging quality control. J Dairy Sci 102:1959–1971. https://doi.org/10.3168/jds.2017-14083

Slattery L, O'Callaghan J, Fitzgerald GF et al (2010) Invited review: Lactobacillus helveticus-A thermophilic dairy starter related to gut bacteria. J Dairy Sci 93:4435–4454. https://doi.org/10.3168/jds.2010-3327

Sosa-Castañeda J, Hernández-Mendoza A, Astiazarán-García H et al (2015) Screening of Lactobacillus strains for their ability to produce conjugated linoleic acid in milk and to adhere to the intestinal tract. J Dairy Sci 98:6651–6659. https://doi.org/10.3168/jds.2014-8515

Sun C, Wu X, Chen X et al (2020) Production and characterization of okara dietary fiber produced by fermentation with Monascus anka. Food Chem 316:126243. https://doi.org/10.1016/j.foodchem.2020.126243

Szajewska H, Skórka A, Ruszczyński M, Gieruszczak-BiaŁek D (2007) Meta-analysis: Lactobacillus GG for treating acute diarrhoea in children. Aliment Pharmacol Ther 25:871–881. https://doi.org/10.1111/j.1365-2036.2007.03282.x

Tajabadi N, Baradaran A, Ebrahimpour A et al (2015) Overexpression and optimization of glutamate decarboxylase in Lactobacillus plantarum Taj-Apis362 for high gamma-aminobutyric acid production. Microb Biotechnol 8:623–632. https://doi.org/10.1111/1751-7915.12254

Takada M, Nishida K, Kataoka-Kato A et al (2016) Probiotic Lactobacillus casei strain Shirota relieves stress-associated symptoms by modulating the gut–brain interaction in human and animal models. Neurogastroenterol Motil 28:1027–1036. https://doi.org/10.1111/nmo.12804

Tamang JP, Shin DH, Jung SJ, Chae SW (2016a) Functional properties of microorganisms in fermented foods. Front Microbiol 7:1–13. https://doi.org/10.3389/fmicb.2016.00578

Tamang JP, Tamang B, Schillinger U et al (2009) Functional properties of lactic acid bacteria isolated from ethnic fermented vegetables of the Himalayas. Int J Food Microbiol 135:28–33. https://doi.org/10.1016/j.ijfoodmicro.2009.07.016

Tamang JP, Watanabe K, Holzapfel WH (2016b) Review: diversity of microorganisms in global fermented foods and beverages. Front Microbiol 7:377. https://doi.org/10.3389/fmicb.2016.00377

Tan H, Zhai Q, Chen W (2019) Investigations of Bacteroides spp. towards next-generation probiotics. Food Res Int 116:637–644. https://doi.org/10.1016/j.foodres.2018.08.088
Treven M, Koenig X, Assadpour E et al (2015) The anticonvulsant retigabine is a subtype selective modulator of GABAA receptors. Epilepsia 56:647–657. https://doi.org/10.1111/epi.12950
Troy EB, Kasper DL (2010) Beneficial effects of Bacteroides fragilis polysaccharides on the immune system. Front Biosci 15:25–34. https://doi.org/10.2741/3603
Urbaniak C, Angelini M, Gloor GB, Reid G (2016) Human milk microbiota profiles in relation to birthing method, gestation and infant gender. Microbiome 4:1–9. https://doi.org/10.1186/s40168-015-0145-y
Van Kessel JS, Karns JS, Gorski L et al (2004) Prevalence of salmonellae, Listeria monocytogenes, and fecal coliforms in bulk tank milk on US dairies. J Dairy Sci 87:2822–2830. https://doi.org/10.3168/jds.S0022-0302(04)73410-4
Van Nieuwenhove CP, Oliszewski R, González SN, Pérez Chaia AB (2007) Conjugated linoleic acid conversion by dairy bacteria cultured in MRS broth and buffalo milk. Lett Appl Microbiol 44:467–474. https://doi.org/10.1111/j.1472-765X.2007.02135.x
Van Wyk J, Witthuhn RC, Britz TJ (2011) Optimisation of vitamin B12 and folate production by Propionibacterium freudenreichii strains in kefir. Int Dairy J 21:69–74. https://doi.org/10.1016/j.idairyj.2010.09.004
Verdier-Metz I, Gagne G, Bornes S et al (2012) Cow teat skin, a potential source of diverse microbial populations for cheese production. Appl Environ Microbiol 78:326–333. https://doi.org/10.1128/AEM.06229-11
Verna EC, Lucak S (2010) Use of probiotics in gastrointestinal disorders: what to recommend? Ther Adv Gastroenterol 3:307–319. https://doi.org/10.1177/1756283X10373814
Vinderola CG, Bailo N, Reinheimer JA (2000) Survival of probiotic microflora in Argentinian yoghurts during refrigerated storage. Food Res Int 33:97–102. https://doi.org/10.1016/S0963-9969(00)00011-9
Walther B, Philip Karl J, Booth SL, Boyaval P (2013) Menaquinones, bacteria, and the food supply: the relevance of dairy and fermented food products to vitamin K requirements. Adv Nutr 4:463–473. https://doi.org/10.3945/an.113.003855
Wei CL, Wang S, Yen JT et al (2019) Antidepressant-like activities of live and heat-killed Lactobacillus paracasei PS23 in chronic corticosterone-treated mice and possible mechanisms. Brain Res 1711:202–213. https://doi.org/10.1016/j.brainres.2019.01.025
Whittington R, Donat K, Weber MF et al (2019) Control of paratuberculosis: who, why and how. A review of 48 countries. BMC Vet Res 15:1–29. https://doi.org/10.1186/s12917-019-1943-4
Yadav D, Papachristou GI, Whitcomb DC (2007) Alcohol-associated pancreatitis. Gastroenterol Clin N Am 36:219–238. https://doi.org/10.1016/j.gtc.2007.03.005
Yamatsu A, Yamashita Y, Maru I et al (2015) The improvement of sleep by oral intake of GABA and apocynum venetum leaf extract. J Nutr Sci Vitaminol (Tokyo) 61:182–187. https://doi.org/10.3177/jnsv.61.182
Yang B, Chen H, Stanton C et al (2015) Review of the roles of conjugated linoleic acid in health and disease. J Funct Foods 15:314–325. https://doi.org/10.1016/j.jff.2015.03.050
Yong SJ, Tong T, Chew J, Lim WL (2020) Antidepressive mechanisms of probiotics and their therapeutic potential. Front Neurosci 13:1361. https://doi.org/10.3389/fnins.2019.01361
Zhang D, Palmer J, Teh KH et al (2019) 16S rDNA high-throughput sequencing and MALDI-TOF MS are complementary when studying psychrotrophic bacterial diversity of raw cows' milk. Int Dairy J 97:86–91. https://doi.org/10.1016/j.idairyj.2019.06.001
Zhong Z, Hou Q, Kwok L et al (2016) Bacterial microbiota compositions of naturally fermented milk are shaped by both geographic origin and sample type. J Dairy Sci 99:7832–7841. https://doi.org/10.3168/jds.2015-10825
Zivkovic AM, Lewis ZT, German JB, Mills DA (2013) Establishment of a Milk-Oriented Microbiota (MOM) in early life: how babies meet their MOMs. Funct Foods Rev 5:3–12. https://doi.org/10.2310/6180.2009.00035

Remarkable Metabolic Versatility of the Commensal Bacteria *Eubacterium hallii* and *Intestinimonas butyriciproducens*: Potential Next-Generation Therapeutic Microbes

5

Jos F. M. L. Seegers, Thi Phuong Nam Bui, and Willem M. de Vos

Abstract

Our gastrointestinal tract is host to a wide variety of bacteria, together referred to as the microbiota. These bacteria influence our health and well-being through many different mechanisms. Most of these effects are the result of metabolites that are being produced by these bacteria or through triggering the expression of metabolites by the host. In this chapter we will highlight two bacterial species that have remarkable metabolic features that make them prime candidates for the development as next-generation probiotics. The first is *Eubacterium hallii*, a bacterium that is capable of producing two important short-chain fatty acids (SCFAs), propionate and butyrate. The other bacterium is *Intestinimonas butyriciproducens*, a bacterium that is capable of producing butyrate from not only sugars but also lysine and even glycated lysine. Both species also can produce pseudovitamin B12. We will discuss conditions that can result in the production of specific metabolites and the implications this can have on human health.

J. F. M. L. Seegers (✉)
Caelus Health, Zegveld, The Netherlands
e-mail: j.seegers@caelushealth.com

T. P. N. Bui
Caelus Health, Zegveld, The Netherlands

Laboratory of Microbiology, Wageningen University & Research, Wageningen, The Netherlands

W. M. de Vos
Caelus Health, Zegveld, The Netherlands

Laboratory of Microbiology, Wageningen University & Research, Wageningen, The Netherlands

Human Microbiome Research Program, Faculty of Medicine, University of Helsinki, Helsinki, Finland

N. Mojgani, M. Dadar (eds.), *Probiotic Bacteria and Postbiotic Metabolites: Role in Animal and Human Health*, Microorganisms for Sustainability 2,
https://doi.org/10.1007/978-981-16-0223-8_5

5.1 Introduction

The human body is host to trillions of microorganisms, mostly bacteria. These microorganisms can be found on every surface that is in contact with the outer world such as our skin, our digestive tract, the lungs, and even the eye (Willcox 2013). By far the largest bacterial community is found in our gut. The importance of these bacteria in relation to our health has long been overlooked, but even Hippocrates, who by many is seen as the founding father of modern medicine, apparently already claimed "all disease begins in the gut."

The first person in more recent times who recognized that bacteria might play an important role in health was the Russian biologist Elie Metchnikoff. While his theory on how to improve health by manipulating the intestinal microbiota with host-friendly bacteria in yoghurt (probiotics) caught the attention for a while (Metchnikoff 1907), in the coming years it would drift out of attention, largely due also to the difficulty of growing gut bacteria in the lab.

Someone whose contribution should not go unmentioned is Carl Woese, an American microbiologist who, together with George Fox, defined the Archaea as a separate kingdom from bacteria, based on 16S ribosomal RNA sequence (Woese and Fox 1977). With the emergence of modern technologies such as polymerase chain reaction and next-generation DNA sequencing came the possibility to quickly distinguish different bacteria on the basis of 16S rRNA sequence and detect and identify bacteria that could not be detected before by classical growth experiments. Not only the identification of bacteria has made considerable progress, also cultivation conditions have evolved and as a result over 1000 different, mainly anaerobic species from our gut can now be cultivated in vitro (Rajilić-Stojanović and de Vos 2014).

In the wake of research that followed, it became increasingly clear that the bacterial populations that surround us produce many substances essential to our bodies and thus play a major role in health and disease. From that came the realization that there is an intricate relationship between the food that we eat, the microbes in our gut, the metabolites they produce, and how they affect our body (Fig. 5.1, Holmes et al. 2012; Patterson et al. 2014). The main metabolites produced by the bacteria in our gut are the short-chain fatty acids acetate, propionate, and butyrate (Den Besten et al. 2013). Acetate, the most abundant SCFA from our gut, is a primary carbon source for other gut bacteria, who convert this to either propionate or butyrate (Bui et al. 2014; Moens et al. 2017; Schwab et al. 2017). Of all SCFA acetate is the one that is systemically most available (Boets et al. 2017). Propionate plays an important role in glucose metabolism through the liver and is thought to lower lipogenesis and serum cholesterol levels, although most studies for this have been conducted in rodents and would need confirmation for humans (Lin et al. 1995). Butyrate is taken up directly by the colonocytes that line our gut for which it serves as a direct source of energy. As a result, butyrate directly contributes to a healthy gut. In addition, these SCFA have an important role as signaling molecules, thereby affecting many factors such as satiety, secretion of hormones, and glucose

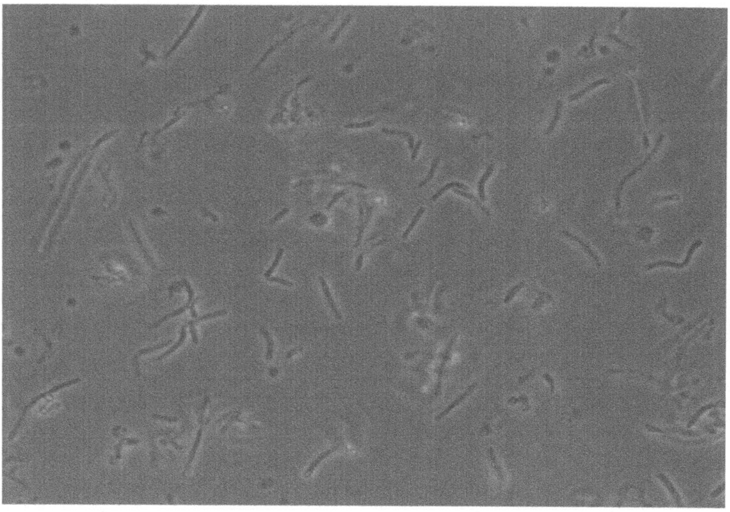

Fig. 5.1 Microscopic picture of *E. hallii*. Swellings, appearing in the form of vacuole-like structures under certain growth conditions, can clearly be observed

metabolism (Den Besten et al. 2013; Bolognini et al. 2016; Morrison and Preston 2016).

Other important metabolites that can be produced by gut bacteria are vitamins (notably K and B) and several neuroactive molecules that can either directly serve as or induce the expression of neurotransmitters and hormones (Oleskin and Shenderov 2019).

Following the elucidation of the human genome, analysis of the genetic composition of the microbiota, the microbiome, revealed its vast genetic potential. Whereas the human genome encodes approximately 23,000 genes, the gut microbiome encodes over ten million genes (Li et al. 2014). This not only reflects genetic power, but also a high level of flexibility. The ten million genes are divided in over 1000 different bacterial species, countless viruses and fungi and yeasts. Each individual will carry an estimate amount of 400 different bacteria and the composition of this is highly diverse between individuals (Qin et al. 2010). Remarkably, the genetic composition is less variable, allowing the microbiota to execute all of its functions, regardless of its species composition.

Within this high variety there are a number of key species that are shared by all individuals. These bacteria perform functions that serve our body, but also contribute to the stability of the highly complex ecosystem which they are part of. Here we describe one of those key species, *Eubacterium hallii* that has recently been renamed *to Anaerobutyricum* spp. but for simplicity we keep the original name. *E. hallii* has a remarkable metabolic versatility and produces metabolites that do not only serve its host but are also important for maintenance of a stable microbiota. We also include another species, *Intestinimonas butyriciproducens* that is highly specialized and adapted to the nutrients that are provided by its hosts.

5.2 *Eubacterium hallii*

5.2.1 General Description

Eubacterium hallii is a member of the *Lachnospiraceae* and based on 16S rRNA sequence analysis belongs to Clostridium cluster XIVa (Collins et al. 1994; Harmsen et al. 2002), a cluster that comprises many of the most prominent butyrate-producing bacteria in the gut. It has been detected as common commensal in human microbiota and recognized as a core species due to its frequent presence in all humans (Shetty et al. 2017). It is a common gut bacterium that can be detected in at least 63–81% of the population (Engels et al. 2016). With a diameter of 0.8–2.4 μm and 4.7 to more than 25 μm in length (De Vos et al. 2009) these bacteria are among the larger bacteria that can be found in our gut. Subterminal and terminal swellings can be observed, but cultures do not survive heating at 80 °C for 10 min (Fig. 5.1).

Because of its high metabolic flexibility *E. hallii* can easily adapt to the different conditions that are found in the gastrointestinal tract. As a result, this bacterium can be found in the small intestine as well as in the colon. Its metabolites can be used by other bacteria and as such *E. hallii* takes a central role in the microbiota, defining it as a key species. Also, it produces metabolites that are essential to human health. The role of butyrate and propionate in this respect are well documented. Moreover, *E. hallii* plays an important role in insulin resistance. In a double blinded study that was performed in The Netherlands, where volunteers that were diagnosed with metabolic syndrome received either fecal matter from lean donors or fecal matter of their own through a nasal duodenal tube, an improvement in peripheral insulin sensitivity could be observed in the former group. This improved insulin sensitivity was accompanied by an increase in the abundance of *E. hallii* that was observed in small intestinal biopsies (Vrieze et al. 2012). In subsequent animal trials, where mice were fed live *E. hallii*, a similar improvement of insulin sensitivity could be observed (Udayappan et al. 2016). Studies on the mode of action to explain this improvement are ongoing.

The omnipresence of this bacterium and its involvement with insulin resistance shows the importance of these bacteria. Through its metabolic diversity it is able to interact both with other members of the microbiota and its host. In the next section, the remarkable versatility of this bacterium is further explained.

5.3 Metabolic Diversity

E. hallii is a metabolically versatile species in the gastrointestinal tract. As *E. hallii* is not able to grow on complex polysaccharides this bacterium is mainly involved in secondary fermentation of simple compounds that arise as metabolites from other gut bacteria (Duncan et al. 2004; Schwab et al. 2017). *E. hallii* can grow very well in dietary-derived sugars such as glucose, fructose, galactose, sucrose, maltose, mannose, and sorbitol with butyrate as the major end metabolite of fermentation. In addition, *E. hallii* has been shown to metabolize the mucin-derived substrate

N-acetylglucosamine (Belzer et al. 2017), thus facilitating the interaction with mucin degrading bacteria. Compared to other butyrate-producing bacteria, *E. hallii* has a relatively broad spectrum of substrates for its growth which is one of several advantages for *E. hallii* to survive in the gut where competition is high, especially for simple substrates. In addition, *E. hallii* is able to efficiently convert both D- and L-lactate to butyrate in the presence of acetate (Duncan et al. 2004; Louis et al. 2010). This can be a of mechanism to prevent lactate accumulation in the gut (Shetty et al. 2018). Accumulation of lactate has been observed in patients with gastrointestinal conditions (Hove et al. 1994) and the capability of removing access of lactate by *E. hallii* can therefore be an important factor in maintaining intestinal health.

E. hallii has been proposed to contribute to propionate production in the gut via a conversion of 1,2-propanediol (Engels et al. 2016). *E. hallii* is capable of converting 1,2-propanediol to propionate using 1,2-propanediol metabolic pathway. Although this metabolic pathway has been detected in several other bacteria, including *Flavonifractor plautii, Intestinimonas butyriciproducens,* and *Veillonella* spp. (Engels et al. 2016). The conversion of 1,2-propanediol to propionate has been demonstrated experimentally only for *E. hallii* and *Lactobacillus reuteri* (Gänzle 2015; Engels et al. 2016).

5.3.1 Butyrate Pathway

Genomic analysis of *E. hallii* revealed the presence of a glycolytic pathway for conversion of sugars to pyruvate while employing butyryl-CoA transferase pathway for butyrate production (Fig. 5.2a). This pathway differentiates from the other butyrogenic pathways at the terminal step which involves butyryl-CoA:acetate CoA transferase for butyrate production. This CoA transferase transfers CoA group from butyryl-CoA to acetate to form butyrate and acetyl-CoA as end products. Either CO_2/H_2 or formate is also produced along this pathway which might confer an opportunity for cross-feed with hydrogenotrophic microbes. In addition, in order to use lactate *E. hallii* first converts it to pyruvate and follow all other steps in butyryl-CoA:acetate CoA transferase to make butyrate. The energy is mainly conserved via the conversion from crotonyl-CoA to butyryl-CoA which involves butyryl-CoA dehydrogenase electron-transferring flavoprotein complex that generates a proton gradient via a membrane-associated NADH-ferredoxin oxidoreductase (Li et al. 2008).

5.3.2 Propionate Pathway

E. hallii is known to not only produce butyrate but also propionate. An entire adenosylcobalamin-dependent dehydratase PduCDE operon was found in the genome which confers the ability to convert 1,2-propanediol to propionate (Fig. 5.2b). This conversion is relatively fast and obtains only a small amount of energy via a last step from propionyl-phosphate to propionate. Remarkably, the

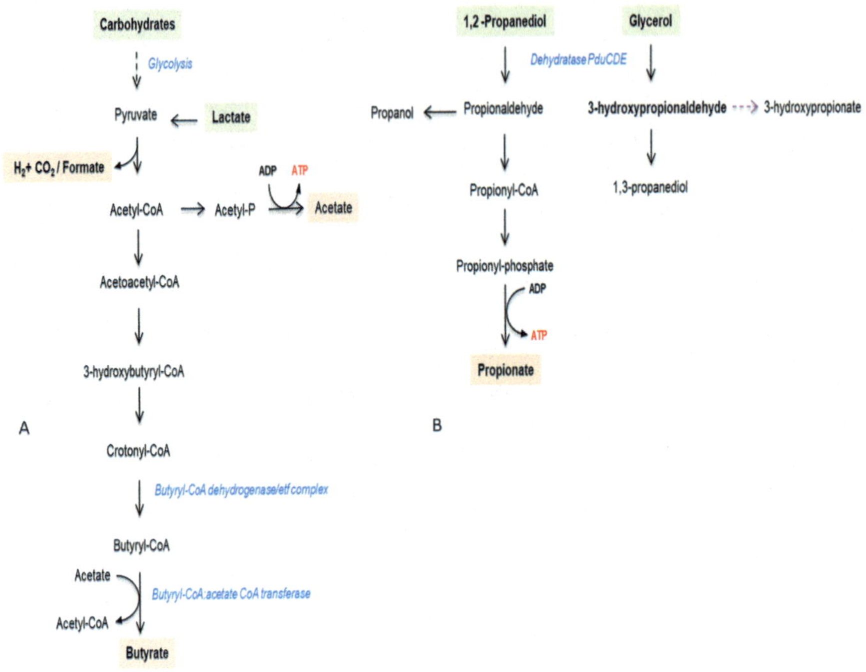

Fig. 5.2 Metabolic pathways of *E. hallii* for butyrate (**a**) and propionate (**b**) production

propanediol pathway was found to be widely spread in several intestinal taxa and contribute to approximately 31% of propionate turnover in the gut (Engels et al. 2016). This is indicative of the notable role of *E. hallii* on formation of propionate in the gut.

5.3.3 Additional Metabolites

Next to the generation of two significant short-chain fatty acids, *E. hallii* is capable of producing several other interesting metabolites. Strikingly, it was found that *E. hallii* was able to convert glycerol to 3-hydroxypropionaldehyde (reuterin) via the same dehydratase PduCDE as is used for the conversion of 1,2-propanediol to propionate (Fig. 5.3b). Reuterin is known as antimicrobial compound and toxic to bacteria. In spite of that, *E. hallii* was able to metabolize a small amount of produced reuterin relatively fast (Engels et al. 2016), which might be important to eliminate the toxicity of this compound at small quantities in the gut.

In addition, it was found that *E. hallii* was capable of producing pseudovitamin B12 (Belzer et al. 2017). Vitamin B12 is known as a modulator in shaping the structure and function of human gut microbial community (Degnan et al. 2014). It is believed that pseudovitamin B12 cannot be used efficiently by humans and other animals. Moreover, the receptors necessary for vitamin B12 absorption are only

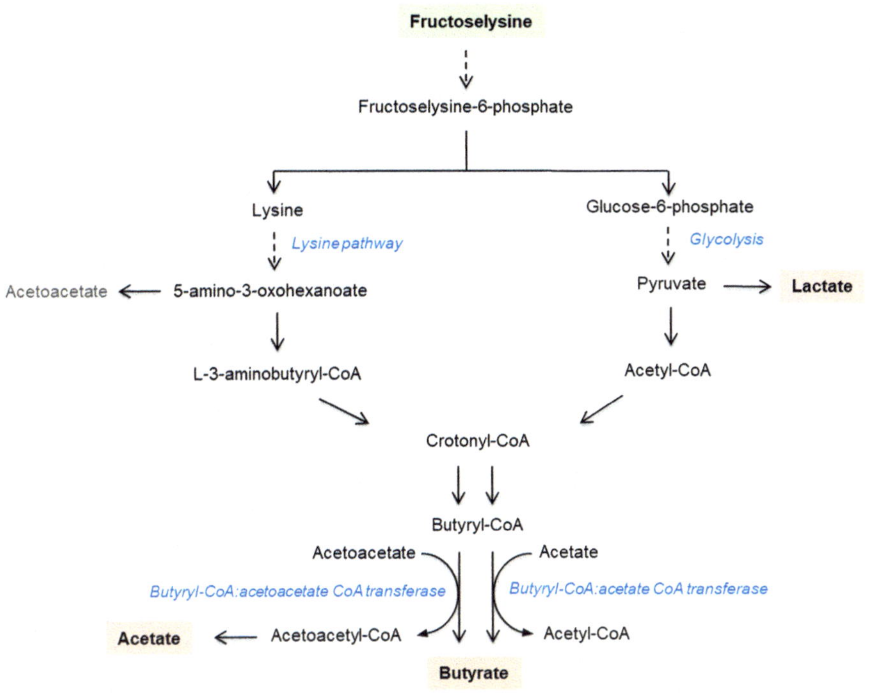

Fig. 5.3 Metabolic pathway of *Intestinimonas butyriciproducens* for butyrate production

found in the small intestine (Seetharam and Alpers 1982). However, animal experiments suggest that pseudovitamin B12 delivered orally may be bioavailable.

Finally, *E. hallii* was found to be able to transform the carcinogenic heterocyclic amine 2-amino-1-methyl-6-phenylimidazo[4,5-b]pyridine (PhIP) to noncarcinogenic PhIP-M1 (Fekry et al. 2016). PhIP is a component that is found in well-cooked meats and is believed to contribute to the carcinogenicity of processed meats.

5.3.4 Interaction with Other Commensals

E. hallii has been reported to interact with many different commensal bacteria. This includes the production of butyrate by *E. hallii* from the metabolites resulting from starch/fructose-oligosaccharides degradation by *Bifidobacterium adolescentis* (Duncan et al. 2004; Belenguer et al. 2006). Another example is the utilization by *E. hallii* of sugars, released from mucin degradation by *Akkermansia muciniphila* to produce butyrate (Belzer et al. 2017). Conversely, *E. hallii* is capable of producing pseudovitamin B12, which is used for metabolizing 1,2 propanediol to propionate but can also be used by *A. muciniphila* to activate the methylmalonyl-CoA pathway, converting succinate to propionate. It has been shown that *E. hallii* might compete with *Desulfovibrio* for lactate consumption in the gut (Marquet et al. 2009) as lactate is mainly fermented to butyrate in the gut (Bourriaud et al. 2005). It was observed that *E. hallii* converted 1,2-propanediol produced by infant bifidobacteria from L-fucose and fucosyllactose to propionate (Schwab et al. 2017). In parallel, *E. hallii* also used intermediate lactate/acetate produced by bifidobacteria to produce butyrate.

5.4 *Intestinimonas butyriciproducens*

5.4.1 General Description

In a bid to identify commensals that are capable of utilizing noncarbohydrate sources for their carbon needs, mouse fecal samples were prepared and cultured on reduced agar medium, containing yeast extract, rumen fluid, and lactic acid as main energy and carbon sources. One specific strain was identified that was only distantly related to any at the time know recognized species. This species was close related to the butyrate-producing bacterium *Flavonifractor plautii* (>94.5% similarity for 16S rRNA sequence), but clustered as a separate genus and was designated *Intestinimonas butyriciproducens* (Kläring et al. 2013). It was the first species that was found to produce butyrate from lysine. Moreover, it was found that *I. butyriciproducens* was able to convert the Amadori product fructoselysine to butyrate. Amadori products such as fructose-lysine have become part of our food since we are able to cook our foods and are formed through heating of reducing sugars with amino acids in a nonenzymatic Maillard reaction. An interesting question that arises from this is if this bacterium has evolved as part of our microbiota since we started consuming cooked foods. Hence it would be of interest to see if other mammals (apart from human and laboratory mice), who do not normally consume cooked foods, would also harbor *I. butyriciproducens*. To date no examples are known.

5.4.2 Metabolic Diversity

I. butyriciproducens is not able to degrade polysaccharides or disaccharides. This species grew poorly in hexose sugars, but the growth was much enhanced in the presence of acetate. *I. butyriciproducens* was able to ferment glucose, galactose, and arabinose to mainly butyrate and minor amounts of ethanol and lactate while no growth was observed on mannitol, cellobiose, raffinose, xylose, D-mannose, sucrose, or sorbitol (Bui et al. 2016). Growth on lactate and acetate was also observed (unpublished data). No hydrogen was detected on any substrate. Genomic analysis showed the presence of a complete glycolysis and butyryl-CoA transferase pathway. In accordance to this, it was shown that associated proteins were overproduced when growing on glucose (Bui et al. 2015). Of this pathway, the conversion of crotonyl-CoA to butyryl-CoA involved a butyryl-CoA dehydrogenase

(Bcd) electron-transferring flavoprotein (Etf) complex that generates a proton motive force via a membrane-integrated Rnf complex (Li et al. 2008).

Remarkably, *I. butyriciproducens* grew much better on lysine as a primary carbon source, converting it to equimolar amounts of butyrate and acetate. The entire lysine pathway was detected in the genome. It has been reported that the lysine to butyrate pathway is the second abundant pathway for butyrate synthesis in the gastrointestinal tract (Vital et al. 2014), suggesting a key role of *Intestinimonas* in colonic butyrate turnover.

The lysine metabolic pathway consists of a sequence of reactions from lysine to L-3-aminobutyryl-CoA which are performed by proteins of which the coding genes are located in a single operon. The conversion from crotonyl-CoA to butyryl-CoA is a major step for generation of energy. Generally, for butyrate formation from butyryl-CoA the butyryl-CoA:acetate CoA transferase pathway is used. For the butyrate formation pathway from lysine, however, it was found that butyryl-CoA: acetoacetate CoA transferase is also used, which transfers CoA from butyryl-CoA to acetoacetate to form butyrate. When growing on fructoselysine, these two pathways are operating simultaneously (Bui et al. 2015). These pathways are depicted in Fig. 5.3. Fructoselysine is first converted to fructoselysine-6-phosphate and subsequently cleaved to lysine and glucose-6-phosphate. While glucose-6-phosphate is used via glycolysis and butyryl-CoA pathway, lysine is further metabolized via lysine pathway. Butyrate, acetate, and ammonium were all detected as major end products. Lactate was formed in small amounts. Several strains of I. *butyriciproducens* have been isolated from different hosts, all of which shared the same metabolic activities for lysine and fructoselysine.

Similar to *E. hallii*, *I. butyriciproducens* produces pseudovitamin B12, which is beneficial for intestinal microbes and, either directly or indirectly, for the host.

Interestingly, an entire dehydratase PduCDE operon was also found on the genome of *I. butyriproducens*, indicating the potential capability of converting 1,2-propanediol to propionate (Engels et al. 2016). This still needs to be proven experimentally.

5.5 Probiotic Potential

Both *I. butyriciproducens* and *E. hallii* show features that are reminiscent of human lifestyle. As mentioned previously, the conversion of fructoselysine to butyrate could be an adaptation to the consumption of Amadori products, which are the result of thermo treatment of food products. Similarly, the capability of *E. hallii* to convert PhIP, a component that is derived from well-cooked meat, could well be an adaptation of our microbiota to food that is normally confined to humans. The Amadori products have been associated with the aging process and chronic diseases (Deppe et al. 2011). Fructoselysine is among the most common Amadori products and also a precursor of Advanced Glycation Endproducts (AGEs). AGEs are implicated in the development of cancer and diabetic complications (Brownlee 1994) and therefore the removal of (predecessors of) AGEs could be an important step in reducing the risk of cancer and diabetics. Similarly, reducing levels of PhIP could have a reducing effect on the risk of cancer development (Fekry et al. 2016).

These anticarcinogenic properties are just two reasons why *E. hallii* and I. *butyriciproducens* are prime candidates for the development as potential therapeutic strains. Another reason is that both strains produce butyrate, a short-chain fatty acid that has also been associated with lowering the risk of (colonic) cancer and a number of other beneficial health traits, including maintenance of blood glucose levels (McNabney and Henagan 2017). Indeed, *E. hallii* has been identified as a bacterium that can have a positive influence on insulin resistance (Vrieze et al. 2012; Udayappan et al. 2016). Simultaneously propionate is also linked to glucose metabolism and insulin production (Chambers et al. 2015; Pingitore et al. 2017), suggesting that the effects on glucose metabolism observed with *E. hallii* could also result from its ability to produce propionate.

E. hallii is also a keystone species as it has vast interactions with other bacteria from the microbiota. On the one hand, it uses the metabolites acetate and lactate that are produced by other bacteria from the fermentation of complex carbohydrates; on the other hand, it produces certain metabolites that can have a profound influence on the stability of the microbial ecosystem. As mentioned previously, (pseudo) vitamin B12 is essential for metabolic processes of several other bacteria such as *Akkermansia muciniphila* (Belzer et al. 2017), while reuterin is a bacteriocin that affects the composition of its close surroundings by killing bacteria.

A complicating factor in developing human gut-derived bacteria as potential probiotics is the ubiquitous presence of specific antibiotic resistance genes. These have emerged in the bacterial population as a result of widespread use of antibiotics over the last decades and bear the inherent risk of being spread to invading pathogens, complicating the treatment of infections (Thiemann et al. 2016). Preselecting of strains that have a preferred antibiotic resistance profile can therefore be a tedious step and might not always be successful. Notably tetracycline resistance genes are omnipresent in the human microbiota of all geographic regions (Hu et al. 2013). Indeed, *E. hallii* L2-7 also carries an active TetO gene. To be able to use this strain for probiotic purposes we used a mutagenic approach for the selection of naturally occurring tetracycline-sensitive strains (unpublished results).

Concluding it can be stated that the high versatility of these strains makes them prime candidates to be developed as next-generation therapeutic strains. First and foremost, both bacteria produce butyrate which has proven health implications. In addition, *Intestinimonas* has a high potential because of its ability to reduce the burden of Amadori products that could potentially develop into carcinogenic AGEs and turn it into a beneficial compound (butyrate). *E. hallii* has an even wider scope of use as it can influence the health of the host through a direct interaction by producing butyrate and propionate as well as its ability to transform the carcinogenic dietary compound PhIP to PhIP-M1. Indirectly, it can influence the health of its host by influencing the health state of the microbiota through the production of pseudovitamin B12 and reuterin.

More recently another potential application became apparent where the presence of *E. hallii* was associated with protection against *Clostridioides difficile* infection (Crobach et al. 2020).

References

Belenguer A et al (2006) Two routes of metabolic cross-feeding between Bifidobacterium adolescentis and butyrate-producing anaerobes from the human gut. Appl Environ Microbiol 72(5):3593–3599. https://doi.org/10.1128/AEM.72.5.3593

Belzer C et al (2017) Microbial metabolic networks at the mucus layer lead to diet-independent butyrate and vitamin B 12 production by intestinal symbionts. MBio 8(5):e00770–e00717. https://doi.org/10.1128/mBio.00770-17

Boets E et al (2017) Systemic availability and metabolism of colonic-derived short-chain fatty acids in healthy subjects: a stable isotope study. J Physiol 595(2):541–555. https://doi.org/10.1113/JP272613

Bolognini D et al (2016) The pharmacology and function of receptors for short-chain fatty acids. Mol Pharmacol 89(3):388–398. https://doi.org/10.1124/mol.115.102301

Bourriaud C et al (2005) Lactate is mainly fermented to butyrate by human intestinal microfloras but inter-individual variation is evident. J Appl Microbiol 99(1):201–212. https://doi.org/10.1111/j.1365-2672.2005.02605.x

Brownlee M (1994) Lilly Lecture 1993. Glycation and diabetic complications. Diabetes 43 (6):836–841. https://doi.org/10.2337/diab.43.6.836

Bui TPN, de Vos WM, Plugge CM (2014) Anaerostipes rhamnosivorans sp. nov., a human intestinal, butyrate-forming bacterium. Int J Syst Evol Microbiol 64(PART 3):787–793. https://doi.org/10.1099/ijs.0.055061-0

Bui TPN et al (2015) Production of butyrate from lysine and the Amadori product fructoselysine by a human gut commensal. Nat Commun 6:1–10. https://doi.org/10.1038/ncomms10062

Bui TPN et al (2016) Comparative genomics and physiology of the butyrate-producing bacterium Intestinimonas butyriciproducens. Environ Microbiol Rep 8(6):1024–1037. https://doi.org/10.1111/1758-2229.12483

Chambers ES et al (2015) Effects of targeted delivery of propionate to the human colon on appetite regulation, body weight maintenance and adiposity in overweight adults. Gut 64 (11):1744–1754. https://doi.org/10.1136/gutjnl-2014-307913

Collins MD et al (1994) The phylogeny of the genus clostridium: proposal of five new genera and eleven new species combinations. Int J Syst Bacteriol 44(4):812–826. https://doi.org/10.1099/00207713-44-4-812

Crobach MJT et al (2020) The bacterial gut microbiota of adult patients infected, colonized or noncolonized by clostridioides difficile. Microorganisms 8(5):1–13. https://doi.org/10.3390/microorganisms8050677

De Vos P et al (eds) (2009) Bergey's manual of Systematic Bacteriology; volume three, the firmicutes. second. Springer International Publishing, Dordrecht Heidelberg London New York. https://doi.org/10.1007/b92997

Degnan PH, Taga ME, Goodman AL (2014) Vitamin B12as a modulator of gut microbial ecology. Cell Metab 20(5):769–778. https://doi.org/10.1016/j.cmet.2014.10.002

Den Besten G et al (2013) The role of short-chain fatty acids in the interplay between diet, gut microbiota, and host energy metabolism. J Lipid Res 54(9):2325–2340. https://doi.org/10.1194/jlr.R036012

Deppe VM et al (2011) Enzymatic deglycation of Amadori products in bacteria: mechanisms, occurrence and physiological functions. Appl Microbiol Biotechnol 90(2):399–406. https://doi.org/10.1007/s00253-010-3083-4

Duncan SH, Louis P, Flint HJ (2004) Lactate-utilizing bacteria, isolated from human feces, that produce butyrate as a major fermentation product lactate-utilizing bacteria, isolated from human feces, that produce butyrate as a major fermentation product. Appl Environ Microbiol 70 (10):5810–5817. https://doi.org/10.1128/AEM.70.10.5810

Engels C et al (2016) The common gut microbe Eubacterium hallii also contributes to intestinal propionate formation. Front Microbiol 7:1–12. https://doi.org/10.3389/fmicb.2016.00713

Fekry MI et al (2016) The strict anaerobic gut microbe Eubacterium hallii transforms the carcinogenic dietary heterocyclic amine 2-amino-1-methyl-6-phenylimidazo[4,5-b]pyridine (PhIP). Environ Microbiol Rep 8(2):201–209. https://doi.org/10.1111/1758-2229.12369

Gänzle MG (2015) Lactic metabolism revisited: metabolism of lactic acid bacteria in food fermentations and food spoilage. Curr Opin Food Sci 2:106–117. https://doi.org/10.1016/j.cofs.2015.03.001

Harmsen HJMM et al (2002) Extensive set of 16S rRNA-based probes for detection of bacteria in human feces. Appl Environ Microbiol 68(6):2982–2990. https://doi.org/10.1128/AEM.68.6.2982-2990.2002

Holmes E et al (2012) Therapeutic modulation of microbiota-host metabolic interactions. Sci Transl Med 4(137). https://doi.org/10.1126/scitranslmed.3004244

Hove H, Nordgaard-Andersen I, Mortensen PB (1994) Faecal DL-lactate concentration in 100 gastrointestinal patients. Scand J Gastroenterol 29(3):255–259. https://doi.org/10.3109/00365529409090473

Hu Y et al (2013) Metagenome-wide analysis of antibiotic resistance genes in a large cohort of human gut microbiota. Nat Commun 4:2151. https://doi.org/10.1038/ncomms3151

Kläring K et al (2013) Intestinimonas butyriciproducens gen. nov., sp. nov., a butyrate-producing bacterium from the mouse intestine. Int J Syst Evol Microbiol 63(PART 12):4606–4612. https://doi.org/10.1099/ijs.0.051441-0

Li F et al (2008) Coupled ferredoxin and crotonyl coenzyme A (CoA) reduction with NADH catalyzed by the butyryl-CoA dehydrogenase/Etf complex from Clostridium kluyveri. J Bacteriol 190(3):843–850. https://doi.org/10.1128/JB.01417-07

Li J et al (2014) An integrated catalog of reference genes in the human gut microbiome. Nat Biotechnol 32(8):834–841. https://doi.org/10.1038/nbt.2942

Lin Y et al (1995) Differences in propionate-induced inhibition of cholesterol and triacylglycerol synthesis between human and rat hepatocytes in primary culture. Br J Nutr 74(2):197–207. https://doi.org/10.1079/BJN19950123

Louis P et al (2010) Diversity of human colonic butyrate-producing bacteria revealed by analysis of the butyryl-CoA:acetate CoA-transferase gene. Environ Microbiol 12(2):304–314. https://doi.org/10.1111/j.1462-2920.2009.02066.x

Marquet P et al (2009) Lactate has the potential to promote hydrogen sulphide formation in the human colon. FEMS Microbiol Lett 299(2):128–134. https://doi.org/10.1111/j.1574-6968.2009.01750.x

McNabney SM, Henagan TM (2017) Short chain fatty acids in the colon and peripheral tissues: a focus on butyrate, colon cancer, obesity and insulin resistance. Nutrients 9(12):1–28. https://doi.org/10.3390/nu9121348

Metchnikoff É (1907) The prolongation of life, optimistic studies. In: Mitchell PC (ed) The prolongation of life: optimistic studies. G.P. Putnam's Sons, New York & London. https://archive.org/details/prolongationofli00metciala

Moens F, Verce M, De Vuyst L (2017) Lactate- and acetate-based cross-feeding interactions between selected strains of lactobacilli, bifidobacteria and colon bacteria in the presence of inulin-type fructans. Int J Food Microbiol 241:225–236. https://doi.org/10.1016/j.ijfoodmicro.2016.10.019

Morrison DJ, Preston T (2016) Formation of short chain fatty acids by the gut microbiota and their impact on human metabolism. Gut Microbes 7(3):189–200. https://doi.org/10.1080/19490976.2015.1134082

Oleskin AV, Shenderov BA (2019) Probiotics and psychobiotics: the role of microbial neurochemicals. Probiotics Antimicrobial Proteins 11(4):1071–1085. https://doi.org/10.1007/s12602-019-09583-0

Patterson E et al (2014) Gut microbiota, the pharmabiotics they produce and host health. Proc Nutr Soc 73(4):477–489. https://doi.org/10.1017/S0029665114001426

Pingitore A et al (2017) The diet-derived short chain fatty acid propionate improves beta-cell function in humans and stimulates insulin secretion from human islets in vitro. Diabetes Obes Metab 19(2):257–265. https://doi.org/10.1111/dom.12811

Qin J et al (2010) A human gut microbial gene catalogue established by metagenomic sequencing. Nature 464(7285):59–65. https://doi.org/10.1038/nature08821

Rajilić-Stojanović M, de Vos WM (2014) The first 1000 cultured species of the human gastrointestinal microbiota. FEMS Microbiol Rev 38(5):996–1047. https://doi.org/10.1111/1574-6976.12075

Schwab C et al (2017) Trophic interactions of infant bifidobacteria and eubacterium hallii during L-fucose and fucosyllactose degradation. Front Microbiol 8:1–14. https://doi.org/10.3389/fmicb.2017.00095

Seetharam B, Alpers DH (1982) Absorption and transport of cobalamin (Vitamin B12). Annu Rev Nutr 2(1):343–369. https://doi.org/10.1146/annurev.nu.02.070182.002015

Shetty SA et al (2017) Intestinal microbiome landscaping: insight in community assemblage and implications for microbial modulation strategies. FEMS Microbiol Rev 41(2):182–199. https://doi.org/10.1093/femsre/fuw045

Shetty SA et al (2018) Reclassification of Eubacterium hallii as Anaerobutyricum hallii gen. nov., comb. nov., and description of Anaerobutyricum soehngenii sp. nov., a butyrate and propionate-producing bacterium from infant faeces. Int J Syst Evol Microbiol 68(12):3741–3746. https://doi.org/10.1099/ijsem.0.003041

Thiemann S, Smit N, Strowig T (2016) Antibiotics and the intestinal microbiome: individual responses, resilience of the ecosystem, and the susceptibility to infections. Curr Top Microbiol Immunol 398:123–146. https://doi.org/10.1007/82_2016_504

Udayappan S et al (2016) Oral treatment with Eubacterium hallii improves insulin sensitivity in db/db mice. NPJ Biofilms Microbiomes 2:16009. https://doi.org/10.1038/npjbiofilms.2016.9

Vital M, Howe A, Tiedje J (2014) Revealing the bacterial butyrate synthesis pathways by analyzing (meta) genomic data. MBio 5(2):1–11. https://doi.org/10.1128/mBio.00889-14.Editor

Vrieze A et al (2012) Transfer of intestinal microbiota from lean donors increases insulin sensitivity in individuals with metabolic syndrome. Gastroenterology 143(4):913–916.e7. https://doi.org/10.1053/j.gastro.2012.06.031

Willcox MDP (2013) Characterization of the normal microbiota of the ocular surface. Exp Eye Res 117:99–105. https://doi.org/10.1016/j.exer.2013.06.003

Woese CR, Fox GE (1977) Phylogenetic structure of the prokaryotic domain: the primary kingdoms (archaebacteria/eubacteria/urkaryote/16S ribosomal RNA/molecular phylogeny). Proc Natl Acad Sci U S A 74(11):5088–5090

Anticarcinogenic Potential of Probiotic, Postbiotic Metabolites and Paraprobiotics on Human Cancer Cells

6

Elham Noroozi, Majid Tebianian, Morteza Taghizadeh, Maryam Dadar, and Naheed Mojgani

Abstract

The performance of probiotic bacteria and their metabolites in the control and treatment of various cancers has been shown by a number of clinical studies. Among probiotic bacteria, lactic acid bacteria are well known for their beneficial role in colonic health, where they also exert anticarcinogenic effects. These beneficial bacteria can inhibit the occurrence of cancer by (1) lowering PH, (2) reducing the level of pro-carcinogenic enzymes, (3) enhancing cell proliferation by inhibiting normal cell apoptosis and by promoting cell differentiation and cytoprotective activities, (4) suppressing inflammation-induced cell apoptosis, (5) enhancing innate immunity, (6) promoting various gut homeostasis, and (7) displaying antioxidant activity. Several research findings showed that probiotic metabolites (postbiotics) can regulate cell proliferation in colorectal cancer and might be considered a therapeutic alternative for treating chemoresistant colorectal cancer. These metabolites including short-chain fatty acids, exopolysaccharides, vitamins, bacteriocin, H_2O_2, etc. are known to be involved in decreasing the viability of cancer cells and the induction of apoptosis by influencing different signaling pathway. Despite the general definition that probiotics are live microorganisms, a variety of biological responses have been reported from administering dead and frequently heat-killed (Paraprobiotic)

E. Noroozi
Department of Cellular Molecular Biology, Science and Research Branch, Islamic Azad University, Tehran, Iran

Department of Medical Genetics, Islamic Azad University, Shahrood, Iran

M. Tebianian · M. Taghizadeh · M. Dadar · N. Mojgani (✉)
Agriculture Research, Education and Extension Organization (AREEO), Razi Vaccine and Serum Research Institute (RVSRI), Karaj, Iran

N. Mojgani, M. Dadar (eds.), *Probiotic Bacteria and Postbiotic Metabolites: Role in Animal and Human Health*, Microorganisms for Sustainability 2,
https://doi.org/10.1007/978-981-16-0223-8_6

probiotic bacterial cultures to various mammalians. The preparations of dead cells have also been fractionated and various cellular components shown to produce a range of biological responses. Many of the biological responses found with the heat-killed probiotic bacterial cells are not antimicrobial effects but are, rather, immunomodulating effects. Owing to the fact that probiotics, paraprobiotics and metabiotics or postbiotics metabolites are the most widely studied biological therapeutic alternatives for the treatment of cancer; hence in this chapter their functions and mechanism of action would be elucidated.

Keywords

Probiotics · Paraprobiotics · Postbiotics · Metabiotic · Bacteriocin · Exopolysaccharide · Short-chain fatty acids · Biosurfactants · Cancer

6.1 Introduction

Despite the fact that cancer risk indisputably depends on genetic factors, and immunological conditions of the host, but most important of all the gut microbiome has known to play considerable role in cancer cases. According to reports, any imbalance in the gut microbiome compositions might result in disorders such as cancer, malignancy, inflammatory bowel disease (IBD), irritable bowel syndrome (IBS), fatty liver diseases, obesity, type 2 diabetes mellitus, asthma, cardiovascular, psychiatric disorders, and immune-mediated diseases (Barteneva et al. 2017; Marques et al. 2017). Such modification of the gut microbiota is referred to as dysbiosis is of paramount importance as they play potential role in initiation and progression of several diseases in humans and animals (Azad et al. 2018). Vast population of these bacteria present in the digestive tract of the host could either prevent cancer cell growth or slow down the process.

Gut microbiota is occupied by members of bacteria belonging to the family Bacteroidetes and *Firmicutes*. Among these, lactic acid bacteria (LAB) are the most dominant flora residing in the gut of a healthy individual that are recognized as GRAS (generally recognized as safe). Majority of the LAB species are widely used as probiotics in a number of food products and supplements. The term probiotic refers to the live microorganisms that provide health benefits to the host when consumed in sufficient amounts (Liu et al. 2015; Lüke et al. 2016; Maghsood et al. 2018; Voigt et al. 2009). Some of the therapeutic effects of probiotic LAB are their antitumor activities that inhibit the carcinogens present in the gastrointestinal tract by stimulating the immune response. Owing to their immune modulating role, these bacteria are also known as "immunobiotics" (Bedada et al. 2020; Ghanavati et al. 2020; Kumar et al. 2012; Sharma and Shukla 2016). According to reports, the occurrence of cancer is usually prevented by these bacteria by (1) lowering pH, (2) reducing the level of pro-carcinogenic enzymes (Kahouli et al. 2013), (3) enhancing cell proliferation by inhibiting normal cell apoptosis and by promoting cell differentiation and cytoprotective activities (Sivamaruthi et al. 2020),

(4) suppressing inflammation-induced cell apoptosis (Kumar et al. 2010), (5) enhancing innate immunity, (6) promoting various gut homeostasis (dos Reis et al. 2017), and (7) displaying antioxidant activity (Kaur and Kaur 2015).

Apart from probiotic bacteria, the dead cells of these bacteria are shown to demonstrate variety of biological responses. These nonviable probiotic bacterial cells are regarded as "Paraprobiotics" or "Probiotic ghost cells" (Sharma and Singh Saharan 2014). Similar to live probiotic cells, the dead cells of probiotic bacteria are known to bring about a number of biological responses in the hosts. Although their exact mechanism of action is yet not fully explored, they are believed to provide health benefits by the ability of their cell wall and other cellular components to boost the immune system, and inhibit the pathogens by adherence to the intestinal walls, etc. (Fujiki et al. 2012). Furthermore, the responses exerted by these live and dead bacteria might also be due to the secretory metabolites released in the cell free supernatant fluids by either the live bacteria or released after the cell lysis, respectively (Aguilar-Toalá et al. 2018). These secretory metabolites released by the probiotic bacteria are often termed as "postbiotics" or "metabiotics" and are known to exert beneficial effects in the gastrointestinal tract of the host (Sharma and Shukla 2016). Organic acids, bacteriocin, and H_2O_2 are some of these metabolites from probiotic bacteria that have significant role in decreasing the viability of colorectal cancer cells and the induction of apoptosis by influencing different signaling pathway (Jacouton et al. 2017).

6.2 Cancers

Abnormal growth of cells that harms an organ of the body is defined as cancerous cells. Cancer usually arises from the transformation of normal cells into tumor cells in a multistage process that generally progresses from a pre-cancerous lesion to a malignant tumor. The most common types of cancers include the following:

- Lung (2.09 million cases)
- Breast (2.09 million cases)
- Colorectal (1.80 million cases)
- Prostate (1.28 million cases)
- Skin cancer (nonmelanoma) (1.04 million cases)
- Stomach (1.03 million cases)

While the most common causes of cancer death include cancers of:

- Lung (1.76 million deaths)
- Colorectal (862,000 deaths)
- Stomach (783,000 deaths)
- Liver (782,000 deaths)
- Breast (627,000 deaths)

These changes are the result of the interaction between a person's genetic factors and three categories of external agents, including:

- Physical carcinogens, such as ultraviolet and ionizing radiation
- Chemical carcinogens, such as asbestos, components of tobacco smoke, aflatoxin (a food contaminant), and arsenic (a drinking water contaminant)
- Biological carcinogens, such as infections from certain viruses, bacteria, or parasites.

Aging is another fundamental factor for the development of cancer. The incidence of cancer rises dramatically with age, most likely due to a build-up of risks for specific cancers that increase with age. The overall risk accumulation is combined with the tendency for cellular repair mechanisms to be less effective as a person grows older (WHO).

Use of tobacco and alcohol, unhealthy diet, and physical inactivity are major cancer risk factors worldwide and are also the four shared risk factors for other noncommunicable diseases. Some chronic infections are risk factors for cancer and have major relevance in low- and middle-income countries. Approximately 15% of cancers diagnosed in 2012 were attributed to carcinogenic infections, including Helicobacter pylori, Human papillomavirus (HPV), Hepatitis B virus, Hepatitis C virus, and Epstein-Barr virus.

According to WHO reports, Hepatitis B and C virus and some types of HPV increase the risk for liver and cervical cancer, respectively. Infection with HIV substantially increases the risk of cancers such as cervical cancer.

A correct cancer diagnosis is essential for adequate and effective treatment because every cancer type requires a specific treatment regimen that encompasses one or more modalities such as surgery, radiotherapy, and chemotherapy. The primary goal is generally to cure cancer or to considerably prolong life. Improving the patient's quality of life is also an important goal. This can be achieved by supportive or palliative care and psychosocial support and most important of all by healthy diet. In this context, probiotic food products are of high importance as the beneficial bacteria in these products can manipulate the microbiome of the gut in a manner leading to desired health outcomes. Hence, the use of these bacteria for the prevention and treatment of various types of cancers has been of key research interest (Dicks et al. 2018; Zhong et al. 2014).

6.3 Anticancer Effects of Probiotic Bacteria

Probiotics are nonpathogenic live microorganisms that provide health benefits when are consumed in sufficient amounts (Mehra et al. 2012). Probiotic bacteria and yeasts are known to colonize, multiply, and produce variety of bioactive substances that accounts for their beneficial effects in the gastrointestinal tract of the host (Forsyth et al. 2009). Probiotics may be highly beneficial to the host as it has been described that they can maintain epithelial integrity, compete for adhesion and nutrition with pathogens, and stimulate cell-mediated immunity, IgA production, and gut associated lymphoid tissue (Goldin and Gorbach 1980).

These beneficial bacteria are vastly studied for their anti-inflammatory effects (Jacouton et al. 2017), playing significant role in the treatment of variety of cancer types (Fig. 6.1). The underlying mechanisms for the anticancer effects of probiotic bacteria are versatile including suppression of the growth of microbiota implicated in the production of mutagens and carcinogens, alteration in carcinogen metabolism, and protection of DNA from oxide damage as well as regulation of immune system (Jacouton et al. 2017). In addition, they have been shown to change expression of different genes participating in cell death and apoptosis, invasion and metastasis, cancer stem cell maintenance, as well as cell cycle control. Probiotic actions such as adhesion of lactic acid bacteria or their components to epithelial cells as well as release of soluble factors have been proposed to be important for the suppression of neoplastic cells (Oelschlaeger 2010).

Diet is known to play a major role in the pathogenesis of cancer especially colon cancer, among which red meat and animal fats are the main enemies. On the other hand, reports have indicated that fruits and vegetables might have preventive effects on such types of cancers. With the recognition of importance of diet in the control and prevention of a number of diseases, the demand for functional foods that are claimed to have health benefits are highly sought for.

Goldin and Gorbach (1980) were among the first to demonstrate the association between a diet enriched with Lactobacillus and a reduced incidence of colon cancer (40% vs. 77% in controls). In another study, a traditional fermented milk product was shown to inhibit *in vitro* proliferation of MCF-7 breast cancer cells, but not normal mammary epithelial cells.

Generally, the importance of probiotics has been shown in different *in vitro, in vivo* animals and clinical trials in humans (Tables 6.1 and 6.2). In a recent study we were able to demonstrate *in vitro* and *in vivo* anticancer effects of a live and heat-killed *L. casei* strain isolated from local dairy product (Noorozi et al. 2021). While, previously, Orlando et al., 2009 had reported antiproliferative effects of L.GG on gastric and colon cancer cells. They found that the highest concentrations of L.GG homogenate and cytoplasm extracts reduced the percentage of cell viability to nearly 55% and 65% in DLD-1(colon) and HGC-27 (gastric) cancer cell lines (Oelschlaeger 2010).

Special attention has been given to the effects of probiotics in reduction of invasion and metastasis in cancer cells. Invasion and metastasis have been regarded as important hallmarks of malignant cells which are endowed to them through diverse and complex genetic or epigenetic aberrations as well as extrinsic signals, such as those relayed from their microenvironment (Górska et al. 2019).

While, during in vivo studies conducted by Jacouton et al. it was shown that dairy strain of probiotic *L. casei* BL23 possessed potential anti-inflammatory and antitumor effects when administered orally to 6–8 weeks old female mice. The protective effect demonstrated by this probiotic strain was through reduction in cell proliferations and apoptosis induction. Apoptosis or programmed cell death is necessary in the treatment of cancer (Jacouton et al. 2017).

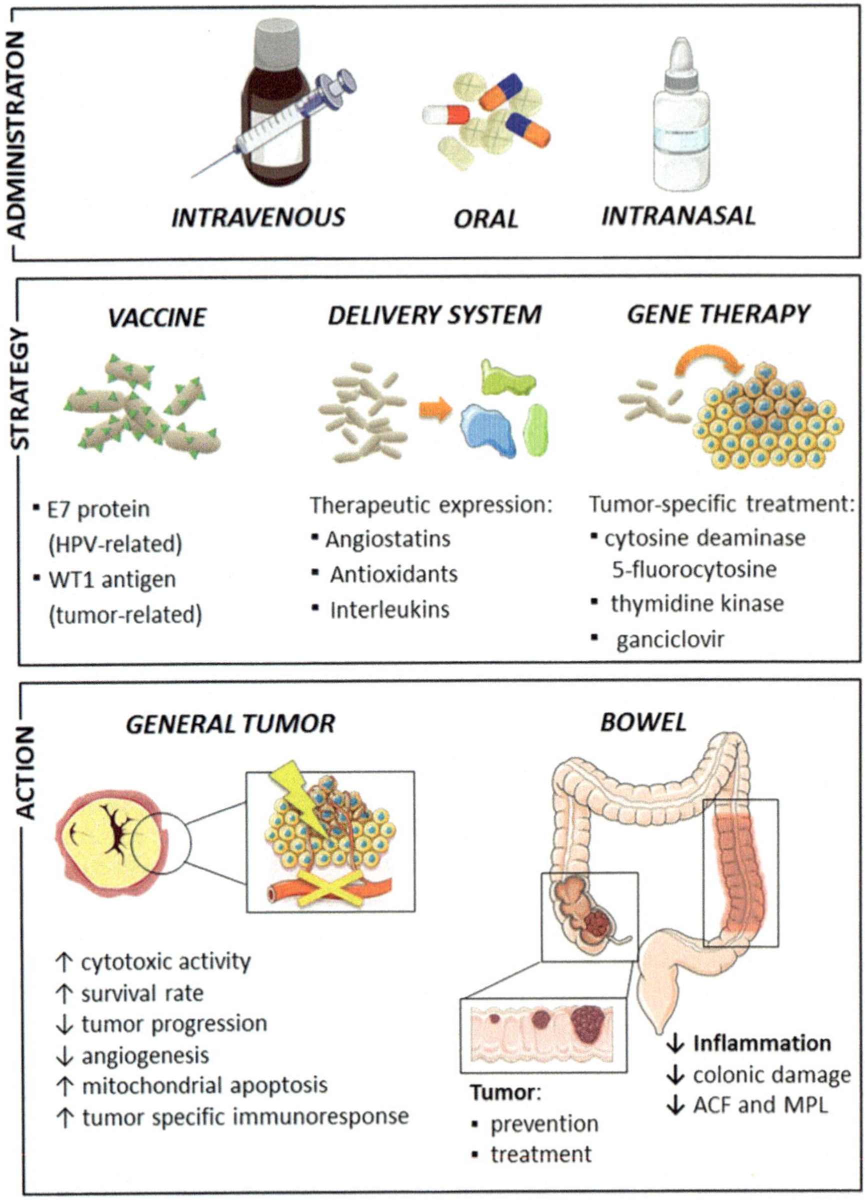

Fig. 6.1 Summary of the possible applications of probiotic bacteria in the treatment and prevention of cancer (Górska et al. 2019), *downwards arrow depicts decrease, upwards arrow depicts an increase; ACF aberrant crypt foci; MPL multiple plaque lesions*

Table 6.1 General effects of probiotics on cancer cells (in vitro studies)

Probiotic strain/details of experiment	Cell line	Effect	Reference
Lactobacillus rhamnosus GG, Bifidobacterium lactis Bb12	Caco-2	↑ Apoptosis	Altonsy et al. (2010)
Lactobacillus casei ATCC 393	HT29 and CT26	Induction of apoptosis	Tiptiri-Kourpeti et al. (2016)
Lactococcus lactis NK34	HT-29, LoVo, AGS	>80% ↓ Cell proliferation	
Bifidobacterium infantis, Lactobacillus paracasei, Bifidobacterium bifidum	MCF7	↓ Cell proliferation	Han et al. (2015)
Lactobacillus paracasei IMPC2.1, Lactobacillus rhamnosus GG/heat killed/	DLD-1	↓ Cell proliferation Induction of apoptosis	Orlando et al. (2012)
Lactobacillus pentosus B281, Lactobacillus plantarum B282/cell free supernatant used/	Caco-2 and HT-29	↓ Cell proliferation Cell cycle arrest (G1)	Oelschlaeger (2010)
Lactobacillus casei CRL431		↓ Cell proliferation	Saxami et al. (2016)
Bacillus polyfermenticus/AOM stimulation/	NMC460	↓ Cell colony formation in cancer cells(N/E on normal colonocytes)	Ma et al. (2010)
Bacillus polyfermenticus KU3	LoVo, HT-29, AGS	↓ Cell proliferation	Lee et al. (2015)
Lactobacillus rhamnosus, Lactobacillus crispatus/cell-free supernatant used/	HT-29	↑ Apoptosis	
Lactobacillus gasseri and *Lactobacillus crispatus*/cell-free supernatant used/	Hela, HNCF	↓ Cell proliferation	Motevaseli et al. (2013)
Lactobacillus lactis IL-17A	TH17	↑ Apoptosis	Jacouton et al. (2017)

↓ Decrease; ↑ increase; N/E no effect. Human colonic cancer cells: Caco-2, HT-29, SW1116, HCT116, SW480, DLD-1, LoVo, Human colonic epithelial cells: NMC460. Human gastric adenocarcinoma cells: AGS Mus musculus colon carcinoma cells: CT26.Cervical cancer:Hela. Head and neck cancer:HNCF.Lung cancer:TH17

The immunomodulatory potential of *L. casei* BL23 is mediated through IL-22 cytokine downregulation, and an antiproliferative property, mediated through Bik, caspase-7, and caspase-9 upregulation (Tiptiri-Kourpeti et al. 2016).

During an *in vitro* and *in vivo* study, it was shown that live *L. casei* ATCC393 and its components exert potent antiproliferative, growth inhibitory, and pro-apoptotic effects. These researchers reported that oral administration of live *L. casei* ATCC 393 and its components to the mice displayed antiproliferative effects, and suggested

Table 6.2 General effects of probiotics in tumor-induced animal models (in vivo studies)

Probiotic strain	Animal model	Induction	Treatment	Result	Reference
Lactobacillus acidophilus, *Lactobacillus casei*	Rat	DMH	40 weeks	↓ TI ↓ TV ↓ TM	Arvind and Sinha (2009)
Bifidobacterium lactis KCTC 5727	SPF C57BL rat		19 weeks	↓ TI ↓ TV	Kim et al. (2010)
Bacillus polyfermenticus	CD-1 mice	DLD-1 cells injection	20 weeks	↓ TI ↓ TV	Ma et al. (2010)
Lactobacillus plantarum	BALB/c mice	CT26 cells injection	14 weeks	↓TV, Induction of necrosis	Hu et al. (2015), Walia et al. (2015)
Lactobacillus plantarum (AdF10), *Lactobacillus rhamnosus GG*	SD rats	DMH 4 weeks	12 weeks	↓ TI ↓ TV ↓ TM	
Lactobacillus casei BL23	C57BL/6 mice	DMH	10 weeks	↓ TI	Tiptiri-Kourpeti et al. (2016)

↓ Decrease, TI tumor incidence, TV tumor volume, TM tumor multiplicity, DMH 1,2 dimethylhydrazine dihydrochloride

that the mechanisms underlying this effect were tumor necrosis factor related apoptosis-inducing ligand TRAIL upregulation and surviving downregulation (Tiptiri-Kourpeti et al. 2016)·

Lactobacilli can stimulate immune cells of the host including dendritic or natural killer (NK) cells or T helper type 1 (TH1) response, which participates in precancerous or anticancerous cell. Oral administration of probiotic *L. acidophilus* isolated from traditional homemade yogurt and neonatal stool reduces tumor growth by immune response modulation or changing the cytokine milieu reducing growth rate of tumor, increasing proliferation of lymphocyte, protecting TH cells, and activating antitumoral cell in in vivo breast cancer murine model, 8–10-week-old Balb/C female mice (Fujiki et al. 2012).

Notably, gut *L. acidophilus* activates NK cells, a major source of interferon (IFN)-γ and play vital role in antitumor immunity. Thus, the mechanism by which *L. acidophilus* prevents tumor growth is by innate anticancer cells activation. *L. acidophilus* produces IFN-γ from splenocyte to increase anticancer property, antiangiogenesis, and NK activity (Fujiki et al. 2012). Additionally, probiotic *Lactococcus lactis* (*L. lactis*) NK34 with a dose 10^6 CFU was shown to possess strong anticancer and anti-inflammatory effects by inhibiting the proliferation of cancer cells such as human lung carcinoma cell line (SK-MES-1), human colon adenocarcinoma cell line (DLD-1, HT-29), human colon adenocarcinoma cell line (LoVo), human stomach adenocarcinoma cell line (AGS), and human breast adenocarcinoma cell line (MCF-7 cells). *L. lactis* NK34 demonstrated anti-inflammatory property by inhibiting lipopolysaccharide-induced RAW 264.7 cells that produce nitric oxide, and proinflammatory cytokines such as interleukin-18, tumor necrosis factor-α, and cyclooxygenase-2 were decreased (Tiptiri-Kourpeti et al. 2016). Probiotic *L. lactis* has been used as a fermentation starter in dairy or fermented foods and is considered as a safe microbe with GRAS (Generally recognized as safe) status.

Fast acidifying lactic acid bacteria, *Streptococcus thermophilus* (*S. thermophilus*) M17PTZA496 and *S. thermophilus TH982* have been reported to possess probiotic properties, anticancer activity, and folate-producing ability *in vitro*. Of most commercially available strains of probiotics, thermophilic *S. thermophiles* is extensively used as starter culture for many dairy products next to *L. lactis* (Bedada et al. 2020). According to these studies, *S. thermophilus* MTH17CL396, TH982, and M17PTZA496 inhibited HT-29 cells significantly. The significant antiproliferative potential of these strains on HT-29 cancer cells was concluded to be the result of lactic acid produced by these bacteria. Various mechanisms are revealed as to how lactic acid bacteria prevent colon cancer, such as carcinogens binding and degrading, immune response increment, antimutagenic compounds production, and physicochemical conditions change in the colon (Bedada et al. 2020; Sanders et al. 2018). Probiotics are used to fight against cancer by enhancing immune response or protecting against gastrointestinal infections. A pro-inflammatory cytokine, interleukin-17A is produced by TH17-cells and used in autoimmune disease and host defense. Recombinant *L. lactis* IL-17A produced and secreted cytokine, Interleukin-17A in murine fibroblasts 3 T3 L1 cells line and human papilloma virus induced cancer in mouse allograft model. This indicates the role of IL 17A in cancer (Kumar et al. 2010; Sanders et al. 2018). Figure 6.1 summarizes most significant findings from in vitro and *in vivo* studies regarding anticancer effects of probiotic bacteria and the therapeutic options (Górska et al. 2019).

6.4 Anticancer Effects of Postbiotic Metabolites Produced by Probiotic Bacteria

Probiotics colonize, multiply, and produce variety of bioactive substances termed "metabiotics," accounting for their beneficial effects in gastrointestinal tract (GIT) diseases. These metabolites produced by probiotics help in maintaining homeostasis in the gut and enhance the growth of friendly bacteria that inhibit the conversion of procarcinogens into carcinogens by decreasing harmful enzyme levels such as nitroreductase, β-glucuronidase, and β-glucosidase enzymes (Sharma 2019). Postbiotics are defined as the soluble factors (products or metabolic byproducts), secreted by live bacteria during metabolism, like hydrogen peroxide, active ribosomal proteins like bacteriocins, exopolysaccharides, etc. or released after bacterial lysis, such as enzymes, peptides, teichoic acids, peptidoglycan-derived muropeptides, polysaccharides, cell surface proteins, and organic acids. A variety of these metabolites, such as plantaricin, exopolysaccharides (EPS), lactic acid, acetic acid, and γ-aminobutyric acid, have been shown to possess the ability to enhance body immunity, antitumor, and antisepsis activity (Dicks et al. 2018; Kaur and Kaur 2015; Sharma 2019). Apart from these, the short-chain fatty acids (SCF) in

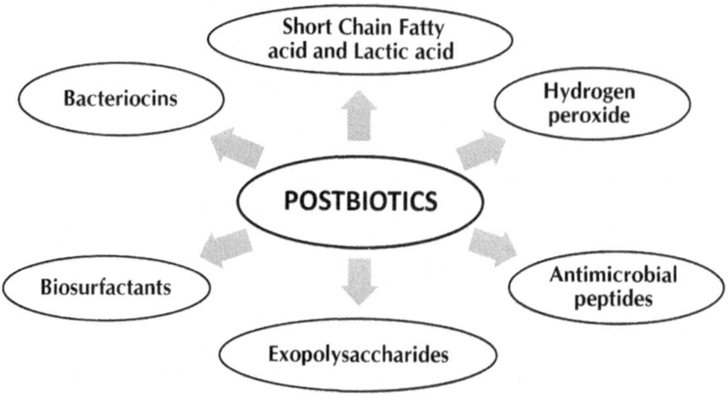

Fig. 6.2 Some biological functions of postbiotic metabolites produced by probiotic bacteria

the postbiotics are known to induce chemopreventive enzymes glutathione S transferase and glutathione transferase pi and impart genetic stability to colon cells. Some most important biological functions of postbiotic metabolites are shown in Fig. 6.2.

Molecules and metabolites derived from probiotic bacteria can prevent tumor development through modulation of immune systems of the host. For instance, bacterial lipopolysaccharide (LPS), a key component of gram-negative bacteria outer membrane, activates toll-like receptor 4, consequently activating immune T cell-mediated response against tumor cells (Sanders et al. 2018; Sharma 2019).

In the last decade, the postbiotic metabolites extracted from beneficial bacteria especially LAB have gained immense importance owing to their clear chemical structure, safety dose parameters, long shelf life, and the content of various signaling molecules that might have anti-inflammatory, immunomodulatory, anti-obesogenic, antihypertensive, hypocholesterolemic, antiproliferative, and antioxidant activities. As pointed out by Zhang et al., *L. acidophilus* and *L. casei* produce compounds that inhibit the growth of breast cancer cell line, MCF7. *L. acidophilus* 606 prevents the proliferation of human pancreatic tumor cell line by soluble polysaccharides production. These properties suggest that postbiotics may contribute to the improvement of host health by improving specific physiological functions, even though the exact mechanisms have not been entirely elucidated (Dicks et al. 2018; Kaur and Kaur 2015).

6.4.1 Short-Chain Fatty Acids (SCFAs)

The potential therapeutic role of probiotic bacteria in the gut is linked to their ability to produce a number of metabolites including short-chain fatty acids (SCFA) like lactic acid, acetic acid, butyric acid, propionic acids, etc. (Kahouli et al. 2013; Sharma and Shukla 2016).

SCFAs, conjugated linoleic acid and other anticarcinogenic products produced by Lactobacilli extracts induce apoptosis in cancer cells. In a report, the antiproliferative

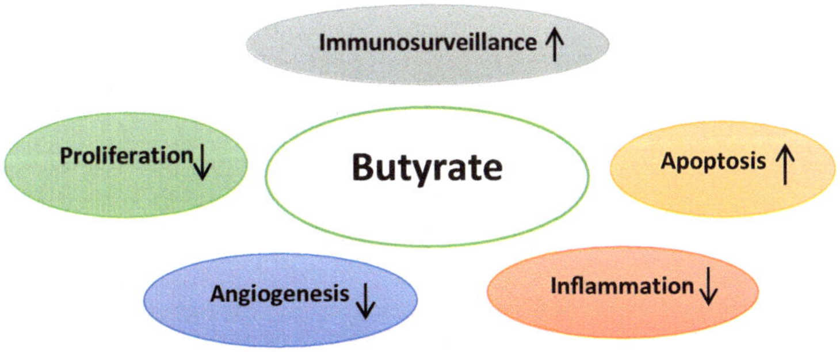

Fig. 6.3 Potential mechanisms by which the antineoplastic actions of butyrate may be mediated (Williams et al. 2003)

potential of probiotics *L. fermentum* NCIMB 5221, 2797, and 8829 was linked to their ability to produce SCFAs *in vitro* and their ability to persist in an intestinal fluid (Kahouli et al. 2013).

In another research report, SCFAs metabolites produced by probiotic *Propionibacterium freudenreichii* was shown to damage colorectal adenocarcinoma cells by producing apoptosis *in vitro*. Similarly, probiotics *Pediococcus pentosaceus* FP3, *Lactobacillus salivarius* FP35, *Lactobacillus salivarius* FP25, and *E. faecium* FP51 also inhibit proliferation of colon cancer cells by SCFAs bioproduction, mostly propionic and butyric acids. Similarly, conjugated linoleic acids produced by probiotic bacteria have the ability to form anticarcinogenic effects *in vitro* and *in vivo*. Another probiotic strain, *L. reuteri* NCIMB 701,359 has demonstrated anticarcinogenic effects owing to its ability to produce propionate (Kahouli et al. 2013; Sharma and Shukla 2016; Sivamaruthi et al. 2020).

Butyrate produced by fermentation of high amylose starch was reported to reduce the overall oxidative stress in gut and may also activate different procarcinogen metabolizing enzymes to aid in colon cancer prevention (Kahouli et al. 2013; Sharma and Shukla 2016; Sivamaruthi et al. 2020). Butyrate acts as the preferred source of energy for colonocytes and has anti-inflammatory and anticancerous properties (Fig. 6.3). Butyrate participates in the mobility of the colon, reduces inflammation, increases visceral irrigation, induces apoptosis, inhibits the progression of tumor cells, and contributes with the prevention of colorectal cancer (Williams et al. 2003). Unlike butyrate, acetate has been reported to be an instigator of cancers including liver, brain, prostate, and breast cancer. In cancer cells, acetate can serve as a source of nutrition required for lipid biosynthesis and can acetylate histones, leading to epigenetic modifications. It can also lead to the considerable posttranslational modification of proteins, altering their functions (Schug et al. 2016).

6.4.2 Bacteriocins

The potential use of bacteriocins in anticancer therapy is due to their inhibition of DNA and membrane protein synthesis, inducing apoptosis or cytotoxicity in tumor cells. Bacteriocins are ribosomally synthesized cationic peptides that are produced by almost all groups of bacteria. The first bacteriocin was discovered in the year 1925 by Gratia from *Escherichia coli* and later named as colicin. Since then, large number of bacteriocins have been identified from a diverse group of bacterial strains. Their physiological functions in bacteria seem to inhibit the growth of competing microorganisms in a particular biological niche by killing them. Microorganisms colonizing the gut may produce bacteriocins in an attempt to outcompete pathogens. Production of bacteriocins in a harsh and complex environment such as the gastrointestinal tract (GIT) may be below minimal inhibitory concentration (MIC) levels. At such low levels, the stability of bacteriocins may be compromised. Despite this, most bacteria in the gut have the ability to produce bacteriocins, distributed throughout the GIT. Most bacteriocins are extremely potent, exhibiting antimicrobial activity at nanomolar concentrations, as opposed to the peptide antimicrobials produced by eukaryotic cells, which normally have 102–103-fold lower activities. Interestingly, the producer cells are immune to their own bacteriocins. The classification of bacteriocins has been revised from time to time. The latest classification arranges bacteriocins into three major classes based on their structural and physicochemical properties (Aguilar-Toalá et al. 2018; Kaur and Kaur 2015).

Several studies have shown that some bacteriocins have anticancer properties and demonstrate selective action toward cancer cells. Although the exact mechanism of the cancer cell specificity has not been studied, the various factors that could account for the selective action of these bacteriocins could be explained based on the generalized cell surface variations of cancer cells from the normal cells. Bacteriocins have a higher affinity for cancer cells due to the general negative charge of cancer cells. This could be explained by the fact that, the bi-layered phospholipid membrane of normal mammalian cells is asymmetric with respect to the distribution of phospholipids on the inner and outer surface. However, in cancer cells there is loss in asymmetry with respect to phospholipid types. Cancer cell membrane is known to carry a predominantly negative charge due to high levels of the anionic phosphatidylserine, O-glycosylated mucins, sialylated gangliosides, and heparin sulfates. Bacteriocins are cationic peptides by nature and thus they preferentially bind to negatively charged cell membrane of cancer cells as compared to normal cell membranes which are neutral in charge. Secondly, the selective binding of bacteriocins to cancer cells can be explained due to differences in the membrane fluidity of cancer cells. Cancer cells have higher membrane fluidity as compared to normal cells and this facilitates easy membrane destabilization. Lastly, the membranes of cancer cells contain a significantly higher number of microvilli compared to normal cells that increases the surface area of cancer cells, which results in the binding of a greater number of antimicrobial peptides to the cancer cell membrane as compared to normal cells (Dicks et al. 2018; Kaur and Kaur 2015).

Table 6.3 Some well-known bacteriocins having anticancer activities against various cancer cell lines

Bacteriocin	Producer organism	Cancer cell lines	References
Colicin E3	*E. coli*	HeLa, HS913T	Fuska et al. (1979), Šmarda et al. (1978)
Colicin A	*E. coli*	HS913T, SKUT-1, BT474, ZR75, SKBR3, MRC5	Chumchalova and Šmarda (2003)
Colicin E1	*E. coli*	MCF7, HS913T	Chumchalova and Šmarda (2003)
MicrocinE492	*K. pneumonia*	Hela, Jurkat, RJ2.25	Hetz Flores et al. (2002)
Pediocin PA-1	*P. acidilactici PAC1.0*	A-549, DLD-1	Beaulieu et al. (2007)
Pediocin K2a2-3	*P. acidilactici K2a2-3*	HT2a, HeLa	Villarante et al. (2011)
Pediocin CP2	*P. acidilactici*	HeLa, MCF7, Sp2/0-Ag14, HepG2	Kumar et al. (2012)
Pyocin S2	*P. aeruginosa 42A*	HepG2, Im9HeLa, AS-II, mKS-ATU-7	Abdi-Ali et al. (2004), Watanabe and Saito (1980)
Nisin	*L. lactis*	MCF7, HepG2	Paiva et al. (2011)
Bovicin	*S. bovis*	HC5 MCF7, HepG2	Paiva et al. (2012)
Plantaricin A	*L. plantarum C11*	Jurkat, GH4, Reh, Jurkat, PC12, N2A, GH4	Sand et al. (2007, 2010, 2013), Zhao et al. (2006)

Nisin is a food grade bacteriocin that has been used as bio-preservative in dairy products. In a study, head and neck squamous cell carcinoma (HNSCC) cells treated with nisin showed that this agent induced DNA fragmentation and apoptosis on three different cancer cell lines. Apoptosis in NHSCC cells, caused by nisin, is associated with calcium influx and upregulation of CHAC1 (cation transport regulator and apoptosis mediator). In another study, the size of tumors in mice with oral cancer was reduced when treated with nisin. The authors concluded that the selective action of nisin was due to structural differences in the composition of the plasma membranes between HNSCC cells and primary keratinocytes (Dicks et al. 2018). Table 6.3 depicts anticancer effects of some bacteriocins produced by gram-negative and gram-positive bacteria.

The class IIc human defensins like bacteriocin, laterosporulin 10, displays cytotoxic effects against several cell lines and causes necrotic and apoptotic cell death at high and low concentrations, respectively. According to these studies, at high concentrations (10 mM), more than 95% of normal prostate epithelial cells remained viable, whereas 80% of cancer cells lost their viability. As with cytotoxicity against normal cells, the concentrations used to be effective against cancerous cells may be higher than the levels crossing the GBB (gut–blood barrier). However, the higher affinity for cancerous cells may result in bacteriocins targeting these cells. Immune priming by bacteriocins may also assist in the elimination of cancer cells. The possibility of bacteriocins crossing the GBB is intriguing, and from the literature, it is clear that they are capable of effecting the host if they do cross. However, if they do cross and if they exert an effect requires further investigation (Dicks et al. 2018; Kaur and Kaur 2015).

6.4.3 Exopolysaccharides

Exopolysaccharides (EPS) are biological high-molecular long-chain extracellular polysaccharides surrounding the envelope of most bacteria. EPS are mainly involved in cell adhesion and protection, and often covalently bound to the cell surface in the form of capsules, or secreted into the extracellular environment in the form of slime (Sivakumar et al. 2012). EPS constitutes rhamnose, galactose, glucose, arabinose, and mannose. The diversity in the sugar composition, chain linkage, and molecular weight of EPS are known to be responsible for their antiproliferative activities (Ismail and Nampoothiri 2013; Wang et al. 2014a). Possible mechanism by which EPS exert their anticancer activity includes: (1) prevention of tumorigenesis, (2) induction of cancer cell apoptosis, and (3) immune modulation.

During last decade, several LAB have been investigated for their EPS-producing ability and their health benefits such as immunomodulatory, antitumor, antibiofilm, and antioxidant activity analyzed (Angelin and Kavitha 2020; Degeest et al. 2002). EPS extracted from probiotics plays a fundamental role in prevention and treatment of cancer. As Fig. 6.4 depicts, EPS of probiotic LAB have antimicrobial, immunomodulatory, anti-inflammatory, antioxidant, antitumor, antiviral, antidiabetic, anti-ulcer, and cholesterol lowering activities (Hussain et al. 2017; Patel et al. 2012; Angelin and Kavitha 2020)).

Several scientific data indicate that lactic acid bacteria found in the gut have a role in regression of cancer through their effect on immunomodulation. These bacteria

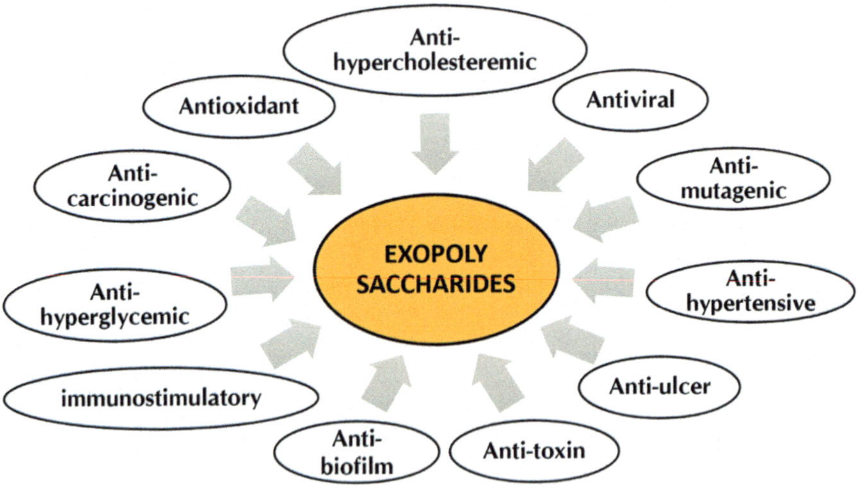

Fig. 6.4 Biological activities of exopolysaccharides produced by LAB

can activate phagocytes to remove early-stage tumor cells. In a recent report, Wu and his colleagues (2021) reported that EPS extracted from LAB are noncytotoxic to normal cells while promote tumor cell apoptosis and cell cycle arrest (Wu et al. 2020). The antitumor and antioxidant properties of EPS produced by *L. acidophilus*, *L. gasseri*, *L. plantarum*, and *L. rhamnosus* have been reported earlier (Adesulu-Dahunsi et al. 2018; Sungur et al. 2017).

In a study, the EPS of *L. fermentum* Lf2 given orally to BALB/c mice mixed with yoghurt samples was shown to increase SCFA concentrations such as acetate and butyrate. As mentioned in earlier section, these fatty acids are volatiles produced by gut microbiota and have intestinal anti-inflammatory properties (Ale et al. 2019).

Reports from Sungur and his colleagues (2017) showed that 400 μg/mL of EPS G10 extracted from *L. gasseri* could significantly inhibit HeLa cell line proliferations. According to these researchers, the enhanced antiproliferative activity observed was related to upregulation of BAX in HeLa cells and an increase in Caspase 3 protein expression that activates apoptosis (Sungur et al. 2017). Similar to these findings, EPS (MSR101) extracted from *L. kefiri* showed apoptotic effect on HT-29 cancerous cells via upregulation of the expression of cytochrome-c, BAX, BAD, caspase-3, caspase-8, and caspase-9 (Rajoka et al. 2019). *L. acidophilus* 20,079 EPS has a direct cytotoxic effect on the tumor cells by mechanisms of apoptosis, immune response stimulation, and NF-κB inflammatory pathway inactivation. The effect of *L. acidophilus* 20,079 extracted EPS on colon is a promising therapeutic target for cancer. Cell wall components of *L. acidophilus* and *L. casei* act as anticancer substances. *L. plantarum* 70,810 extracted EPS prevents the proliferation of hepatocellular carcinoma cell line. Additionally, EPS produced by probiotic lactic acid bacteria such as *L. plantarum* GD2, *L. rhamnosus* E9, *L. brevis* LB63 isolated from healthy infant feces and *Lactobacillus delbrueckii sp. bulgaricus* B3 isolated from yogurt has shown anticancer effect on colon cancer cells (HT-29) (Sharma and Shukla 2016). Lactobacilli EPS induces apoptosis in CRC in vitro through Caspase 3 and 9 and BAX increased expression and decreased Bcl-2 and survivin. Vital molecular pathways target and different forms of cell death induction by components of probiotic yeasts are considered as potential therapeutic tools against CRC. EPSs produced by probiotic *Kluyveromyces marxianus (K. marxianus)* and *Pichia kudriavzevii (P. kudriavzevii)* inhibit various colon cancer cell lines (Dicks et al. 2018; Sharma and Shukla 2016).

Moreover, probiotic *Bifidobacterium breve (B. breve)* lw01 EPS improves immune development and possesses anticancer and anti-inflammation effects (Sharma and Shukla 2016). According to these researchers EPS shows anticancer property against head and neck squamous cell carcinoma cell line by controlling apoptosis and cell cycle arrest. The investigators suggested that *B. breve* lw01 EPS can be used to assist genetic and metabolic engineering and might play role in application of functional food or drug industries.

6.4.4 Biosurfactants

Several studies reported the prospective of LABS as biosurfactant producers and their potential role in biomedical and food research (Fariq and Saeed 2016; Fracchia et al. 2012; Thavasi et al. 2011). Biosurfactants (BS) are the diverse polymeric molecules synthesized during the late log or early stationary phase of the growth cycle of an organism, secreted extracellularly or cell wall-bound. These compounds are generally localized on the microbial surface and made of amphiphilic molecule, comprising both hydrophobic and hydrophilic moieties (Banat et al. 2010). As shown in Table 6.4, the major group of biosurfactants comprises phospholipids,

Table 6.4 Structural composition of biosurfactants derived from various LAB strains

No	Bacteria	Biosurfactant produced	References
1.	*L. acidophilus*RC14	Rich in protein, high amount of polysaccharides and Phosphate content	Velraeds et al. (1996)
2.	*S. thermophilus*	Glycolipid	Busscher and Van der Mei (1997)
3.	*L. acidophilus*	Surlactin	Velraeds et al. (1996)
4.	*S. mutans* NS	Rhamnolipid like	van Hoogmoed et al. (2004)
5.	*S. thermophiles* A	Glycolipid	Rodrigues et al. (2006)
6.	*L. casei*	Glycoprotein	Gol·ek et al. (2009)
7.	*L. lactis*	Xylolipids	Saravanakumari and Mani (2010)
8.	*L. acidophilus*	Glycoprotein	Tahmourespour et al. (2011)
9.	*L. plantarum*	Glycolipids	Sauvageau et al. (2012)
10.	*L. plantarum*	Glycoprotein	Madhu and Prapulla (2014)
11.	*L. pentosus*	Glycolipids	Vecino et al. (2014)
12.	*L. casei* MRTL3	Glycolipids	Sharma et al. (2014)
13.	*E. faecium* MRTL9	Xylolipids	Sharma et al. (2015)
14.	*L. helveticus* MRTL91	Glycolipids (Xylolipids)	Sharma et al. (2015)
15.	*L. pentosus*	Glycolipopeptide	Vecino et al. (2015)
16	*L. gasseri* P_{65} and *L. jensenii* P_{6A}	Glycolipioproteins	Morais et al. (2017)
17	*L.lactis53*	Glycoprotein	Rodrigues et al. (2006)
18	*L. paracasei*	Glycoprotein	Gudina et al. (2010)
19	*L. pentosus*	Glycoprotein	Moldes et al. (2013)
20	*Bacillus subtilis ATCC 6633*	Lipopeptide	Dehghan-Noudeh et al. (2005)

fatty acids, glycolipids, lipopeptides, lipoproteins, polymeric surfactants, and particulate surfactants.

These bioactive compounds have recently emerged as promising molecules for their structural versatility, novelty, and diverse properties that are potentially useful for many therapeutic applications. One of the most important therapeutic effects of biosurfactants is their anticancer actions and their ability to regulate cancer progression processes (Gudiña et al. 2013).

The ability of these biomolecules to interact with cell membranes of several organisms and/or with the surrounding environments can be viewed as potential cancer therapeutics (Rodrigues 2011). The glycoproteins derived from *L. paracasei* were shown to have antitumor activity against breast cancer cell lines. According to the results of these researchers, the biocompound produced by the mentioned probiotic LAB was able to decrease cell viability after 48 h and reported cell cycle arrest in the tested cell lines. Lipopeptides have also been extensively studied for their potential antitumor activity. In another report, growth inhibition activity of mannosylerythritol lipids against human leukemia cells was stated (Isoda and Nakahara 1997).

Cao et al. (2010, 2011) demonstrated that surfactin induces apoptosis in human breast cancer MCF7 cells through a ROS/JNK-mediated mitochondrial/caspase pathway (Cao et al. 2010; Cao et al. 2011). While Kim et al. (2007) evaluated the effect of surfactin on the human colon carcinoma cell line LoVo and showed that the lipopeptide presents a strong growth inhibitory activity by inducing apoptosis and cell cycle arrest (Kim et al. 2010). Lee et al. (2012) demonstrated that surfactin inhibited the growth of MCF7 human breast cancer cells in a dose-dependent manner (Lee et al. 2012). In addition, several other lipopeptides (isoforms of surfactin and fengycin) were also found to have potent cytotoxic effects against the human colon cancer cell lines HCT15 and HT29 (Sivapathasekaran et al. 2010). While Durate and his colleagues were able to show the effect of biosurfactants on viability and proliferation of human breast cancer cells (Duarte et al. 2014).

However, an important drawback of using surfactin as a chemotherapeutic agent is its hemolytic activity (Dehghan-Noudeh et al. 2005) that has been reported for concentrations above 0.05 g/L. Since surfactin has never been tested in humans, to prevent future complications several strategies have been explored envisaging its use as a safe therapeutic agent. Symmank et al. (2002) reported several minor modifications of the surfactin molecule by altering surfactin synthetase. These modifications changed the molecule toxicity profile, resulting in a "new" lipopeptide with improved activity and not revealing any signs of toxicity or hemolytic activity. Another interesting approach consists in the incorporation of surfactin in nanoparticles in order to provide a directed as in order to provide a directed administration and in situ release of the cyclic peptide (Symmank et al. 2002).

Since there is an enormous diversity of microbial surfactants, the attention of the scientific community in the search for new molecules with interesting antitumor activities is continuously increasing, as well as in looking deeply into their mechanisms of action. Therefore, clinical applications of specifically probiotic biosurfactants yet need to be explored. In addition, further research is required to unravel the mechanism of their action in human body which would assist in their target sited applications without interfering with other body microflora. Moreover, investigation on genetics of probiotic bacteria in regulation of biosurfactants is vital for their optimum production and potential applications in health sector.

6.5 Anticancer Effects of Paraprobiotics (Dead Probiotic Cells)

In contrast to probiotics, paraprobiotics are defined as dead or nonviable probiotic bacterial cells or cell components that confer health benefits when administered in adequate doses. They are also referred to as ghost or inactivated probiotics and mainly constitute ruptured cell components of probiotic cells such as teichoic acids, peptidoglycan-derived muropeptides, pili, fimbriae, flagella, polysaccharides, biosurfactants, etc. Some paraprobiotic products have been commercialized and present in the market with the trade name of Lactéol Fort® from PUMC Pharmaceutical Co., Ltd. and Fermenti Lattici Tindalizzati® from Frau, AF United Spa (Taverniti and Guglielmetti 2011).

The anticancerous activity of paraprobiotics has been reported by many; in fact, the suppressive potentials of dead probiotic have been shown to be superior to live probiotic. According to these statements, administration of high doses of dead probiotic reduced a number of tumors considerably compared with pure live probiotic. Dead probiotics showed fewer colonic tumors, longer colons, and less weight loss compared with pure live probiotic *L. plantarum* (Taverniti and Guglielmetti 2011; Wang et al. 2014b). This property was due to the effects of inflammation suppression, apoptosis, and enhanced IgA secretion. Figure 6.5 dictates some proposed mechanism of action of paraprobiotics (Fig. 6.5).

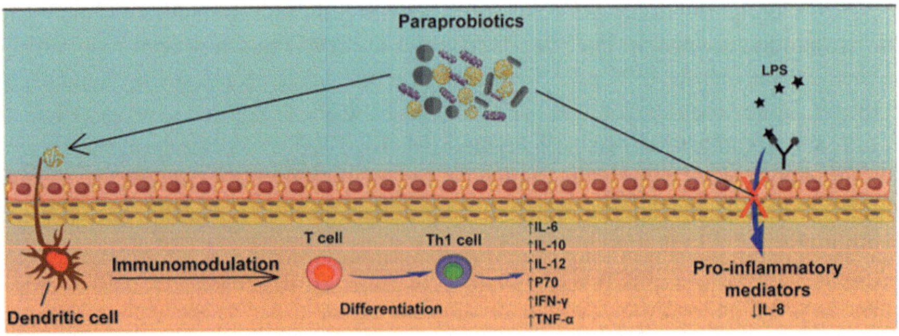

Fig. 6.5 Diagrammatic representation of mechanism of action of paraprobiotic; the inactivated or dead probiotic bacterial cells are known to exert their effects by immunomodulation of T cells by dendritic cells, stimulating their differentiation into Th1 cells, promoting the production of anti-inflammatory cytokines. Another proposed mechanism includes inhibition of signaling pathways related to LPS stimulation, resulting in a reduction of proinflammatory mediators, like IL-8 (Batista et al. 2020)

In a study, combined administration of live and dead probiotics was shown to significantly reduce proinflammatory cytokines and inflammatory genes overexpression, and suppressive potentials than separate administration of either groups. All the experimental AOM/DSS control animals group possessed colon tumors, but administration with dead *L. plantarum* suppressed the development of neoplasty significantly. The mechanism by which dead probiotic sustains the status of mucosal immune system is by increasing levels of secretory IgA. This indicates that the antitumor property of dead probiotic is related with the easier uptake of dead probiotic by more cells than pure live probiotic and the stronger secretory immune responses (Sharma and Shukla 2016).

Recent report indicated that, the application of heat-inactivated probiotic *Enterococcus faecalis* (*E. faecalis*) protects against dextran sodium sulfate-induced CRC and ameliorates intestinal inflammation severity in wild-type mice. *E. faecalis* paraprobiotic fractions provided protection to experimental animals against dextran sodium sulfate-induced colitis and CRC by reducing intestinal inflammation severity through phagocytosis attenuation. It was concluded that heat-killed probiotic *E. faecalis* is safe and useful for inflammation-associated colon carcinogenesis attenuation by inhibition of IL-1β secondsretion induction in macrophages (Sharma and Shukla 2016).

6.6 Conclusions

Probiotics are known to exert various health benefits such as immunomodulation, inactivation of carcinogens, and maintenance of gut integrity, but the present review represents their role in cancer prevention and treatment. However, owing to their viable status these beneficial bacteria might impose some adverse effects in certain immunosuppressed individuals, which in turn might limit their use. Thus, attempts are being made to the dead counterpart (paraprobiotics) of these viable bacteria or their secretory substances (postbiotic metabolites/metabiotics) that might be safer alternative and effective bio-interventions. The significant role of these metabolites has been shown to possess remarkable antimutagenic, anti-inflammatory, antiproliferative potentials attributed to their epigenetic effects in one or the other way, and may target cancer at different stages. Hence, paraprobiotic and metabiotics independently or in conjunction with other approaches could be considered as a potent prophylactic/or therapeutic modulator for cancer or other diseases in the post-antibiotic era. However, there is a high need for human/clinical trials focusing on the validation of health claims of these bioactive molecules. The trials in immunocompromised subjects would be further augmentable to investigate the tolerance of immunocompromised subjects on these biomolecules. On the other hand, there is a lack of knowledge regarding the stability of paraprobiotics and postbiotics under *in vitro* and *in vivo* digestive conditions to comprehend specific mechanistic actions by interacting with the ligands. These biomolecules may eliminate the adverse effects and reduce the difficulties in the maintenance of viability of probiotics, but yet enough information regarding their human clinical trials is not available that could elucidate their mechanism of action and propose their safe use in human especially for the treatment of variety of cancer.

References

Abdi-Ali A, Worobec E, Deezagi A, Malekzadeh F (2004) Cytotoxic effects of pyocin S2 produced by Pseudomonas aeruginosa on the growth of three human cell lines. Can J Microbiol 50:375–381

Adesulu-Dahunsi AT, Jeyaram K, Sanni AI, Banwo K (2018) Production of exopolysaccharide by strains of Lactobacillus plantarum YO175 and OF101 isolated from traditional fermented cereal beverage. Peer J 6:e5326

Aguilar-Toalá J, Garcia-Varela R, Garcia H, Mata-Haro V, González-Córdova A, Vallejo-Cordoba-B, Hernández-Mendoza A (2018) Postbiotics: an evolving term within the functional foods field. Trends Food Sci Technol 75:105–114

Ale EC, Bourin MJ-B, Peralta GH, Burns PG, Ávila OB, Contini L, Reinheimer J, Binetti AG (2019) Functional properties of exopolysaccharide (EPS) extract from Lactobacillus fermentum Lf2 and its impact when combined with Bifidobacterium animalis INL1 in yoghurt. Int Dairy J 96:114–125

Altonsy MO, Andrews SC, Tuohy KM (2010) Differential induction of apoptosis in human colonic carcinoma cells (Caco-2) by Atopobium, and commensal, probiotic and enteropathogenic bacteria: mediation by the mitochondrial pathway. Int J Food Microbiol 137:190–203

Angelin J, Kavitha M (2020) Exopolysaccharides from probiotic bacteria and their health potential. Int J Biol Macromol 162:853–865

Arvind NKS, Sinha PR (2009) Inhibition of 1, 2 dimethylhydrazine induced genotoxicity in rats by the administration of probiotic curd. Int J Probiotics Prebiotics 4:201–204

Azad M, Kalam A, Sarker M, Li T, Yin J (2018) Probiotic species in the modulation of gut microbiota: an overview. Biomed Res Int 2018:9478630. https://doi.org/10.1155/2018/9478630

Banat IM, Franzetti A, Gandolfi I, Bestetti G, Martinotti MG, Fracchia L, Smyth TJ, Marchant R (2010) Microbial biosurfactants production, applications and future potential. Appl Microbiol Biotechnol 87:427–444

Barteneva NS, Baiken Y, Fasler-Kan E, Alibek K, Wang S, Maltsev N, Ponomarev ED, Sautbayeva Z, Kauanova S, Moore A (2017) Extracellular vesicles in gastrointestinal cancer in conjunction with microbiota: on the border of Kingdoms. Biochim Biophys Acta 1868:372–393

Batista, V.L., Da Silva, T.F., De Jesus, L.C.L., Dias Coelho-Rocha, N., Barroso, F.A.L., Tavares, L. M., Azevedo, V.A.D.C., Mancha-Agresti, P.D.C., Drumond, M.M., 2020. Probiotics, prebiotics, synbiotics, and paraprobiotics as a therapeutic alternative for intestinal mucositis running head: alternative treatment for intestinal mucositis. Front Microbiol 11, 2246

Beaulieu L, Tolkatchev D, Jette J-F, Groleau D, Subirade M (2007) Production of active pediocin PA-1 in Escherichia coli using a thioredoxin gene fusion expression approach: cloning, expression, purification, and characterization. Can J Microbiol 53:1246–1258

Bedada TL, Feto TK, Awoke KS, Garedew AD, Yifat FT, Birri DJ (2020) Probiotics for cancer alternative prevention and treatment. Biomed Pharmacother 129:110409

Busscher H, Van der Mei H (1997) Physico-chemical interactions in initial microbial adhesion and relevance for biofilm formation. Adv Dent Res 11:24–32

Cao X-H, Wang A-H, Wang C-L, Mao D-Z, Lu M-F, Cui Y-Q, Jiao R-Z (2010) Surfactin induces apoptosis in human breast cancer MCF-7 cells through a ROS/JNK-mediated mitochondrial/caspase pathway. Chem Biol Interact 183:357–362

Cao X-H, Zhao S-S, Liu D-Y, Wang Z, Niu L-L, Hou L-H, Wang C-L (2011) ROS-Ca2+ is associated with mitochondria permeability transition pore involved in surfactin-induced MCF-7 cells apoptosis. Chem Biol Interact 190:16–27

Chumchalova J, Šmarda J (2003) Human tumor cells are selectively inhibited by colicins. Folia Microbiol 48:111–115
Degeest B, Mozzi F, De Vuyst L (2002) Effect of medium composition and temperature and pH changes on exopolysaccharide yields and stability during Streptococcus thermophilus LY03 fermentations. Int J Food Microbiol 79:161–174
Dehghan-Noudeh G, Housaindokht M, Bazzaz BSF (2005) Isolation, characterization, and investigation of surface and hemolytic activities of a lipopeptide biosurfactant produced by Bacillus subtilis ATCC 6633. J Microbiol 43:272–276
Dicks LM, Dreyer L, Smith C, Van Staden AD (2018) A review: the fate of bacteriocins in the human gastro-intestinal tract: do they cross the gut–blood barrier? Front Microbiol 9:2297
dos Reis SA, da Conceição LL, Siqueira NP, Rosa DD, da Silva LL, Maria do Carmo GP (2017) Review of the mechanisms of probiotic actions in the prevention of colorectal cancer. Nutr Res 37:1–19
Duarte C, Gudiña EJ, Lima CF, Rodrigues LR (2014) Effects of biosurfactants on the viability and proliferation of human breast cancer cells. AMB Express 4:40
Fariq A, Saeed A (2016) Production and biomedical applications of probiotic biosurfactants. Curr Microbiol 72:489–495
Forsyth CB, Farhadi A, Jakate SM, Tang Y, Shaikh M, Keshavarzian A (2009) Lactobacillus GG treatment ameliorates alcohol-induced intestinal oxidative stress, gut leakiness, and liver injury in a rat model of alcoholic steatohepatitis. Alcohol 43:163–172
Fracchia L, Cavallo M, Martinotti MG, Banat IM (2012) Biosurfactants and bioemulsifiers biomedical and related applications–present status and future potentials. Biomed Sci Eng Technol 14:326–335
Fujiki T, Hirose Y, Yamamoto Y, Murosaki S (2012) Enhanced immunomodulatory activity and stability in simulated digestive juices of Lactobacillus plantarum L-137 by heat treatment. Biosci Biotechnol Biochem 76:918–922
Fuska J, Fusková A, Šmarda J, Mach J (1979) Effect of colicin E3 on leukemia cells P 388 in vitro. Experientia 35:406–407
Ghanavati R, Asadollahi P, Shapourabadi MB, Razavi S, Talebi M, Rohani M (2020) Inhibitory effects of Lactobacilli cocktail on HT-29 colon carcinoma cells growth and modulation of the Notch and Wnt/β-catenin signaling pathways. Microb Pathog 139:103829
Goldin BR, Gorbach SL (1980) Effect of Lactobacillus acidophilus dietary supplements on 1, 2-dimethylhydrazine dihydrochloride-induced intestinal cancer in rats. J Natl Cancer Inst 64:263–265
Gołek P, Bednarski W, Brzozowski B, Dziuba B (2009) The obtaining and properties of biosurfactants synthesized by bacteria of the genusLactobacillus. Ann Microbiol 59:119–126
Górska A, Przystupski D, Niemczura MJ, Kulbacka J (2019) Probiotic bacteria: a promising tool in cancer prevention and therapy. Curr Microbiol:1–11
Gudina EJ, Teixeira JA, Rodrigues LR (2010) Isolation and functional characterization of a biosurfactant produced by Lactobacillus paracasei. Colloids Surf B: Biointerfaces 76:298–304
Gudiña EJ, Rangarajan V, Sen R, Rodrigues LR (2013) Potential therapeutic applications of biosurfactants. Trends Pharmacol Sci 34:667–675
Han KJ, Lee N-K, Park H, Paik H-D (2015) Anticancer and anti-inflammatory activity of probiotic Lactococcus lactis NK34. J Microbiol Biotechnol 25:1697–1701
Hetz Flores C, Bono Merino MR, Barros LF, Lagos Mónaco R (2002) Microcin E492, a channel-forming bacteriocin from Klebsiella pneumoniae, induces apoptosis in some human cell lines. Proc Natl Acad Sci 99:2696–2971
Hu J, Wang C, Ye L, Yang W, Huang H, Meng F, Shi S, Ding Z (2015) Anti-tumour immune effect of oral administration of Lactobacillus plantarum to CT26 tumour-bearing mice. J Biosci 40:269–279
Hussain A, Zia KM, Tabasum S, Noreen A, Ali M, Iqbal R, Zuber M (2017) Blends and composites of exopolysaccharides; properties and applications: a review. Int J Biol Macromol 94:10–27

Ismail B, Nampoothiri K (2013) Exposition of antitumour activity of a chemically characterized exopolysaccharide from a probiotic Lactobacillus plantarum MTCC 9510. Biologia 68:1041–1047

Isoda H, Nakahara T (1997) Mannosylerythritol lipid induces granulocytic differentiation and inhibits the tyrosine phosphorylation of human myelogenous leukemia cell line K562. Cytotechnology 25:191–195

Jacouton E, Chain F, Sokol H, Langella P, Bermudez-Humaran LG (2017) Probiotic strain Lactobacillus casei BL23 prevents colitis-associated colorectal cancer. Front Immunol 8:1553

Kahouli I, Tomaro-Duchesneau C, Prakash S (2013) Probiotics in colorectal cancer (CRC) with emphasis on mechanisms of action and current perspectives. J Med Microbiol 62:1107–1123

Kaur S, Kaur S (2015) Bacteriocins as potential anticancer agents. Front Pharmacol 6:272

Kim SW, Kim HM, Yang KM, Kim S-A, Kim S-K, An MJ, Park JJ, Lee SK, Kim TI, Kim WH (2010) Bifidobacterium lactis inhibits NF-κB in intestinal epithelial cells and prevents acute colitis and colitis-associated colon cancer in mice. Inflamm Bowel Dis 16:1514–1525

Kumar M, Kumar A, Nagpal R, Mohania D, Behare P, Verma V, Kumar P, Poddar D, Aggarwal P, Henry C (2010) Cancer-preventing attributes of probiotics: an update. Int J Food Sci Nutrit 61:473–496

Kumar RS, Kanmani P, Yuvaraj N, Paari K, Pattukumar V, Thirunavukkarasu C, Arul V (2012) Lactobacillus plantarum AS1 isolated from south Indian fermented food Kallappam suppress 1, 2-dimethyl hydrazine (DMH)-induced colorectal cancer in male Wistar rats. Appl Biochem Biotechnol 166:620–631

Lee JH, Nam SH, Seo WT, Yun HD, Hong SY, Kim MK, Cho KM (2012) The production of surfactin during the fermentation of cheonggukjang by potential probiotic Bacillus subtilis CSY191 and the resultant growth suppression of MCF-7 human breast cancer cells. Food Chem 131:1347–1354

Lee N-K, Son S-H, Jeon EB, Jung GH, Lee J-Y, Paik H-D (2015) The prophylactic effect of probiotic Bacillus polyfermenticus KU3 against cancer cells. J Funct Foods 14:513–518

Liu Z, Li C, Huang M, Tong C, Zhang X, Wang L, Peng H, Lan P, Zhang P, Huang N (2015) Positive regulatory effects of perioperative probiotic treatment on postoperative liver complications after colorectal liver metastases surgery: a double-center and double-blind randomized clinical trial. BMC Gastroenterol 15:34

Lüke J, Vukoja V, Brandenbusch T, Nassar K, Rohrbach JM, Grisanti S, Lüke M, Tura A (2016) CD147 and matrix-metalloproteinase-2 expression in metastatic and non-metastatic uveal melanomas. BMC Ophthalmol 16:74

Ma EL, Choi YJ, Choi J, Pothoulakis C, Rhee SH, Im E (2010) The anticancer effect of probiotic Bacillus polyfermenticus on human colon cancer cells is mediated through ErbB2 and ErbB3 inhibition. Int J Cancer 127:780–790

Madhu AN, Prapulla SG (2014) Evaluation and functional characterization of a biosurfactant produced by Lactobacillus plantarum CFR 2194. Appl Biochem Biotechnol 172:1777–1789

Maghsood F, Mirshafiey A, Farahani MM, Modarressi MH, Jafari P, Motevaseli E (2018) Dual effects of cell free supernatants from Lactobacillus acidophilus and Lactobacillus rhamnosus GG in regulation of MMP-9 by up-regulating TIMP-1 and down-regulating CD147 in PMA-differentiated THP-1 cells. Cell J (Yakhteh) 19:559

Marques FZ, Nelson E, Chu P-Y, Horlock D, Fiedler A, Ziemann M, Tan JK, Kuruppu S, Rajapakse NW, El-Osta A (2017) High-fiber diet and acetate supplementation change the gut microbiota and prevent the development of hypertension and heart failure in hypertensive mice. Circulation 135:964–977

Mehra N, Majumdar R, Kumar S, Dhewa T (2012) Probiotics: preventive and clinical applications. Biosci Res Bull 1:15–20

Moldes A, Paradelo R, Vecino X, Cruz J, Gudiña E, Rodrigues L, Teixeira J, Domínguez JM, Barral M (2013) Partial characterization of biosurfactant from Lactobacillus pentosus and comparison with sodium dodecyl sulphate for the bioremediation of hydrocarbon contaminated soil. Biomed Res Int 2013

Morais I, Cordeiro A, Teixeira G, Domingues V, Nardi R, Monteiro A, Alves R, Siqueira E, Santos V (2017) Biological and physicochemical properties of biosurfactants produced by Lactobacillus jensenii P 6A and Lactobacillus gasseri P 65. Microb Cell Factories 16:155

Motevaseli E, Shirzad M, Akrami SM, Mousavi A-S, Mirsalehian A, Modarressi MH (2013) Normal and tumour cervical cells respond differently to vaginal lactobacilli, independent of pH and lactate. J Med Microbiol 62:1065–1072

Noorozi E, Mojgani N, Motaveselli E, Tebianian M, Modaressi S (2021) Prophylactic role of probiotic, paraprobiotic and postbiotic of Lactobacillus casei strains against colorectal cancer cell; invitro studies. Braz J Pharm Res (in press)

Oelschlaeger TA (2010) Mechanisms of probiotic actions—a review. Int J Med Microbiol 300:57–62

Orlando A, Refolo M, Messa C, Amati L, Lavermicocca P, Guerra V, Russo F (2012) Antiproliferative and proapoptotic effects of viable or heat-killed Lactobacillus paracasei IMPC2. 1 and Lactobacillus rhamnosus GG in HGC-27 gastric and DLD-1 colon cell lines. Nutr Cancer 64:1103–1111

Paiva AD, Breukink E, Mantovani HC (2011) Role of lipid II and membrane thickness in the mechanism of action of the lantibiotic bovicin HC5. Antimicrob Agents Chemother 55:5284–5293

Paiva AD, de Oliveira MD, de Paula SO, Baracat-Pereira MC, Breukink E, Mantovani HC (2012) Toxicity of bovicin HC5 against mammalian cell lines and the role of cholesterol in bacteriocin activity. Microbiology 158:2851–2858

Patel S, Majumder A, Goyal A (2012) Potentials of exopolysaccharides from lactic acid bacteria. Indian J Microbiol 52:3–12

Rajoka MSR, Mehwish HM, Fang H, Padhiar AA, Zeng X, Khurshid M, He Z, Zhao L (2019) Characterization and anti-tumor activity of exopolysaccharide produced by Lactobacillus kefiri isolated from Chinese kefir grains. J Funct Foods 63:103588

Rodrigues LR (2011) Inhibition of bacterial adhesion on medical devices. Adv Exp Med Biol 715:351–367

Rodrigues L, Banat IM, Teixeira J, Oliveira R (2006) Biosurfactants: potential applications in medicine. J Antimicrob Chemother 57:609–618

Sand SL, Haug TM, Nissen-Meyer J, Sand O (2007) The bacterial peptide pheromone plantaricin A permeabilizes cancerous, but not normal, rat pituitary cells and differentiates between the outer and inner membrane leaflet. J Membr Biol 216:61–71

Sand SL, Oppegård C, Ohara S, Iijima T, Naderi S, Blomhoff HK, Nissen-Meyer J, Sand O (2010) Plantaricin A, a peptide pheromone produced by Lactobacillus plantarum, permeabilizes the cell membrane of both normal and cancerous lymphocytes and neuronal cells. Peptides 31:1237–1244

Sand SL, Nissen-Meyer J, Sand O, Haug TM (2013) Plantaricin A, a cationic peptide produced by Lactobacillus plantarum, permeabilizes eukaryotic cell membranes by a mechanism dependent on negative surface charge linked to glycosylated membrane proteins. Biochim Biophys Acta 1828:249–259

Sanders M, Merenstein D, Merrifield C, Hutkins R (2018) Probiotics for human use. Nutr Bull 43:212–225

Saravanakumari P, Mani K (2010) Structural characterization of a novel xylolipid biosurfactant from Lactococcus lactis and analysis of antibacterial activity against multi-drug resistant pathogens. Bioresour Technol 101:8851–8854

Sauvageau J, Ryan J, Lagutin K, Sims IM, Stocker BL, Timmer MS (2012) Isolation and structural characterisation of the major glycolipids from Lactobacillus plantarum. Carbohydr Res 357:151–156

Saxami G, Karapetsas A, Lamprianidou E, Kotsianidis I, Chlichlia A, Tassou C, Zoumpourlis V, Galanis A (2016) Two potential probiotic lactobacillus strains isolated from olive microbiota exhibit adhesion and anti-proliferative effects in cancer cell lines. J Funct Foods 24:461–471

Schug ZT, Voorde JV, Gottlieb E (2016) The metabolic fate of acetate in cancer. Nat Rev Cancer 16:708–717

Sharma A (2019) Importance of probiotics in cancer prevention and treatment. In: Recent developments in applied microbiology and biochemistry. Elsevier, pp 33–45

Sharma M, Shukla G (2016) Metabiotics: one step ahead of probiotics; an insight into mechanisms involved in anticancerous effect in colorectal cancer. Front Microbiol 7:1940

Sharma D, Singh Saharan B (2014) Simultaneous production of biosurfactants and bacteriocins by probiotic Lactobacillus casei MRTL3. Int J Microbiol 2014

Sharma D, Saharan BS, Chauhan N, Bansal A, Procha S (2014) Production and structural characterization of Lactobacillus helveticus derived biosurfactant. Sci World J 2014

Sharma D, Saharan BS, Chauhan N, Procha S, Lal S (2015) Isolation and functional characterization of novel biosurfactant produced by Enterococcus faecium. Springerplus 4:1–14

Sivakumar T, Sivasankara Narayani S, Shankar T, Vijayabaskar P (2012) Optimization of cultural conditions for exopolysaccharides production by Frateuria aurentia. Int J Appl Biol Pharm Technol 3:133–144

Sivamaruthi BS, Kesika P, Chaiyasut C (2020) The role of probiotics in colorectal cancer management. Evid Based Complement Alternat Med:3535982

Sivapathasekaran C, Das P, Mukherjee S, Saravanakumar J, Mandal M, Sen R (2010) Marine bacterium derived lipopeptides: characterization and cytotoxic activity against cancer cell lines. Int J Pept Res Ther 16:215–222

Šmarda J, Obdržálek V, Táborský I, Mach J (1978) The cytotoxic and cytocidal effect of colicin E3 on mammalian tissue cells. Folia Microbiol 23:272–277

Sungur T, Aslim B, Karaaslan C, Aktas B (2017) Impact of Exopolysaccharides (EPSs) of Lactobacillus gasseri strains isolated from human vagina on cervical tumor cells (HeLa). Anaerobe 47:137–144

Symmank H, Franke P, Saenger W, Bernhard F (2002) Modification of biologically active peptides: production of a novel lipohexapeptide after engineering of Bacillus subtilis surfactin synthetase. Protein Eng 15:913–921

Tahmourespour A, Salehi R, Kermanshahi RK (2011) Lactobacillus acidophilus-derived biosurfactant effect on gtfB and gtfC expression level in Streptococcus mutans biofilm cells. Braz J Microbiol 42:330–339

Taverniti V, Guglielmetti S (2011) The immunomodulatory properties of probiotic microorganisms beyond their viability (ghost probiotics: proposal of paraprobiotic concept). Genes Nutr 6:261–274

Thavasi R, Jayalakshmi S, Banat IM (2011) Application of biosurfactant produced from peanut oil cake by Lactobacillus delbrueckii in biodegradation of crude oil. Bioresour Technol 102:3366–3372

Tiptiri-Kourpeti A, Spyridopoulou K, Santarmaki V, Aindelis G, Tompoulidou E, Lamprianidou EE, Saxami G, Ypsilantis P, Lampri ES, Simopoulos C (2016) Lactobacillus casei exerts anti-proliferative effects accompanied by apoptotic cell death and up-regulation of TRAIL in colon carcinoma cells. PLoS One 11:e0147960

van Hoogmoed CG, van der Mei HC, Busscher HJ (2004) The influence of biosurfactants released by S. mitis BMS on the adhesion of pioneer strains and cariogenic bacteria. Biofouling 20:261–267

Vecino X, Devesa-Rey R, Moldes A, Cruz J (2014) Formulation of an alginate-vineyard pruning waste composite as a new eco-friendly adsorbent to remove micronutrients from agroindustrial effluents. Chemosphere 111:24–31

Vecino X, Barbosa-Pereira L, Devesa-Rey R, Cruz JM, Moldes AB (2015) Optimization of extraction conditions and fatty acid characterization of Lactobacillus pentosus cell-bound biosurfactant/bioemulsifier. J Sci Food Agric 95:313–320

Velraeds MM, van der Mei HC, Reid G, Busscher HJ (1996) Physicochemical and biochemical characterization of biosurfactants released by Lactobacillus strains. Colloids Surf B: Biointerfaces 8:51–61

Villarante KI, Elegado FB, Iwatani S, Zendo T, Sonomoto K, de Guzman EE (2011) Purification, characterization and in vitro cytotoxicity of the bacteriocin from Pediococcus acidilactici K2a2-3 against human colon adenocarcinoma (HT29) and human cervical carcinoma (HeLa) cells. World J Microbiol Biotechnol 27:975–980

Voigt H, Vetter-Kauczok CS, Schrama D, Hofmann UB, Becker JC, Houben R (2009) CD147 impacts angiogenesis and metastasis formation. Cancer Investig 27:329–333

Walia S, Kamal R, Kanwar SS, Dhawan DK (2015) Cyclooxygenase as a target in chemoprevention by probiotics during 1, 2-dimethylhydrazine induced colon carcinogenesis in rats. Nutr Cancer 67:603–611

Wang K, Li W, Rui X, Chen X, Jiang M, Dong M (2014a) Characterization of a novel exopolysaccharide with antitumor activity from Lactobacillus plantarum 70810. Int J Biol Macromol 63:133–139

Wang S-M, Zhang L-W, Fan R-B, Han X, Yi H-X, Zhang L-L, Xue C-H, Li H-B, Zhang Y-H, Shigwedha N (2014b) Induction of HT-29 cells apoptosis by lactobacilli isolated from fermented products. Res Microbiol 165:202–214

Watanabe T, Saito H (1980) Cytotoxicity of pyocin S2 to tumor and normal cells and its interaction with cell surfaces. Biochim Biophys Acta 633:77–86

Williams EA, Coxhead JM, Mathers JC (2003) Anti-cancer effects of butyrate: use of micro-array technology to investigate mechanisms. Proc Nutr Soc 62:107–115

Wu J, Zhang Y, Ye L, Wang C (2020) The anti-cancer effects and mechanisms of lactic acid bacteria exopolysaccharides in vitro: a review. Carbohydr Polym 253:117308

Zhao H, Sood R, Jutila A, Bose S, Fimland G, Nissen-Meyer J, Kinnunen PK (2006) Interaction of the antimicrobial peptide pheromone Plantaricin A with model membranes: implications for a novel mechanism of action. Biochim Biophys Acta 1758:1461–1474

Zhong L, Zhang X, Covasa M (2014) Emerging roles of lactic acid bacteria in protection against colorectal cancer. World J Gastroenterol: WJG 20:7878

Postbiotic Metabolites of Probiotics in Animal Feeding

7

Teck Chwen Loh, Hooi Ling Foo, and Hui Mei Chang

Abstract

The gut is consistently exposed to broad harmful pathogens from the external environment. It also acts as a barrier against infections, involves in digestion, nutrient absorption, and immune response. Any disruption and dysfunction in the gut microenvironment can cause huge implications on the general well-being of the host. Hence, a good gut health, which is an integral component of the diet, gut microbiota, and intestinal mucus layer, is vital to maintain the normal function of physiological and metabolic activities in the body. In this chapter, we highlighted the positive effects of supplementation of dietary postbiotic on growth performance, gut microbiota, intestinal morphology, immune response, meat quality, and expression of genes related to the barrier function and immunity in livestock. The inclusion of antibiotic growth promoters to achieve good intestinal health and performance which led to the emergence of antibiotic resistance gene and

T. C. Loh (✉)
Department of Animal Science, Faculty of Agriculture, Universiti Putra Malaysia, Serdang, Selangor, Malaysia

Institute of Tropical Agriculture and Food Security, Faculty of Agriculture, Universiti Putra Malaysia, Serdang, Selangor, Malaysia
e-mail: tcloh@upm.edu.my

H. L. Foo (✉)
Department of Bioprocess Technology, Faculty of Biotechnology and Biomolecular Science, Universiti Putra Malaysia, Serdang, Selangor, Malaysia

Institute of Bioscience, Universiti Putra Malaysia, Serdang, Selangor, Malaysia
e-mail: hlfoo@upm.edu.my

H. M. Chang
Department of Animal Science, Faculty of Agriculture, Universiti Putra Malaysia, Serdang, Selangor, Malaysia

N. Mojgani, M. Dadar (eds.), *Probiotic Bacteria and Postbiotic Metabolites: Role in Animal and Human Health*, Microorganisms for Sustainability 2,
https://doi.org/10.1007/978-981-16-0223-8_7

antibiotic residue level in the food product as well as beneficial compounds detected in the postbiotic are discussed in more detail.

Keywords

Probiotics · Postbiotics · Gut microbiota · Animal feeding

7.1 Introduction

The gut is a vital organ involved in nutrient digestion and absorption, metabolism, and immunity. It is highly exposed to the extrinsic pathogens (Yegani and Korver 2008). Failure or disruption on the digestion and nutrient absorption could affect the optimum performance and health of the birds. The intestinal mucosal is one of the critical determinants in the gut health and performance of the birds; it protects the internal environment against various harmful agents from the external environment, aids nutrient absorption, and acts as lubricant and site for microflora colonization (Rinttila and Apajalahti 2013).

The main factor that is responsible for the gut health and performance in a poultry flock is the gut microbiota. Even though it helps to shape the gut structure and morphology, which aid in the digestion and nutrient absorption, a good colonization of commensal microflora also protects against the pathogenic invasion and immune response (Sugiharto 2016). Also, a well-balanced intestinal mucosal barrier function is associated with the internal homeostasis between the mucus layer, epithelial cells, gut microflora, and immune cells (Schenk and Mueller 2008). The diet, nutrition, and infectious disease agents have been identified as the prominent factors that interrupt with the mucin dynamics in the gut. In turn, it will affect the well-being and performance of the birds (Yegani and Korver 2008).

Antibiotic growth promoter (AGP) has been included sub-therapeutically in the feed extensively to achieve optimum gut health. The extensive application of AGP in livestock nutrition may produce public health consequences as the resistance develops in many pathogenic bacterial species in exposed animals. It was reported that *E. coli* isolated from pigs' faeces were resistant to various commonly used antibiotics such as neomycin, oxytetracycline, nalidixic acid, and chloramphenicol (Loh et al. 2013). Similar findings were also reported in the Netherlands (Van de Bogaard et al. 2001). AGP has been widely used to maintain the equilibrium of the gut ecosystem as well as to improve the growth performance of birds (Huyghebaert et al. 2011).

Nonetheless, the usage of AGP for long periods has led to the emergence of antibiotic resistance gene (Shazali et al. 2014) and exceeded permitted residue levels in animal products (Van de Bogaard et al. 2001). Genes encoding for this resistance can also be transferred to other formerly susceptible bacteria, thereby posing a threat to both animal and human health (Montagne et al. 2003). Due to the rising concern of food safety and security from the public, the inclusion of AGP in animal feed is either restricted or outright banned in many countries (Ohimain and Ofongo 2012).

However, with the ban and restriction to the use of AGPs in poultry nutrition, other natural and safe alternatives are required to maintain the well dynamic gut microbiota in chickens (Sugiharto 2016).

Another natural alternative to replace AGP is the postbiotic. Unlike probiotic which deals with living cells, postbiotic is the secondary metabolic or the molecules secreted by the probiotic during the metabolic activity (Tsilingiri and Rescigno 2013). The postbiotic contains bacteriocin, enzymes, and other unelucidated protein compounds. It also contains organic acids such as lactic acid, acetic acid, and other short-chain fatty acids (Lee et al. 2019). It has been proven that postbiotics have antimicrobial and anti-inflammatory functions, prevent the proliferation of pathogens by lowering the pH in the gut environment (Kareem et al. 2014), stimulate cytotoxicity effect on cancer cells (Chuah et al. 2019) and act as an antioxidant agent (Humam et al. 2019; Izuddin et al. 2020). In the in vitro studies, the metabolites produced from *Lactiplantibacillus plantarum (L. plantarum)* were able to inhibit the growth of pathogenic bacteria such as *Listeria monocytogenes, Salmonella typhimurium, Escherichia coli,* and Vancomycin-Resistant *Enterococci* and *Paediococcus acidilactici* (Kareem et al. 2014; Loh et al. 2010). Furthermore, enhancement in the growth performance, faecal lactic acid bacteria, intestinal morphology and immune status were observed when metabolite combinations were added to the feed of broilers (Thanh et al. 2009), laying hens (Choe et al. 2012), pigs (Thu et al. 2011), and lambs (Izuddin et al. 2020).

7.2 Postbiotics

Postbiotics, the metabiotics or metabolites are the cell-free extracts or soluble factors (metabolic by-products) secreted by live bacteria or released after bacterial lysis (Aguilar-Toala et al. 2018). The hosts are benefitted physiologically by the capacity of the metabolites to induce additional bioactivity (Cicenia et al. 2014). Substances such as organic acids, bacteriocins, enzymes, cell surface proteins, peptides, polysaccharides (endo- and exo-), and plasmalogens are few examples of soluble factors that are commonly found in postbiotic metabolite (Lee et al. 2019; Thu et al. 2011; Aguilar-Toala et al. 2018; Lim et al. 2019; Malashree et al. 2019; Toe et al. 2019). Furthermore, their broad inhibitory activities against several pathogenic bacteria have been reported. For instance, reduced colonization of pathogenic bacteria such as *E. coli*, *Salmonella*, and *Enterobacteriaceae* in the gastrointestinal tract was reported in poultry following supplementation with various lactic acid bacteria (LAB) metabolites (Humam et al. 2019; Izuddin et al. 2020; Thanh et al. 2009; Choe et al. 2012; Loh et al. 2009; Kareem et al. 2016a).

Although the underlining mechanisms regarding the beneficial implications of postbiotics on livestock health are not well understood, there is scientific-based evidence to show that these metabolites possess various functional characteristics such as antimicrobial, immunomodulatory antioxidant, and other unconfirmed properties. Currently, these properties are known to influence gut microbiota and metabolic pathways associated with physiological, immunological, and hormonal functions (Foo et al. 2019; Yang et al. 2019).

7.3 Implications of Postbiotic Supplementation on Growth Performance

The growth performance in poultry is measured using parameters such as total feed intake, feed conversion ratio (FCR), body weight gain (BWG), as well as factors relating to egg production and meat quality. Postbiotic efficacy is determined by various mechanisms involved in their ability to increase the beneficial bacteria population to protect mucosal surfaces from pathogens, which include direct competition between them for adhesion and the rate of regeneration (Aguilar-Toala et al. 2018). Also, postbiotics have both bactericidal and bacteriostatic properties, thus reducing the proliferation of pathogenic bacteria in the gastrointestinal tract (Loh et al. 2010; Rosyidah et al. 2011; Loh et al. 2014; Kareem et al. 2016b). The *L. plantarum* postbiotics have been documented to exert an antimicrobial effect on different pathogenic bacteria (Thanh et al. 2009). Thu et al. (2011) used the inhibitory activity test to assess such a relationship with the application of *P. acidilactici* as an indicator microbe. Their findings showed that postbiotic combinations (TL1, RG14, RG11, and RS5) have stronger inhibitory activity as compared with a single strain. Presence of a large amount of lactic acid and acetic acid, as well as bacteriocins inhibitory compound, is suggestive of contributing to such event. These mechanisms of actions enhance better growth performance in poultry.

Based on the inhibitory activities observed on the pathogens using postbiotics from *L. plantarum* (Thanh et al. 2010), an experiment was conducted to investigate the effect of a different combination of postbiotics on growth performance in broiler chickens (Thanh et al. 2009). Postbiotic treatment groups had greater BW, WG, ADG, and the best FCR compared to the negative control (fed basal diet only). Based on the improved growth performance found after postbiotic supplementation in piglets (Thu et al. 2011), the authors also evaluated if such beneficial effects could be replicated in laying hens (Loh et al. 2014). There was a significant reduction in faecal pH and specific pathogenic bacteria and improvement in faecal LAB following the addition of postbiotics, whereas there was no significant difference in overall feed intake, egg mass, and FCR among the treatment groups. The use of postbiotics combinations of *L. plantarum* strains in laying hens resulted in significantly higher laying performance and other parameters relating to gut microbiota and small intestine histomorphology (Choe et al. 2012). Thus, the latter study indicated the need for further evaluation on the potential benefits of various postbiotics combination on growth performance and immune response in poultry. Metabolite combinations and single strains of postbiotics could confer varying impact on the growth performance of broiler chickens. Kareem (2016) investigated the growth performance in broiler chickens after supplemented with a combination of postbiotic and inulin. Through the study, it showed that such combination could lower the FCR compared to the group without adding any feed additive. Another benefit of postbiotic is shown in the ruminal fermentation and growth performance of lambs (Izuddin et al. 2018; Izuddin et al. 2019a; Izuddin et al. 2019b). Majority of the research on postbiotics in poultry have established their positive impact on growth performance.

7.4 Implications of Postbiotic Supplementation on Gut Microbiota and pH

The gut microbiota in poultry chickens is vital for general health and productivity (Yang et al. 2019; Shang et al. 2018). The beneficial effects of postbiotic supplementation on the gut microbiota have been determined in several studies (Thanh et al. 2009; Choe et al. 2012; Thu et al. 2011; Loh et al. 2014; Kareem et al. 2016b). In poultry, such studies assess the population of pathogenic and non-pathogenic microorganisms following postbiotic supplementation in feeds (Thanh et al. 2009; Choe et al. 2012; Loh et al. 2014). Broilers supplemented with combinations of basal diet, inulin, and varying concentration of postbiotic *L. plantarum* RG14 promoted the proliferation of total caecal bacteria and *Bifidobacterium* compared to the negative control group (Kareem et al. 2016b). Also, the non-treatment group had greater counts of *E. coli* and *Enterobacteriaceae*, whereas no effect of diet on *Lactiplantibacillus*, *Enterococcus* (ENT), and *Salmonella* population in broilers fed postbiotic with prebiotic. Similar findings were reported following the supplementation of broiler feed with the combination of various postbiotics and inulin (a prebiotic) (Kareem et al. 2016b). Thanh et al. (2009) observed increased faecal LAB population after supplementing broilers feed with different strains of *L. plantarum* derived metabolites. The LAB counts and mesophilic microbial population increased significantly following dietary supplementation with combined fermentative products of LAB (bacteriocins and organic acids) in Ross broilers (Fajardo et al. 2012). A similar result was reported using the postbiotics in layer hens (Choe et al. 2012). These works reinstate that intestinal microbiota maturation index could serve as an important parameter when comparing the efficacy of feed additives on poultry microecology.

The feeding of *L. plantarum* postbiotic to livestock has been proven to shift the gut microbiome towards the proliferation of good bacteria. Kareem (2016) revealed that supplementation of *L. plantarum* RG14 combined with inulin enhanced the colonization of *Bifidobacterium* while lowered the population of *E. coli* and *Enterobacteriaceae* compared to control groups. Similarly, broiler and layer birds fed with a combination of *L. plantarum* postbiotics had a higher faecal LAB population (Thanh et al. 2009; Choe et al. 2012). Humam et al. (2019) reported that the supplementation of postbiotic to broiler chicken reared under heat stress environment recorded higher counts of caecum total bacteria, *Lactiplantibacillus* and *Bifidobacterium* but with a significantly lower population of pathogenic bacteria such as *Enterobacteriaceae*, *E. coli,* and Salmonella compared to the control groups. It also has been shown that caecal pH correlates with the bacteria counts. The presence of various organic acids in the postbiotic lowers the pH of the caecum.

Subsequently, this inhibits the growth of the low acidic tolerant pathogenic bacteria such as *Enterobacteriaceae*, *E. coli,* and Salmonella and stimulates the proliferation of beneficial bacteria such as *Lactiplantibacillus* and *Bifidobacterium*. In lambs, the inclusion of *L. plantarum* postbiotics in feed improved fibre degrading bacteria but with a reduction in total protozoa and methanogens in the rumen (Izuddin et al. 2019b).

7.5 Implications of Postbiotic Supplementation on Gut Morphology and Immune Response

Maintenance of the integrity of the GIT is a prerequisite for optimal physiological function. Such maintenance enables a stable microbial population and conditions to protect against offending substances (Yang et al. 2009; Adedokun and Olojede 2019; Dudek-Wicher et al. 2018). Majority of the studies investigate the relationship between postbiotics and gut morphology by focussing on the height and depth of the intestinal villi and crypts, respectively, since they are the portals for the absorption of nutrients (Choe et al. 2012; Gao et al. 2019). Accordingly, gut health can be assessed based on the condition of the villus height and crypt depth (Uni et al. 1995).

Postbiotic supplementation in laying hens was reported to improve nutrient digestibility based on the maintenance of villus height and crypt depth of mucosal architecture (Choe et al. 2012). Increment in villi height and reduced crypt depth are optimal for efficient nutrients absorption to take place across the intestine into the systemic circulation (Markovi et al. 2009). The addition of a combination of postbiotics obtained from *L. plantarum* strains also increased small intestinal villus height in broilers (Loh et al. 2010; Thanh et al. 2009; Kareem et al. 2016a) and pigs (Loh et al. 2013; Thu et al. 2011). Likewise, the addition of six per cent liquid metabolites obtained from the same bacteria showed enhancement on villus height in layers (Choe et al. 2012). Moreover, Izuddin et al. (2019a) reported that addition of postbiotic postweaning lamb feed had increased the villus height. Improvement was also detected in the small intestine morphology (higher villi height and lower crypt depth) when the broiler chickens were supplemented with postbiotic (Humam et al. 2019). Based on the information generated by Markovi et al. (2009) the positive effect of probiotic supplementation on intestinal villi and crypt depth could be one of the reasons for better health and growth in broiler chickens when supplemented with postbiotic.

It is assumed that associations between postbiotic supplementation and improved health and productivity, as shown in several studies, are connected to underlying mechanisms that activate a protective immune response against offending pathogens. Many studies have reported that inclusion of postbiotic in the diets improves the immune response in the broilers (Kareem et al. 2016b) and lambs (Izuddin et al. 2019a) and the gut health for animals.

7.6 Implications of Postbiotic Supplementation on Meat Quality

There is limited published data on the meat quality of broiler chickens fed with postbiotics. However, one study showed that broiler chickens fed a combination of postbiotics and inulin showed reduction of drip loss and better breast muscle lightness (Kareem et al. 2015). Other parameters, such as shear force, cooking loss, and related bone attributes, were not affected by the in-feed supplementation of the combination, as mentioned earlier (Kareem et al. 2015). Another study conducted by Humam et al. (2020) revealed that when postbiotic was added to the feed of heat-stressed broiler chickens, there was an increased of breast meat pH but decreased in the shear force, lightness, drip loss, cooking loss, and yellowness. The prevention of pH drop might contribute to the results, improved antioxidant activity and reduced corticoid hormone levels (Sato et al. 2010; Hao and Gu 2014; Zaboli et al. 2019).

7.7 Implications of Postbiotic Supplementation and Gene Expression Related to the Gut Barrier Function and Immunity

Exploring expression of gene information regarding postbiotic supplementation could assist in better understanding of the mechanisms underlining their effect in poultry birds. The gene expression includes immune response and reactions relating to the impacts on the gut microbiota and barrier function, as well as processes involved in arresting pathogenic bacteria. Another study conducted by Kareem et al. (2016a) evaluated the ileal cytokine expression in birds fed with postbiotic (RG14) and prebiotic (inulin). The birds fed with diet not containing the metabolites had higher expression of interferon (IFN) and the tumour necrosis factor alpha (TNF-α) when compared with all the other treatment groups. Also, the mRNA expression of IL-6 (interleukin-6) had significantly higher expression in the group fed various percentages of the metabolite combination than the basal diet and antibiotic-treated groups. In another study, the liver of broilers fed with various postbiotics (metabolites from the LAB) combined with inulin had significantly up-regulated the IGF-1 expression compared to other treatments groups (Kareem et al. 2016b). Moreover, the group fed RG14 and inulin had greater GHR mRNA expression compared to other treatments. Izuddin et al. (2019b) found out the addition of in feed postbiotic of postweaning lambs led to an increase in the IL-6 mRNA and decrease of IL-1β, IL-10, TNF gene expression. Postbiotics increased the gene expression of IGF-1 in the liver and MCT-1 in the rumen of the postweaning lambs (Izuddin et al. 2019b).

There are intrinsic associations between gut microbiota and IGF-1 production as established in experiments conducted in mice (Yan et al. 2016). Accordingly, mice supplemented with short-chain fatty acids (SCFA) recorded increased secretion of IGF-1 in both the adipose and liver. This effect (increased SCFA production) was successfully found in broiler chickens following the combined dietary supplementation using *L. plantarum* metabolites and prebiotics (Kareem et al. 2016a) and heat stress broiler chickens (Humam et al. 2019).

The effects on antioxidant enzyme-related gene expression due to supplementation of *L. plantarum* postbiotics can be detected in postweaning lambs. The study carried out by Izuddin et al. (2020) showed that the postweaning lambs fed with postbiotics had up-regulated the expression of hepatic glutathione peroxidase (GPX1 and GPX4) and Cu/ZN SOD genes. Besides, the same group of lambs received postbiotic supplementation affected the gene expression of TJP1, OCLD, CLDN1, and CLDN 4 in the rumen and zonula occludin-1 in broiler chickens under heat stress (Humam et al. 2020). As mucin secretion affects the gut permeability, immune status, and nutrient absorption in the intestine, the postbiotic of *L. plantarum* has been proven to improve the expression of intestinal mucin in livestock. Humam et al. (2020) showed that postbiotic RI11 up-regulated the expression of MUC2 with the increased inclusion level in feed.

7.8 Implications of Postbiotic Supplementation on Antioxidant Activities

During heat stress condition, there is an increased production of reactive oxygen species such as hydroxyl free radical and superoxide anions which posed harmfully and even damaging effect on the animal tissues (Akbarian et al. 2016; Altan et al. 2003). In poultry, antioxidants are substances synthesized by related enzymes such as superoxide dismutase, glutathione peroxidase, and catalase as defensive products to mitigate the effects of reactive oxygen species by making them non-toxic (Surai 2015). Humam et al. (2020) revealed that postbiotic supplementation RI11 improved the activity of total-antioxidant capacity, catalase, and glutathione while lowered the level of alpha-1-acid-glycoprotein and ceruloplasmin in the blood plasma when the broiler chickens were raised under heat stress condition. However, no significant changes were observed in plasma corticosterone and heat shock protein 70 between all the treatments. On the other hand, Izuddin et al. (2020) revealed that the glutathione peroxidase and thiobarbituric acid reactive substance activities in the blood serum were affected due to the supplementation of postbiotic.

7.9 Conclusions

A good and efficient gastrointestinal tract is essential for nutrient digestion and absorption besides protecting the animals from various infections and stress by secreting immune-related compounds. The feed additives such as antibiotics were included since a long time in broilers to enhance the gut health and boilers productivity. However, excessive usage of antibiotics contributed to the emergence of antibiotic-resistant bacteria and residual effects in the ecosystem and food product, which could have a harmful influence on both broiler and human health. Several alternatives to antibiotics such as ascorbic acid and probiotics were used to promote the broilers health and productivity. However, some probiotics own antimicrobial resistance genes which can be transferred to other living organisms. Therefore, this limitation may compromise the expected health consequences exhibited by the probiotic. For these consequences, it becomes necessary to find safe and effective alternatives to antibiotics and probiotic. Postbiotics are the metabolites produced by probiotic and exert antimicrobial activity due to the presence of antimicrobial compounds, for example, bacteriocins and organic acids. This characteristic leads to better gut health. Postbiotics have been documented to enhance gut health and growth performance and production in livestock such as poultry, lambs, and pigs.

Acknowledgment The authors would like to thank the Ministry of Higher Education, Malaysia for the financial support under the Fundamental Research Grant Scheme (FRGS/1/2017/WAB01/UPM/01/1) and Higher Institution Centers of Excellence (HICoE) research grant (HICoE-ITAFoS/2017).

References

Adedokun SA, Olojede OC (2019) Optimizing gastrointestinal integrity in poultry: the role of nutrients and feed additives. Front Vet Sci 5:348

Aguilar-Toala JE, Garcia-Varela R, Garcia HS, Mata-Haro V, Gonzalez-Cordova AF, Vallejo-Cordoba B, Hernandez-Mendoza A (2018) Postbiotics: an evolving term within the functional foods field. Trends Food Sci Technol 75:105

Akbarian A, Michiels J, Degroote J, Majdeddin M, Golian A, De Smet S (2016) Association between heat stress and oxidative stress in poultry; mitochondrial dysfunction and dietary interventions with phytochemicals. J Anim Sci Biotechnol 7(1):37

Altan O, Pabuccuoglu A, Altan A, Konyalioglu S, Bayraktar H (2003) Effect of heat stress on oxidative stress, lipid peroxidation and some stress parameters in broilers. Br Poult Sci 44 (4):545

Choe DW, Loh TC, Foo HL, Hair-Bejo M, Awis QS (2012) Egg production, faecal pH and microbial population, small intestine morphology and plasma and yolk cholesterol in laying hens given liquid metabolites produced by *Lactobacillus plantarum* strains. Br Poult Sci 53 (1):106

Chuah LO, Foo HL, Loh TC, Alitheen NBM, Yeap SK, Mutalib NEA, Yusoff K (2019) Postbiotic metabolites produced by *Lactobacillus plantarum* strains exert selective cytotoxicity effects on cancer cells. BMC Complement Altern Med 19(1):114

Cicenia A, Scirocco A, Carabotti M, Pallotta L, Marignani M, Severi C (2014) Postbiotic activities of lactobacilli-derived factors. J Clin Gastroenterol 48(1):18

Dudek-Wicher RK, Junka A, Bartoszewicz M (2018) The influence of antibiotics and dietary components on gut microbiota. Przegl Gastroenterol 13(2):85

Fajardo P, Pastrana L, Méndez J, Rodríguez I, Fuciños C, Guerra NP (2012) Effects of feeding of two potentially probiotic preparations from lactic acid bacteria on the performance and faecal microflora of broiler chickens. Sci World J 2012(562635):1

Foo HL, Loh TC, Abdul Mutalib NE, Rahim RA (2019) The myth and therapeutic potentials of postbiotics, Chapter 21. In: Microbiome and metabolome in diagnosis, therapy, and other strategic applications. Academic Press, New York, pp 201–211

Gao J, Li Y, Wan Y, Hu T, Liu L, Yang S, Yang W (2019) A novel postbiotic from *Lactobacillus rhamnosus* GG with a beneficial effect on intestinal barrier function. Front Microbiol 10:477

Hao Y, Gu X (2014) Effects of heat shock protein 90 expression on pectoralis major oxidation in broilers exposed to acute heat stress. Poult Sci 93(11):2709

Humam AM, Loh TC, Foo HL, Samsudin AA, Mustapha NM, Zulkifli I, Izuddin WI (2019) Effects of feeding different postbiotics produced by *Lactobacillus plantarum* on growth performance, carcass yield, intestinal morphology, gut microbiota composition, immune status and growth gene expression in broilers under heat stress. Animals 9(9):644

Humam AM, Loh TC, Foo HL, Izuddin WI, Zulkifli I, Samsudin AA, Mustapha NM (2020) Supplementation of postbiotic RI11 improves antioxidant enzymes activity, up-regulated gut barrier genes and reduced cytokines acute phase proteins an HSP70 gene expression levels in heat-stressed. Poult Sci. https://doi.org/10.21203/rs.3.rs-22778/v1

Huyghebaert G, Ducatelle R, Van Immerseel F (2011) An update on alternatives to antimicrobial growth promoters for broilers. Vet J 187(2):182

Izuddin WI, Loh TC, Samsudin AA, Foo HL (2018) In vitro study of postbiotics from *Lactobacillus plantarum* RG14 on rumen fermentation and microbial population. Rev Bras Zootec 47: e20170255

Izuddin WI, Loh TC, Foo HL, Samsudin AA, Humam AM (2019a) Postbiotic *L. plantarum* RG14 improves ruminal epithelium growth, immune status and up-regulates the intestinal barrier function in postweaning lambs. Sci Rep 9(1):9938

Izuddin WI, Loh TC, Samsudin AA, Foo HL, Humam AM, Shazali N (2019b) Effects of postbiotic supplementation on growth performance, ruminal fermentation and microbial profile, blood metabolite and G.H.R., IGF-1 and MCT-1 gene expression in postweaning lambs. BMC Vet Res 15(1):1

Izuddin WI, Humam AM, Loh TC, Foo HL, Samsudin AA (2020) Dietary postbiotic *Lactobacillus plantarum* improves serum and ruminal antioxidant activity and up-regulates hepatic antioxidant enzymes and ruminal barrier function in postweaning lambs. Antioxidants 9(3):250

Kareem KY (2016) Effect of postbiotic and inulin supplements on growth performance, gut morphology, gene expression and faecal characteristics of broiler chickens. Unpublished Doctoral Thesis, Universiti Putra Malaysia

Kareem KY, Ling FH, Chwen LT, Foong OM, Asmara SA (2014) Inhibitory activity of postbiotic produced by strains of *Lactobacillus plantarum* using reconstituted media supplemented with inulin. Gut Pathog 6(1):23

Kareem KY, Loh TC, Foo HL, Asmara SA, Akit H, Abdulla NR, Ooi MF (2015) Carcass, meat and bone quality of broiler chickens fed with postbiotic and prebiotic combinations. Int J Probiotics Prebiotics 10(1):23

Kareem KY, Loh TC, Foo HL, Asmara SA, Akit H (2016a) Influence of postbiotic RG14 and inulin combination on cecal microbiota, organic acid concentration, and cytokine expression in broiler chickens. Poult Sci 96(4):966

Kareem KY, Loh TC, Foo HL, Akit H, Samsudin AA (2016b) Effects of dietary postbiotic and inulin on growth performance, IGF1 and G.H.R. mRNA expression, faecal microbiota and volatile fatty acids in broilers. BMC Vet Res 12(1):163

Lee FH, Wan SY, Foo HL, Loh TC, Mohamad R, Abdul Rahim R, Idrus Z (2019) Comparative study of extracellular proteolytic, cellulolytic, and hemicellulolytic enzyme activities and biotransformation of palm kernel cake biomass by lactic acid bsacteria isolated from Malaysian foods. Int J Mol Sci 20(20):4979

Lim YH, Foo HL, Loh TC, Mohamad R, Abdullah N (2019) Comparative studies of versatile extracellular proteolytic activities of lactic acid bacteria and their potential for extracellular amino acid productions as feed supplements. J Anim Sci Biotechnol 10(1):15

Loh T, Chong S, Foo H, Law F (2009) Effects on growth performance, faecal microflora and plasma cholesterol after supplementation of spray-dried metabolite to postweaning rats. Czeh J Anim Sci 54(1):10–16

Loh TC, Thanh NT, Foo HL, Hair-Bejo M, Azhar BK (2010) Feeding different levels of metabolite combinations on growth performance, faecal microflora, volatile fatty acid and villi height in broilers. Anim Sci J 81(2):205

Loh TC, Thu TV, Foo HL, Bejo MH (2013) Effects of different levels of metabolite combination produced by *Lactobacillus plantarum* on growth performance, diarrhoea, gut environment and digestibility of postweaning piglets. J Appl Anim Res 41(2):200
Loh TC, Choe DW, Foo HL, Sazili AQ, Bejo MH (2014) Effects of feeding different postbiotic metabolite combinations produced by *Lactobacillus plantarum* strains on egg quality and production performance, faecal parameters and plasma cholesterol in laying hens. BMC Vet Res 10(1):149
Malashree L, Angadi V, Yadav KS, Prabha R (2019) Postbiotics - one step ahead of probiotics. Int J Curr Microbiol App Sci 8(1):2049
Markovi R, Sefer D, Krsti M, Petrujki B (2009) Effect of different growth promoters on broiler performance and gut morphology. Archivos De Medicina Veterinaria 41(2):163–169
Montagne L, Pluske JR, Hampson DJ (2003) A review of interactions between dietary fibre and the intestinal mucosa, and their consequences on digestive health in young non-ruminant animals. Anim Feed Sci Technol 108(1–4):95
Ohimain EI, Ofongo RT (2012) The effect of probiotic and prebiotic feed supplementation on chicken health and gut microflora: a review. Int J Anim Vet Adv 4(2):135
Rinttila T, Apajalahti J (2013) Intestinal microbiota and metabolites-implications for broiler chicken health and performance. J Appl Poult Res 22(3):647
Rosyidah M, Loh T, Foo H, Cheng X, Bejo M (2011) Effect of feeding metabolites and acidifier on growth performance, faecal characteristics and microflora in broiler chickens. J Anim Vet Adv 10(21):2758
Sato H, Takahashi T, Sumitani K, Takatsu H, Urano S (2010) Glucocorticoid generates R.O.S. to induce oxidative injury in the hippocampus, leading to impairment of cognitive function of rats. J Clin Biochem Nutr 47(3):224
Schenk M, Mueller C (2008) The mucosal immune system at the gastrointestinal barrier. Best Pract Res Clin Gastroenterol 22(3):391
Shang Y, Kumar S, Oakley B, Kim WK (2018) Chicken gut microbiota: importance and detection technology. Front Vet Sci 5:254.29
Shazali N, Foo HL, Loh TC, Choe DW, Rahim RA (2014) Prevalence of antibiotic resistance in lactic acid bacteria isolated from the faeces of broiler chicken in Malaysia. Gut Pathogens 6(1):1
Sugiharto S (2016) Role of nutraceuticals in gut health and growth performance of poultry. J Saudi Soc Agric Sci 15(2):99
Surai P (2015) Antioxidant systems in poultry biology: heat shock proteins. J Sci 5(12):1188
Thanh NT, Loh TC, Foo HL, Hair-Bejo M, Azhar BK (2009) Effects of feeding metabolite combinations produced by *Lactobacillus plantarum* on growth performance, faecal microbial population, small intestine villus height and faecal volatile fatty acids in broilers. Br Poult Sci 50 (3):298
Thanh NT, Chwen LT, Foo HL, Hair-Bejo M, Kasim AB (2010) Inhibitory activity of metabolites produced by strains of *Lactobacillus plantarum* isolated from Malaysian fermented food. Int J Probiotics Prebiotics 5(1):37
Thu TV, Loh TC, Foo HL, Yaakub H, Bejo MH (2011) Effects of liquid metabolite combinations produced by *Lactobacillus plantarum* on growth performance, faeces characteristics, intestinal morphology and diarrhoea incidence in postweaning piglets. Trop Anim Health Prod 43(1):69
Toe CJ, Foo HL, Loh TC, Mohamad R, Abdul Rahim R, Idrus Z (2019) Extracellular proteolytic activity and amino acid production by lactic acid Bacteria isolated from Malaysian foods. Int J Mol Sci 20(7):1777
Tsilingiri K, Rescigno M (2013) Postbiotics: what else? Benefic Microbes 4(1):101
Uni Z, Noy Y, Sklan D (1995) Posthatch changes in morphology and function of the small intestines in heavy- and light-strain chicks. Poult Sci 74(10):1622
Van de Bogaard AE, London N, Driessen CAGG, Stobberingh EE (2001) Antibiotic resistance of faecal *Escherichia coli* in poultry, poultry farmers and poultry slaughterers. J Antimicrob Chemother 47(6):763

Yan J, Herzog JW, Tsang K, Brennan CA, Bower MA, Garrett WS, Charles JF (2016) Gut microbiota induce IGF-1 and promote bone formation and growth. Proc Natl Acad Sci 113 (47):7554

Yang PC, Tu YH, Perdue MH, Oluwole C, Struiksma S (2009) Regulatory effect of heat shock protein 70 in stress-induced rat intestinal epithelial barrier dysfunction. N Am J Med Sci 1(1):9

Yang J, Wang C, Liu L, Zhang M (2019) *Lactobacillus reuteri* KT260178 supplementation reduced morbidity of piglets through its targeted colonization, improvement of cecal microbiota profile, and immune functions. Probiotics Antimicrob Proteins 1:10

Yegani M, Korver DR (2008) Factors affecting intestinal health in poultry. Poult Sci 87(10):2052

Zaboli G, Huang X, Feng X, Ahn DU (2019) How can heat stress affect chicken meat quality?–a review. Poult Sci 98(3):1551

Probiotics Application: Implications for Sustainable Aquaculture

8

Milad Adel and Mahmoud A. O. Dawood

Abstract

Probiotics, known as beneficial microorganisms, are being proposed as an effective and eco-friendly alternative to antibiotics. They were first applied in aquaculture species more than three decades ago, but considerable attention had been given only in the early 2000s. Probiotics defined as live, dead, or a component of the microorganisms which act under different modes of action in conferring beneficial effects to the host or its environment. Several probiotics have been characterized and applied in fish, and a number of them are of host origin. Unlike some disease control alternatives being adopted and proposed in aquaculture where actions are unilateral, the immense potential of probiotics lies on their multiple mechanisms in conferring benefits to the host fish and the rearing environment. The staggering number of probiotics papers in aquaculture highlights the multitude of advantages from these microorganisms and conspicuously position them in the dynamic search for health-promoting alternatives for cultured fish. The present review provides an update on the use of probiotics in finfish aquaculture, mainly focusing on their modes of action. It explores the contemporary understanding of their spatial and nutritional competitiveness, inhibitory metabolites, environmental modification capability, immunomodulatory potential, and stress-alleviating mechanism.

M. Adel
Department of Food Hygiene and Quality Control, Faculty of Veterinary Medicine, Shahrekord University, Shahrekord, Iran

M. A. O. Dawood (✉)
Department of Animal Production, Faculty of Agriculture, Kafrelsheikh University, Kafrelsheikh, Egypt
e-mail: Mahmoud.dawood@agr.kfs.edu.eg

N. Mojgani, M. Dadar (eds.), *Probiotic Bacteria and Postbiotic Metabolites: Role in Animal and Human Health*, Microorganisms for Sustainability 2,
https://doi.org/10.1007/978-981-16-0223-8_8

Keywords

Probiotics · Fish · Shellfish aquaculture · Positive effects · Antibiotic resistance

8.1 Introduction

The aquaculture industry is rapidly growing and is now considered a significant contributor to global food production. According to the United Nations Food and Agriculture Organization, the growth of the aquaculture sector is higher than any other types of animal food production systems (www.fao.org). To meet the global demand, aquaculture production practices have been intensified to a greater extent both in technological and practical measures (Tuan et al. 2013; Dawood et al. 2020a). However, the growth of the aquaculture industry is hampered by unpredictable mortalities which are caused by pathogenic microorganisms. Bacterial diseases have been attributed as biological production bottlenecks in intensive aquaculture, hence necessitating the use of chemicals such as drugs and antibiotics in health management strategies (Newaj-Fyzul and Austin 2014; Ringø 2020). The antibiotic application had been an effective strategy in the beginning, but the residuals remaining in the rearing environment exerted selective pressure for long periods and became a big challenge (Lakshmi et al. 2013; Soltani et al. 2019). The indiscriminate use resulted in the emergence of antibiotic-resistant bacteria in aquaculture environments, in the increase of antibiotic resistance in fish pathogens, in the transfer of these resistance determinants to bacteria of land animals and human pathogens, and in alterations of the bacterial flora both in sediments and in the water column (Dawood et al. 2018). These alarming disadvantages prompted the aquaculture industry to explore and develop strategies that are as equally effective as antibiotics, eco- and consumer-friendly, and most importantly sustainable (Standen et al. 2013; Lazado et al. 2015a, b).

Probiotics are one of the identified alternatives that can lessen the dependence of the aquaculture industry to antibiotics (Verschuere et al. 2000; Nayak 2010; Lazado and Caipang 2014a, b; Akhter et al. 2015). Probiotics have several mechanisms in conferring their benefits to the host fish (Fig. 8.1). Such a feature makes probiotic research in aquatic animals a very dynamic field. The results demonstrating the multitude of ways in delivering benefits to the host have immensely expanded the traditional understanding of probiotics as a modifier of the microbial community in the host. This paper discusses the immense potential of probiotics as a health-promoting alternative through the identified different modes of action of probiotics following their application in finfish aquaculture. It focuses more on how they improve the quality of the rearing environment, protect fish from biological hazards, and modulate physiological processes that eventually promote the health and welfare status of fish in culture. The synthesis provided here collates our current understanding of how probiotics are beneficial to fish and how we can utilize these microorganisms in fostering a more sustainable aquaculture practice.

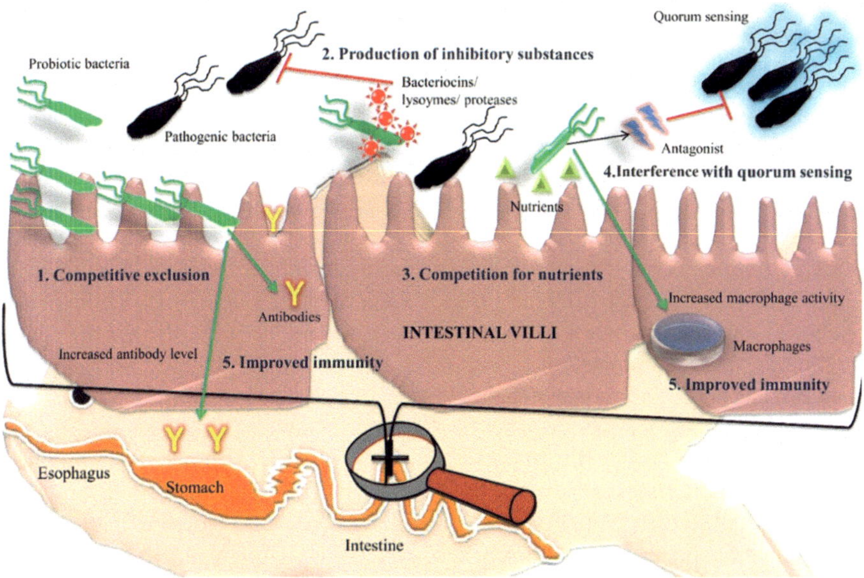

Fig. 8.1 General mechanism of action of probiotics. 1. Competitive exclusion—probiotic organism colonizes the gut, thereby inhibiting colonization of pathogenic bacteria. 2. Probiotic organisms produce certain inhibitory substances which hinder pathogenic organism. 3. Competition for nutrients—probiotic body utilizes the nutrients causing unavailability of nutrients to the pathogens. 4. Substances produced by probiotics act as an antagonist for quorum sensing mechanism. 5. Improved immunity—increase macrophage activity and antibody level

8.2 Probiotic Definition

Several definitions of probiotics have been put forward since the first definition was given by Lilly and Stillwell (1965), but the most widely used is the definition by World Health Organisation's (WHO); "live microorganisms that when administrated in adequate amounts, confer a health benefit to the host." The word probiotic originated from the Greece words "pro" and "bios" which collectively mean "for life," hence being widely regarded as beneficial microorganisms. For some time, Fuller's definition of probiotics as "a live microbial feed supplement which beneficially affects the host animal by improving microbial balance" was the adapted understanding of the probiotic concept in many cultured animals (Fuller 1989). Interestingly, the results of probiotics research in aquaculture have opened numerous possibilities on the benefits of this group of microorganisms. Recently, Lazado and Caipang (2014a, b) proposed that probiotics under an aquaculture understanding be defined as "alive or dead, or even a component of the microorganisms that act under different modes of action in conferring beneficial effects to the host or its environment." This contemporary definition reflects all the advances in probiotics research in aquaculture for over three decades since its first application.

8.3 Regular Probiotics in Aquaculture

In the last three decades, several probiotic microorganisms have been identified, characterized, and applied in aquaculture. These beneficial microorganisms can be of host or non-host origin (Lazado et al. 2015a; Lazado and Caipang 2014b). In a recent review paper, it was highlighted that host-associated microorganisms offer a great prospect as a source of probiotics with diverse biochemical features (Dawood and Koshio 2016; Lazado et al. 2015a). Bacteria obtained from intestine of aquatic as well as terrestrial animals are commonly used as probiotics in aquaculture (Hai and Fotedar 2010). Several bacterial species such as *Vibrio* and *Pseudomonas* spp. isolated from marine fishes are being proposed as probiotics. Different species of probiotics used in aquaculture and their beneficial effects are enumerated in Table 8.1. There is no united stand as to what the best source of probiotics is to be applied for fish. Probiotics from the terrestrial environment have been documented conferring numerous benefits to the cultured animals. On the other hand, probiotics of host origin offer several advantages as well, uniquely leverage in some biotechnical concerns (i.e. temperature, salinity, the familiarity of the environment).

Various factors impose a decisive role in the selection of a suitable probiotic for aquatic species. Different features like the type of probiotic (i.e. bacteria, fungi or algae), host from which they are derived (i.e. host or non-host), single strain probiotic or multi-strain, viable or non-viable organisms as probiotic and also use of spore formers or non-spore formers (Nayak 2010; Dawood et al. 2019). These are some of the reasons why having probiotics of universal application seem impractical.

The most commonly used probiotic species include genera *Lactobacillus, Bifidobacterium, Aeromonas, Plesiomonas, Bacteroides, Fusobacterium, Alteromonas, Carnobacterium and Eubacterium and strains of Bacillus, Enterococcus, Bacteroides, Clostridium, Agrobacterium, Pseudomonas, Paenibacillus, Brevibacterium, Microbacterium, Staphylococcus, Streptomyces, Micrococcus, Vibrio, Psychrobacter, Carnobacterium, Phaeobacter, Pediococcus, Pseudoalteromonas, Rhodosporidium, Saccharomyces, Debaryomyces, Aeromonas, Tetraselmis, Roseobacter, Weissella* and *Aspergillus* (Balcazar et al. 2006; Nayak 2010; Lakshmi et al. 2013; Tuan et al. 2013; Lazado et al. 2015b; Zorriehzahra et al. 2016; Dawood et al. 2019; Ringø 2020).

8.4 Modes of Action

8.4.1 Competition for Space

Many of the pathogenic bacteria require attachment to the mucosal layer of the host gastrointestinal tract to initiate the development of a disease (Zorriehzahra et al. 2016). An essential mechanism of action in probiotic bacteria is competition for adhesion sites, also known as "competitive exclusion" (Boaventura et al. 2012). The ability of bacteria to colonize the gut and adhere to the epithelial surface and

Table 8.1 The application of different species of probiotics in aquaculture

Probiotics	Species	Positive effects	References
Aeromonas hydrophila	Rainbow trout (Oncorhynchus mykiss)	*Aeromonas salmonicida* infection reduced	Irianto and Austin (2002a, b)
Aeromonas media A199	*Crassostrea gigas* (Pacific oyster)	Reduced *Vibrio tubiashii* infection	Gibson (1999)
Aeromonas sobria GC2	Rainbow trout	Protection against *Lactococcus garvieae* and *Streptococcus iniae*	Pieters et al. (2008); Brunt and Austin (2005)
Agarivorans albus F1-UMA	*Haliotis rufescens* (abalone)	Survivability increased	Silva-Aciares et al. (2011)
Alteromonas CA2	Pacific oyster	Survivability increased	Douillet and Langdon (1994)
Alteromonas macleodii 0444	*Perna canaliculus* (Greenshell mussel)	Controls *Vibrio splendidus* infection	Kesarcodi-Watson et al. (2010); Kesarcodi-Watson et al. (2012)
	Pecten maximus (scallop)	Controls *Vibrio coralliilyticus* and *V. splendidus*	
Burkholderia cepacia Y021	*Crassostrea corteziensis* (Cortez oyster), *Nodipecten subnodosus* (lions-pay scallop)	Increased growth and survival	Granados-Amores et al. (2012)
Enterobacter amnigenus	Rainbow trout	Increased resistance towards *Flavobacterium psychrophilum*	Burbank et al. (2011)
Neptunomonas 0536	*Perna P. canaliculus* (Greenshell mussel)	*V. splendidus* infection controlled	Kesarcodi-Watson et al. (2010, 2012)
Pseudomonas aeruginosa, P. synxantha	*Penaeus latisulcatus* (Western king prawns)	General health and immune status improved	Hai et al. (2009)
Pseudomonas sp. (GP21) and *Psychrobacter* sp.	Atlantic cod (*Gadus morhua*)	Immune response	Lazado and Caipang (2014b)
Shewanella putrefaciens	*Sparus aurata L.*, (Gilthead Sea bream)		De la Banda et al. (2012)
Gordonia bronchialis	*Rainbow trout (Oncorhynchus mykiss)*	Growth performance, intestinal histology and biochemical parameters	Shabanzadeh et al. (2016)
Arthrobacter XE-7	*L. vannamei* (Pacific white shrimp)	Alters intestinal microbes	Li et al. (2008)
Bacillus circulans PB7	*Labeo rohita* (Rohu)	Act as immune stimulant and protects against *A. hydrophila*	Ghosh et al. (2003); Bandyopadhyay and Das

(continued)

Table 8.1 (continued)

Probiotics	Species	Positive effects	References
			Mohapatra (2009)
Bacillus subtilis and *Bacillus licheniformis*	*Rainbow trout*	Protects against *Yersinia ruckeri*, FCR and growth improved	Raida et al. (2003)
Bacillus coagulans	Pacific white shrimp	Stress tolerance and disease resistance	Cai et al. (2019)
Bacillus subtilis	*Labeo rohita* (Indian major carp)	Controls *A. hydrophila*	Kumar et al. (2006)
	White shrimp	Immunity increased and resistance against *V. harveyi* increased	Zokaeifar et al. (2012)
	Ictalurus punctatus (channel cat fish) and *Pangasianodon hypophthalmus* (striped cat fish)	Decreased mortality rate due to *Edwardsiella ictaluri*	Ran et al. (2012)
Bacillus subtilis UTM 126	*Litopenaeus vannamei* (White shrimp)	Protection against vibriosis	Das et al. (2006)
Bacillus subtilis E20	*Litopenaeus vannamei* (White shrimp)	Growth improved; mortality reduced	Liu et al. (2010); Tsai et al. (2019)
Bacillus megaterium	Shrimp	Immunity improved, intestinal microbes altered and resistant to white spot syndrome virus	Li et al. (2009)
Bacillus pumilus	*P. japonicus*	Improved larval survival	El-Sersy et al. (2006)
	O. niloticus (Tilapia)	Immunity increased and survivability increased against *A. hydrophila* challenge	Aly et al. (2008a)
Bacillus P64	*L. vannamei* (white shrimp)	Immunostimulant	Gullian et al. (2004)
Bacillus 48	*Centropomus undecimalis* (common snook)	Growth improved	Kennedy et al. (1998)
Brevibacillus brevis	*Dicentrarchus labrax* (sea bass)	Prevent vibriosis and improve growth	Mahdhi et al. (2012)
Brochothrix thermosphacta BA211	Rainbow trout	Protect against *A. Bestiarum*	Pieters et al. (2008)
Clostridium butyricum	Rainbow trout	Protect against vibriosis and also from *A. hydrophila* and *V. anguillarum* infections	Sakai et al. (1995)
	Miichthys miiuy (Chinese drum)	Increased immunity and disease resistance	Pan et al. (2008)

(continued)

Table 8.1 (continued)

Probiotics	Species	Positive effects	References
Carnobacterium divergens	*Gadus morhua* (Atlantic cod), Atlantic salmon (*Salmo salar)* and and rainbow trout	Protects against *V. anguillarum* infection	Gildberg et al. (1997); Robertson et al. (2000)
Enterococcus faecium SF 68	*Anguilla Anguilla* (European eel)	Prevents against Edwardsiellosis	Chang and Liu (2002)
E. faecium MC13	*P. monodon* (shrimp)	Protects against *V. harveyi* and *V. parahaemolyticus*	Swain et al. (2009)
Enterococcus faecalis	*P. monodon* (shrimp)	Improved growth performance and levels of glutathione peroxidase (GPs)	Guzmán-Villanueva et al. (2020)
Enterococcus casseliflavus	Rainbow trout (Oncorhynchus mykiss)	Improved resistance against *Streptococcus iniae* infection	Safari et al. (2016)
Kocuria SM1	Rainbow trout	Protects against *V. anguillarum* and *V. ordalii*	Sharifuzzaman and Austin (2010)
Lactobacillus acidophilus	Nile tilapia	Immunity increased and protects against *P. fluorescens* and *S. iniae*	Aly et al. (2008b)
Lactobacillus acidophilus	*Clarias gariepinus* (African catfish)	Growth performance, haematological parameters and immunoglobulin concentration	Al-Dohail et al. (2009)
Lactobacillus paracasei spp. *paracasei* (06TCa22)	Japanese pufferfish (*Takifugu rubripes*)	Disease resistance	Biswas et al. (2013)
Lactobacillus rhamnosus ATCC 53101	Rainbow trout	Reduction in mortality caused by A. salmonicida	Nikoskelainen et al. (2001)
L. rhamnosus	*O. niloticus*	Protects against *E. tarda* infection	Pirarat et al. 2006
Lactobacillus fructivorans and *L. plantarum*	*S. aurata* (sea bream)	Increase in production of HSP70 thereby increasing heat tolerance	Carnevali et al. (2004); Rollo et al. (2006)
Lactococcus lactis	*Litopenaeus vannamei*	Improved growth and immunity	Adel et al. (2017b)
Lactococcus lactis AR21	Rotifers	Improved growth and protects against *V. anguillarum* infection	Harzevili et al. (1998)

(continued)

Table 8.1 (continued)

Probiotics	Species	Positive effects	References
Lactobacillus sporogenes	*Macrobrachium rosenbergii* (freshwater prawn)	Boosts the survival, growth and levels of biochemical constituents	Seenivasan et al. (2012)
Leuconostoc mesenteroides CLFP 196 and L. plantarum CLFP 238	Rainbow trout	Mortality due to *L. garvieae* was reduced	Vendrell et al. (2008)
Lactobacillus brevis	*Macrobrachium rosenbergii*	Growth promotion	Karthik and Bhavan (2018).
Micrococcus luteus	*O. mykiss* (rainbow trout)	Infection due to *A. salmonicida* was reduced	Irianto and Austin (2002a)
Micrococcus MCCB 104	*M. rosenbergii* (fresh water prawn)	Different bacteria inhibited	Jayaprakash et al. (2005)
Pediococcus pentosaceus	Red sea bream (Pagrus major)	Growth performance, feed utilisation and blood characteristics	Dawood et al. (2016a, b, c)
Pediococcus pentosaceus	White shrimp (*Litopenaeus vannamei*)	Increase the amylase, protease and lipase activities	Adel et al. (2017a)
Pediococcus acidilactici	Rainbow trout fry	Vertebral column compression syndrome (VCCS) was reduced	Aubin et al. (2005)
Rhodococcus SM2	Rainbow trout	Immunity improved and protection against *V. anguillarum*	Sharifuzzaman et al. (2011)
Streptococcus phocae P180	*P. monodon*	Growth increased and protects against *V. harveyi* infection	Swain et al. (2009)
Streptococcus faecium	*Oreochromis niloticus* (Nile tilapia)	As growth promoters	Lara-Flores et al. (2003)
Streptococcus faecium	*Cyprinus carpio* (carp)	Improves growth and intestinal micro flora	Bogut et al. (1998)
Streptomyces	*P. monodon*	Growth improved and water quality was also increased	Das et al. (2006); Newaj-Fyzul et al. (2014).
Vagococcus fluvialis	Sea bass	Protection against *V. anguillarum* infection	Sorroza et al. (2012)
Weissella hellenica DS-12	–	Protects against several fish pathogens	Byun et al. (1997); Cai et al. (1998)
Phages of family Myoviridae and Podoviridae	*Plecoglossus altivelis*	Protection against *Pseudomonas plecoglossicida*	Park et al. (2000)

(continued)

Table 8.1 (continued)

Probiotics	Species	Positive effects	References
Microaglae *Tetraselmis suecica*	Penaeids	Protection against bacterial pathogen	Austin and Day (1990)
Dunaliella tertiolecta	*Artemia*	Protection against *Vibrio campbellii* and *V. proteolyticus*	Marques et al. (2006)
Phaffia rhodozyma, Saccharomyces cerevisiae and *Saccharomyces exiguous*)	Penaeids	Protection against vibriosis	Scholz et al. (1999)
Vibrionaceae 51M6 and Pdp11	Rainbow trout (Oncorhynchus mykiss)	Immune responses	Choi and Yoon (2008)

consequently interfere with the adhesion of pathogens is a desirable criterion in the selection of probiotics (Lazado et al. 2011; Balcazar et al. 2006). Non-pathogenic intestinal microbes such as *Lactobacilli* compete with the pathogens for adhesion sites on the intestinal surfaces, particularly on intestinal villus and enterocytes (Brown 2011).

Probiotic addition is being suggested as an early stage husbandry practice in larviculture because the feature of competitive exclusion for attachment sites could provide favourable rearing conditions (Irianto and Austin 2002a). Attachment of probiotics may be non-specific based on the physicochemical agents or specific based on the adhesion of the probiotics on the surface of the adherent bacteria and receptor molecules on the epithelial cells (Salminen et al. 1996; Lazado et al. 2015a).

8.4.2 Production of Inhibitory Substances

Probiotic bacteria produce substances with bactericidal or bacteriostatic effects such as bacteriocins, hydrogen peroxide, siderophores, lysozymes, and proteases (Panigrahi and Azad 2007; Servin 2004; Tinh et al. 2007). Besides, some bacteria produce organic acid and volatile fatty acids (e.g. lactic, acetic, butyric, and propionic acids) that can result into the reduction of pH in the gastrointestinal lumen, thus preventing the growth of opportunistic pathogenic microorganisms (Tinh et al. 2007; Boaventura et al. 2012).

Recently, a compound called indole (s,3-benzopyrrole) with potent inhibitory activity against pathogens was identified in some bacteria known to have anti-bacterial and anti-fungal activities (Gibson 1999; Lategan et al. 2006).

8.5 Probiotics vs. Antibiotics

One of the most vulnerable points of aquaculture is the fish's mortality related to infectious diseases (10–20% of total mortality). Infectious diseases are often a consequence of stress conditions as an excessive density of fish in tank or basin, hypoxia, high nitrite, ammonia concentrations, etc. So, antibiotics are used in the aquaculture industry to prevent or treat bacterial diseases in farmed fish and shellfish (Adel et al. 2018).

Today, with the increasing population, access to healthy food resources has become one of the main concerns for human beings (Zaineldin et al. 2018). Protein from different sources and especially red and white meats plays a vital role in human daily nutrition. However, due to the prevalence of various cardiovascular diseases in today's society, which occurs following the excessive consumption of unhealthy foods, including red meat, consumers are increasingly inclined to white meat consumption like seafood. On the other hand, natural fish and shellfish products are restricted, and therefore aquaculture tries to meet this demand. But intensive and highly intensive aquaculture has caused many infectious diseases and force the farmers to use antibiotics and chemical compounds (Dawood et al. 2016c). The presence of drug residues in fish and fish products threatens public health. Moreover, antibiotic misuse and overuse can promote antibiotic resistance.

The use of healthy or pathogen-free fish, balanced diet, high water quality, biosecurity plans in managing fish farms can substantially prevent disease entrance and transmission into the farm. Despite the above points, antibiotics are widely used to reduce the complications and symptoms of diseases. In some countries, antibiotics are widely used as a routine procedure for treating different types of diseases in aquaculture. Therefore, residues of antibiotics in fish meat and roe will have harmful and destructive impacts on the health of consumers. Studies have shown that antibiotic residues can change the resistance of normal microbial flora in the human body to a specific group according to the antibiotics type. Also, it may cause resistance to the antibiotics used in the treatment of humans and other animals, which is why attempts are now being made to use alternatives to antibiotics.

8.5.1 Antibiotics Action

The antibiotics mechanism of action seems to be based on the selective action on intestinal bacteria (Fig. 8.2). Antibiotics are used in the aquaculture industry to accelerate the growth performance, treat diseases, reduce mortality, increase meat production, increase the absorption of nutrient in the gastrointestinal tract, and increase the neutralization of toxins produced by harmful intestinal bacteria. For optimal effectiveness of antibiotics, they should be available in high concentrations in the body for a few days.

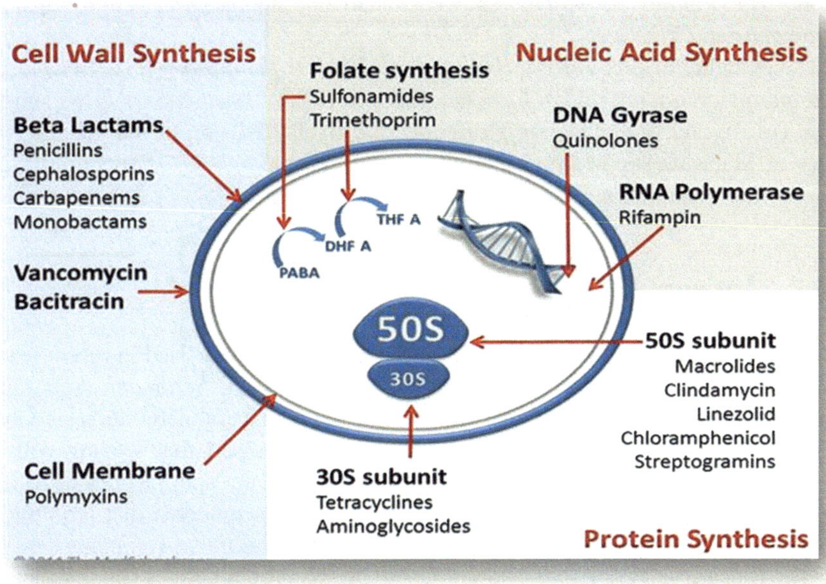

Fig. 8.2 Mechanism of action of antibiotic

8.6 Overview of the Effect of Probiotic on Aquatic Animals

8.6.1 Antibacterial Activity

Several probiotics in aquaculture have been documented possessing antibacterial activity against known pathogens. For example, probiotic *L. lactis* RQ516 that is being used in tilapia (*Oreochromis niloticus*) exhibited inhibitory activity against *A. hydrophila* (Zhou et al. 2010). It was also shown by Balcázar et al. (2008) that probiotic *L. lactis* had antibacterial activity towards two fish pathogens, namely, *A. salmonicida* and *Yersinia ruckeri.*

Zapata and Lara-Flores (2013) found that *Leuconostoc mesenteroides* was able to inhibit the growth of pathogenic fish bacteria in Nile tilapia (*O. niloticus*). Ghosh et al. (2008) found that *B. subtilis* significantly reduced the amount of motile *Aeromonads*, presumptive *Pseudomonads,* and total Coliforms in ornamental fishes (Newaj-Fyzul and Austin 2014). Moosavi-Nasab et al. (2014) also reported that lactic acid bacteria (*Lactobacillus buchneri, Lactococcus lactis, Lactobacillus acidophilus, Lactobacillus fermentum,* and *Streptococcus salivarius*) isolated from the intestine of Spanish mackerel (*Scomberomorus commerson*) were able to inhibit the growth of *Listeria innocua.* Dhanasekaran et al. (2008) reported that several *Lactobacilli* isolated from *the* intestine of catfish (*Clarias orientalis*), Hari fish (*Anguilla* sp.), Rohu fish (*Labeo rohita*), Jillabe fish (*Oreochromis* sp.), and

Gende fish (*Puntius carnaticus*) showed remarkable antibacterial activity against *Aeromonas* and *Vibrio* sp.

The potential of probiotics including *Lactobacillus plantarum* (LP1, LP2), *Saccharomyces cerevisiae* (SC3), *Candida glabrata* (CG2), *Lactococcus lactis* subsp. lactis (LL2), and *Staphylococcus arlettae* (SA) isolated from an indigenous fish sauce in Malaysia showed high inhibitory activity on *Staphylococcus aureus* and *Listeria monocytogenes*.

8.6.2 Antiviral Activity

The knowledge of the antiviral activity of probiotics has been raised in recent years (Lakshmi et al. 2013). For example, *Pseudomonas*, *Vibrio*, *Aeromonas* spp., and *Coryneforms* had antiviral activity against infectious hematopoietic necrosis virus (IHNV) (Kamei et al. 1988). Li et al. (2009) demonstrated that feeding with a *B. megaterium* strain increased the resistance to white spot syndrome virus (WSSV) in the shrimp *Litopenaeus vannamei*. It was documented that probiotics, like *Bacillus* and *Vibrio* sp., positively protect shrimp *Litopenaeus vannamei* against WSSV (Balcazar 2003). Application of *Lactobacillus* probiotics as a single strain or mixed with Sporlac improved disease resistance against lymphocystis viral disease in olive flounder (*Paralichthys olivaceus*) (Harikrishnan et al. 2010).

8.6.3 Antifungal Activity

There are few studies regarding the antifungal effect of probiotics. Lategan et al. (2004) isolated *Aeromonas media* (strain A199) from eel (*Anguilla australis*) culture water and was observed to have a strong inhibitory activity against *Saprolegnia* sp. In a separate study, *Pseudomonas* sp. M162, *Pseudomonas* sp. M174, and *Janthinobacterium* sp. M169 enhanced immunity against saprolegniasis in rainbow trout (Heikkinen 2013). Atira et al. (2012) demonstrated that *Lactobacillus plantarum* FNCC 226 exhibited inhibitory activity against *Saprolegnia parasitica* A3 in catfish (*Pangasius hypophthalmus*).

8.6.4 Competition for Chemicals or Available Energy

The existence of any microbial population depends on its ability to compete for chemicals and available energy with the other microbes in the same environment (Verschuere et al. 2000). Many microorganisms, including the known probiotic group lactic acid bacteria, consume the nutrients that are essential for the growth of several pathogens (Dawood et al. 2016a, b; Brown 2011).

For example, siderophores are low-molecular-weight ferric iron chelating agents that can dissolve precipitated iron or extract it from iron complexes, then making it available for bacterial growth (Neilands 1981). Siderophore-producing bacteria can be used as probiotics because they can sequester ferric iron in an iron-low environment, hence making it unavailable for the growth of pathogenic bacteria (Tinh et al. 2007). Gram et al. (1999) showed that a culture supernatant of *Pseudomonas fluorescens*, grown in iron-limited conditions, inhibited growth of *Vibrio anguillarum*. It has been revealed that *P. fluorescens* can competitively inhibit the growth of the fish pathogen *Aeromonas salmonicida*, by competing for free iron (Smith and Davey 1993; Gram et al. 1999). It was also revealed that GP12 and GP21, candidate probiotics from Atlantic cod, are capable of releasing siderophores, and this ability had been implicated for their beneficial use (Lazado et al. 2011).

8.6.5 Positive Effects on Rearing Water Quality

Interactive effects between aquaculture environment and aquatic species have been confirmed for sustainable aquaculture (Dawood et al. 2019). Application of Gram-positive bacteria, such as *Bacillus* spp., is beneficial in improving the quality of the water system. *Bacillus* spp. have a more efficient ability in converting organic matter into carbon dioxide in comparison to the Gram-negative bacteria, which converts a higher proportion of organic matter into bacterial biomass or slime (Mohapatra et al. 2012; Balcazar et al. 2006). Certain probiotic bacteria possess significant algicidal effect as well, particularly on several species of microalgae (Fukami et al. 1997). Ammonia and nitrite toxicity can be eliminated by the application of nitrifying cultures into the fish environment (Mohapatra et al. 2012). Besides, probiotics are beneficial as they can increase microbial species' composition in the water and modify its quality (Elsabagh et al. 2018; Mohapatra et al. 2012). The temperature, pH, dissolved oxygen, NH_3, and H_2S in rearing water were found to be of higher quality when probiotics were added, hence maintaining a positive, healthy environment for shrimp and prawn larval in green water system (Aguirre-Guzman et al. 2012; Banerjee et al. 2010; Dawood et al. 2020b). Dalmin et al. (2001) reported that the using of an indigenous *Bacillus* spp. in the rearing water of giant tiger prawn was able to maintain optimum transparency and low organic carbon of the pond. Mohamed et al. (2013) approved that applications of commercial probiotics to saline tilapia (*Oreochromis mossambicus*) could improve the growth performance, phytoplankton production, and water quality.

8.6.6 Nutrients and Enzymatic Contribution

Some microorganisms have a positive effect on the digestive processes of aquatic animals (Dawood et al. 2017; Balcazar et al. 2006). It has been shown that some bacteria contribute to the digestion process by producing extracellular enzymes, such as proteases, lipases, as well as growth-promoting factors (Wang et al. 2000).

Reports are demonstrating that some probiotics, especially from *Bacteroides* and *Clostridium* sp., are capable of supplying vitamins, fatty acids, and essential amino acids to the host (Balcazar et al. 2006; Tinh et al. 2007). Gnotobiotic oyster larvae

(*Crassostrea gigas*), fed with auxenic algae (*Isochrysis galbana*) supplemented with a bacterial strain CA2, showed not only improved growth performance but efficient nutrient utilization as well (Douillet and Langdon 1994). Yeasts are well known in animal nutrition because they can produce polyamines, which enhance intestinal maturation. Besides bacterial probiotics, many strains of yeast have been used as dietary supplements in several fish species (Tinh et al. 2007).

8.6.7 Interference of Quorum Sensing

Quorum sensing is defined as the regulation of gene expression in response to fluctuations in cell-population density. Many bacteria are using this system to communicate and regulate a diverse array of physiological activities (Miller and Bassler 2001). The disruption of quorum sensing (QS) is considered a potential anti-infective strategy in aquaculture (Defoirdt et al. 2004).

Halogenated furanones, which are produced by the marine red alga *Delisea pulchra* (Manefield et al. 1999), have been investigated as a promising QS antagonist. These compounds, added at adequate concentrations, protected Brachionus, Artemia, and rainbow trout from the harmful effects of pathogenic *Vibrios* (Rasch et al. 2004; Defoirdt et al. 2006; Tinh et al. 2007). Also, some probiotic bacteria such as *Lactobacillus*, *Bifidobacterium*, and *B. cereus* strains degrade the signal molecules of pathogenic bacteria by enzymatic secretion or production of autoinducer antagonists (Brown 2011). It was demonstrated by Medellin-Pena et al. (2007) that *L. acidophilus* secretes a molecule that inhibits the QS or interacts with a bacterial transcription of *E. coli* O157 gene.

8.6.8 Immunostimulants

8.6.8.1 Fish

Probiotics by stimulation of immune system of hosts, including the stimulation of pro-inflammatory cytokines on the activity of immune cells, increasing the phagocytic activity of leucocytes (Pirarat et al. 2006), increasing the levels of antibodies, acid phosphatase, lysozymes (Lara-Flores and Aguirre-Guzman 2009), complement (Balcazar et al. 2007), cytokines (interleukin-1 (IL-1), IL-6, IL-12, tumor necrosis factor α (TNF- α), gamma interferon (IFN-γ), IL-10 and transforming growth factor b) (Nayak 2010) and antimicrobial peptides (Mohapatra et al. 2012), and also by improving the intestinal microbial balance, inhibition of the colonization of fish pathogens in the digestive tract, production of inhibitory compounds such as bacteriocins, siderophores, lysozymes, proteases, hydrogen peroxides (Saurabh et al. 2005), increasing the digestive enzymes activity (amylase, protease and lipase) (Ringø 2020) and production of fatty acids, vitamins (Sakata 1990) and essential amino acids that are useful for lactic acid bacteria (Ringø and Gatesoupe 1998) could improve the growth performance, immune system and increased resistance on common pathogens in fish (Fig. 8.3) and shrimp (Lakshmi et al. 2013).

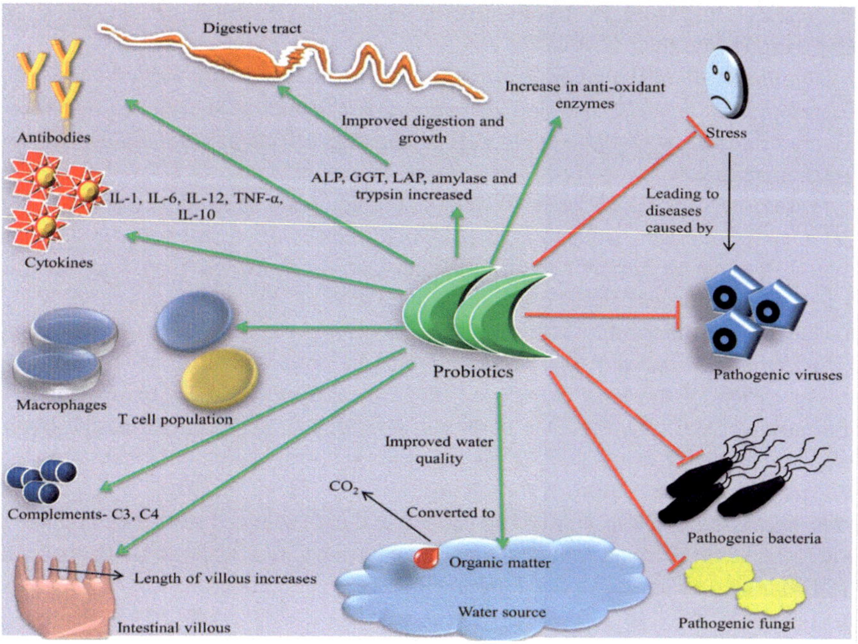

Fig. 8.3 Overall beneficial effects of probiotics in aquaculture. The green arrow indicates additive effects. Red lines indicate inhibitory effect

The administration of probiotics in tilapia (*Oreochromis niloticus*) caused an increase in lysozyme activity, neutrophil migration, bactericidal action, and finally enhanced resistance of fish to infection of *Edwardsiella tarda* (Taoka et al. 2006b). Also, Gomez et al. (2007) used *Vibrio alginolyticus* strains as probiotics in white shrimp (*Litopenaeus vannamei*) and observed increased survival and growth in shrimp (Zhou et al. 2009).

Harikrishnan et al. (2011a) reported that administration of probiotics (*Lactobacillus sakei* BK19) with herb (*Scutellaria baicalensis*) in tilapia (*O. fasciatus*) reduces the mortality, altered haematological parameters, and enhances innate immunity against *Edwardsiella tarda*. The same researchers repeated this experiment in olive flounder (*Paralichthys olivaceus*) against *Streptococcus parauberis* and found growth, blood biochemical constituents, and non-specific immunity improved in the groups treated with probiotics and herbals mixture supplementation diet (Harikrishnan et al. 2011b). Irianto and Austin (2002a) reported that feeding with Gram-positive and Gram-negative probiotics resulted in the stimulation of cellular rather than humoral (serum of mucus antibodies) immunity. There was an increase in the number of erythrocytes, macrophages, and lymphocytes, and enhanced lysozyme activity during feeding with probiotics. Feeding with diets containing single or mixed isolated probiotic bacteria for *O. niloticus* showed different results in survival rates and highest with fish fed diets supplemented with

B. pumilus, followed by a mixture of probiotics (*B. firmus*, *B. pumilus*, and *C. freundii*), and then *C. freundii*.

Avella et al. (2010) used a mixture of *Bacillus* probiotic bacteria including *B. subtilis*, *B. licheniformis* and *B. pumilus* in the diet of the gilthead seabream (*Sparus aurata*) larviculture and observed apparent effects on survival, growth, and general welfare.

Assessment of mRNA levels of several immune parameters like cytokine IL-8 in the intestine of the control and *L. plantarum* groups by using real-time PCR showed that IL-8 gene expression was significantly up-regulated by *L. plantarum* after *Lactococcus garvieae* infection (Pérez-Sánchez et al. 2011). Standen et al. (2013) evaluated the probiotic effect of *Pediococcus acidilactici* on Nile tilapia (*Oreochromis niloticus*) and suggested that the probiotic treatment may cause up-regulation of the gene expression of the proinflammatory cytokine TNF-α in the probiotic fed fish. Presence of *Bacillus subtilis* C-3102 in the diets of hybrid tilapia juvenile (*O. niloticus* × *O. aureus*) caused up-regulation of cytokines such as IL-1β, TGF-β, and TNF-α in the intestine of fish (He et al. 2013). *Lactobacillus delbrueckii* ssp. *delbrueckii* (AS13B) added in diet of gilthead sea bream resulted in lower transcription of proinflammatory cytokine genes such as IL1β, IL10, cox2, and TGF-β in the intestine of treated group (Picchietti et al. 2009).

8.6.8.2 Shrimp

Use of probiotics in different species of shrimps has improved the innate immunity (natural or non-specific immunity) (Fig. 8.4). Several studies have demonstrated that by using probiotics the production of cellular components such as phagocytosis, encapsulation, the formation of nodules and humoral components including

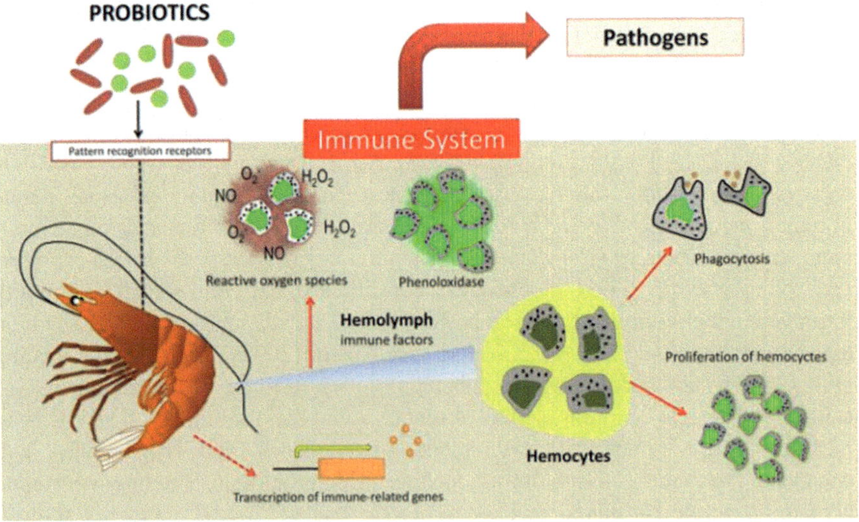

Fig. 8.4 Model of immunomodulation in shrimp by probiotics (Lazado et al. 2015b)

anticoagulant proteins, agglutinins, phenoloxidase enzyme (Lakshmi et al. 2013; Song et al. 2014), antimicrobial peptides (defensins and chemokines), an antiapoptotic protein, free radicals, bacteriocins, siderophores, monostatin, lysozymes, proteases, hydrogen peroxide, gramicidin, polymyxin, tyrotricidin, competitive exclusion and organic acid was increased. The transcription of several immune-related genes can also be modulated by probiotic treatment (Antony et al. 2011). These stimulated immune defences to play a crucial role in the responses and eventual protection during pathogen exposure. Probiotics have an essential role in enhancing the resistance of shrimps against common diseases such as vibriosis, white spot disease and *Aeromonas hydrophila* infection (Ahilan et al. 2004; Ma et al. 2007; Harikrishnan et al. 2009; Liu et al. 2010; Zokaeifar et al. 2014).

It was also confirmed by RNA interference (RNAi) assay that the immunity of shrimps was increased against viral diseases using probiotics (Kawai and Akira 2006). Rangpipat et al. (2000) showed that *Bacillus* sp. (strain S11) protected against infection by activating the *Penaeus monodon* immune system.

8.6.8.3 Immunomodulation of the Gut Immune System

The immune system of the gut is related to gut-associated lymphoid tissue (GALT) (Nayak 2010; Lazado and Caipang 2014a, b), and there are some differences in respect of Peyer's patches, secretory IgA, and antigen-transporting M cells in the intestine of piscine and mammal gut immune system (Nayak 2010). Although lymphoid cells, macrophages, granulocytes, and mucus IgM were observed in the intestine of fish (Bakke-McKellep et al. 2007).

There is limited knowledge about the application of probiotics and their ability in stimulating the piscine gut immune system (Nayak 2010; Lazado and Caipang 2014a, b; Mamun et al. 2019). The present knowledge is mostly associated with humans and terrestrial vertebrates (Lazado and Caipang 2014a, b). However, studies indicated that probiotics could stimulate the piscine gut immune system, increasing the number of Ig^+-cells and acidophilic granulocytes (AGs) (Picchietti et al. 2007, 2008, 2009; Salinas et al. 2008). For example, it has been reported that the supplementation of LAB (*Lactobacillus rhamnosus* GG, human origin) in the diet of tilapia, *Oreochromis niloticus* could modulate the population of the intestinal immune cells. Also, the amount of intraepithelial lymphocytes and acidophilic granulocytes (AGs) enhanced significantly in the probiotic-fed group (Pirarat et al. 2011). Addition of probiotic-containing *Lactobacillus fructivorans* (host origin) and *Lactobacillus plantarum* (human origin) to the diet of larval gilthead sea bream, *Sparus aurata* by live vectors affected the extent of Ig^+-cells and acidophilic granulocytes mostly the MAb G7(+) phagocytic population in the gut (Picchietti et al. 2007).

Picchietti et al. (2009) used rotifers and artemia in the administration of *Lactobacillus delbrueckii* ssp. *delbrueckii* (AS13B) as the live vectors to the larval sea bass, *Dicentrarchus labrax*. They observed that the population of T cells and acidophilic granulocytes in the intestinal mucosa significantly increased in probiotic-fed fish.

In a study, rainbow trout (*Oncorhynchus mykiss*) were fed by diets supplemented with probiotics such as *L. Lactis* spp. *lactis*, *L. mesenteroides*, and *L. sakei*. In the end, an enhancement was observed in the phagocytic activity of mucosal leucocytes by the LAB group (Balcazar et al. 2006). *Pediococcus acidilactici* was used by Standen et al. (2013) in the feeding of Nile tilapia (*Oreochromis niloticus*).

8.7 Amelioration of the Effects of Stress

Stress might be regarded as a physical or chemical agent, causing reactions that may result in disease and death (Rottmann et al. 1992). Any change in water parameters may have a side effect on the physiological and behavioural aspect of aquatic animals (Dawood 2021). Different types of stress that may have adverse effects on fish include thermal (Das et al. 2005; Logan and Somero 2011), nutritional, high density (Lupatsch et al. 2010), anoxia, hypoxia, chemical, and toxins (DeMicco et al. 2010). Many harmful agents for fish exist in their environments like the water, soil, air, or even their own body (Smith et al. 2012). In intensive systems of aquaculture where the high density is an essential factor for the outbreak, in stressful conditions, aquatic animals are more susceptible than wild fishes. Application of probiotic bacteria, both as a feed supplement and water quality, can prevent stressful conditions, enhancing the immune system and therefore reducing the harmful effects of various stressors (Taoka et al. 2006a).

Any situation that enhances reactive oxygen species (ROS) concentration is called oxidative stress that can lead to disturbing cellular metabolism and its regulation, thereby damaging cellular constituents (Jia et al. 2011; Lushchak 2011). ROS production is nearly related to antioxidant responses (Lesser 2006; Bidhan et al. 2014). The alterations of temperature and other environmental parameters can severely affect the physiological activities of aquatic animals (Wabete et al. 2008). Also, a wide range of contaminants (xenobiotics), UV-radiation, hypoxia, and other environmental physicochemical parameters may cause oxidative stress in the animal (Mohapatra et al. 2012). Feeding with probiotics may ameliorate the effects of these oxidative stress factors by increasing the antioxidant status (Mohapatra et al. 2012).

Blood glucose, cortisol, and the RNA/DNA ratio of the different tissues is used as valid biochemical stress indicators to study the fish stresses, growth, and health status (Sivaraman et al. 2012). Another way to assess stress tolerance in fish involves subjecting them to heat shock (Cruz et al. 2012).

Taoka et al. (2006a) grew flounder (*Paralichthys olivaceus*) under stress conditions and evaluated the effects of probiotics on growth, stress tolerance, and non-specific immune response in fish. Plasma lysozyme activity in the probiotic diet group and the water supply group was significantly higher than in the control group. In heat shock stress tests, flounder in the probiotics-treated groups showed higher heat tolerance. Koninkx and Malago (2008) demonstrated that under stress conditions, normal intestinal microflora taken as probiotics were able to enhance the defence system by increasing specifically the putative heat shock protein (HSP).

Some probiotic bacteria have been found to decrease several biochemical stress indicators. There is a report regarding the decrease in cortisol level on supplementation of *Lactobacillus delbrueckii* ssp. *delbrueckii* in the diet of European sea bass

(*Dicentrarchus labrax*) compared to the controls during temperature stress (Carnevali et al. 2006).

Gomes et al. (2008, 2009) found that administration of *Bacillus* spp. during transport reduced handling stress by influencing the cortisol level. Varela et al. (2010) carried out probiotic administration studies on gilthead seabream (*Sparus auratus*) and concluded that there was improved tolerance to stress with this treatment under high stocking density. Castex et al. (2009) evaluated the antioxidative effect of *Pediococcus acidilactici* MA 18/5 in shrimp, *Litopenaeus stylirostris*. Results showed the modulation of the activities of antioxidant enzymes such as superoxide dismutase and catalase. It has been reported that administration of *Lactobacillus plantarum* could enhance the antioxidant state in shrimp *Litopenaeus vannamei* and consequently improve resistance to *V. alginolyticus* infection (Chiu et al. 2007).

Chai et al. (2016) investigated the effects of *Bacillus* bacteria isolated from the intestine of healthy, wild shrimps on the growth of Pacific white shrimp and showed that probiotics reduced shrimp culture risks from stressful conditions.

8.8 Side Effects

Probiotics are generally considered safe and well tolerated (Boyle et al. 2006). One theoretical concern associated with probiotics includes the potential for these viable organisms to move from the gastrointestinal tract and cause systemic infections (Snydman 2008). Another theoretical risk associated with probiotics involves the possible transfer of antibiotic resistance from probiotic strains to pathogenic bacteria; however, this has not yet been observed (Martin et al. 2013). Also, the possibilities of change in intestinal microflora, emerging diseases, mutagenesis, or recombination of DNA of bacteria may result into systemic infections and economical losses in fish farms (Ringø et al. 2010).

8.9 Maximizing the Benefit of Probiotics

To maximize the competitive advantage of probiotics, early delivery notably before first feeding improved the chances of producing persistent fish population (Ringo et al. 1997). Choosing the right probiotic, appropriate concentrations of probiotic, sufficient feeding time, and feeding status are several important parameters that are necessary for maximizing the benefit of probiotics.

8.10 Conclusion Remarks and Future Consequences

Despite doing many studies about efficiency and mechanisms of probiotics, many questions are not yet clear. Additional and future studies can be directed to transcriptome and proteome profiling of gut microbiota, host/microbe interactions,

interactions between gut microbes, the intestinal epithelium, gut immune system, antioxidant status, lipid level of hosts, antagonistic and synergist activity or probably side effects of probiotics.

Use of probiotics is a useful alternative sustainable source of beneficial microbes with bactericidal or bacteriostatic effect on pathogenic bacteria, with anti-bacterial, anti-viral and anti-fungal activity, immunomodulatory capabilities of promoting health and welfare to improve the growth performance, augment the immune system, disruption of quorum sensing (QS) as a new anti-infective strategy, ameliorate the harmful effects of oxidative stress factors and increased resistance for common pathogens in fishes for controlling potential fish pathogens. An interactive approach among academicians, scientists, producers and fish sector owners is required to focus and explore the specific aspects of bacterial host interactions conferring the possible favourable changes in diverse immune responses elicited by different bacterial strains to propose clinically useful, bacteria-based strategies to promote the health, production and economic growth of the aquaculture industry.

In future studies on probiotics in shellfish aquaculture, bio-floc culture system using probiotics should be investigated on growth performance, immune response, gut microbiome and disease resistance as only some information are available on this topic. Also, more studies are needed to investigate the fate of probiotic organisms in the environment and the shellfish.

The probiotic formulation should be viable on a large scale at low operational cost. They should not be treated as "elixir of life"; instead they should be used as a supplement to balance the diet to avail and maintain the sound health free of infections and disease-causing microorganisms.

References

Adel M, Yeganeh S, Dawood MAO, Safari R, Radhakrishnan S (2017a) Effects of *Pediococcus pentosaceus* supplementation on growth performance, intestinal microflora and disease resistance of white shrimp, *Litopenaeus vannamei*. Aquac Nutr 23:1401–1409

Adel M, El-Sayed AFM, Yeganeh S, Dadar M, Giri SS (2017b) Effect of potential probiotic *Lactococcus lactis* subsp. lactis on growth performance, intestinal microbiota, digestive enzyme activities, and disease resistance of *Litopenaeus vannamei*. Probiotics Antimicrob Proteins 9 (2):150–156

Adel M, Dadar M, Oliveri Conti G (2018) Antibiotics and malachite green residues in farmed rainbow trout (*Oncorhynchus mykiss*) from the Iranian markets: a risk assessment. Int J Food Prop 20(2):402–408

Aguirre-Guzman G, Lara-Flores M, Sanchez-Martinez JG, Campa-Cordova AI, Luna-Gonzalez A (2012) The use of probiotics in aquatic organisms: a review. Afr J Microbiol Res 6 (23):4845–4857

Ahilan B, Shine G, Santhanam R (2004) Influence of probiotics on thegrowth and gut microflora load of juvenile gold fish (*Carassiusauratus*). Asian Fish Sci 17:271–278

Akhter N, Wu B, Memon AM, Mohsin M (2015) Probiotics and prebiotics associated with aquaculture: a review. Fish Shellfish Immunol 45(2):733–741

Al-Dohail MA, Hashim R, Aliyu-Paiko M (2009) Effects of the probiotic, *Lactobacillus acidophilus*, on the growth performance, haematologyparameters and immunoglobulin concentration in African catfish (*Clarias gariepinus*, Burchell 1822) fingerling. Aquac Res 40:1642–1652

Aly SM, Mohamed MF, John G (2008a) Effect of probiotics on the survival, growth and challenge infection in *Tilapia nilotica* (*Oreochromis niloticus*). Aquac Res 39:647–656

Aly SM, Ahmed YAG, Ghareeb AAA, Mohamed MF (2008b) Studies on *Bacillus subtilis* and *Lactobacillus acidophilus*, as potential probiotics, on the immune response and resistance of *Tilapia nilotica* (*Oreochromis niloticus*) to challenge infections. Fish Shellfish Immunol 25:128–136

Antony SP, Singh ISB, Jose RM, Kumar PRA, Philip R (2011) Antimicrobial peptide gene expression in tiger shrimp, *Penaeus monodon* in response to gram positive bacterial probionts and white spot virus challenge. Aquaculture 316:6–12

Atira NJ, Aryantha INP, Kadek IDG (2012) The curative action of *Lactobacillus plantarum* FNCC 226 to *Saprolegnia parasitica* A3 on catfish (*Pangasius hypophthalamus*, Sauvage). Int Food Res J 19(4):1723–1727

Aubin J, Gatesoupe FJ, Labbé L, Lebrun L (2005) Trial of probiotics to prevent the vertebral column compression syndrome in rainbow trout (*Oncorhynchus mykiss*, Walbaum). Aquac Res 36:758–767

Austin B, Day JG (1990) Inhibition of prawn pathogenic *Vibrio* spp. by a commercial spray-dried preparation of *Tetraselmis suecica*. Aquaculture 90:389–392

Avella MA, Gioacchini G, Decamp O, Makridis P, Bracciatelli C, Carnevali O (2010) Application of multi-species of *Bacillus* in sea bream larviculture. Aquaculture 305:12–19

Bakke-McKellep AM, Froystad MK, Lilleeng E, Dapra F, Refstie S, Krogdahl A (2007) Response to soy: T-cell-like reactivity in the intestine of Atlantic salmon, *Salmo salar* L. J Fish Dis 30:13–25

Balcazar JL (2003) Evaluation of probiotic bacterial strains in *Litopenaeus vannamei*. Final Report, National Center for Marine and Aquaculture Research, Guayaquil

Balcazar JL, de Blas I, Ruiz-Zarzuela I, Cunningham D, Vendrell D, Mu'zquiz JL. (2006) The role of probiotics in aquaculture. Vet Microbiol 114:173–186

Balcazar JL, de Blas I, Ruiz-Zarzuela I, Vendrell D, Calvo AC, Marquez I, Gironés O, Muzquiz JL (2007) Changes in intestinal microbiota and humoral immune response following probiotic administration in brown trout (*Salmo trutta*). Br J Nutr 97:522–527

Balcázar JL, Vendrell D, de Blas I, Ruiz-Zarzuela I, Gironés O, Múzquiz JL (2008) In vitro competitive adhesion and production of antagonistic compounds by lactic acid bacteria against fish pathogens. Vet Microbiol 122(3–4):373–380

Bandyopadhyay P, Das Mohapatra PK (2009) Effect of a probiotic bacterium *Bacillus circulans* PB7 in the formulated diets: on growth, nutritional quality and immunity of *Catla catla* (ham.). Fish Physiol Biochem 35:467–478

Banerjee S, Khatoon H, Shariff M, Yusoff F Md. (2010) Enhancement of *Penaeus monodon* shrimp post larvae growth and survival without water exchange using marine *Bacillus pumilus* and periphytic microalgae. Fish Sci 76:481–487

Bidhan CD, Meena DK, Behera BK, Das P, Das Mohapatra PK, Sharma AP (2014) Probiotics in fish and shellfish culture: immunomodulatory and ecophysiological responses. Fish Physiol Biochem 40(3):921–971

Biswas G, Korenaga H, Nagamine R, Kawahara S, Takeda S, Kikuchi Y et al (2013) Elevated cytokine responses to *Vibrio harveyi* infection in the Japanese pufferfish (*Takifugu rubripes*) treated with *Lactobacillus paracasei* spp. paracasei (06TCa22) isolated from the Mongolian dairy product. Fish Shellfish Immunol 35:756–765

Boaventura C, Azevedo R, Uetanabaro A, Nicoli J, Braga LG (2012) The benefits of probiotics in human and animal nutrition. In: Brzozowski T (ed) New advances in the basic and clinical gastroenterology. Intech, Rijeka, pp 1–26

Bogut I, Milakovic Z, Bukvic Z, Brkicand S, Zimmer R (1998) Influenceof probiotic *Streptococcus faecium* M74 on growth and content ofintestinal microfllora in carp *Cyprinus carpio*. Czech J Anim Sci 43:231–235

Boyle RJ, Robins-Browne RM, Tang MLK (2006) Probiotic use in clinical practice: what are the risks? Review articles. Am J Clin Nutr 83:1256–1264

Brown M (2011) Modes of action of probiotics: resent developments. J Anim Vet Adv 10 (14):1895–1900

Brunt J, Austin B (2005) Use of a probiotic to control lactococcosis and streptococcosis in rainbow trout, *Oncorhynchus mykiss* (Walbaum). J Fish Dis 28:693–701

Burbank DR, Shah DH, La Patra SE, Fornshell G, Cain KD (2011) Enhanced resistance to Coldwater disease following feeding of probiotic bacterial strains to rainbow trout (*Oncorhynchus mykiss*). Aquaculture 321:185–190

Byun JW, Park SC, Benno Y, Oh TK (1997) Probiotic effect of *Lactobacillus* sp. DS-12 in flounder (*Paralichthys olivaceus*). J Gen Appl Microbiol 43:305–308

Cai YM, Benno Y, Nakase T, Oh TK (1998) Specific probiotic characterization of *Weissella hellenica* DS-12 isolated from flounder intestine. J Gen Appl Microbiol 44:311–316

Cai Y, Yuan W, Wang S, Guo W, Li A, Wu Y (2019) *In vitro* screening of putative probiotics and their dual beneficial effects: to white shrimp (*Litopenaeus vannamei*) postlarvae and to the rearing water. Aquaculture 498:61–71

Carnevali O, Zamponi MC, Sulpizio R, Rollo A, Nardi M, Orpianesi C, Silvi S, Caggiano M, Polzonetti AM, Cresci A (2004) Administration of probiotic strain to improve sea bream wellness during development. Aquac Int 12:377–386

Carnevali O, de Vivo L, Sulpizio R (2006) Growth improvement by probiotic in European sea bass juveniles (*Dicentrarchus labrax*, L), with particular attention to IGF-1, myostatin and cortisol gene expression. Aquaculture 258:430–438

Castex M, Lemaire P, Wabete N, Chim L (2009) Effect of dietary probiotic *Pediococcus acidilacticion* antioxidant defences and oxidative stress status of shrimp *Litopenaeus stylirostris*. Aquaculture 294(3–4):306–313

Chai P-C, Song X-L, Chen G-F, Xu H, Huang J (2016) Dietary supplementation of probiotic *Bacillus* PC465 isolated from the gut of *Fenneropenaeus chinensis* improves the health status and resistance of *Litopenaeus vannamei* against white spot syndrome virus. Fish Shellfish Immunol 54:602–611

Chang CI, Liu WY (2002) An evaluation of two probiotic bacterial strains, *Enterococcus faecium*SF68 and *Bacillus toyoi*, for reducing edwardsiellosis in cultured European eel, *Anguilla anguilla, L.* J Fish Dis 25:311–315

Chiu CH, Guu YK, Liu CH, Pan TM, Cheng W (2007) Immune responses and gene expression in white shrimp (*Litopenaeus vannamei*), induced by *Lactobacillus plantarum*. Fish Shellfish Immunol 23:364–377

Choi SH, Yoon TJ (2008) Non-specific immune response of rainbow trout (*Oncorhynchus mykiss*) by dietary heat-inactivated potential probiotics. Immune Netw 8(3):67–74

Cruz PM, Ibanez AL, Monroy Hermosillo OA, Ramirez Saad HC (2012) Use of probiotics in aquaculture. ISRN Microbiol 2:1–13

Dalmin G, Kathiresan K, Purushothaman A (2001) Effects of probiotics on bacterial population and health status of shrimp in culture pond system. Indian J Exp Biol 39:939–942

Das T, Pal AK, Chakraborty SK, Manush SM, Sahu NP, Muk-herjee SC (2005) Thermal tolerance, growth and oxygen consumption of *Labeo rohitafry* (Hamilton, 1822) acclimated to four temperatures. J Therm Biol 30:378–383

Das S, Lyla PS, Khan SA (2006) Application of Streptomyces as a probiotic in the laboratory culture of *Penaeus monodon* (Fabricius). Isr J Aquac 58(3):198–204

Dawood MA, Koshio S (2016) Recent advances in the role of probiotics and prebiotics in carp aquaculture: a review. Aquaculture 454:243–251

Dawood MA, Koshio S, Ishikawa M, Yokoyama S, El Basuini MF, Hossain MS, Nhu TH, Dossou S, Moss AS (2016a) Effects of dietary supplementation of *Lactobacillus rhamnosus* or/and *Lactococcus lactis* on the growth, gut microbiota and immune responses of red sea bream, *Pagrus major*. Fish Shellfish Immunol 49:275–285

Dawood MA, Koshio S, Ishikawa M, El-Sabagh M, Esteban MA, Zaineldin AI (2016b) Probiotics as an environment-friendly approach to enhance red sea bream, *Pagrus major* growth, immune response and oxidative status. Fish Shellfish Immunol 57:170–178

Dawood MAO, Koshio S, Ishikawa M, Yokoyama S (2016c) Effects of dietary inactivated *Pediococcus pentosaceus* on growth performance, feed utilization and blood characteristics of red sea bream, *Pagrus major* juvenile. Aquac Nutr 22(4):923–932

Dawood MAO, Koshio S, Ishikawa M, El-Sabagh M, Yokoyama S, Wang WL et al (2017) Physiological response, blood chemistry profile and mucus secretion of red sea bream (*Pagrus major*) fed diets supplemented with *Lactobacillus rhamnosus* under low salinity stress. Fish Physiol Biochem 43(1):179–192

Dawood MA, Koshio S, Esteban MÁ (2018) Beneficial roles of feed additives as immunostimulants in aquaculture: a review. Rev Aquac 10(4):950–974

Dawood MA, Koshio S, Abdel-Daim MM, Van Doan H (2019) Probiotic application for sustainable aquaculture. Rev Aquac 11(3):907–924

Dawood MA, Abo-Al-Ela HG, Hasan MT (2020a) Modulation of transcriptomic profile in aquatic animals: probiotics, prebiotics and synbiotics scenarios. Fish Shellfish Immunol 97:268–282

Dawood MA, Eweedah NM, Moustafa EM, Farahat EM (2020b) Probiotic effects of *Aspergillus oryzae* on the oxidative status, heat shock protein, and immune related gene expression of Nile tilapia (*Oreochromis niloticus*) under hypoxia challenge. Aquaculture 520:734669

Dawood MAO (2021) Nutritional immunity of fish intestines: important insights for sustainable aquaculture. Rev Aquacult 13(1):642–663

De la Banda IG, Lobo C, Chabrillon M, Leon-Rubio JM, Arijo S, Pazos G, Lucas LM, Morinigo MA (2012) Influence of dietary administration of a probiotic strain *Shewanella putrefaciens* on Senegalese sole (*Solea senegalensis*, Kaup 1858) growth, body composition and resistance to *Photobacterium damselae* subsp *piscicida*. Aquac Res 43:662–669

Defoirdt T, Boon N, Boosier P, Verstraete W (2004) Disruption of bacterial quorum sensing: an unexplored strategy to fight infections in aquaculture. Aquaculture 240:69–88

Defoirdt T, Crab R, Wood TK, Sorgeloos P, Verstraete W, Bossier P (2006) Quorum sensing disrupting brominated furanones protect the gnotobiotic brine shrimp *Artemia franciscana* from pathogenic *Vibrio harveyi*, *Vibrio campbellii*, and *Vibrio parahaemolyticus* isolates. Appl Environ Microbiol 72:6419–6423

DeMicco A, Cooper KR, Richardson JR, White LA (2010) Developmental neurotoxicity of pyrethroid insecticides in zebra fish embryos. Toxicol Sci 113:177–186

Dhanasekaran D, Saha S, Thajuddin N, Panneerselvam A (2008) Probioticeffect of *Lactobacillus* isolates *against* bacterial pathogens in *Clarias orientalis*. Med Biol 15(3):97–102

Douillet PA, Langdon CJ (1994) Use of probiotic for the culture of larvae of the Pacific oyster (*Crassostrea gigas Thurnberg*). Aquaculture 119:25–40

Elsabagh M, Mohamed R, Moustafa EM, Hamza A, Farrag F, Decamp O, Dawood MA, Eltholth M (2018) Assessing the impact of *Bacillus* strains mixture probiotic on water quality, growth performance, blood profile and intestinal morphology of Nile tilapia, *Oreochromis niloticus*. Aquac Nutr 24(6):1613–1622

El-Sersy NA, Abdelrazek FA, Taha SM (2006) Evaluation of various probiotic bacteria for the survival of *Penaeus japonicus* larvae. Fresenius Environ Bull 15:1506–1511

Fukami K, Nishijima T, Ishida Y (1997) Stimulative and inhibitory effects of bacteria on the growth of microalgae. Hydrobiol 358:185–191

Fuller R (1989) Probiotics in man and animals. J Appl Bacteriol 66:365–378

Ghosh K, Kumar SK, Kumar RA (2003) Supplementation of an isolatedfish gut bacterium *Bacillus circulans*, in: formulated diets for Rohu, Labeo rohita, fingerlings. Isr J Aquac Bamidgeh 55 (1):13–21

Ghosh S, Sinha A, Sahu C (2008) Dietary probiotic supplementation on growth and health of live-bearing ornamental fishes. Aquac Nutr 14(4):289–299

Gibson LF (1999) Bacteriocin activity and probiotic activity of Aeromonas media. J Appl Microbiol 85:243–248

Gildberg A, Mikkelsen H, Sandaker E, Ringø E (1997) Probiotic effect of lactic acid bacteria in the feed on growth and survival of fry of Atlantic cod (*Gadus morhua*). Hydrobiol 352:279–285

Gomez GD, Balcázar JL, Shen MA (2007) Probiotics as control agents in aquaculture. J Ocean Univ China 6(1):76–79

Gram L, Melchiorsen J, Spanggaard B, Huber I, Nielsen TF (1999) Inhibition of *Vibrio anguillarum* by *Pseudomonas fluorescens* AH2, a possible probiotic treatment of fish. Appl Environ Microbiol 65(3):969–973

Granados-Amores A, Campa-Cordova AI, Araya R, Mazon-Suastegui JM, Saucedo PE (2012) Growth, survival and enzyme activity of lions-paw scallop (*Nodipecten subnodosus*) spat treated with probiotics at the hatchery. Aquac Res 43:1335–1343

Gullian M, Thompson F, Rodriguez J (2004) Selection of probiotic bacteria and study of their immunostimulatory effect in *Penaeus vannamei*. Aquaculture 233:1–14

Guzmán-Villanueva LT, Escobedo-Fregoso C, Barajas-Sandoval DR, Gomez-Gil B, Peña-Rodríguez A, Martínez-Diaz SF, Balcázar JL, Quiroz-Guzmán E (2020) Assessment of microbial dynamics and antioxidant enzyme gene expression following probiotic administration in farmed Pacific white shrimp (*Litopenaeus vannamei*). Aquaculture 519:734907

Hai NV, Fotedar R (2010) A review of probiotics in shrimp aquaculture. J Appl Aquac 22:251–266

Hai NV, Buller N, Fotedar R (2009) Effects of probiotics (*Pseudomonas synxantha* and *Pseudomonas aeruginosa*) on the growth, survival and immune parameters of juvenile western king prawns (*Penaeus latisulcatus* Kishinouye, 1896). Aquac Res 40:590–602

Harikrishnan R, Balasundaram C, Heo MS (2009) Effect of chemotherapy, vaccines and immunostimulants on innate immunity of goldfish infected with *Aeromonas hydrophila*. Dis Aquat Org 88(1):45–54

Harikrishnan R, Balasundaramb C, Heo MS (2010) Effect of probiotics enriched diet on *Paralichthys olivaceus* infected with lymphocystis disease virus (LCDV). Fish Shellfish Immunol 29:868–874

Harikrishnan R, Kim MC, Kim JS, Balasundaram C, Heo MS (2011a) Probiotics and herbal mixtures enhance the growth, blood constituents, and nonspecific immune response in *Paralichthys olivaceus* against *Streptococcus parauberis*. Fish Shellfish Immunol 31:310–317

Harikrishnan R, Kim MC, Kim JS, Balasundaramb C, Heo MS (2011b) Protective effect of herbal and probiotics enriched diet on haematological and immunity status of *Oplegnathus fasciatus* (Temminck & Schlegel) against *Edwardsiella tarda*. Fish Shellfish Immunol 30:886–893

Harzevili ARS, Van Duffel H, Dhert P, Swings J, Sorgeloos P (1998) Use of a potential probiotic *Lactococcus lactis* AR21 strain for the enhancement of growth in the rotifer *Brachionus plicatilis* (Muller). Aquac Res 29:411–417

He S, Zhang Y, Xu L, Yang Y, Marubashi T, Zhou Z (2013) Effects of dietary *Bacillus subtilis* C-3102 on the production, intestinal cytokine expression and autochthonous bacteria of hybrid tilapia *Oreochromis niloticus* × *Oreochromis aureus*. Aquaculture 412-413:125–130

Heikkinen J (2013) Novel applications of Pseudomonas sp. bacterial strains in rainbow trout aquaculture. University of Eastern Finland, pp 128–134

Irianto A, Austin B (2002a) Use of probiotics to control furunculosis in rainbow trout, *Oncorhynchus mykiss* (Walbaum). J Fish Dis 25:1–10

Irianto A, Austin B (2002b) Probiotics in aquaculture. J Fish Dis 25:633–642

Jayaprakash NS, Pai SS, Anas A, Preetha R, Philip R, Singh ISB (2005) A marine bacterium, *Micrococcus* MCCB 104, antagonistic to vibrios in prawn larval rearing systems. Dis Aquat Org 68:39–45

Jia X, Zhang H, Liu X (2011) Low levels of cadmium exposure induce DNA damage and oxidative stress in the liver of Oujiang colored common carp (*Cyprinus carpio* var color). Fish Physiol Biochem 37:97–103

Kamei Y, Yoshimizu M, Ezura Y, Kimura T (1988) Screening of bacteria with antiviral activity from fresh water salmonid hatcheries. Microbiol Immunol 32:67–73

Karthik M, Bhavan PS (2018) Supplementation of *Lactobacillus brevis* for growth promotion of the freshwater prawn *Macrobrachium rosenbergii* post larvae and identification of gut microflora through 16S rDNA. Res J Biotechnol 13(1):34–50

Kawai T, Akira S (2006) Innate immune recognition of viral infection. Nat Immunol 7:131–137

Kennedy SB, Tucker JW, Neidic CL, Vermeer GK, Cooper VR, Jarrell JL, Sennett DG (1998) Bacterial management strategies for stock enhancement of warm water marine fish: a case study with common Snook (*Centropomus undecimalis*). Bull Mar Sci 62:573–588

Kesarcodi-Watson A, Kaspar H, Lategan MJ, Gibson L (2010) Alteromonas macleodii 0444 and *Neptunomonas* sp. 0536, two novel probiotics for hatchery-reared Greenshell (TM) mussel larvae, *Perna canaliculus*. Aquaculture 309:49–55

Kesarcodi-Watson A, Miner P, Nicolas JL, Robert R (2012) Protective effect of four potential probiotics against pathogen-challenge of the larvae of three bivalves: Pacific oyster (*Crassostrea gigas*), flat oyster (*Ostrea edulis*) and scallop (*Pecten maximus*). Aquaculture 344:29–34

Koninkx JFJG, Malago JJ (2008) The protective potency of probiotic bacteria and their microbial products against enteric infections-review. Folia Microbiol 53:189–194

Kumar R, Mukherjee SC, Prasad KP, Pal AK (2006) Evaluation of *Bacillus subtilis* as a probiotic to Indian major carp *Labeo rohita* (ham). Aquac Res 37:1215–1221

Lakshmi B, Viswanath B, Sai Gopal DVR (2013) Probiotics as antiviral agents in shrimp aquaculture. J Pathogens 2013:Article ID 424123

Lara-Flores M, Aguirre-Guzman G (2009) The use of probiotic in fish and shrimp aquaculture. A review: chapter 4. In: Perez-Guerra N, Pastrana-Castro L (eds) Probiotics: production, evaluation and uses in animal feed. Research Signpost, Kerala, pp 4–16

Lara-Flores M, Olvera-Novoa MA, Guzman-Mendez BE, Lopez MW (2003) Use of bacteria *Streptococcus faecium* and *Lactobacillus acidophilus,* and the yeast *Saccharomyces cerevisiae* as growthpromoters in the Nile tilapia (*Oreochromis niloticus*). Aquaculture 216:193–201

Lategan MJ, Torpy FR, Gibson LF (2004) Control of saprolegniosis in the eel (*Anguilla australis*, Richardson) by *Aeromonas media* strain A199. Aquaculture 240:19–27

Lategan MJ, Booth W, Shimmon R, Gibson LF (2006) An inhibitory substance produced by *Aeromonas* media A199, an aquatic probiotic. Aquaculture 254:115–124

Lazado CC, Caipang CMA (2014a) Mucosal immunity and probiotics in fish. Fish Shellfish Immunol 39:78–89

Lazado CC, Caipang CM (2014b) Bacterial viability differentially influences the immunomodulatory capabilities of potential host-derived probiotics in the intestinal epithelial cells of Atlantic cod *Gadus morhua*. J Appl Microbiol 116(4):990–998

Lazado CC, Caipang CMA, Brinchmann MF, Kiron V (2011) In vitro adherence of two candidate probiotics from Atlantic cod and their interference with the adhesion of two pathogenic bacteria. Vet Microbiol 148:252–259

Lazado CC, Caipang CM, Estante EG (2015a) Prospects of host-associated microorganisms in fish and penaeids as probiotics with immunomodulatory functions. Fish Shellfish Immunol 45 (1):2–12

Lazado CC, Lacsamana JI, Caipang CM (2015b) Mechanisms of probiotic actions in shrimp: implications to tropical aquaculture. In: Biotechnological advances in shrimp health management in the Philippines. Research Signpost, Kerala

Lesser MP (2006) Oxidative stress in marine environments: biochemistry and physiological ecology. Annu Rev Physiol 68:253–278

Li JQ, Tan BP, Mai KS, Ai QH, Zhang WB, Liufu ZG, Xu W (2008) Immune responses and resistance against *Vibrio parahaemolyticus* induced by probiotic bacterium Arthrobacter XE-7 in Pacific white shrimp, *Litopenaeus vannamei*. J World Aquac Soc 39:477–489

Li J, Tan B, Mai K (2009) Dietary probiotic *Bacillus* OJ and isomalto oligosaccharides influence the intestine microbial populations, immune responses and resistance to white spot syndrome virus in shrimp (*Litopenaeus vannamei*). Aquaculture 291:35–40

Liu KF, Chiu CH, Shiu YL, Cheng W, Liu CH (2010) Effects of the probiotic, *Bacillus subtilis* E20, on the survival, development, stress tolerance, and immune status of white shrimp, *Litopenaeus vannamei* larvae. Fish Shellfish Immunol 28(5–6):837–844

Logan CA, Somero GN (2011) Effects of thermal acclimation on transcriptional responses to acute heat stress in the eurythermal fish *Gillichthys mirabilis* (Cooper). Am J Physiol Regul Integr Comp Physiol 300(6):1373–1383

Lupatsch GA, Santos JW, Schrama JA, Verreth J (2010) Effect of stocking density and feeding level on energy expenditure and stress responsiveness in European sea bass (*Dicentrarchus labrax*). Aquaculture 298:245–250

Lushchak VI (2011) Adaptive response to oxidative stress: bacteria, fungi, plants and animals. Comp Biochem Physiol Toxicol Pharmacol 153:175–190

Ma THT, Tiu SHK, He JG, Chan SM (2007) Molecular cloning of a C-type lectin (LvLT) from the shrimp *Litopenaeus vannamei*: early gene down-regulation after WSSV infection. Fish Shellfish Immunol 23(2):430–437

Mahdhi A, Kamoun F, Messina C, Santulli A, Bakhrouf A (2012) Probiotic properties of Brevibacillus brevis and its influence on sea bass (*Dicentrarchus labrax*) larval rearing. Afr J Microbiol Res 6:6487–6495

Mamun MAA, Nasren S, Rathore SS, Sidiq MJ, Dharmakar P, Anjusha KV (2019) Assessment of probiotic in aquaculture: functional changes and impact on fish gut. Microbiol Res J Int 29:1–10

Manefield M, de Nys R, Kumar N, Read R, Givskov M, Steinberg P, Kjelleberg SA (1999) Evidence that halogenated furanones from *Delisea pulchra* inhibit acylated homoserine lactone (AHL)-mediated gene expression by displacing the AHL signal from its receptor protein. Microbiol 145:283–291

Marques A, Thanh TH, Sorgeloos P, Bossier P (2006) Use ofmicroalgae and bacteria to enhance protection of gnotobiotic artemia against different pathogens. Aquaculture 258:116–126

Martin H, Floch MD, Macg A (2013) Probiotic safety and risk factors. J Clin Gastroenterol 47 (5):375–376

Medellin-Pena MJ, Wang H, Johnson R, Anand S, Griffiths MW (2007) Probiotic effects virulence related gene expression in *Escherichia coli O157*:H7. Appl Environ Microbiol 73:4259–4267

Miller MM, Bassler BL (2001) Quorum sensing in bacteria. Annu Rev Microbiol 55:165–199

Mohamed AH, Traifalgar RFM, Serrano A (2013) Assessment of probiotic application on natural food, water quality and growth performance of saline tilapia *Oreochromis mossambicus* L. cultured in concrete tanks. Fisheries Aquacult J 2013:FAJ-75

Mohapatra S, Chakraborty T, Kumar V, De Boeck G, Mohanta KN (2012) Aquaculture and stress management: a review of probiotic intervention. J Anim Physiol Anim Nutr 14:1–26

Moosavi-Nasab M, Abedi E, Moosavi-Nasab S, Eskandari MH (2014) Inhibitory effect of isolated lactic acid bacteria from *Scomberomorus commerson* intestines and their bacteriocin on *Listeria innocua*. Iran Agric Res 33(1):43–52

Nayak SK (2010) Probiotics and immunity: a fish perspective. Fish Shellfish Immunol 29:2–14

Neilands JB (1981) Iron absorption and transport in microorganisms. Annu Rev Nutr 1:27–46

Newaj-Fyzul A, Austin B (2014) Probiotics, immunostimulants, plant products and oral vaccines, and their role as feed supplements in the control of bacterial fish diseases. J Fish Dis 14:12313–12318

Newaj-Fyzul A, Al-Harbi AH, Austin B (2014) Review: developments in the use of probiotics for disease control in aquaculture. Aquaculture 431:1–11

Nikoskelainen S, Ouwehand A, Salminen S, Bylund G (2001) Protection of rainbow trout (*Oncorhynchus mykiss*) from furunculosis by *Lactobacillus rhamnosus*. Aquaculture 198:229–236

Pan X, Wu T, Zhang L, Song Z, Tang H, Zhao Z (2008) In vitro evaluation on adherence and antimicrobial properties of a candidate probiotic *Clostridium butyricum* CB2 for farmed fish. J Appl Microbiol 105:1623–1629

Panigrahi A, Azad IS (2007) Microbial intervention for better fish health in aquaculture: the Indian scenario. Fish Physiol Biochem 33:429–440

Park SC, Shimamura I, Fukunaga M, Mori K, Nakai T (2000) Isolation of bacteriophages specific to a fish pathogen, *Pseudomonas plecoglossicida*, as a candidate for disease control. Appl Environ Microbiol 66:1416–1422

Pérez-Sánchez T, Balcázar JL, Merrifield DL, Carnevali O, Gioacchini G, de Blas I (2011) Expression of immune-related genes in rainbow trout (*Oncorhynchusmykiss*) induced by probiotic bacteria during *Lactococcus garvieae* infection. Fish Shellfish Immunol 31:196–201

Picchietti S, Mazzini M, Taddei AR, Renna R, Fausto AM, Mulero V (2007) Effects of administration of probiotic strains on GALT of larval gilthead sea bream: immunohistochemical and ultrastructural studies. Fish Shellfish Immunol 22:57–67

Picchietti S, Guerra L, Selleri L, Buonocore F, Abelli L, Scapigliati G (2008) Compartmentalisation of T cells expressing CD8a and TCRb in developing thymus of sea bass *Dicentrarchus labrax* (L.). Dev Comp Immunol 32:92–99

Picchietti S, Fausto AM, Randelli E, Carnevali O, Taddei AR, Buonocore F (2009) Early treatment with *Lactobacillus delbrueckii* strain induces an increase in intestinal T-cells and granulocytes and modulates immune-related genes of larval *Dicentrarchus labrax* (L.). Fish Shellfish Immunol 26:368–376

Pieters N, Brunt J, Austin B, Lyndon AR (2008) Efficacy of in-feed probiotics against *Aeromonas bestiarum* and *Ichthyophthirius multifiliis* skin infections in rainbow trout (*Oncorhynchus mykiss, Walbaum*). J Appl Microbiol 105:723–732

Pirarat N, Kobayashi T, Katagiri T, Maita M, Endo M (2006) Protective effects and mechanisms of a probiotic bacterium *Lactobacillus rhamnosus* against experimental *Edwardsiella tarda* infection in tilapia (*Oreochromis niloticus*). Vet Immunol Immunopathol 113:339–347

Pirarat N, Pinpimai K, Endo M, Katagiri T, Ponpornpisit A, Chansue N (2011) Modulation of intestinal morphology and immunity in Nile tilapia (*Oreochromis niloticus*) by *Lactobacillus rhamnosus* GG. Res Vet Sci 91:9–97

Raida MK, Larsen JL, Nielsen ME, Buchmann K (2003) Enhanced resistance of rainbow trout, *Oncorhynchus mykiss* (Walbaum), against *Yersinia ruckeri* challenge following oral administration of *Bacillus subtilis* and *B. licheniformis*. J Fish Dis 26:495–498

Ran C, Carrias A, Williams MA, Capps N, Dan BCT, Newton JC, Kloepper JW, Ooi EL, Browdy CL, Terhune JS, Liles MR (2012) Identification of Bacillus strains for biological control of catfish pathogens. PLoS One 7:9–14

Rangpipat S, Rukpratanporn S, Piyatiratitivorakul S, Menasaveta P (2000) Immunity enhancement in black tiger shrimp (*Penaeus monodon*) by a probiont bacterium (*Bacillus* S11). Aquaculture 191:271–288

Rasch M, Buch C, Austin B, Slierendrecht WJ, Ekmann KS, Larsen JL, Johansen C, Riedel K, Eberl L, Givskov M, Gram L (2004) An inhibitor of bacterial quorum sensing reduces mortalities caused by vibriosis in rainbow trout (*Oncorhynchus mykiss*, Walbaum). Syst Appl Microbiol 27:350–359

Ringø E (2020) Probiotics in shellfish aquaculture. Aquac Fish 5(1):1–27

Ringø E, Gatesoupe FJ (1998) Lactic acid bacteria in fish: a review. Aquaculture 160:177–203

Ringo E, Olsen RE, Overli JO, Lovik F (1997) Effect of dominance hierarchy formation on aerobic microbiota associated with the epithelial mucosa of subordinate and dominant individuals of Arctic char, *Salvelinus alpinus* (L.). Aquac Res 28:901–904

Ringø E, Olsen RE, Gifstad T, Dalmo RA, Amlund H, Hemre G, Bakke AM (2010) Probiotics in aquaculture: a review. Aquac Nutr 16:117–136

Robertson PAW, Dowd O, Burrells C, Williams C, Austin B (2000) Use of *Carnobacterium* sp. as probiotic for Atlantic salmon (*Salmo salar*) and rainbow trout (*Oncorhynchus mykiss*, Walbaum). Aquaculture 185:235–243

Rollo A, Sulpizio R, Nardi M, Silvi S, Orpianesi C, Caggiano M, Cresci A, Carnevali O (2006) Livemicrobial feed supplement in aquaculture for improvement of stress tolerance. Fish Physiol Biochem 32:167–177

Rottmann RW, Francis-Floyd R, Durborow R (1992) The role of stress in fish disease. SRAC Publication Number, pp 470–474

Safari R, Adel M, Lazado CC, Caipang CMA, Dadar M (2016) Host-derived probiotics *Enterococcus casseliflavus* improves resistance against *Streptococcus iniae* infection in rainbow trout (*Oncorhynchus mykiss*) via immunomodulation. Fish Shellfish Immunol 52:198–205

Sakai M, Yoshida T, Atsuta S, Kobayashi M (1995) Enhancement of resistance to vibriosis in rainbow trout, *Oncorhynchus mykiss* Walbaum, by oral administration of *Clostridium butyrium* bacterin. J Fish Dis 18:187–190

Sakata T (1990) Microflora in the digestive tract of fish and shellfish. In: Lesel R (ed) Microbiology in Poecilotherms. Elsevier, Amsterdam, pp 171–176

Salinas I, Myklebust R, Esteban MA, Olsen RE, Meseguer J, Ringø E (2008) In vitro studies of *Lactobacillus delbrueckii* subsp. *lactis* in Atlantic salmon (*Salmo salar* L.) foregut: tissue responses and evidence of protection against *Aeromonas salmonicida* subsp. *salmonicida* epithelial damage. Vet Microbiol 128:167–177

Salminen S, Isolauri E, Salminen E (1996) Clinical uses of probiotics for stabilizing the gut mucosal barrier: successful strains for future challenges. Anton van Leeuwenhoek 70:347–358

Saurabh S, Choudhary AK, Sushma GS (2005) Concept of probiotics in aquaculture. Fish Chimes 25(4):19–22

Scholz U, Garcia Diaz G, Ricque D, Cruz Suarez LE, Vargas Albores F, Latchford J (1999) Enhancement of vibriosis resistance in juvenile *Penaeus vannamei* by supplementation of diets with different yeast products. Aquaculture 176:271–283

Seenivasan C, Saravana BP, Radhakrishnan S, Shanthi R (2012) Enrichment of *Artemia nauplii* with *Lactobacillus sporogenes* forenhancing the survival, growth and levels of biochemical constituents in the post- larvae of the freshwater prawn *Macrobrachium rosenbergii*. Turk J Fish Aquat Sci 12:23–31

Servin A (2004) Antagonistic activities of *lactobacilli* and *bifidobacteria* against microbial pathogens. Microbiol Rev 28:405–440

Shabanzadeh S, Shapoori M, Sheikhzadeh N, Nofouzi K, Oushani AK, Enferadi MHN et al (2016) Growth performance, intestinal histology, and biochemical parameters of rainbow trout (*Oncorhynchus mykiss*) in response to dietary inclusion of heat-killed *Gordonia bronchialis*. Fish Physiol Biochem 42(1):65–71

Sharifuzzaman SM, Austin B (2010) Kocuria SM1 controls vibriosis in rainbow trout (*Oncorhynchus mykiss*, Walbaum). J Appl Microbiol 108:2162–2170

Sharifuzzaman SM, Abbass A, Tinsley JW, Austin B (2011) Subcellular components of probiotics Kocuria SM1 and *Rhodococcus* SM2 induce protective immunity in rainbow trout (*Oncorhynchus mykiss*, Walbaum) against *Vibrio anguillarum*. Fish Shellfish Immunol 30:347–353

Silva-Aciares FR, Carvajal PO, Mejias CA, Riquelme CE (2011) Use of macroalgae supplemented with probiotics in the *Haliotis rufescens* (Swainson, 1822) culture in northern Chile. Aquac Res 42:953–961

Sivaraman GK, Barat A, Ali S, Mahanta PC (2012) Prediction of fish growth rate and food availability in the Himalayan water bodies by estimation of RNA/DNA ratios. IUP J Genet Evol 4(3):15–19

Smith P, Davey S (1993) Evidence for the competitive exclusion of *Aeromonas salmonicida* from fish with stress-inducible furunculosis by *Pseudomonas fluorescens*. J Fish Dis 16:521–524

Smith KF, Schmidt V, Rosen GE, Amaral-Zettler L (2012) Microbial diversity and potential pathogens in ornamental fish aquarium water. PLoS One 7(9):39971

Snydman DR (2008) The safety of probiotics. Clin Infect Dis 46(2):104–111

Soltani M, Lymbery A, Song SK, Hosseini Shekarabi P (2019) Adjuvant effects of medicinal herbs and probiotics for fish vaccines. Rev Aquac 11(4):1325–1341

Song SK, Beck BR, Kim D, Park J, Kim J, Kim HD, Ringø E (2014) Prebiotics as immunostimulants in aquaculture: a review. Fish Shellfish Immunol 40(1):40–48

Sorroza L, Padilla D, Acosta F, Roman L, Grasso V, Vega J, Real F (2012) Characterization of the probiotic strain Vagococcus fluvialis in the protection of European sea bass (*Dicentrarchus labrax*) against vibriosis by *Vibrio anguillarum*. Vet Microbiol 155:369–373

Standen BT, Rawling MD, Davies SJ, Castex M, Foey A, Gioacchini G (2013) Probiotic *Pediococcus acidilactici* modulates both localised intestinal- and peripheral-immunity in tilapia (*Oreochromis niloticus*). Fish Shellfish Immunol 35:1097–1104

Swain SM, Singh C, Arul V (2009) Inhibitory activity of probiotics *Streptococcus phocae* PI80 and *Enterococcus faecium* MC13 against vibriosis in shrimp *Penaeus monodon*. World J Microbiol Biotechnol 25:697–703

Taoka Y, Maeda H, Jo JY (2006a) Growth, stress tolerance and non-specific immune response of Japanese flounder *Paralichthys olivaceusto* probiotics in a closed recirculating system. Fish Sci 72(2):310–321

Taoka Y, Maeda H, Jo JY (2006b) Use of live and dead probiotic cells in tilapia (*Oreochromis niloticus*). Fish Sci 72:755–766

Tinh NTN, Dierckens K, Sorgeloos P, Bossier P (2007) A review of the functionality of probiotics in the larviculture food chain. Mar Biotechnol 10:1–12

Tsai C-Y, Chi CC, Liu CH (2019) The growth and apparent digestibility of white shrimp, *Litopenaeus vannamei*, are increased with the probiotic, *Bacillus subtilis*. Aquac Res 50:1475–1481

Tuan TN, Duc PM, Hatai K (2013) Overview of the use of probiotics in aquaculture. Int J Res Fish Aquac 3(3):89–97

Varela JL, Ruız IR, Vargas L (2010) Dietary administration of probiotic Pdp11 promotes growth and improves stress tolerance to high stocking density in gilthead sea bream *Sparus auratus*. Aquaculture 309(1–4):265–271

Vendrell D, Balcazar JL, de Blas I, Ruiz-Zarzuela I, Girones O, Muzquiz JL (2008) Protection of rainbow trout (*Oncorhynchus mykiss*) from lactococcosis by probiotic bacteria. Comp Immunol Microbiol Infect Dis 31:337–345

Verschuere L, Rombaut G, Sorgeloos P, Verstraete W (2000) Probiotic bacteria as biological control agents in aquaculture. Microbiol Mol Biol Rev 64(4):655–671

Wabete N, Chim L, Lemaire P, Massabuau JC (2008) Life on the edge: physiological problems in penaeid prawns *Litopenaeus stylirostris*, living on the low side of their thermo preferendum. Mar Biol 154:403–412

Wang X, Li H, Zhang X, Li Y, Ji W, Xu H (2000) Microbial flora in the digestive tract of adult penaeid shrimp (*Penaeus chinensis*). J Ocean Univ Qingdao 30:493–498

Zaineldin AI, Hegazi S, Koshio S, Ishikawa M, Bakr A, El-Keredy AM, Dawood MA, Dossou S, Wang W, Yukun Z (2018) *Bacillus subtilis* as probiotic candidate for red sea bream: growth performance, oxidative status, and immune response traits. Fish Shellfish Immunol 79:303–312

Zapata AA, Lara-Flores M (2013) Antimicrobial activities of lactic acid bacteria strains isolated from Nile tilapia (*Oreochromis niloticus*) intestine. J Biol Life Sci 4(1):123–129

Zhou XX, Wang YB, Li WF (2009) Effect of probiotic on larvae shrimp (*Penaeus vannamei*) based on water quality, survival rate and digestive enzyme activities. Aquaculture 287:349–353

Zhou X, Wang Y, Yao J, Li W (2010) Inhibition ability of lactic-acid bacteria *Lactococcus lactis*, against a. hydrophila and study of its immunostimulatory effect in tilapia (*Oreochromis niloticus*). Int J Engineer Sci Technol 2(7):73–80

Zokaeifar H, Balcazar JL, Saad CR, Kamarudin MS, Sijam K, Arshad A, Nejat N (2012) Effects of *Bacillus subtilis* on the growth performance, digestive enzymes, immune gene expression and disease resistance of white shrimp, *Litopenaeus vannamei*. Fish Shellfish Immunol 33:683–689

Zokaeifar H, Babaei N, Saad CR, Kamarudin MS, Sijam K, Balcazar JL (2014) Administrationof *Bacillus subtilis* strains in the rearing water enhances the water quality, growth performance, immune response, and resistance against *Vibrio harveyi* infection in juvenile white shrimp, *Litopenaeus vannamei*. Fish Shellfish Immunol 36:68–74

Zorriehzahra MJ, Delshad ST, Adel M, Tiwari R, Karthik K, Dhama K et al (2016) Probiotics as beneficial microbes in aquaculture: an update on their multiple modes of action: a review. Vet Q 36:228–241

Honeybee Gut: Reservoir of Probiotic Bacteria

9

Samira Tootiaie, Mojtabah Moharrami, and Naheed Mojgani

Abstract

Honeybees are the most important crop pollinators that contribute significantly to agricultural productivity and profitability worldwide. Microbiota accounts for up to 1–10% of the insect's biomass. The intestine of European Honeybees, *Apis mellifera,* have diverse microbiota and are known to be occupied by approximately 70% Gram-negative bacteria, 27% Gram-positive bacteria, and 1% yeast. The native microbiota of the honeybees is known to contribute to their nutrition, growth, digestion, pathogens defense, and insecticide resistance. As with other humans and animals, intestinal dysbiosis might greatly influence these insects' health status posing a threat to their safe existence. Lactic acid bacteria (LAB) have been discovered in abundance in the honeybee gut and are believed to be of great importance to the honeybee health. Among several symbiotic LAB species isolated from the digestive tract of honeybees, it is found that some of them have the potential to be developed as probiotics. One of the most important health benefits of probiotic LAB in honeybees is their ability to protect against several bee pathogens and contribute to honey's antimicrobial properties. Hence, the use of probiotics in beekeeping could prevent diseases, enhance bee health, and consequently increase honey production. Although probiotic bacteria isolated from different sources could be used for honeybees, using the host bacteria, i.e., the bacteria from the honeybees' gut microbiome community would be more desirable for their own health. In this review study, we discuss the important aspects related to *Apis mellifera* gut microbiome such as composition, perturbation, fermentation, and most important of all, the probiotic bacterial community, mainly LAB species residing in the gut of these insects.

S. Tootiaie · M. Moharrami · N. Mojgani (✉)
Agriculture Research, Education and Extension Organization (AREEO), Razi Vaccine and Serum Research Institute (RVSRI), Karaj, Iran

N. Mojgani, M. Dadar (eds.), *Probiotic Bacteria and Postbiotic Metabolites: Role in Animal and Human Health*, Microorganisms for Sustainability 2,
https://doi.org/10.1007/978-981-16-0223-8_9

Keywords

Apis mellifera · Gut microbiome · Lactic acid bacteria · Probiotic · Dysbiosis

9.1 Introduction

Apis mellifera (A. mellifera) or European honeybee could colonize virtually all habitable biomes on Earth and adapt to diverse bioclimatic conditions. *A. mellifera* has been classified as a member of the order Hymenoptera and the superfamily Apoidea (Alatawy et al. 2020). They have been divided into six evolutionary lineages including A lineage (Africa), M lineage (western and northern Europe), C lineage (southern and eastern Europe), O lineage (Caucasus, Turkey, Middle East, Cyprus, Crete), Y lineage (Ethiopia), and S lineage (Syria and Lebanon) (Tihelka et al. 2020). As *A. mellifera* is an essential pollinator species for natural ecosystems and agricultural production, its health status and, consequently, continuous existence is of great importance.

As obvious from numerous research reports, *Apis mellifera* depends on its gut microbiome to perform its basic functions and survival. Bacterial communities living in symbiosis with their hosts, also known as probiotics, are essential factors in maintaining host health (Zeinali et al. 2020). Additionally, a close association between Honeybee colony productivity and increased bacterial diversity was discovered recently. Hence, an improved understanding of the honeybees' gut microbiome can help manage modern challenges to these insects' health and production.

The gut of the honeybee is a continuous tube starting from mouth to anus and demarcated into foregut (stomodeum), midgut (mesenteron), and the hindgut (proctodeum). In many insects, the hindgut is the gut region bearing the largest microbial populations. In particular, the ileum (the region between the proximal pylorus and distal rectum) is a relatively benign environment, in that it lacks the digestive enzymes of the midgut and, for many terrestrial insects, the desiccation stress of the distal hindgut, where water is actively resorbed from the lumen into insect tissues. Microbial function and growth may also be favored by the ions and metabolites delivered to the hindgut in the filtrate from the Malpighian tubules (Huang et al. 2010). On the contrary, due to midgut epithelium actively secreting immunologically active enzymes as well as several antimicrobial peptides, the midgut shows a hostile environment for microorganisms. Besides, the midgut also contains a region of $pH < 3$ that mediate many microbial cells degradation (Engel and Moran 2013; Shanbhag and Tripathi 2009).

9.2 Gut Microbiome Composition

In the last couple of years, marked deterioration in honeybee hives' colony health has been reported that has raised worldwide concerns (Meixner 2010). One of the major reasons for such depurations of honeybee colonies is due to the effect of several honeybee diseases (Genersch et al. 2010). In this context, the honeybee gut's microbial ecosystem has known to play an essential role in maintaining their health and survival. Thus, understanding the microbial community residing in the gut of different honeybee species could lead us to better health management of the bees that would consequently result in enhanced agriculture productivity and human well-being.

The bee gut microbial community is far simpler than the mammalian microbiota and contains a distinctive community of bacterial species. The composition of the gut bacterial communities of these social insect insects has been shaped by coevolution. These insects' social behavior provides favorable conditions for the exchange of the symbiont microbes, and a number of these microorganisms are efficiently transmitted between bee colony members and their different generations (Engel and Moran 2013).

The composition of microbial communities in the honeybee gut varies enormously within and between species. Honeybees acquire gut microorganisms from the natural environment via foods, such as nectar, pollen, and water. Hence, honeybees' gut flora varies according to seasonal or geographical differences in food sources, even among individual honeybees from the same colony (Mohr and Tebbe 2006; Moran et al. 2012).

A.mellifera gut microbiota is dominated by only nine bacterial species clusters that are specific to bees and are transmitted through social interactions between individuals (Fig. 9.1). According to available information, there are five main bacterial groups in the honeybees gut including (i) Gram-negative bacteria group (*Snodgrassella alvi* and *Gilliamella apicola*), (ii) phylum Proteobacteria to Gram-positive bacteria, Firmicutes (Lactobacillus Firm-4 and Lactobacillus Firm 5 groups), (iii) phylum Actinobacteria (*Bifidobacterium asteroids*), (iv) a small number of Proteobacteria species (*Frischella perrara, Bartonella apis, Parasaccharibacter apium*), and (v) Gluconobacter-related species group designated Alpha2.1 (Bottacini et al. 2012; Kwong and Moran 2016).

As seen in Table 9.1, honeybee microbiota occupies distinct metabolic niches in the *A. mellifera* gut. According to reports, the abundance of *Snodgrassella alvi, Frischella perrara, Gilliamella apicola, and Bartonella apis* is seen in the ileum while *Lactobacillus Firm-4*, *Lactobacillus Firm-5*, as well as Bifidobacterium predominantly reside in the rectum (Kwong and Moran 2016).

Many factors are known to affect the gut microbiota composition and profile of the honeybees, including the age and physiological condition of honeybees. Martinson and Moran in 2012 reported that newly emerged honeybee workers have no or very few gut bacteria, while they uptake bacteria later via contact with the collected honey and bee bread and through trophallactic exchange with nestmates (Martinson et al. 2012). In other research findings, it was stated that

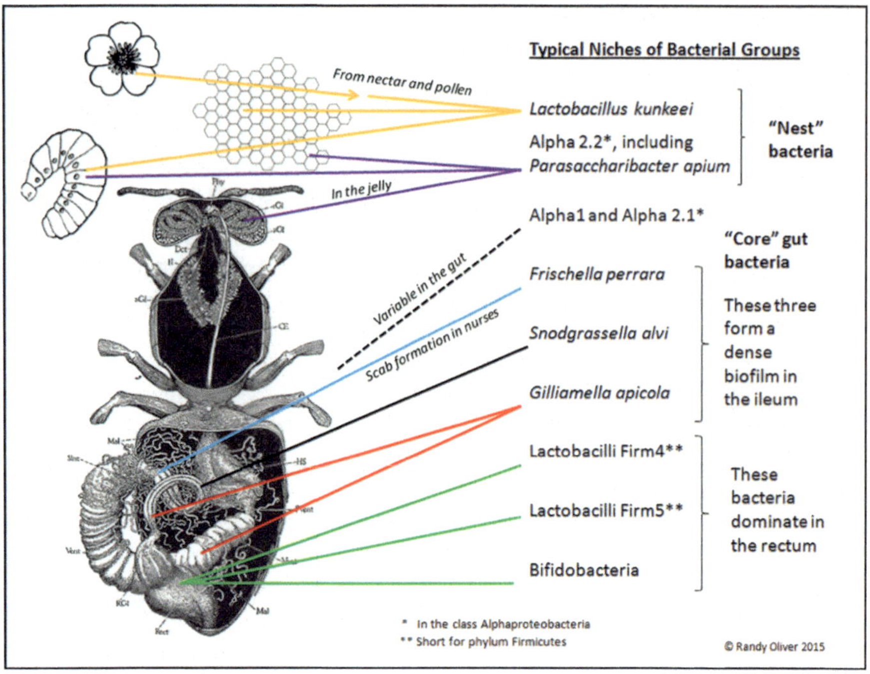

Fig. 9.1 Bacterial groups present in honeybees

larvae's gut microbial composition differs vastly due to the differences in the bee bread microbial communities (Martinson et al. 2012; Mohr and Tebbe 2006). The observation that certain gut bacteria are maintained in all of the developmental stages of an adult bee, irrespective of differences between species, colonies, and individuals, suggests that distinctive gut bacteria are transferred between generations by eusocial behaviors, such as food exchange between the honeybee populations in a hive (Martinson et al. 2011; Martinson et al. 2012; Vásquez et al. 2012).

Zhi-Xiang Dong and his colleagues, through the use of 16 s rRNA gene sequencing analysis, found that 0-day postemergence (dpe) did not harbor core gut flora in the gut, and the critical points for colonization of the core gut flora were around 1–3 dpe. For example, colonization of Frischella, Gilliamella, and Snodgrassella occurred at 1dpe, while Bifidobacterium, Commensalibacter, and Lactobacillus colonization were significantly detected at 3 dpe (Dong et al. 2020). It is worth mentioning that type of overwintering sugar also influences honeybee gut microbiota. Wang et al. in 2020 via 16 s rRNA sequencing determined bacterial communities in honeybee midguts and hindguts before winter and after bees were fed honey, sucrose, and high-fructose syrup as winter-food. In the midgut, the sucrose group's microbial diversity was higher than that of the honey and high-fructose syrup groups, but in the hindgut, the microbial diversity of the honey and high-fructose groups was higher than that in the sucrose group. Sucrose increased the relative abundance of Actinobacteria (Bifidobacteriales) and

Table 9.1 *A. mellifera* gut microbiome composition and their main functions

A.mellifera microbiome composition	Location	Main function	Reference
Snodgrassella alvi	Ileum	Activation of the innate immune system	Horak et al. (2020)
Gilliamella apicola	Ileum	Degradation of pectin	Kwong and Moran (2016)
Lactobacillus Firm-4	Rectum	Degradation of pectin	Lee et al. (2018)
Lactobacillus Firm-5	Rectum	Degradation of pectin	Lee et al. (2018)
Bifidobacterium asteroids	Rectum	Through glycosidase and pectinesterase could degrade pollen	Lee et al. (2018)
Frischella perrara	Ileum	Activation of the innate immune system	Emery et al. (2017)
Bartonella apis	Ileum	Positive effects on disease resistance	(Cornman et al. (2012)
Parasaccharibacter apium and a Gluconobacter- related species group designated Alpha2.1	Rectum	Oxidative fermentation/perform gluconeogenesis	Bonilla-Rosso et al. (2019)

Alphaproteobacteria of honeybee midgut, and honey enriched the Bacteroidetes and Gammaproteobacteria in honeybee hindgut. High-fructose syrup increased the relative abundance of Betaproteobacteria of the midgut. Furthermore, they introduce sucrose as an appropriate overwintering food for honeybee. In this manner, the relative abundance of the dominant microbiota significantly altered with the different sugar types and seasons (Wang et al. 2020).

In another research report, Christina Geldert et al. investigated the effect of phytochemical supplementations on the microbiome diversity and abundance of *A. mellifera*. They disclosed that phytochemical supplementations are able to enhance gut microbial diversity and significantly increase the abundance of the most represented bacterial genera such as *Snodgrassella* spp. and *Lactobacillus* spp. (Geldert et al. 2020).

9.3 Gut Microbiome Perturbation

Exposure to an environmental stressor, including antibiotics as well as the herbicide, is one of the major sources of perturbation to the microbiome that has a detrimental effect on *A. mellifera* health. Antibiotic treatment of bee colonies has been widely used for over 50 years to prevent bee larvae's bacterial disease. Some of the most frequently used antibiotics by beekeepers include tetracycline, fumagillin, and tylosin (Genersch et al. 2010). However, these chemical drugs are known to have many side effects, including the emergence of antibiotic resistance. Hence, more attention has been paid to evaluate the effect of antibiotics on the survival and growth of honeybees to identify and pinpoint the main disadvantages of these drugs on honeybee and their environment's health. Raymann et al., in 2017, assessed the relationship between tetracycline exposure and the size and composition of honeybee gut communities. Their results showed that treatment with tetracycline greatly influenced both the honeybee gut microbiome's composition and size. According to their observations, tetracycline induced dysbiosis in these insects, which resulted in increased susceptibility to opportunistic pathogens and subsequently led to a significant reduction in bee survival rate (Raymann et al. 2017).

Apart from antibiotics, some agrochemicals such as herbicides can perturb the honeybee gut microbiota and therefore compromise bee health. Shikimate pathway that is found in the bacterial community residing in the bee gut, such as in *Snodgrassella alvi, Gilliamella* spp., and *Bifidobacterium* spp., is known to play a key role in the production of essential aromatic compounds such as the amino acids phenylalanine, tryptophan, and tyrosine. In this context, glyphosate, the primary herbicide, inhibits 5-enolpyruvyl-shikimate-3-phosphate synthase (EPSPS) in the shikimate pathway (Motta et al. 2018). Thus, glyphosate via shikimate pathway inhibition and subsequently essential nutrients depletion play a crucial role in bacterial death and reducing beneficial bacteria in bee gut.

Propolis is another critical factor involved in honeybee gut microbiome consistency. In this context, Saelao et al. investigate the association between propolis and microbial community consistency in the honeybee microbiome. They disclosed that propolis insufficiency contributes to significant perturbation in the abundance of several key gut microbiota members. These authors proposed that propolis, via restricting alterations in the microbial community, play a key role in honeybee colony microbial health (Saelao et al. 2020).

9.4 Probiotic Potential of Honeybee Gut-Associated Bacteria

Inappropriate and misuse of antibiotics has led to a rise in antibacterial resistance and diminished the efficacy of these once considered miracle drugs. Since the alarming rise of antibiotic resistance, many strategies and investigations have been carried out to explore other safer ways to treat human ailments without harming the natural immunity of the host, and replacing or augmenting these antibiotics.

In the late nineteenth century, microbiologists identified microflora in healthy individuals' gastrointestinal tracts that differed from those found in diseased individuals. The beneficial microflora found in the gastrointestinal tract was termed probiotics. FAO and WHO experts defined the term probiotics as "Live microorganisms which when administered in adequate amounts confer a health benefit on the host" (Joint 2002). In other words, probiotics are living microorganisms used to restore gut health by maintaining the intestinal microbiota (Manzanares et al. 2016). Similar to humans and animals, the gut-associated bacterial flora in honeybees has been reported to have the ability to provide health benefits, most important of all, which is the capability to protect them from several honeybee diseases (Li et al. 2017; Schwarz et al. 2016).

Below we discuss some of the beneficial functions carried out by the gut microbiome of honeybees.

9.4.1 Antimicrobial Effects of Honeybee Gut-Associated Bacteria

As stated earlier, the gut microbial community in honeybee *A. mellifera* protects the host from infection. Schwarz et al. provide experimental support linking parasite susceptibility of honeybee to dysbiosis of their core microbiota. They disclosed that honeybee, in a dysbiosis state, lose their ability to control encounter protozoan *Lotmaria passim* and lead to *L. passim* infection in these insects (Schwarz et al. 2016). *Nosema ceranae* is a gut intracellular parasite of honeybees that destroys epithelial cells and gut tissue integrity. In this context, Jiang Hong Li in 2017 revealed that disruption of bacteria in the honeybee by antibiotic treatment mediates honeybee's susceptibility to Nosema infection (Li et al. 2017).

Moreover, Huang and Evans in 2020 investigated the effect of Nosema on the gut microbiome via suppression of *N. ceranae* with specific siRNAs. They found that suppressing *N. ceranae* led to significant positive effects on gut microbial abundance. These researchers concluded that *N. ceranae* is negatively correlated with the abundance of 15 identified bacteria (Huang and Evans 2020). In a study conducted by a group of researchers, it was found that the members of the gut microbiome, by lowering the local intestinal pH with the production of lactic acid, antimicrobial metabolites, as well as induction of innate immunity, interfere with the growth of Nosema infection (El Khoury et al. 2018). Furthermore, Streptomycin is an aminoglycoside antibiotic function in protein synthesis inhibition in Gram-negative bacteria. In 2015, through a metagenomics approach, Saraiva identified genes involved in streptomycin biosynthesis in *A. mellifera* microbiome. The presence of such genes raises the hypothesis about the possible role of normal microbiota in protecting *Apis mellifera* against pathogenic bacteria and in maintaining the healthy status of the hive (Saraiva et al. 2015).

Paenibacillus larvae, a Gram-positive sporulated bacterium that causes the American foulbrood disease, is an extremely contagious and dangerous pathogen of honeybees. In 2009, Sabate et al. aimed to explore the biological control capability of Bacillus strains associated with the bee intestine and evaluate their influence against *P. larvae*. They found that Bacillus strains through surfactin synthesis inhibit the growth of *P. larvae* (Sabaté et al. 2009). These novel findings collectively emphasize the importance of *A. mellifera* gut bacteria in modulating honeybees' susceptibility to various infections.

9.4.2 Gut Microbiome Role in Immune Function

Various predators, including parasites, parasitoids, and pathogens, threaten insect health during their life cycle. A complex immune system has evolved in insects for protection against these threats. Several studies have illustrated that gut bacteria are key mediators in immune modulation and are essential for a healthy immune system (Kaltenpoth and Engl 2014). Hemocyte (immune cells), as a crucial element in the innate immune system through phagocytosis, plays a key role in hemolymph pathogen clearance. Vitellogenin (Vg) is a protein engaged in honeybee worker's stress tolerance, and behavior. Vg is the main zinc carrier in honeybee workers, and zinc deficiency is associated with hemocyte pycnosis (cell death). Thereby, Vg is considered a critical mediator in honeybee immunity and lead to a longer life span. In an experimental investigation by Zheng et al. in 2017, it was found that normal microbiota compared to germ-free bees increase vitellogenin expression almost fivefold (Kaltenpoth and Engl 2014; Zheng et al. 2017). Overall, based on these findings, we can consider the gut microbiome as a major contributing factor for honeybee immune activation.

Furthermore, the Scab phenotype as a prominent immune response factor is triggered by reminiscent of melanization and develops 5–7 days after adult worker bees have emerged. Scab phenotype is characterized by a dark brown to black deposit forming a localized thin band in the pylorus at the midgut-hindgut boundary, in close proximity to the Malpighian tubules of the honeybees. Emery et al. identified significant host gene expression alteration in the pylorus region following *Frischella perrara* colonization compared to non-colonized bee. Using gene ontology (GO) enrichment analysis, they disclosed that immune-related genes, including irp30, cdc2c, abaecin, apid73, b-guc2, and def-1, were increased in the pylorus region of the screened honeybees. In this manner, *Frischella perrara* via colonization in a restricted region in the pylorus, as well as immune-related genes activation, play a key role in scab phenotype induction (Emery et al. 2017). Additionally, Horak and his research team investigate the beneficial effect of symbiont Snodgrassella alvi on honeybee immune gene expression. They illustrate that Snodgrassella alvi via expression of host antimicrobial peptides as well as Toll pathway upregulation aid in the clearance of opportunistic pathogen Serratia marcescens from the honeybees gut (Horak et al. 2020).

9.4.3 Gut Microbiome Role in Food Fermentation

Fermentation products such as short-chain fatty acids (SCFAs) are highly beneficial for host energy metabolism. In the fermentation process, *A. mellifera* gut microbiota members play an important role in breaking saccharides into an array of alcohols, SCFAs, gases, and other organic acids such as acetate and lactate. Acetate kinase (ackA) and L-lactate dehydrogenase (ldh) are the main enzymes responsible for acetate and lactate production, respectively. In turn, acetate production and lactate through increased sucrose sensitivity play a crucial role in honeybee weight gaining

(Lee et al. 2018; Zheng et al. 2017). Therefore, *A. mellifera* gut microbiota through organic acid production plays an important role in honeybee weight gain.

9.4.4 Gut Microbiome Role in Detoxification

Gut microbiota strongly promotes the expression of key enzymes of the honeybee xenobiotic detoxification pathway. Three important enzymes responsible for insect detoxification, including carboxylesterases (COEs), Cytochrome P450 monooxygenases (CYPs, also called P450s), and glutathione S-transferases (GSTs), have been identified recently. Thereby, honeybee gut microbiota enhance host detoxification capability and manipulate host metabolism (Wu et al. 2020). Furthermore, some monosaccharide sugars, including xylose, mannose, rhamnose, and arabinose, have been reported to endorse toxic effects on *A. mellifera* and decrease their life span. Recently, genes responsible for mannose metabolism, including phosphotransferase systems (PTSs) and mannose-6-phosphate isomerase (MPI), were identified in the *Gilliamella apicola* genome. Additionally, several genes associated with catabolism of rhamnose, xylose, and arabinose have also been detected in the genome of *Gilliamella apicola*. Hence, it is concluded that *Gilliamella apicola* is able to metabolize xylose, mannose, rhamnose, and arabinose and subsequently boost *A. mellifera* life span (Zheng et al. 2016).

9.4.5 Probiotic Properties of Honeybee-Specific Lactic Acid Bacteria

Lactic acid bacteria (LAB) are a group of Gram-positive lactic acid-producing bacteria present in diverse habitats. LAB belongs to phylum Firmicutes with low G + C in the genome. These bacteria are well known for their role in food fermentation, and a wide variety of strains are routinely employed as starter cultures in the manufacture of dairy, meat, vegetable, and bakery products. Additionally, they have a significant role as starter cultures for cheese and yogurts. One of the factors that make LAB of high importance, especially for human and animal use, is their "generally recognized as safe" (GRAS) status that make these food-grade microorganisms to be employed as probiotics (Åvall-Jääskeläinen and Palva 2005; Choi et al. 2005). While the European Food Safety Authority (EFSA) proclaimed the LAB strain to have QPS (Qualified presumption of safety) status (EFSA 2008). In several findings, the presence of LAB in the gut of honeybees has been reported, where they are shown to provide beneficial effects to their host, and thereby they are potential probiotic candidates.

Majority of honeybee-specific LAB has found significant importance owing to their probiotic potentials. Up to date, thirteen genetically distinct lactic acid-producing bacteria have been identified from the honeybee crop, of which nine are Lactobacilli and four are Bifidobacteria (Olofsson et al. 2014; Olofsson and Vásquez 2008).

Table 9.2 Lactic acid bacteria isolated from different species of honeybees

Honeybee species	LAB isolated	Country of isolation	Reference
Apis mellifera	*Micrococcus, Bifidobacterium asteroids, Fructobacillus fructosus*	Iran	Sharifpour et al. (2016)
	Lactobacillus johnsonii, Enterococcus faecium, Lactobacillus kunkeei	Argentina, Egypt	Audisio et al. (2011), Elzeini et al. (2020)
	Lactobacillus brevis, Lactobacillus casei	Egypt	Elzeini et al. (2020)
	L. melliventris, L. kimbladii, L. mellis, L. apinorum, L. kullabergensis, L. helsingborgensis	Sweden	Olofsson et al. (2014)
Apis mellifera jemenitica	*L. kunkeei, Lact. Lactis, Enterococcus faecalis*	Saudi Arabia	Khan et al. (2017)
Apis cerena	*Bifidobacterium indicum, Bifidobacterium asteroids, Fructobacillus fructosus, L. apinorum, L. apis, L. helsingborgensis, L. kimbladii, L. kullabergensis, L. kunkeei*	Vietnam	Duong et al. (2020)
Apis dorsata	*Bifidobacterium indicum, Lactobacillus kunkeei, Lactobacillus vermiform, Lactobacillus sp.*	Malaysia, Indonesia	
Apis florea fabricius	*L. kunkeei, L. plantarum, L. apis*	Iran	Parichehreh et al. (2018)

Lactic acid bacteria has been isolated from the gut of several honeybee species, including *Apis mellifera, A. dorsata, A. florea, A. nigrocincta, Apis nuluenis, Apis laboriosa, A. cerana indica, Melipona beecheii, Meliponula bacandei*, and *Trigona* sp. (Mathialagan et al. 2018; Niode et al. 2020; Vásquez et al. 2012). Lactobacillus is one of the most important genera within the LAB, which at present includes 175 listed species (Euzéby 1997). Among this group of bacteria, genus *Lactobacillus* is the most frequent Gram-positive bacteria isolated from different honeybee species' gut. While *Lactobacillus kunkeei* has been reported to be one of the most dominant species of this genus residing in their gut (Niode et al. 2020). Table 9.2 depicts a variety of LAB species isolated from different honeybee species around the world.

9.4.6 Antimicrobial Effect of Honeybee Gut-Associated LAB Against Honeybee Diseases

LAB comprises a group of Gram-positive, catalase-negative, non-motile, non-spore-forming facultative anaerobic bacteria that are commonly found as both exogenous and endogenous microbes in healthy individuals. Similar to the LAB found within humans and animals, the honeybee-specific LAB defends their hosts from invasion and colonization of several pathogenic bacteria via the production of a variety of antimicrobial metabolites and modulation of the host immune response (Huang and Evans 2020; Mathialagan et al. 2018; Niode et al. 2020). Hence, honeybee-specific LAB can also provide protection against several honeybee diseases by production of these metabolites.

The antimicrobial effects exerted by these bacteria are owing to their metabolite-producing abilities. These metabolites, also referred to as postbiotic metabolites, includes organic acids (lactic acid, acetate acid, and formic acid) (Olofsson et al. 2016), extracellular proteins, benzoate, bacteriocins, hydrogen peroxide (H_2O_2), lipopolysaccharides, and lipoteichoic acid volatile compounds (Butler et al. 2013; Olofsson et al. 2016; Olofsson and Vásquez 2008) etc.

Hence, due to the stated health-promoting functions of LAB in honeybees, they are considered safe alternative therapeutic strategy for the control of a number of honeybee diseases, including *Paenibacillus larvae* (infective bacterial agent of American foulbrood disease), *Melissococcus pluton* (infective bacterial agent of European foulbrood), Nosemosis, and varroosis (Audisio 2017; Forsgren et al. 2010). In a study conducted by a group of researchers, an organic acid-producing *L. johnsonii* was shown to inhibit the growth of *Nosema ceranae* and harbored fumigillin activity (60).

Paratransgenesis has come to mean a Trojan horse strategy, where endogenous microorganisms via effector molecules production inhibit pathogen development. Candidate microorganisms to being practical in honeybee, they should possess several criteria including (1) candidate microorganism should be genetically modifiable for effector molecules expression; (2) ideally the candidate microorganism must be ecologically and functionally fit with other nonpathogenic bee-associated microorganisms, and (3) following reintroduction the modified organism should have no negative impact on honeybee health. In this context, Rangberg et al. investigated *L. kunkeei* potency in honeybee paratransgenesis. They concluded that *L. kunkeei* complies with the three criteria required for being a suitable paratransgenic candidate (Rangberg et al. 2015). Similar to these findings, Maddaloni and his co-investigators demonstrated that *Fructobacillus fructosus* can be used as a powerful tool for honeybee paratransgenesis to control diseases and expand nutrition repertoire (Maddaloni et al. 2014).

9.5 Commonly Used Methods for Microbiome Analysis

Researcher frequently utilizes full-length 16S rRNA gene sequences with nine hypervariable regions (V1–V9) to infer phylogenetic relationship among the microbiome. Therefore, a full-length 16S rRNA amplicon sequencing approach with high accuracy and efficiency can be used for microbial diversity detection in various biological samples. Nanopore DNA sequencer (MinION) containing several significant advantages including rapid library construction, low cost, real-time detection and small size that made it a suitable tool for identifying microbiome composition at the species levels (Shin et al. 2016). However, it was disclosed that bacteria with almost identical 16S rRNA sequences could exhibit high sequence divergence levels at other loci and very different gene repertoires. Thereby, it is difficult to gain insight into intraspecific diversification of bacterial lineages in the gut with 16S rRNA sequencing. Single-cell genomics and transcriptomics can provide reliable context for assembled genome fragments and gene expression activity on the level of individual prokaryotic genomes. In this manner single cell genomics, through allowing direct access to information from individual microorganisms, has the potential to elucidate processes of bacterial diversification (Engel et al. 2014). However, low DNA and mRNA content restrict the yield of reasonable amounts of genetic material for sequencing analysis from a single cell.

Furthermore, the lack of polyadenylation of bacterial mRNA limits its separation from rRNA. Additionally, cell walls and membranes diversity induce a challenge to consistent lysis or permeabilization required for single-cell RNA sequencing (scRNA-seq). These problems impede the characterization of microbes by traditional single-cell sequencing methods (Sharma and Thaiss 2020).

9.6 Microbiome Engineering as a Future Perspective

Engineering of microbiomes is used to modify structures of the microbiota and restore ecological balance. Synthetic biology and engineering principles are frequently applied in microbiome engineering to improve microbiome function. Thereby, microbiome engineering could lead to a breakthrough in agriculture and medicine. In medicine, microbiome engineering enables exploring individual microbes' contribution and generating potential therapies against metabolic (e.g., phenylketonuria and chronic kidney disease), inflammatory, and immunological diseases, among others. In the case of honeybee, due to their agricultural importance as well as the simple gut microbiome, they are a promising testbed for the nascent field of microbiome engineering (Foo et al. 2017; Leonard 2020). There are several approaches to honeybee microbiome engineering. A plasmid toolkit by combining a broad-host-range (BHR) replicon with a set of modular genetic parts can be applied to bacteria from the *A. mellifera* gut microbiome. It was disclosed that plasmids constructed using bee microbiome toolkit (BTK) act faithfully in various species of Proteobacteria detected in the *A. mellifera* gut microbiome. The BTK can be used to express heterologous genes or to repress or disrupt genes in the bacterial chromosome (Leonard 2020). Consequently, microbiome engineering could be employed as a powerful tool for improving *A. mellifera* health and subsequently agricultural productivity.

9.7 Conclusions

The economic value of commercial honeybee pollination is estimated at over US $220 billion worldwide. Any damage to these insects leads to detrimental consequences not only to our agriculture and production values that ultimately would result in economic losses but might also threaten and endanger our lives on the planet. Hence, intensive research has been done and is still ongoing to find solutions to prevent colony losses and find ways to increase their survival and control the pathogens from harming their viability.

A.mellifera digestive tract is a reservoir of a diverse variety of bacterial communities that play a significant role in these insects' growth and survival. Recent studies with gut microbiome disclosed the honeybee gut-associated microbial in immune system activation, carbohydrate fermentation, and inhibition of disease in the host. This suggests that the gut bacterial community structure may be considered as an indicator of honeybee health. Since related microbiotas are found across bee species, it strongly suggests a close evolutionary relationship between bacteria and hosts, as well as underscoring the importance of LAB symbionts for bees. Not only are LAB symbionts involved in honeybee food production and preservation, but they are also of importance in host defense against pathogen and transient microbes intercepted during foraging. Hence preserving the balance of these gut bacteria is crucial for maintaining honeybee health and vigor. Tools to engineer a microbial member of these honeybees might play a significant role in beekeeping management issues such as increased colony survival.

References

Alatawy M, Al-Attas SG, Assagaf AI, Al-shehri A, Alghamdi KM, Bahieldin A (2020) Gut microbial communities of adult honey bee workers (Apis Mellifera). Biosci Biotechnol Res Asia 17

Audisio MC (2017) Gram-positive bacteria with probiotic potential for the Apis mellifera L. honey bee: the experience in the northwest of Argentina. Probiotics Antimicrob Proteins 9:22–31

Audisio MC, Torres MJ, Sabaté DC, Ibarguren C, Apella MC (2011) Properties of different lactic acid bacteria isolated from Apis mellifera L. bee-gut. Microbiol Res 166:1–13

Åvall-Jääskeläinen S, Palva A (2005) Lactobacillus surface layers and their applications. FEMS Microbiol Rev 3:511–529

Bonilla-Rosso G, Paredes JC, Das S, Ellegaard KM, Emery O, Garcia-Garcera M, Glover N, Hadadi N, van der Meer JR, Tagini F (2019) Acetobacteraceae in the honey bee gut comprise two distant clades with diverging metabolism and ecological niches. bioRxiv:861260

Bottacini F, Milani C, Turroni F, Sánchez B, Foroni E, Duranti S, Serafini F, Viappiani A, Strati F, Ferrarini A (2012) Bifidobacterium asteroides PRL2011 genome analysis reveals clues for colonization of the insect gut. PLoS One 7:e44229

Butler È, Alsterfjord M, Olofsson TC, Karlsson C, Malmström J, Vásquez A (2013) Proteins of novel lactic acid bacteria from Apis mellifera mellifera: an insight into the production of known extra-cellular proteins during microbial stress. BMC Microbiol 13:235

Choi SS, Kang BY, Chung MJ, Kim SD, Park SH, Kim JS, Kang CY, Ha NJ (2005) Safety assessment of potential lactic acid bacteria Bifidobacterium longum SPM1205 isolated from healthy Koreans. J Microbiol 43:493–498

Cornman RS, Tarpy DR, Chen Y, Jeffreys L, Lopez D, Pettis JS, Evans JD (2012) Pathogen webs in collapsing honey bee colonies. PLoS One 7:e43562

Dong Z-X, Li H-Y, Chen Y-F, Wang F, Deng X-Y, Lin L-B, Zhang Q-L, Li J-L, Guo J (2020) Colonization of the gut microbiota of honey bee (Apis mellifera) workers at different developmental stages. Microbiol Res 231:126370

Duong BTT, Lien NTK, Thu HT, Hoa NT, Lanh PT, Yun B-R, Yoo M-S, Cho YS, Van Quyen D (2020) Investigation of the gut microbiome of Apis cerana honeybees from Vietnam. Biotechnol Lett 42:2309–2317

EFSA (2008) Opinion of the scientific panel on biological hazards on the maintenance of the list of QPS microorganisms intentionally added to food or feed. EFSA J 923:1–48

El Khoury S, Rousseau A, Lecoeur A, Cheaib B, Bouslama S, Mercier P-L, Demey V, Castex M, Giovenazzo P, Derome N (2018) Deleterious interaction between honeybees (Apis mellifera) and its microsporidian intracellular parasite Nosema ceranae was mitigated by administrating either endogenous or allochthonous gut microbiota strains. Front Ecol Evol 6:58

Elzeini HM, Ali A-RA-A, Nasr NF, Elenany YE, Hassan AAM (2020) Isolation and identification of lactic acid bacteria from the intestinal tracts of honey bees, Apis mellifera L., in Egypt. J Apicult Res:1–9

Emery O, Schmidt K, Engel P (2017) Immune system stimulation by the gut symbiont Frischella perrara in the honey bee (Apis mellifera). Mol Ecol 26:2576–2590

Engel P, Moran NA (2013) The gut microbiota of insects–diversity in structure and function. FEMS Microbiol Rev 37:699–735

Engel P, Stepanauskas R, Moran NA (2014) Hidden diversity in honey bee gut symbionts detected by single-cell genomics. PLoS Genet 10:e1004596

Euzéby JP (1997) List of Bacterial Names with Standing in nomenclature: a folder available on the Internet. Int J Syst Evol Microbiol 47:590–592

Foo JL, Ling H, Lee YS, Chang MW (2017) Microbiome engineering: current applications and its future. Biotechnol J 12:1600099

Forsgren E, Olofsson TC, Váasquez A, Fries I (2010) Novel lactic acid bacteria inhibiting Paenibacillus larvae in honey bee larvae. Apidologie 41:99–108

Geldert C, Abdo Z, Stewart JE, HS A (2020) Dietary supplementation with phytochemicals improves diversity and abundance of honey bee gut microbiota. J Appl Microbiol

Genersch E, Von Der Ohe W, Kaatz H, Schroeder A, Otten C, Büchler R, Berg S, Ritter W, Mühlen W, Gisder S (2010) The German bee monitoring project: a long term study to understand periodically high winter losses of honey bee colonies. Apidologie 41:332–352

Horak RD, Leonard SP, Moran NA (2020) Symbionts shape host innate immunity in honeybees. Proc R Soc B 287:20201184

Huang Q, Evans JD (2020) Targeting the honey bee gut parasite Nosema ceranae with siRNA positively affects gut bacteria. BMC Microbiol 20:1–6

Huang SW, Zhang HY, Marshall S, Jackson TA (2010) The scarab gut: a potential bioreactor for bio-fuel production. Insect Sci 17:175–183

Joint F (2002) WHO working group report on drafting guidelines for the evaluation of probiotics in food. London, Ontario, Canada 30

Kaltenpoth M, Engl T (2014) Defensive microbial symbionts in H ymenoptera. Funct Ecol 28:315–327

Khan KA, Ansari MJ, Al-Ghamdi A, Nuru A, Harakeh S, Iqbal J (2017) Investigation of gut microbial communities associated with indigenous honey bee (Apis mellifera jemenitica) from two different eco-regions of Saudi Arabia. Saudi J Biol Sci 24:1061–1068

Kwong WK, Moran NA (2016) Gut microbial communities of social bees. Nat Rev Microbiol 14:374–384

Lee FJ, Miller KI, McKinlay JB, Newton IL (2018) Differential carbohydrate utilization and organic acid production by honey bee symbionts. FEMS Microbiol Ecol 94:fiy113

Leonard SP (2020) Engineering the gut microbiome of honey bees

Li JH, Evans JD, Li WF, Zhao YZ, DeGrandi-Hoffman G, Huang SK, Li ZG, Hamilton M, Chen YP (2017) New evidence showing that the destruction of gut bacteria by antibiotic treatment could increase the honey bee's vulnerability to Nosema infection. PLoS One 12:e0187505

Maddaloni M, Hoffman C, Pascual D (2014) Paratransgenesis feasibility in the honeybee (A pis mellifera) using F ructobacillus fructosus commensal. J Appl Microbiol 117:1572–1584

Manzanares W, Lemieux M, Langlois PL, Wischmeyer PE (2016) Probiotic and synbiotic therapy in critical illness: a systematic review and meta-analysis. Crit Care 20:262
Martinson VG, Danforth BN, Minckley RL, Rueppell O, Tingek S, Moran NA (2011) A simple and distinctive microbiota associated with honey bees and bumble bees. Mol Ecol 20:619–628
Martinson VG, Moy J, Moran NA (2012) Establishment of characteristic gut bacteria during development of the honeybee worker. Appl Environ Microbiol 78:2830–2840
Mathialagan M, Johnson Y, Thangaraj E (2018) Isolation, characterization and identification of probiotic lactic acid bacteria (LAB) from honey bees. Int J Curr Microbiol App Sci 7:849–906
Meixner MD (2010) A historical review of managed honey bee populations in Europe and the United States and the factors that may affect them. J Invertebr Pathol 103:S80–S95
Mohr KI, Tebbe CC (2006) Diversity and phylotype consistency of bacteria in the guts of three bee species (Apoidea) at an oilseed rape field. Environ Microbiol 8:258–272
Moran NA, Hansen AK, Powell JE, Sabree ZL (2012) Distinctive gut microbiota of honey bees assessed using deep sampling from individual worker bees. PLoS One 7:e36393
Motta EV, Raymann K, Moran NA (2018) Glyphosate perturbs the gut microbiota of honey bees. Proc Natl Acad Sci 115:10305–10310
Niode NJ, Salaki CL, Rumokoy LJ, Tallei TE (2020) Lactic acid bacteria from honey bees digestive tract and their potential as probiotics, International Conference and the 10th Congress of the Entomological Society of Indonesia (ICCESI 2019). Atlantis Press, pp. 236–241
Olofsson TC, Vásquez A (2008) Detection and identification of a novel lactic acid bacterial flora within the honey stomach of the honeybee Apis mellifera. Curr Microbiol 57:356–363
Olofsson TC, Alsterfjord M, Nilson B, Butler È, Vásquez A (2014) Lactobacillus apinorum sp. nov., Lactobacillus mellifer sp. nov., Lactobacillus mellis sp. nov., Lactobacillus melliventris sp. nov., Lactobacillus kimbladii sp. nov., Lactobacillus helsingborgensis sp. nov. and Lactobacillus kullabergensis sp. nov., isolated from the honey stomach of the honeybee Apis mellifera. Int J Syst Evol Microbiol 64:3109
Olofsson TC, Butler È, Markowicz P, Lindholm C, Larsson L, Vásquez A (2016) Lactic acid bacterial symbionts in honeybees–an unknown key to honey's antimicrobial and therapeutic activities. Int Wound J 13:668–679
Parichehreh S, Tahmasbi G, Sarafrazi A, Imani S, Tajabadi N (2018) Isolation and identification of Lactobacillus bacteria found in the gastrointestinal tract of the dwarf honey bee, Apis florea Fabricius, 1973 (Hymenoptera: Apidae). Apidologie 49:430–438
Rangberg A, Mathiesen G, Amdam G, Diep D (2015) The paratransgenic potential of Lactobacillus kunkeei in the honey bee Apis mellifera. Benefic Microbes 6:513–523
Raymann K, Shaffer Z, Moran NA (2017) Antibiotic exposure perturbs the gut microbiota and elevates mortality in honeybees. PLoS Biol 15:e2001861
Sabaté DC, Carrillo L, Audisio MC (2009) Inhibition of Paenibacillus larvae and Ascosphaera apis by Bacillus subtilis isolated from honeybee gut and honey samples. Res Microbiol 160:193–199
Saelao P, Borba RS, Ricigliano V, Spivak M, Simone-Finstrom M (2020) Honeybee microbiome is stabilized in the presence of propolis. Biol Lett 16:20200003
Saraiva MA, Zemolin APP, Franco JL, Boldo JT, Stefenon VM, Triplett EW, de Oliveira Camargo FA, Roesch LFW (2015) Relationship between honeybee nutrition and their microbial communities. Antonie Van Leeuwenhoek 107:921–933
Schwarz RS, Moran NA, Evans JD (2016) Early gut colonizers shape parasite susceptibility and microbiota composition in honey bee workers. Proc Natl Acad Sci 113:9345–9350
Shanbhag S, Tripathi S (2009) Epithelial ultrastructure and cellular mechanisms of acid and base transport in the Drosophila midgut. J Exp Biol 212:1731–1744
Sharifpour MF, Mardani K, Ownagh A (2016) Molecular identification and phylogenetic analysis of Lactobacillus and Bifidobacterium spp. isolated from gut of honeybees (Apis mellifera) from West Azerbaijan, Iran, Veterinary Research Forum. Faculty of Veterinary Medicine, Urmia University, Urmia, Iran, p 287
Sharma PV, Thaiss CA (2020) Host-microbiome interactions in the era of single-cell biology. Front Cell Infect Microbiol 10

Shin J, Lee S, Go M-J, Lee SY, Kim SC, Lee C-H, Cho B-K (2016) Analysis of the mouse gut microbiome using full-length 16S rRNA amplicon sequencing. Sci Rep 6:29681
Tihelka E, Cai C, Pisani D, Donoghue PC (2020) Mitochondrial genomes illuminate the evolutionary history of the Western honey bee (Apis mellifera). Sci Rep 10:1–10
Vásquez A, Forsgren E, Fries I, Paxton RJ, Flaberg E, Szekely L, Olofsson TC (2012) Symbionts as major modulators of insect health: lactic acid bacteria and honeybees. PLoS One 7:e33188
Wang H, Liu C, Liu Z, Wang Y, Ma L, Xu B (2020) The different dietary sugars modulate the composition of the gut microbiota in honeybee during overwintering. BMC Microbiol 20:1–14
Wu Y, Zheng Y, Chen Y, Wang S, Chen Y, Hu F, Zheng H (2020) Honey bee (Apis mellifera) gut microbiota promotes host endogenous detoxification capability via regulation of P450 gene expression in the digestive tract. J Microbial Biotechnol 13:1201–1212
Zeinali F, Zarch SMA, Mehrjardi MYV, Kalantar SM, Jahan-Mihan A, Karimi-Nazari E, Fallahzadeh H, Hosseinzadeh-Shamsi-Anar M, Rahmanian M, Fazeli MR (2020) Effects of synbiotic supplementation on gut microbiome, serum level of TNF-α, and expression of microRNA-126 and microRNA-146a in patients with type 2 diabetes mellitus: study protocol for a double-blind controlled randomized clinical trial. Trials 21:1–9
Zheng H, Nishida A, Kwong WK, Koch H, Engel P, Steele MI, Moran NA (2016) Metabolism of toxic sugars by strains of the bee gut symbiont Gilliamella apicola. MBio 7
Zheng H, Powell JE, Steele MI, Dietrich C, Moran NA (2017) Honeybee gut microbiota promotes host weight gain via bacterial metabolism and hormonal signaling. Proc Natl Acad Sci 114:4775–4780

Role of Probiotic Bacteria on Bioavailability of Functional Ingredients Under Fermentation Process

10

Zeinab E. Mousavi and Seyed Mohammad Ali Mousavi

Abstract

Consumer attention to consume healthier foods has been significantly encouraged the food industry to formulate new products within the area of so-called functional foods. Functional foods are defined as whole foods, enriched, enhanced, and fortified foods or dietary compounds that in addition to traditional nutrient contents possess healthy and physiological benefits. Food products containing probiotics compromise the majority of functional food market worldwide. This chapter focuses on the bioactive compounds produced in different probiotic fermented food matrices and investigates how these metabolites and fermentation conditions affect the bioavailability of different compounds in foods.

Keywords

Probiotic · Functional foods · Bioactive compounds · Postbiotics · Dairy products

10.1 Introduction

Probiotics are defined as live microorganisms which when ingested in adequate numbers (at least 10^6_10^7 CFU/ml) impart health benefits to the host and include mainly *Lactobacillus* and *Bifidobacterium* genera but some other bacteria and yeast species are also considered as probiotics (de Melo Pereira et al. 2018; George Kerry et al. 2018; Meira et al. 2015; Morton 2015) (Table 10.1). Probiotics have

Z. E. Mousavi (✉) · S. M. A. Mousavi (✉)
Bio-processing and Bio-detection Lab, Department of Food Science, Technology and Engineering, Faculty of Agricultural Engineering and Technology, University of Tehran, Karaj, Iran
e-mail: zeinab.mosavi@ut.ac.ir; mousavi@ut.ac.ir

N. Mojgani, M. Dadar (eds.), *Probiotic Bacteria and Postbiotic Metabolites: Role in Animal and Human Health*, Microorganisms for Sustainability 2,
https://doi.org/10.1007/978-981-16-0223-8_10

Table 10.1 List of some important probiotic microorganisms (Morton 2015)

Category and genus	Species
Bacteria	
Lactobacillus	*Lb. acidophilus, Lb. amylovorus, Lb. brevis, Lb. casei, Lb. curvatus, Lb. crispatus, Lb. delbrueckii subsp. bulgaricus, Lb. fermentum, Lb. helveticus, Lb. gasseri, Lb. johnsonii, Lb. reuteri, Lb. rhamnosus, Lb. salivarius, Lb. paracasei, Lb. plantarum*
Bifidobacterium	*B. adolescentis, B. animalis, B. bifidum, B. lactis, B. breve, B. infantis, B. longum, B. thermophilum, B. essensis, B. laterosporus*
Streptococcus	*S. cremoris, S. diacetylactis, S. intermedius, S. salivarius*
Propionibacterium	*P. freudenreichii, P. freudenreichii subsp. shermanii, P. jensenii*
Enterococcus	*E. faecalis, E. faecium*
Lactococcus	*L. lactis subsp. cremoris, L. lactis subsp. lactis*
Other bacteria	*Pediococcus acidilactici, Leuconostoc mesenteroides, Bacillus cereus, Clostridium butyricum, Escherichia coli Nissle 1917*
Yeast	*Kluyveromyces lactis, Saccharomyces boulardii, Saccharomyces cerevisiae*

anticarcinogenic antimutagenic activities and are able to suppress cholesterol level and blood pressure. They improve digestive system function, epithelial homeostasis, nutrient uptake, intestinal barrier function, immune modulation, and antagonism action against pathogens (Liptáková et al. 2017; Marhaida et al. 2015).

Traditionally, the effectiveness of probiotics was assumed to be related to cell viability. Apart from probiotic cells, bacterial products may have similar benefits to the host. These products are characterized as postbiotics which have biological activity in the host cell (George Kerry et al. 2018; Wegh et al. 2019). Postbiotics are generally regarded as functional fermentation products and include a wide range of metabolites such as bacteriocins, enzymes, vitamins, amino acids, oligosaccharides, exopolysaccharides, short-chain fatty acids, and immunomodulatory compounds (George Kerry et al. 2018; Zielińska and Kolożyn-Krajewska 2018).

In other words, the functionality of probiotics in fermented foods is accomplished in different ways which eventually affect the nutritional quality of foods which include: 1) increase of nutrient density, mostly due to a decrease of sugar content, 2) hydrolysis of polymers from the raw material and bioactive compounds content, 3) biosynthesis of bioactive molecules, 4) degradation of toxic or anti-nutritional factors, and 5) synthesis of promoters for absorption and uptake (Septembre-Malaterre et al. 2018; Tamang et al. 2016).

10.2 Probiotic Fermentation of Foods

Food fermentation is considered as one of the oldest ways of food processing and preservation. Fermentation results in the enhancement of the flavor and nutritional quality of food and extending its shelf life (Beena Divya et al. 2012). Fermentation is microbe-driven process in which the low value substrates are converted to added-value products (Hussain et al. 2016; Sadh et al. 2018). According to scientific data, both nutritive and non-nutritive components are in fermented foods which could potentially implement specific target functions in the body relevant to well-being and health of the consumers (Tamang et al. 2016). Probiotic bacteria as functional microorganisms, in fermentation process, convert the chemical constituents of raw materials of plant/animal sources leading to the enhancement of the bioavailability of nutrients, enrichment of sensory quality of the food, improvement of food safety, degradation of toxic components and anti-nutritive factors, production of antioxidant and antimicrobial compounds, stimulation of the probiotic functions, and fortification with some health-promoting bioactive compounds (Homayoonfal et al. 2018; Mousavi and Mousavi 2019; Rollán et al. 2019). In fact, the probiotic microorganisms promote beneficial effects in a host which are due to the production of bioactive compounds (Indira et al. 2019).

These bioactive compounds play an important role in bio-preservation of fermented food products including dairy, fish, seaweeds, microalgae, beverages, and fruits and vegetables (Mousavi and Mousavi 2019). Additionally, they show antimicrobial activities against food pathogens such as *Listeria monocytogenes*, *Staphylococcus aureus* and *Enterococcus faecalis*. In addition to their antimicrobial properties, these metabolites can be aromatic which can influence the sensory and organoleptic features of food products . Some peptides with health benefits are also produced as bioactive compound in fermentation of and prevent diseases associated with metabolic syndrome (Indira et al. 2019; Ojha and Tiwari 2016) (Table 10.1).

10.3 Production and Modification of Bioactive Compounds Over Probiotic Fermentation

Bioactive compounds as result of probiotic fermentation have two major sources. The first source is direct synthesis of the compound by the probiotic such as bacteriocins, exopolysaccharides (EPS), or enzymes and they can be found in either supplements or foods. The second source of bioactive is a compound that only appears as a result of the modification of the food matrix itself by the probiotic culture fermentation (Champagne et al. 2018). The following section will discuss the bioactive compounds produced during probiotic fermentation and their effect on food bioavailability.

10.3.1 Bioactive Peptides

Bioactive peptides are short sequences of amino acids generally consisting from 2 and 20 amino acids. Such sequences stay intact and inactive when present in the parental protein, but can be released after protein hydrolysis during gastrointestinal digestion (GID), *in vitro* enzymatic hydrolysis, or microbial fermentation. These peptides have biological activities that may influence human health in addition to basic human nutrition (Erdmann et al. 2008). Cardioprotective functions, modulation of immune system, anti-atherosclerosis, antioxidant, mental health, and general well-being functions are associated with bioactive peptides (Ojha and Tiwari 2016; Septembre-Malaterre et al. 2018). According to various researches, it is concluded that microbial fermentation could be regarded as an appropriate approach improving protein bioavailability and digestibility in different food products (Chi and Cho 2016; Hur et al. 2014; Limon et al. 2015; Wu et al. 2015).

10.3.1.1 Dairy Products

Milk-proteins and associated bioactive peptides released during microbial or enzymatic fermentation of milk offer a broad spectrum of new functional properties, for instance antihypertensive, antimicrobial, antioxidative, immunomodulatory, opioid, and mineral-binding properties (Beermann and Hartung 2013).

Calcium casein phosphopeptides (CCP) are phosphorylated bioactive peptides derived from calcium-sensitive caseins (αs1, αs2, and β caseins). These peptides are inactive fragments entrapped in the sequence of precursor protein, and exhibit biological action after its release during the passage through the gastrointestinal tract. In addition, they are also produced in vitro by the action of specific enzymes during fermentation of a number of dairy-based products such as cheese, yogurts, and fermented milks (Ledesma-Martínez et al. 2019; Mohanty et al. 2016). The main activities of CCP include anticancer, body fat reduction, prevention of cardiovascular diseases through the reduction of atherosclerosis lesions and levels of cholesterol and triacylglycerides, anti-inflammatory, and antioxidant. A great number of studies approved the role of CCP on calcium, iron, and zinc (Ledesma-Martínez et al. 2019).

The effect of peptidases activity of *Lactobacillus delbrueckii* ssp. *bulgaricus* and *Streptococcus thermophiles* on milk proteins resulted in the production of antimicrobial and hypotensive peptides. These small biological peptides can be used as food supplements to improve the health-promoting qualities of liquid and semisolid dairy foods prepared by the yogurt fermentation process (Paul and Somkuti 2009).

Investigations revealed that probiotic LAB such as *Lactobacillus helveticus* produces bioactive peptide like, proline-containing peptides isoleucyl-prolyl-proline (IPP) and valyl-prolyl-proline (VPP) which may induce greater availability of calcium (Dubey and Patel 2018). The study on the level of level of calcium, magnesium, phosphorus, and zinc absorption in a series of fermented goat and cow milk showed that the bioavailability of minerals was significantly higher compared with non-fermented milks (Bergillos-Meca et al. 2013).

Oxidative damage caused by various free radicals which are by-products of physiological reactions within human body can be protected by antioxidants. It has been found that yogurt and fermented milks have a higher antioxidant activity than milk. In fermented milks, bioactive peptides are released following the proteolysis of milk proteins, especially lactalbumin, lactoglobulin, and casein (Melini et al. 2019). Yogurt produced with camel milk by fermentation with *Lactobacillus rhamnosus* strain PTCC 1637 has a higher antioxidant activity than cow milk, because of the higher proline content in camel milk caseins. The presence and position of the amino acids tryptophan, tyrosine, and methionine in the peptides are claimed responsible for the antioxidant activity of fermented milks as well.

10.3.1.2 Fruits, Vegetables, Legumes, and Grains

Various studies showed that probiotic fermentation of nondairy foods including vegetables, fruits, legumes, and grains could enhance the level of protein, peptides, and amino acid in these products (Septembre-Malaterre et al. 2018). Bioactive peptides have been mainly studied from milk or whey hydrolysis during lactic fermentation. However, different studies on fermented soybeans, grapes, and cereal flours also showed a significant increase in their bioactive contents (Septembre-Malaterre et al. 2018). Probiotic LAB are naturally present in legume grains; they have also been traditionally used for legume fermentation. Evidences showed that fermentation of legumes with *Lactobacillus* genera can encourage the production of bioactive compounds, improving health benefits beyond basic nutrition. Fermentation of cowpeas with *Lactobacillus plantarum* resulted in the modification of phenolic compounds and improvement of antioxidant activity (Dueñas et al. 2005; Limon et al. 2015).

L. plantarum B1–6 has been studied for its potential proteolysis effect on mung bean protein during fermentation. Electrophoresis profiles revealed that *L. plantarum B1–6* degraded Mung bean proteins with the hydrolysis percentages between 49 and 64%. In addition, reverse phase high-performance liquid chromatography (RP-HPLC) analysis showed that larger/more hydrophobic peptide contents decrease the amount of smaller/more hydrophilic peptides has substantially augmented after fermentation (Wu et al. 2015). In addition, the degradation of gluten could render the final product to be suitable for celiac consumers (Heredia-Sandoval et al. 2016; Houben et al. 2012; Poutanen et al. 2009; Verni et al. 2019). Cereals are in general good sources of proteins. The proportions of essential amino acids and their digestibility mainly determine protein nutritional quality. Peptidase enzymes produced by LAB convert peptides to amino acids. Specific products of these enzymes are responsible for the aroma and taste of final products (Pessione and Cirrincione 2016; Verni et al. 2019).

Solid-state fermentation of whole soybeans by *Lactobacillus plantarum P-8* mixed with *B. subtilis natto* also resulted in an intensive protein degradation and generation of hydrophilic peptides during fermentation (Pessione and Cirrincione 2016; Zhang et al. 2014).

Various scientific reports stated that hydrolyzed peptides produced by probiotics during fermentation can act as antioxidants (Coda et al. 2012; Raveschot et al. 2018; Taha et al. 2017). Rapeseed proteins are hydrolyzed to amino acids and peptides by proteases produced by probiotic *Bacillus subtilis* during fermentation (Rong et al.

2012). Hydrolysis of peptide bonds enhances the levels of free amino and carboxyl groups, resulting in increased solubility. This enhanced solubility may improve the antioxidant activity of the peptide (Karami and Akbari-Adergani 2019; Sohaib et al. 2017). Low-molecular-weight peptides have been reported to exhibit better radical-scavenging activities than their high-molecular-weight counterparts (Xie et al. 2008). Thus, increasing the low-molecular-weight peptides by enzymatic hydrolysis may influence the antioxidative activity during fermentation. Metal-chelating amino acid residues, such as methionine, glutamic acid, glutamine, lysine or arginine, within the sequences of these peptides contributed to the superior Fe^{2+}-chelating ability of the antioxidant peptides (Hur et al. 2014).

According to different studies, proteolytic activity of probiotic LAB could enhance the level of bioactive peptides in fermented cereals. Antihypertensive properties are attributed to these bioactive peptides. In addition, thanks to the production of flavoring free amino acids and other amino acid derivatives during fermentation which convey tastiness to fermented cereals such as bread, it is possible to decrease salt content in the final product (Melini et al. 2019).

10.3.1.3 Fish

Large quantity of liquid and solid waste generated by fish industry can be regarded as a potential resource for valuable products. Due to their high protein contents, fish waste could be used as a suitable medium for culturing probiotic bacteria. Fermentation of fish waste can partially degrade the protein contents, which could help the absorption from the gut and influence its bioactive properties through the production of bioactive peptides (Venegas-Ortega et al. 2019).

10.4 Digestible Saccharides

10.4.1 Fruits and Vegetables

Fruits and vegetables are a rich source of sugars. During fermentation of fruits and vegetables, monosaccharide are significantly consumed by probiotic bacteria specially LAB species. However, with the help of glucosidases and glycosyl hydrolases produced from the cells, hydrolysis of polysaccharides occurs, which release monomers of sugars. Release of monomers contributes to the increase in nutrient density of the fermented products (Ojha and Tiwari 2016).

10.4.2 Cereals

Lactobacillus species are the predominant organisms involved in the fermentation of cereal-based foods and beverages in African countries (Richard and Jooste 2012). A multiple of researches showed that cereal fermentation is considered as a significant potential in improvement and design of the nutritional quality and health effects of foods and ingredients (Rollán et al. 2019). Cereal grains are primarily a source of carbohydrates, and thus a good source of energy. However, a high proportion of starch in cereals is in the form of amylopectin, which is not completed digested and absorbed in the small intestine. Digestible polysaccharides are produced as a result of probiotic lactic acid fermentation of cereal, which are more accessible to gut microbiota. (Liptáková et al. 2017).

10.5 Exopolysaccharides (EPS)

EPS are secondary metabolites with long chain of homo or hetreo-polysaccharides containing repeated units of sugars or sugar derivatives. These polysaccharides are produced outside of the cell. Depending on the carbon source, LAB belonging to the genera *Lactobacillus, Lactococcus, Leuconostoc, Pediococcus*, and *Weissella* are able of producing a variety of EPS (Zeidan et al. 2017).

10.5.1 Dairy

In dairy-based foods, extracellular polysaccharides (EPS) are produced from ropy probiotic cultures (*Streptococci, lactobacilli, and lactococci* strains) (Prasanna et al. 2012). These EPS could improve physicochemical and rheological properties of foods. In addition, they may also protect cells to against phage attack, desiccation, and osmotic stress, thus behaving as prebiotics and improve immunity to fight against pathogenic organisms (Ruas-Madiedo et al. 2002). In addition, blood cholesterol-lowering, immunostimulatory, antitumoral, and antiulcer activity have been also attributed to EPS produced in fermented probiotic dairy products (Madhuri and Prabhakar 2014; Shao et al. 2014).

10.6 Galacto-Olygosaccharides (GOS)

GOS are non-digestible carbohydrates and comprise a chain of galactose units usually with a terminal glucose unit. They are derived from lactose by the action of β-galactosidase enzyme in a trans-galactosylation reaction that occurs simultaneously with the hydrolysis. These bioactive compounds can be synthesized by probiotic microorganisms in fermented products during processing (Otieno 2010). In the case of use of probiotics as enzyme sources for GOS synthesis, they could provide the double advantage as probiotics as well as in prebiotic. GOS are fermented by the beneficial gut microflora of the large intestine resulting in the inhibition of pathogenic and putrefactive bacteria growth. Therefore, the level of toxic metabolites is significantly decreased which could prevent diarrhea, constipation relief, and lactose tolerance. Also, metabolism of GOS results in the production of short-chain fatty acids which could assist in increased calcium and magnesium absorption, control of serum lipid and cholesterol level, and reduction of cancer risk (Davani-Davari et al. 2019).

10.6.1 Dairy

Milk sugar (lactose) is a component of dairy by-products especially from whey which is half consumed by human and animals and the remaining is generally discarded. Various reports used different probiotic strains specially *Lactobacillus* and *Bifidobacterium* species to produce GOS from milk, cheese, whey, and yogurt which can be used as a suitable substrate for GOS synthesis. Therefore, fermented dairy-based foods could be considered as the main carrier of GOS (Lappa et al. 2019; Sabater et al. 2018; Song et al. 2013). There are many parameters affecting the synthesis of these compounds such as β-galactosidase enzyme source and concentration, type and counts of microorganisms, concentration of substrate (lactose), composition of food matrix, conditions of fermentation and storage, and time/temperature of hydrolysis/transgalactosylation (Morton 2015).

10.7 Conjugated Linoleic Acid (CLA)

CLA is a collective term used to describe a heterogeneous mixture of positional and geometric isomers of octadecadienoic acid or linoleic acid (c9,c12-C18:2) in which double bonds are conjugated (cis-, trans-, or mixed configurations). Biological and biochemical roles attributed to CLA include anticancer, body fat reduction, prevention of cardiovascular diseases through the reduction of atherosclerosis lesions and levels of cholesterol and triacylglycerides, anti-inflammatory and antioxidant. Linoleate isomerase (LAI) enzyme is responsible of CLA synthesis, which is bond to the cell membrane of microorganisms. CLAs exert various health benefits and their effectiveness depends on CLA isomer form. Studies demonstrated that trans-9, trans-11 C18:2 has a much higher inhibitory and antiproliferative effect on the growth of the human colon and breast cancer cells, than cis9, trans-11 CLA isomer (Beppu et al. 2007; El Roz et al. 2013; Park 2009).

In contrast, the results of other studies showed that cis-9, trans-11 CLA has extra beneficial effects, such as anti-inflammatory and antiatherogenic effects (Tricon et al. 2006). However, the mixture of the two CLA isomers (cis-9, trans-11 and trans-9, trans-11 CLA) had a synergistic anti-proliferation effect on a human colorectal carcinoma cell line (Zhong et al. 2012).

10.7.1 Meat Products

CLA is a compound found mainly in the meat of ruminants that is recently the subject of many researches due its health-promoting properties, i.e., antiatherogenic, cancer inhibition, anti-diabetic, obesity lowering, and improved immunity (Mulvihill 2002). In a detoxification mechanism, some probiotic bacteria of *Lactobacillus* and *Bifidobacterium* types are able to change fatty acid profile in meat sausages by converting polyunsaturated fatty acids into CLA through isomerization, hydrogenation, and dehydration (Galgano et al. 2015).

10.7.2 Dairy

In some countries, liquid milk, powdered milk, fermented milk, yogurt, and cheese enriched in CLA are marketed. On the other hand, the known fact that several strains of bacteria possess the ability to synthesize CLA in vitro in the presence of precursor substrate raised the possibility for increasing the production of CLA *in situ* during manufacture of fermented dairy foods. The co-culture of *L. rhamnosus* and yogurt starter in the presence of hydrolyzed soy oil as the lipid source showed that CLA contents significantly increased in the final fermented (Xu et al. 2005). A study performed by Ribeiro et al. (2017) showed that *Lactobacillus plantarum* isolated from Pico cheese exhibited probiotic properties and presented the highest production of both cis-9, trans-11 and trans-9, trans-11 CLA isomers, exhibiting a great potential for application in health-promoting food product.

10.8 Short-Chain Fatty Acids (SCFA)

SCFA such as such as butyrate, acetate, propionate, and lactate are secondary metabolites released from the hydrolysis of food fiber and non-digestible carbohydrates in gut by probiotic bacteria and are used as a source of energy for colon cells. In humans, 10% of the daily caloric requirement is from short-chain fatty acids produced in large intestine. Among all short-chain fatty acids, 60–70% of the energy is from butyrate produced in colonocytes. SCFAs, particularly butyrate, have a therapeutic effect in various diseases such as inflammatory bowel disease, antibiotic-associated diarrhea, colon cancer, and heart diseases (Indira et al. 2019; Septembre-Malaterre et al. 2018).

According to different researches, the increase of Ca bioavailability by probiotics would definitely satisfy the bone health. The mechanism behind the increase in Ca bioavailability and ensure the bone health is that the probiotics produce short-chain fatty acids, which increase the solubility of available calcium. Simultaneously, the level of the para-thyroid hormone level (increased PTH level causes the Bone resorption by stimulating the osteoclasts) decreases and minimizes the bone loss (Dubey and Patel 2018).

10.9 Vitamins

Vitamins play an important role in regulating the intestinal metabolism and absorption of minerals. Calcium absorption is enhanced in the presence of Folate and vitamin C, D, and K (Kiela and Ghishan 2016). Probiotics are associated with the synthesis of vitamins and increase the metabolism and absorption of available calcium (Parvaneh et al. 2014; Whisner and Castillo 2018). Therefore, food fermentation with probiotic bacteria could result in an increased vitamin content of the final product (Richard and Jooste 2012). Probiotic LAB are able of producing B vitamins including niacin (B3), panthothenic acid (B5), folic acid (B9), and also vitamins B1, B2, B6, and B12 (Capozzi et al. 2012; Septembre-Malaterre et al. 2018).

10.9.1 Fruits and Vegetables

Vitamin B12 deficiencies in plant based-diet forced researchers to investigate potential ways to fortify plant-based foods with vitamin B12 (Chamlagain 2016; Melini et al. 2019) Cereal-based products such as Ogi, Mageu, and Kenkey, which are considered as traditional fermented products in Africa, have been reported to have an improved B-vitamin content. Beside probiotic LAB benefits in the enrichment of foods with vitamins, they may lower production costs by eliminating the need to add synthetic vitamins (Rollán et al. 2019). A study performed by Varmanen et al. (2016) showed that *L. reuteri* can be used for vitamin B12 fortification in soy-yogurt.

10.9.2 Dairy

Folate, as an essential vitamin, plays an important role in human life for the synthesis of nucleotides, vitamins, and some amino acids. However, this vitamin could not be synthesized by human and have to be taken by daily diet. Dairy products, especially yogurt, are an appropriate choice for bio-fortification of folate as they contain folate-binding protein which improves folate bioavailability. It is reported that the use of folate-producing probiotic bacteria in combination with *S. thermophilus* and/or *L. bulgaricus* provides the largest increase in folate during the fermentation process of probiotic yogurt compared to original milk and conventional fermented milk (Rad et al. 2016). The level of vitamin B12 is significant in dairy products. This vitamin is necessary for the maintenance of the nervous system and the formation of blood cells. Fermentation by probiotic bacteria could increase its content up to 10-folds (Melini et al. 2019).

10.10 Enzymes: Anti-Nutrient Degradation

Food fermentation is considered as an important part in food detoxification. Probiotics LAB are able to metabolize anti-nutrient compounds including phytates, trypsin inhibitors, saponins, tannins, cyanogens, or phenolic compounds in foods. This effect can be associated with modification of minerals bioavailability (Septembre-Malaterre et al. 2018).

10.10.1 Phytates

According to clinical investigations, it has been found that vegetarians may suffer from nutritional deficiencies and, specially, they have an impaired absorption of

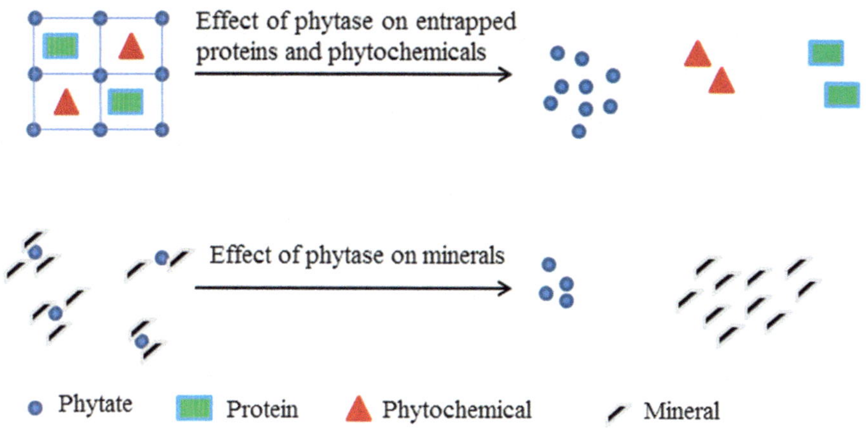

Fig. 10.1 Effect of fermentation on minerals, phytochemicals, and proteins bioavailability of foods

trace minerals, such as zinc, iron, and calcium, proteins, vitamin B_{12}, and folate (Bergillos-Meca et al. 2013; Masum Akond et al. 2011; Popova and Mihaylova 2019; Rekha and Vijayalakshmi 2010). This malabsorption syndrome may cause severe health-threatening diseases ranging from anemia to neurological disorders and immune deficiency (Hunt 2003). It is postulated that this intestinal malabsorption of minerals is due to the high content of phytate in cereals, nuts, legumes, and oilseeds. Furthermore, it accounts from 60% to 90% of total phosphorus content in cereals and is, therefore, the major storage compound for phosphorus (Gupta et al. 2015). Phytate is able of chelating nutritionally important cations such as Ca^{2+}, Mg^{2+}, Fe^{2+}, and Zn^{2+}, thus decreasing the dietary bioavailability of these nutrients.

Intestinal microfloras, especially LAB, are an important source of phytase with high activity. The consequence of phytate hydrolysis by LAB in gut is the release of phosphate, other metal ions and proteins through the degradation of complexes formed by phytate (Dubey and Patel 2018; Famularo et al. 2005) (Fig. 10.1). Various studies have approved an improvement in mineral bioavailability by different probiotic microorganisms used in the fermentation process (Bergillos-Meca et al. 2013). Daily diet enriched with probiotic lactic acid bacteria could minimize phytate or phytic acid in plants. The fermentation of bran with probiotic LAB could provide optimal pH conditions for enzymatic degradation of anti-nutritional factors induced by the degradation of phytate (up to 90%). This results in better bioavailability of minerals (Lopez et al. 2001; Rollán et al. 2019). According to researches, Ca absorption is related to pH in the colon (Diaz de Barboza et al. 2015; Rekha and Vijayalakshmi 2010). Calcium is a divalent cation which salt form is available in food. The soluble and ionized form of Ca is absorbed. Phytate and oxalate in a diet form insoluble salts with calcium and inhibit the calcium absorption (Dubey and Patel 2018). Fermentation of soymilk with five strains of probiotic lactic-acid bacteria (*L. acidophilus* B4496, *L. bulgaricus* CFR 2028, *L. casei* B1922,

L. plantarum B4495, and *L. fermentum* B4655) with the yeast *Saccharomyces boulardii* made Ca more soluble (Parvaneh et al. 2014; Ramsubeik et al. 2014; Rekha and Vijayalakshmi 2010). In a study performed by Lorusso et al. (2017), evaluations showed that the minerals bioavailability in quinoa-based pasta flour fermented by selected LAB with phytase activity substantially augmented. A similar study (Rizzello et al. 2016) reported that phytase activity of quinoa sourdough has increased 2.75 times after fermentation with autochthonous LAB *(L. plantarum T6B10* and *L. rossiae T0A16*).

10.10.2 Phenolic Compounds

Phenolic compounds as secondary metabolites produced by plants are widely used as dietary supplements and have numerous biological and pharmacological effects such as anticancer, antioxidative, antiviral, anti-inflammatory, and antiatherogenic activities (de Souza et al. 2019; Hur et al. 2014; Rollán et al. 2019). Many phenolic compounds occur in food as esters, glycoconjugates, or polymers, which are not directly bioavailable (Rossi et al. 2013). According to estimations,, as little as 5–10% of total ingested phenolic compounds can be absorbed in the small intestine, whereas 90–95% reach the colon because of insufficient gastric residence time, low permeability or solubility in the intestine (de Souza et al. 2019). The evidences showed that the gut microbiota are major responsible of polyphenols biotransformation into more biologically active components (de Souza et al. 2019; Pereira-Caro et al. 2018). Enzymatic activity of intestinal bacteria able to catabolize phenolics could results in the production of various compounds with different bioavailability and biological functions to their parent compounds (Dudonné et al. 2015) As oligo- and polysaccharides bounded to phenolic compounds are the major carbon sources for saccharolytic fermentative bacteria, in the first step of phenolic degradation, aglycones are released from glycol-conjugated forms of polyphenols by microbial enzymes including glycosidases, glucuronidases, and sulfatases (Rossi et al. 2013). These aglycones are further degraded through several functional groups cleavages reactions (dehydroxylation, demethylation, and decarboxylation) and ring-fission. Therefore, the produced microbial metabolites are absorbed from the colon and are also subjected to liver metabolism, resulting in their conjugated derivatives. This intensive microbial metabolism ultimately reduces the structural diversity of native phenolic compounds to a limited number of smaller phenolic acids and derivatives of phenylpropionic and phenyl acetic acids metabolites. Biological activities of phenolic compounds have mostly been attributed to their microbial metabolites, present in higher quantities in circulation than the native compounds (Marín et al. 2015).

Modulating the activity of gut microbiota by the incorporation of appropriate probiotics into daily diet can enhance bioavailability and/or biological activity of these phenolic compounds. In a study performed by Rekha and Vijayalakshmi (2010), investigations showed that soymilk fermentation with LAB in combination with probiotic yeast *Saccharomyces boulardii* could increase the bioactive aglycones form of soy isoflavone (Rekha and Vijayalakshmi 2010). Investigations showed that glucoside conjugates of isoflavones exist principally in soya foods which is poorly absorbed in the body and their biological effect are mainly attributed to their glycosides form (Rekha and Vijayalakshmi 2010; Zubik and Meydani 2003).

Various studies revealed an increase in total phenols after fermentation of different foods, and observed that the increase in antioxidative activity may be due to the increase in the total phenolic compounds (Călinoiu et al. 2019; Hur et al. 2014; Zou et al. 2017). Probiotic LAB are naturally present in legume grains; they have also been traditionally used for legume fermentation. Evidences showed that fermentation of legumes with *Lactobacillus* genera can encourage the production of bioactive compounds, improving health benefits beyond basic nutrition. Fermentation of cowpeas with *Lactobacillus plantarum* resulted in the modification of phenolic compounds and improvement of antioxidant activity (Dueñas et al. 2005; Limon et al. 2015).

A research showed that complex polyphenols were hydrolyzed to simpler and more biologically active compounds during fermentation of cowpea flour, and the concentration of phenolic compounds in fermented has significantly increased (Dueñas et al. 2005). In humans, isoflavones bioavailability depends on the relative ability of gut microflora to degrade these compounds. Variation in the intestinal bacterial community as a result of illnesses, diet, or age could significantly influence isoflavones bioavailability (Rekha and Vijayalakshmi 2010; van der Velpen et al. 2014). A research carried out by Dudonné et al. (2015) consumption of showed thatcranberry extract co-supplemented with probiotic *Bacillus subtilis CU1* resulted in the significant change in the composition of gut microbial communities of high-fat fed diet mice through the inhibition of pathogenic bacteria and stimulation of beneficial bacteria (de Souza et al. 2019). According to a study performed by Parkar et al. (2014), anthocyanin-rich blackcurrant juice stimulated the in vitro growth and adhesion properties of *L. rhamnosus 299*. In contrast, it suppressed the growth and adhesion properties of *Salmonella Typhimurium 450*.

It has been reported that fermentation can significantly improve total phenolic content and antioxidant activity of cereals and pseudocereals, which is highly dependent on the species of microorganism, on the grains types,, fermentation conditions, particularly time, temperature, and pH values (Hur et al. 2014; Rollán et al. 2019). The enzymes involved in the phenolic metabolism by LAB are mainly decarboxylases (PAD), reductases (PAR), esterases, and/or glycosidases (Rollán et al. 2019). Fermentation of cowpeas with *Lactobacillus plantarum* resulted in the modification of phenolic compounds and improvement of antioxidant activity (Dueñas et al. 2005; Limon et al. 2015) (Fig. 10.1).

Catabolic products of orange juice flavanones identified by HPLC–HR–MS showed that probiotication of orange juice by *Bifidobacterium longum R0175* could significantly enhance the aglycone form of flavonones in orange juice which could eventually augment the bioavailability of orange juice flavanones, and, therefore, their potential beneficial effects on health. A study on the effect of probiotic fermentation of pomegranate juice revealed that fermentation of the juice using *L. plantarum* and *L. acidophilus* as probiotic starter organisms increased the antioxidant activity significantly (Mousavi et al. 2013). In a similar study, investigations showed that fermentation of liquorice root extract could effectively improve the antioxidant activity of the extract from 53% to a maximum level of 73% (Mousavi and Mousavi 2019).

10.10.3 Allergens

Hydrolysis of proteins into smaller peptide fragments during lactic acid fermentation by probiotics could also suppress the potential allergenicity of parent proteins in different foods (Verhoeckx et al. 2015; Xiang et al. 2019). For instance, despite the high protein content, balanced amino acid composition, and high level of lysine in comparison with other vegetable protein sources, soybean meals contain anti-nutritional factors (ANFs) and allergens, which cause decrease in protein digestibility and absorption in animals (Gu et al. 2010). The soybean is one of the "Big 8" food allergens. The allergen proteins account for 65–80% of total protein content in the soybean and approximately 30% in soybean. The major allergen proteins are beta conglycinin, the 30-kDa allergen (GlymBd 30), and glycinin. In human subjects, these allergens can induce symptoms ranging from skin, gastrointestinal, or respiratory reactions to anaphylaxis. They also cause hypersensitivity in weaned piglets, with the primary adverse effect being diarrhea (Adachi et al. 2009).

Lactobacillus kefiranofaciens M1 isolated from Kefir grains has an anti-allergic effect. Digestion of caseins during maturation of fermented milk products has shown to facilitate loss of allergenic reactivity (Chen et al. 2012).

Fermentation of soybean meal enhanced the bioavailability of nutritious components and decreased the incidence of diarrhea in weaned pigs due to the degradation of allergens into peptides (Chi and Cho 2016). The absorption of peptides was significantly improved by the animal. In addition, soybean protein hydrolysate also exhibited antioxidative, metal-chelating activity and lipid peroxidation inhibitory activity attributed mainly to the low-molecular-weight (3 kDa) peptide (Chi and Cho 2016).

The probiotic *B. coagulans GBI-30, 6086* has the capacity to produce enzymes degrading proteins and a wide of carbohydrates. These enzymes can increase the amount of digested milk protein available for absorption. *B. coagulans GBI-30*, 6086 could be exploited to improve protein quality in plant protein sources with lower essential amino acid such as Leucine (Jager et al. 2018).

10.10.4 Cyanogenic Glucosides

Galactosidase is recognized to metabolize cyanogenic glucosides present in some vegetal matrixes such as cassava roots, bitter almonds, or whole sorghum. Cyanogenic glycoside linamarin and lotaustralin in cassava tubers can be detoxified by species of *Leuconostoc*, *Lactobacillus*, and *Streptococcus* during traditional method to Gari and Fufu productions to yield hydrocyanic acid (HCN). This compound is volatile and can escape from the dewatered pulp during toasting rendering the product safe for human consumption (Tamang et al. 2016).

10.10.5 Tannins

Tannins are polyphenols widely available in cereals and legumes. They can easily bind to proteins making indigestible complexes with reduced bioaccessibility of nutrients. Various probiotic *Lactobacillus* species such as *L. plantarum*, *L. paraplantarum*, and *L. pentosus* have been confirmed to have tannase activity (Osawa et al. 2000). Therefore, the exploitation of these bacteria in the fermentation of plant-based foods rich in tannins can cleave the protein-tannin complexes rendering protein more available to the cells (Nkhata et al. 2018).

10.11 Conclusion

During food fermentation with probiotic bacteria, a number of chemical changes occur in the structure of components of the raw matrix, which thus results in the improvement of the functional properties of foods. This improvement is resulted from several mechanisms such as the elimination of anti-nutritional factors, production of metabolites with a positive effect (bioactive peptides, exopolysaccharides), improvement of the bioavailability through biopolymers hydrolysis (esters of phenolic compounds), and increased vitamin, mineral, and phenolic compounds, leading to an increase in the antioxidant capacity of the final product.

A higher bioactive molecule content and an improved antioxidant activity were found in fermented milks, cereals, fruit and vegetables, meat and fish. Antihypertensive peptides were detected in fermented milk and cereals. Changes in vitamin content were mainly observed in fermented milk and fruits. The imparted health benefits of probiotic fermentation to consumers make this category of foods worthy of recommending for regular dietary guidelines. However, it seems that molecular mechanisms behind the bioavailability and the potential health effects of the newly formed compounds by probiotic fermentation are not deeply investigated yet. Therefore, development of molecular tool analysis such as metabolomics, proteomics, and transcriptomics would considerably help in that respect. Analysis of food composition and enzyme activity evaluation in the gastrointestinal tract would be helpful to evaluate the extent of molecular changes at each stage. Eventually, clinical trials would be useful to measure the health effect of probiotic-fermented foods on different groups of the population.

References

Adachi A, Horikawa T, Shimizu H, Sarayama Y, Ogawa T, Sjolander S, Tanaka A, Moriyama T (2009) Soybean beta-conglycinin as the main allergen in a patient with food-dependent exercise-induced anaphylaxis by tofu: food processing alters pepsin resistance. Clin Exp Allergy 39:167–173

Beena Divya J, Kulangara Varsha K, Madhavan Nampoothiri K, Ismail B, Pandey A (2012) Probiotic fermented foods for health benefits. Eng Life Sci 12:377–390

Beermann C, Hartung J (2013) Physiological properties of milk ingredients released by fermentation. Food Funct 4:185–199

Beppu F, Hosokawa M, Tanaka L, Kohno H, Tanaka T, Miyashita K (2007) Potent inhibitory effect of trans9, trans11 isomer of conjugated linoleic acid on the growth of human colon cancer cells. J Nutr Biochem 17:830–836

Bergillos-Meca T, Navarro-Alarcon M, Cabrera-Vique C, Artacho R, Olalla M, Gimenez R, Moreno-Montoro M, Ruiz-Bravo A, Lasserrot A, Ruiz-Lopez MD (2013) The probiotic bacterial strain Lactobacillus fermentum D3 increases in vitro the bioavailability of Ca, P, and Zn in fermented goat milk. Biol Trace Elem Res 151:307–314

Călinoiu LF, Cătoi A-F, Vodnar DC (2019) Solid-state yeast fermented wheat and oat bran as a route for delivery of antioxidants. Antioxidants (Basel) 8:372

Capozzi V, Russo P, Dueñas M, López P, Spano G (2012) Lactic acid bacteria producing B-group vitamins: A great potential for functional cereals products. Appl Microbiol Biotechnol 96:1383–1394

Chamlagain B (2016) Fermentation fortification of active vitamin B12 in food matrices using propionibacterium freudenreichii: analysis, production and stability. Dissertation

Champagne CP, da Cruz AG, Daga M (2018) Strategies to improve the functionality of probiotics in supplements and foods. Curr Opin Food Sci 22:160–166

Chen YP, Hsiao PJ, Hong WS, Dai TY, Chen MJ (2012) Lactobacillus kefiranofaciens M1 isolated from milk kefir grains ameliorates experimental colitis in vitro and in vivo. J Dairy Sci 95:63–74

Chi C-H, Cho S-J (2016) Improvement of bioactivity of soybean meal by solid-state fermentation with Bacillus amyloliquefaciens versus Lactobacillus spp. and Saccharomyces cerevisiae. LWT Food Sci Technol 68:619–625

Coda R, Rizzello CG, Pinto D, Gobbetti M (2012) Selected lactic acid bacteria synthesize antioxidant peptides during sourdough fermentation of cereal flours. Appl Environ Microbiol 78:1087–1096

Davani-Davari D, Negahdaripour M, Karimzadeh I, Seifan M, Mohkam M, Masoumi SJ, Berenjian A, Ghasemi Y (2019) Prebiotics: definition, types, sources, mechanisms, and clinical applications. Foods (Basel, Switzerland) 8:92

de Melo Pereira GV, de Oliveira Coelho B, Magalhaes Junior AI, Thomaz-Soccol V, Soccol CR (2018) How to select a probiotic? A review and update of methods and criteria. Biotechnol Adv 36:2060–2076

de Souza EL, de Albuquerque TMR, Dos Santos AS, Massa NML, de Brito Alves JL (2019) Potential interactions among phenolic compounds and probiotics for mutual boosting of their health-promoting properties and food functionalities: a review. Crit Rev Food Sci Nutr 59:1645–1659

Diaz de Barboza G, Guizzardi S, Tolosa de Talamoni N (2015) Molecular aspects of intestinal calcium absorption. World J Gastroenterol 21:7142–7154

Dubey MR, Patel VP (2018) Probiotics: a promising tool for calcium absorption. Open Nutri J 12

Dudonné S, Varin TV, Forato Anhê F, Dubé P, Roy D, Pilon G, Marette A, Levy É, Jacquot C, Urdaci M, Desjardins Y (2015) Modulatory effects of a cranberry extract co-supplementation with Bacillus subtilis CU1 probiotic on phenolic compounds bioavailability and gut microbiota composition in high-fat diet-fed mice. Pharma Nutr 3:89–100

Dueñas M, Fernández D, Hernández T, Estrella I, Muñoz R (2005) Bioactive phenolic compounds of cowpeas (Vigna sinensis L). Modifications by fermentation with natural microflora and with Lactobacillus plantarum ATCC 14917. J Sci Food Agric 85:297–304

El Roz A, Bard J-M, Huvelin JM, Nazih H (2013) The anti-proliferative and pro-apoptotic effects of the trans9,trans11 conjugated linoleic acid isomer on MCF-7 breast cancer cells are associated with LXR activation. Prostaglandins Leukot Essent Fatty Acids 88:265–272

Erdmann K, Cheung BW, Schröder H (2008) The possible roles of food-derived bioactive peptides in reducing the risk of cardiovascular disease. J Nutr Biochem 19:643–654
Famularo G, De Simone C, Pandey V, Sahu AR, Minisola G (2005) Probiotic Lactobacilli: an innovative tool to correct the malabsorption syndrome of vegetarians? Med Hypotheses 65:1132–1135
Galgano F, Condelli N, Caruso M, Colangelo M, Favati F (2015) Beneficial microbes in fermented and functional foods. CRC Press, Boca Raton, pp 189–206
George Kerry R, Patra JK, Gouda S, Park Y, Shin H-S, Das G (2018) Benefaction of probiotics for human health: a review. J Food Drug Anal 26:927–939
Gu C, Pan H, Sun Z, Qin G (2010) Effect of soybean variety on anti-nutritional factors content, and growth performance and nutrients metabolism in rat. Int J Mol Sci 11:1048–1056
Gupta RK, Gangoliya SS, Singh NK (2015) Reduction of phytic acid and enhancement of bioavailable micronutrients in food grains. J Food Sci Technol 52:676–684
Heredia-Sandoval NG, Valencia-Tapia MY, Calderon de la Barca AM, Islas-Rubio AR (2016) Microbial proteases in baked goods: modification of gluten and effects on immunogenicity and product quality. Foods (Basel, Switzerland) 5:59
Homayoonfal M, Mousavi SM, Kiani H, Askari G, Khani M, Rezazad Bari M, Alizadeh M (2018) The use of an innovative inverse numerical modeling method for the evaluation and parameter estimation of barberry anthocyanins ultrasound assisted extraction. Chem Eng Process Process Intensif 133:1–11
Houben A, Höchstötter A, Becker T (2012) Possibilities to increase the quality in gluten-free bread production: an overview. Eur Food Res Technol 235:195–208
Hunt JR (2003) Bioavailability of iron, zinc, and other trace minerals from vegetarian diets. Am J Clin Nutr 78:633s–639s
Hur SJ, Lee SY, Kim Y-C, Choi I, Kim G-B (2014) Effect of fermentation on the antioxidant activity in plant-based foods. Food Chem 160:346–356
Hussain A, Bose S, Wang J-H, Yadav MK, Mahajan GB, Kim H (2016) Fermentation, a feasible strategy for enhancing bioactivity of herbal medicines. Food Res Int 81:1–16
Indira M, Venkateswarulu T, Peele KA, Bobby MN, Krupanidhi S (2019) Bioactive molecules of probiotic bacteria and their mechanism of action: a review. 3 Biotech 9:306
Jager R, Purpura M, Farmer S, Cash HA, Keller D (2018) Probiotic Bacillus coagulans GBI-30, 6086 improves protein absorption and utilization. Probiotics Antimicrob Proteins 10:611–615
Karami Z, Akbari-Adergani B (2019) Bioactive food derived peptides: a review on correlation between structure of bioactive peptides and their functional properties. J Food Sci Technol 56:535–547
Kiela PR, Ghishan FK (2016) Physiology of intestinal absorption and secretion. Best Pract Res Clin Gastroenterol 30:145–159
Lappa I, Papadaki A, Kachrimanidou V, Terpou A, Koulougliotis D, Eriotou E, Kopsahelis N (2019) Cheese whey processing: integrated biorefinery concepts and emerging food applications. Foods (Basel, Switzerland) 8:347
Ledesma-Martínez E, Aguíñiga-Sánchez I, Weiss-Steider B, Rivera-Martínez AR, Santiago-Osorio E (2019) Casein and peptides derived from casein as antileukaemic agents. J Oncol 2019:8150967
Limon RI, Penas E, Torino MI, Martinez-Villaluenga C, Duenas M, Frias J (2015) Fermentation enhances the content of bioactive compounds in kidney bean extracts. Food Chem 172:343–352
Liptáková D, Matejčeková Z, Valík Ľ (2017) Lactic acid bacteria and fermentation of cereals and pseudocereals. Fermentation Processes 10:65459
Lopez HW, Krespine V, Guy C, Messager A, Demigne C, Remesy C (2001) Prolonged fermentation of whole wheat sourdough reduces phytate level and increases soluble magnesium. J Agric Food Chem 49:2657–2662
Lorusso A, Verni M, Montemurro M, Coda R, Gobbetti M, Rizzello CG (2017) Use of fermented quinoa flour for pasta making and evaluation of the technological and nutritional features. LWT 78:215–221

Madhuri K, Prabhakar KV (2014) Microbial exopolysaccharides: biosynthesis and potential applications. Orient J Chem 30:1401–1410

Marhaida S, Chua LS, El Enshasy H, Majid F, Malek RA (2015) A review on fruit juice probiotication: pomegranate. Curr Nutr Food Sci 11:4–11

Marín L, Miguélez EM, Villar CJ, Lombó F (2015) Bioavailability of dietary polyphenols and gut microbiota metabolism: antimicrobial properties. Biomed Res Int 2015:905215–905215

Masum Akond ASMG, Crawford H, Berthold J, Talukder ZI, Hossain K (2011) Minerals (Zn, Fe, Ca and Mg) and antinutrient (phytic acid) constituents in common bean. Am J Food Technol 6:235–243

Meira QGS, Magnani M, de Medeiros Júnior FC, Queiroga RDCR, Madruga MS, Gullón B, Gomes AMP, Pintado MME, de Souza EL 2015 Effects of added Lactobacillus acidophilus and Bifidobacterium lactis probiotics on the quality characteristics of goat ricotta and their survival under simulated gastrointestinal conditions. Food Res Int 76:828–838

Melini F, Melini V, Luziatelli F, Ficca AG, Ruzzi M (2019) Health-promoting components in fermented foods: an up-to-date systematic review. Nutrients 11:1189

Mohanty D, Mohapatra S, Misra S, Sahu P (2016) Milk derived bioactive peptides and their impact on human health–A review. Saudi Jo Biol Sci 23:577–583

Morton AE (2015) Fermented foods: sources, consumption and health benefits. Nova Science Publishers, Incorporated, New York, NY

Mousavi ZE, Mousavi M (2019) The effect of fermentation by Lactobacillus plantarum on the physicochemical and functional properties of liquorice root extract. LWT 105:164–168

Mousavi ZE, Mousavi SM, Razavi SH, Hadinejad M, Emam-Djomeh Z, Mirzapour M (2013) Effect of fermentation of pomegranate juice by lactobacillus plantarum and lactobacillus acidophilus on the antioxidant activity and metabolism of sugars, organic acids and phenolic compounds. Food Biotechnol 27:1–13

Mulvihill B (2002) Ruminant meat as a source of conjugated linoleic acid (CLA). Nutr Bull 26:295–299

Nkhata SG, Ayua E, Kamau EH, Shingiro J-B (2018) Fermentation and germination improve nutritional value of cereals and legumes through activation of endogenous enzymes. Food Sci Nutr 6:2446–2458

Ojha KS, Tiwari BK (2016) Novel food fermentation technologies, Novel Food Fermentation Technologies. Springer Cham, Switzerland, pp 1–5

Osawa R, Kuroiso K, Goto S, Shimizu A (2000) Isolation of tannin-degrading Lactobacilli from humans and fermented foods. Appl Environ Microbiol 66:3093–3097

Otieno DO (2010) Synthesis of β-galactooligosaccharides from lactose using microbial β-galactosidases. Compr Rev Food Sci Food Saf 9:471–482

Park YW (2009) Bioactive components in milk and dairy products. John Wiley & Sons, Hoboken, NJ

Parkar SG, Redgate EL, McGhie TK, Hurst RD (2014) In vitro studies of modulation of pathogenic and probiotic bacterial proliferation and adhesion to intestinal cells by blackcurrant juices. J Funct Foods 8:35–44

Parvaneh K, Jamaluddin R, Karimi G, Erfani R (2014) Effect of probiotics supplementation on bone mineral content and bone mass density. ScientificWorldJournal 2014:595962–595962

Paul M, Somkuti G (2009) Degradation of milk-based bioactive peptides by yogurt fermentation bacteria. Lett Appl Microbiol 49:345–350

Pereira-Caro G, Fernandez-Quiros B, Ludwig IA, Pradas I, Crozier A, Moreno-Rojas JM (2018) Catabolism of citrus flavanones by the probiotics Bifidobacterium longum and Lactobacillus rhamnosus. Eur J Nutr 57:231–242

Pessione E, Cirrincione S (2016) Bioactive molecules released in food by lactic acid bacteria: encrypted peptides and biogenic amines. Front Microbiol 7:876–876

Popova A, Mihaylova D (2019) Antinutrients in plant-based foods: a review. Open Biotechnol J 13:68–76

Poutanen K, Flander L, Katina K (2009) Sourdough and cereal fermentation in a nutritional perspective. Food Microbiol 26:693–699

Prasanna P, Grandison A, Charalampopoulos D (2012) Effect of dairy-based protein sources and temperature on growth, acidification and exopolysaccharide production of Bifidobacterium strains in skim milk. Food Res Int 47:6–12

Rad AH, Khosroushahi AY, Khalili M, Jafarzadeh S (2016) Folate bio-fortification of yoghurt and fermented milk: a review. Dairy Sci Technol 96:427–441

Ramsubeik K, Keuler NS, Davis LA, Hansen KE (2014) Factors associated with calcium absorption in postmenopausal women: a post hoc analysis of dual-isotope studies. J Acad Nutr Diet 114:761–767

Raveschot C, Cudennec B, Coutte F, Flahaut C, Fremont M, Drider D, Dhulster P (2018) Production of bioactive peptides by lactobacillus species: from gene to application. Front Microbiol 9:2354

Rekha CR, Vijayalakshmi G (2010) Bioconversion of isoflavone glycosides to aglycones, mineral bioavailability and vitamin B complex in fermented soymilk by probiotic bacteria and yeast. J Appl Microbiol 109:1198–1208

Ribeiro S, Stanton C, Yang B, Ross R, Silva C (2017) Conjugated linoleic acid production and probiotic assessment of Lactobacillus plantarum isolated from Pico cheese. LWT 90:403–411

Richard N, Jooste P (2012) Cereal-based functional foods. IntechOpen, London, pp 161–196

Rizzello C, Lorusso A, Russo V, Pinto D, Marzani B, Gobbetti M (2016) Improving the antioxidant properties of quinoa flour through fermentation with selected autochthonous lactic acid bacteria. Int J Food Microbiol 241:252–261

Rollán GC, Gerez CL, LeBlanc JG (2019) Lactic fermentation as a strategy to improve the nutritional and functional values of pseudocereals. Front Nutr 6:98–98

Rong H, Ju X, Yuan J, Wang L, Girgih A, Aluko R (2012) Antioxidant activities of rapeseed peptides produced by solid state fermentation. Food Res Int 49:432–438

Rossi M, Amaretti A, Leonardi A, Raimondi S, Simone M, Quartieri A (2013) Potential impact of probiotic consumption on the bioactivity of dietary phytochemicals. J Agric Food Chem 61:9551–9558

Ruas-Madiedo P, Hugenholtz J, Zoon P (2002) An overview of the functionality of exopolysaccharides produced by lactic acid bacteria. Int Dairy J 12:163–171

Sabater C, Fara A, Palacios J, Corzo N, Requena T, Montilla A, Zárate G (2018) Synthesis of prebiotic galactooligosaccharides from lactose and lactulose by dairy propionibacteria. Food Microbiol 77:93–105

Sadh PK, Kumar S, Chawla P, Duhan JS (2018) Fermentation: a boon for production of bioactive compounds by processing of food industries wastes (by-products). Molecules 23:2560

Septembre-Malaterre A, Remize F, Poucheret P (2018) Fruits and vegetables, as a source of nutritional compounds and phytochemicals: changes in bioactive compounds during lactic fermentation. Food Res Int 104:86–99

Shao L, Wu Z, Zhang H, Chen W, Ai L, Guo B (2014) Partial characterization and immunostimulatory activity of exopolysaccharides from Lactobacillus rhamnosus KF5. Carbohydr Polym 107:51–56

Sohaib M, Anjum FM, Sahar A, Arshad MS, Rahman UU, Imran A, Hussain S (2017) Antioxidant proteins and peptides to enhance the oxidative stability of meat and meat products: a comprehensive review. Int J Food Prop 20:2581–2593

Song T-S, Lee K-S, Kang S-B, Yoo S-H, Lee J-I, Yoon S-S (2013) Synthesis of galactooligosaccharides in the cheese whey-based medium by a Lactase from Lactobacillus paracasei YSM0308. Korean J Food Sci Anim Resour 33:565–571

Taha S, El Abd M, De Gobba C, Abdel-Hamid M, Khalil E, Hassan D (2017) Antioxidant and antibacterial activities of bioactive peptides in buffalo's yoghurt fermented with different starter cultures. Food Sci Biotechnol 26:1325–1332

Tamang JP, Shin D-H, Jung S-J, Chae S-W (2016) Functional properties of microorganisms in fermented foods. Front Microbiol 7:578

Tricon S, Burdge G, Jones E, Russell J, El-Khazen S, Moretti E, Hall W, Gerry A, Leake D, Grimble R, Williams C, Calder P, Yaqoob P (2006) Effects of dairy products naturally enriched with cis-9,trans-11 conjugated linoleic acid on the blood lipid profile in healthy middle-aged men. Am J Clin Nutr 83:744–753

van der Velpen V, Hollman PC, van Nielen M, Schouten EG, Mensink M, van't Veer P, Geelen A (2014) Large inter-individual variation in isoflavone plasma concentration limits use of isoflavone intake data for risk assessment. Eur J Clin Nutr 68:1141–1147

Varmanen P, Deptula P, Chamlagain B, Piironen V (2016) Letter to the editor on 'Enhancing vitamin B12 content in soy-yogurt by Lactobacillus reuteri, IJFM. 206:56–59'. Int J Food Microbiol 228

Venegas-Ortega MG, Flores-Gallegos AC, Martínez-Hernández JL, Aguilar CN, Nevárez-Moorillón GV (2019) Production of bioactive peptides from lactic acid bacteria: a sustainable approach for healthier foods. Compr Rev Food Sci Food Saf 18:1039–1051

Verhoeckx KCM, Vissers YM, Baumert JL, Faludi R, Feys M, Flanagan S, Herouet-Guicheney C, Holzhauser T, Shimojo R, van der Bolt N, Wichers H, Kimber I (2015) Food processing and allergenicity. Food Chem Toxicol 80:223–240

Verni M, Rizzello CG, Coda R (2019) Fermentation biotechnology applied to cereal industry by-products: nutritional and functional insights. Front Nutr 6:42

Wegh CAM, Geerlings SY, Knol J, Roeselers G, Belzer C (2019) Postbiotics and their potential applications in early life nutrition and beyond. Int J Mol Sci 20:4673

Whisner CM, Castillo LF (2018) Prebiotics, bone and mineral metabolism. Calcif Tissue Int 102:443–479

Wu H, Rui X, Li W, Chen X, Jiang M, Dong M (2015) Mung bean (Vigna radiata) as probiotic food through fermentation with Lactobacillus plantarum B1-6. LWT Food Sci Technol 63:445–451

Xiang H, Sun-Waterhouse D, Waterhouse GIN, Cui C, Ruan Z (2019) Fermentation-enabled wellness foods: a fresh perspective. Food Sci Human Wellness 8:203–243

Xie Z, Huang J, Xu X, Jin Z (2008) Antioxidant activity of peptides isolated from alfalfa leaf protein hydrolysate. Food Chem 111:370–376

Xu S, Boylston T, Glatz B (2005) Conjugated linoleic acid content and organoleptic attributes of fermented milk products produced with probiotic bacteria. J Agric Food Chem 53:9064–9072

Zeidan AA, Poulsen VK, Janzen T, Buldo P, Derkx PM, Øregaard G, Neves AR (2017) Polysaccharide production by lactic acid bacteria: from genes to industrial applications. FEMS Microbiol Rev 41:S168–S200

Zhang S, Shi Y, Zhang S, Shang W, Gao X, Wang H (2014) Whole soybean as probiotic lactic acid bacteria carrier food in solid-state fermentation. Food Control 41:1–6

Zhong X-F, Luo T, Huang G-D, Deng Z-Y, Lei L (2012) Equimolar mixture of c9,t11 and t9,t11 CLA inhibits the growth and induces apoptosis in Caco-2 cells. Eur J Lipid Sci Technol 114:479–485

Zielińska D, Kołożyn-Krajewska D (2018) Food-origin lactic acid bacteria may exhibit probiotic properties: review. Biomed Res Int 2018:5063185–5063185

Zou B, Wu J, Yu Y, Xiao G, Xu Y (2017) Evolution of the antioxidant capacity and phenolic contents of persimmon during fermentation. Food Sci Biotechnol 26:563–571

Zubik L, Meydani M (2003) Bioavailability of soybean isoflavones from aglycone and glucoside forms in American women. Am J Clin Nutr 77:1459–1465

Quality and Health Aspects of Dairy Foods as Affected by Probiotic Bacteria and Their Metabolites

11

Mahdieh Iranmanesh

Abstract

With the growing demand for healthy food products, there has been increasing scientific and commercial interests for developing foods that besides providing the nutritive values could also improve the overall health status of the consumer. In this context, the probiotic dairy products are of immense interest to both consumers and researchers who are searching for healthy food products with increased health benefits. Administration of live bacteria especially lactic acid bacteria (LAB) to ferment and non-fermented dairy food products are considered a health-promoting strategy that could bestow health benefits on the consumer. Probiotics are mixture of friendly bacteria capable of maintaining and improving intestinal balance and hence boosting immune system effectiveness. Mounting evidence are present on the role of probiotic strains acting as adjuncts to antibiotic therapy by reducing adverse effects, improving antibacterial function and enhancing mucosal immunity. Apart from probiotic bacteria, their nonviable counterparts (paraprobiotics) and the metabolites produced by probiotic bacteria (postbiotic metabolites) are also known to provide physiological health benefits to the consumers, and demonstrate therapeutic actions that are comparable to the actions of probiotics. In this review these concepts will be approached, as well as their potential applications in dairy products, highlighting the functional and technological advantages compared to the use of probiotics.

M. Iranmanesh (✉)
Department of Food science and Technology, Science and Research Branch, Islamic Azad University, Tehran, Iran

N. Mojgani, M. Dadar (eds.), *Probiotic Bacteria and Postbiotic Metabolites: Role in Animal and Human Health*, Microorganisms for Sustainability 2,
https://doi.org/10.1007/978-981-16-0223-8_11

Keywords

Probiotics · Postbiotics · Paraprobiotics · Dairy food · Fermented milk · Non-fermented milk

11.1 Introduction

With the rise in healthy living and integrative medicine, the importance of beneficial bacteria has become prominent and the use of products known as probiotics is becoming more and more common among ordinary people. The importance of probiotic foods especially dairy foods has been emphasized by many researchers in last decades (FAO/WHO 2001, 2002; Gardiner et al. 2002; Moeller and de Vrese 2004; Malcata et al. 2005; Sharma and Ghosh 2006; Shah 2007; Shiby and Mishra 2013; Sharma and Devi 2014; Santiago Lopez et al. 2015; Tunick and van Hekken 2015). The word Probiotic is derived from a Greek word that means "for life" (Kollath 1953). As stated in FAO/WHO reports (2001), Probiotics are "*live microorganisms which when administered in adequate amounts confer a health benefit on the host.*" The most widely used probiotic species includes species from the Genus *Lactobacillus* and *Bifidobacterium*, while some species of *Streptococcus*, *Lactococcus* and *Enterococcus* has also been used. Apart from these, some yeast like *Saccharomyces boulardii* and *Kluyveromyces lactis* are also used as a probiotics (Kumura et al. 2004; Kumar et al. 2015). The health benefit of probiotics and their metabolites has been observed in many probiotic food products. Among different category of food products, fermented dairy products are considered as the most important vehicle for delivering probiotic organisms (Tamime et al. 1995), and many research reports have highlighted their therapeutic effects (Granato et al. 2010; Parmjit 2011; Sánchez et al. 2017). In addition, the postbiotic metabolites produced by these added probiotic bacteria can affect the microbiological and sensory qualities of dairy products (Guzel-Seydim et al. 2005; Hekmat and Reid 2006; Sobrino-López and Martín-Belloso 2008; Allgeyer et al. 2010). In this chapter, the effect of probiotic bacteria and their metabolite on quality of dairy products and the health benefits of consumption of these kinds of products will be reviewed.

11.2 Probiotic, Paraprobiotic, and Postbiotic

Probiotics in fermented dairy products impose beneficial health effects on the host by several mechanisms (Fig. 11.1). According to Oelschlaeger (2010) these effects can be divided into three groups based on their mode of action: (1) modulating hosts defenses through the mucosal barrier function by decreasing the apoptosis of epithelial cells and increasing mucin production (Mattar et al. 2002; Gaudier et al. 2005; Yan and Polk 2006; Caballero-Franco et al. 2007; Gogineni et al. 2013; Saad et al. 2013), (2) direct effect on pathogenic microorganisms by producing antimicrobial substances such as bacteriocins (Alakomi et al. 2000; Penner et al.

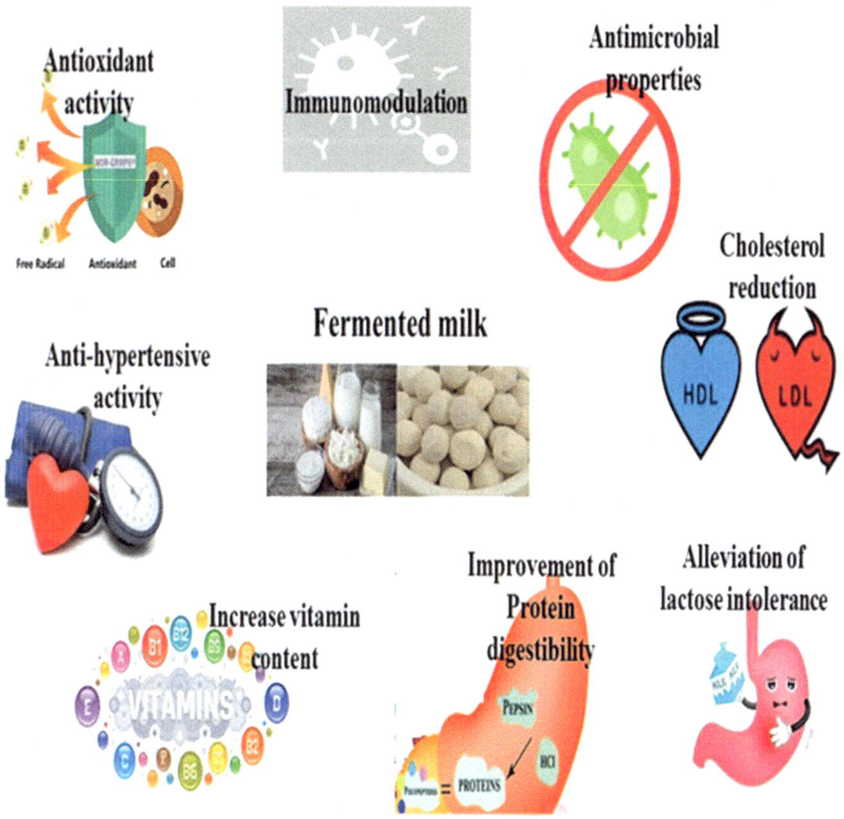

Fig. 11.1 Health benefits of fermented dairy products

2005; Liévin-Le Moal and Servin 2006; Sharma and Devi 2014) and antimicrobial peptides (Schlee et al. 2008; Kelsall 2008; Mondel et al. 2009), and (3) effect on microbial products such as toxins.

Most of the functions performed by probiotic bacteria have been reported to depend on their viability (Sanders 2009), and it was considered essential for a probiotic bacterium to retain its viability at concentrations of approximately10^9 cfu/mL to be effective. However, recent studies have suggested that bacterial viability is not an imperative factor for these beneficial microbes to show their health effects, and dead probiotic bacterial cells are also able to show significant health benefits. The nonviable counterpart of the probiotic bacteria was termed paraprobiotic.

Paraprobiotic or nonviable probiotic can be defined as "inactivated microbial cells or cell fractions that confer health benefits to the consumer" (Taverniti and Guglielmetti 2011). Paraprobiotics include the cell wall components including peptidoglycans, surface proteins, cell wall polysaccharides, etc. (Shin et al. 2010).

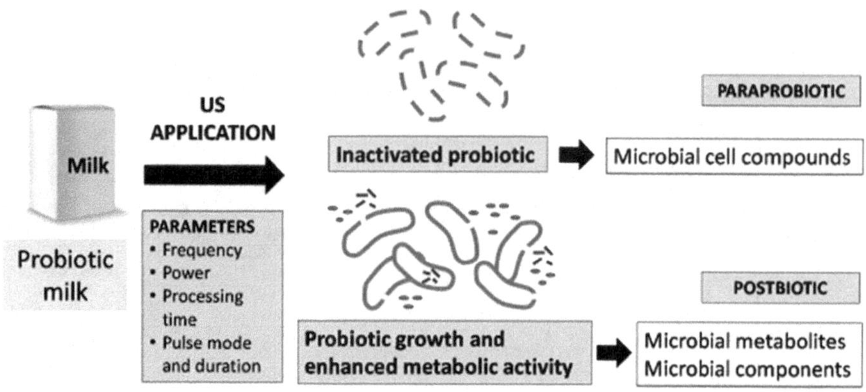

Fig 11.2 High-intensity ultrasound in the development of paraprobiotics and postbiotics (Guimarães et al. 2019)

Various methods like heat, high pressure, sonication, UV irradiation, and other methods such as dehydration, pulsed electric field (PEF), and ohmic heating have been used for inactivating bacterial cells (de Almada et al. 2016). These kinds of methods are also used in dairy products manufactured with probiotics in order to improve the functional activities. Guimarães and his colleagues (2019) used high-intensity ultrasound (HIUS) as a mild preservation technology in dairy products (Fig. 11.2). These researchers concluded that HIUS technology could shorten the processing time, improve probiotic viability, and could be utilized for development of paraprobiotics and improving the production of postbiotics with health effects.

Postbiotic or probiotic metabolic, biogenics, or simply metabolites/CFS (cell-free supernatants) refers to soluble fractions (products or metabolic byproducts) secreted by live probiotic bacteria or released after bacterial lysis (Tsilingiri and Rescigno 2013). Postbiotics are classified differently based on the data available in the literature. As on these reports, postbiotics can be categorized depending on their elemental composition like lipids (e.g., butyrate, propionate, dimethyl acetyl-derived plasmalogen), proteins (e.g., lactocepin, p40 molecule), carbohydrates (e.g., galactose-rich polysaccharides, and teichoic acids), vitamins/cofactors (e.g., B-group vitamins), organic acids (e.g., propionic and 3-phenyllactic acid) enzymes, bacteriocins, and complex molecules such as peptidoglycan-derived muropeptides and lipoteichoic acids (Kostantinov et al. 2013; Tsilingiri and Rescigno 2013). In another words, postbiotics include extracellular and intracellular bacterial cell fractions. The extracellular cell wall components include exopolysaccharide and peptidoglycans, while the intracellular metabolites are organic acids, short-chain fatty acids, and bacteriocins like acidophilin, bifidin, reuterin, peptides, etc. (Matsuguchi et al. 2003). Besides, postbiotics can be divided into different groups according to their function such as immunomodulation, anti-inflammatory, hypocholesterolemic, antiobesogenic, antihypertensive, antiproliferative, and anti-oxidant effects (Nakamura et al. 2016; Shin et al. 2010; Sawada et al. 1990).

11.3 Dairy Products

With increasing consumer demands, there is surge for safe and nutritive foods that would not only provide the required energy and the nutrients for the human, but might also improve the overall health status of the consumer including their physiological and psychological status (Young 2000; Mollet and Rowland 2002). These food products are referred to as functional foods and are known for their ability to improve the general health of the consumers (Stanton et al. 2005). Functional foods enriched with probiotics are termed probiotic functional foods.

Dairy products are an important source of energy as well as micro- and macronutrients and among the highly consumed food product worldwide. Among these dairy products yogurts, fermented dairy products, LAB drinks and mixture of probiotic (fermented) milks and fruit juices are highly reputed for their dietary health benefits and are considered a rich source of beneficial bacteria including lactic acid bacteria. It is a well-known fact that milk and dairy products are a powerful tool and significant vectors for creating probiotic dairy foods (Fig. 11.3).

11.4 Classification of Probiotic Dairy Products

This category of dairy products includes raw milk (fermented and unfermented), cheese, ice cream, and dried dairy products (infant formula and dairy base dried products). Below we discuss the importance of some of these probiotic products.

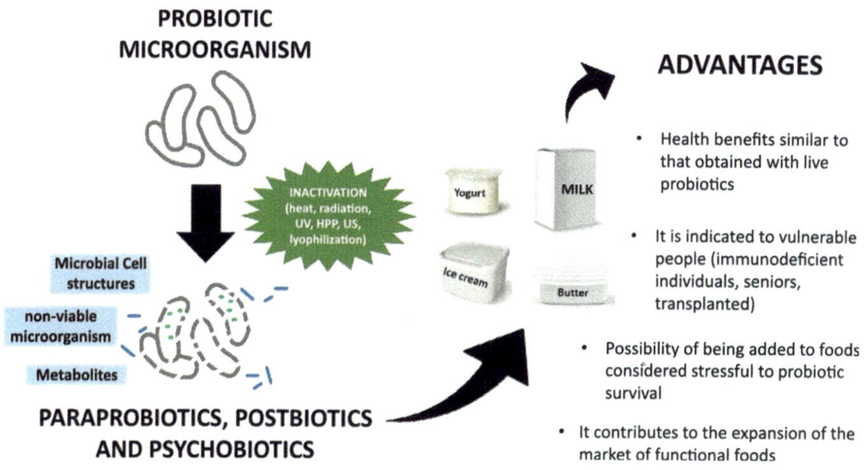

Fig. 11.3 Advantages of probiotics, paraprobiotics, and postbiotics in dairy products (Barros et al. 2020)

11.4.1 Probiotic Fermented Milks

Fermented milk is a dairy product that is produced by fermentation of milk with abundant number of viable and active microorganism that are safe for use (García-Burgos et al. 2020). This milk has been produced in many countries for centuries, and is considered one of the oldest methods for extending the shelf life of milk and also helps to produce various products from milk. Fermented milk products offer vast array of nutritional and health benefits due to the presence of abundant LAB and their metabolites which are produced during the process of fermentation (Granier et al. 2013). These products are produced from different mammal milk like cow, sheep, goat, buffalo, and camel (Tamime 2002), and are a good source of calcium which is vital for bone formation and mineralization (Baba et al. 2014). Fermented milk products are recognized as suitable carrier for probiotic microorganisms.

According to Robinson and Tamime (1990) fermented milks can be classified into three groups based on the dominant microorganisms in the products as follows:

A. Lactic fermentations: (i) mesophilic type like cultured buttermilk, (ii) thermophilic type such as yogurt, Bulgarian buttermilk, zabadi, dahi; and (iii) therapeutic or probiotic type including acidophilus milk
B. Yeast—lactic fermentations like kefir and koumiss
C. Mold—lactic fermentations such as viili

In addition to these groups, we discussed other probiotic dairy product in this section as shown in Fig. 11.4.

Lactic fermentations could be divided into mesophilic probiotic fermented milks and thermophilic probiotic fermented milks.

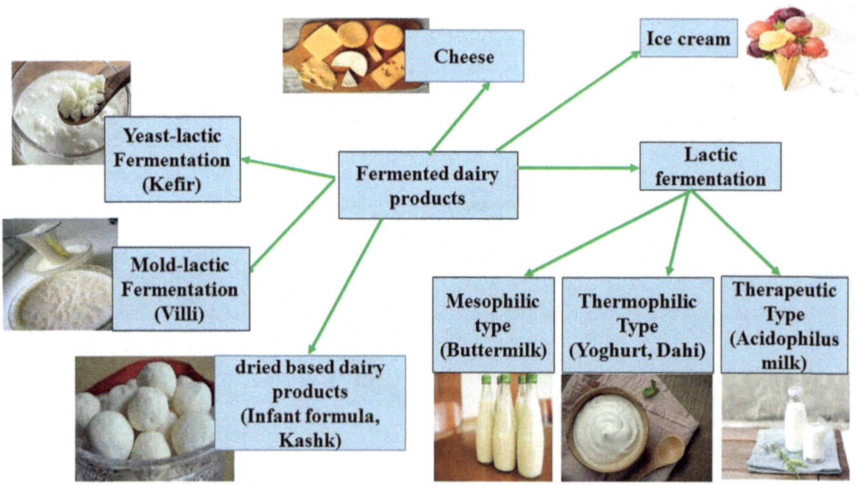

Fig. 11.4 Classification of fermented dairy products

Mesophilic cultures are widely used in fermented dairy products, like in sour creams, cultured buttermilk, kefir, etc. The primary mesophilic starter cultures used in these fermented dairy products belongs to the Genus *Lactobacillus*, *Lactococcus,* and *Leuconostoc* spp. These species are known to produce a variety of biogenic metabolites including proteins, peptides, oligosaccharides, vitamins, and fatty acids (Ebringer et al. 2008). The production of some fatty acids such as pyruvic, acetic, propionic, and especially lactic acid in cultured buttermilk produced during the fermentation by *Lc. lactis* and *Leu. citrovorum* was evaluated by Marsili 1981. According to the results, the content of acetaldehyde was shown to increase initially and then a decline was observed followed by formation of ethanol, whereas the content of acetone and uric acid remained constant during fermentation. In our previous study, some probiotic LAB species like *Lb. brevis, Lb. pentosus, Ped. acidilactici*, and *Lb. paracasei* were isolated from a traditional buttermilk in Iran (Iranmanesh et al. 2012). The paraprobiotic of the isolates also showed interesting health characteristics and were able to lower the cholesterol concentrations *in vitro*. Among the tested species, *L. brevis* demonstrated the highest level of cholesterol removal (Iranmanesh et al. 2014). Our results also showed that *Lb. pentosus, Lc. lactis, Lb. paracasei, and Ped. acidilactici* were able to produce protein metabolite (bacteriocins) that inhibited the growth of *L. monocytogenes* and *S. aureus*. The bacteriocin produced by the mentioned species was produced in the culture medium in the early logarithmic phase and continued to the end of exponential phase (Iranmanesh et al. 2015). Similar to our results, *Lc. lactis* subsp. *Hordniae* and *Lc. lactis* subsp. *Lactis* isolated from homemade buttermilk was shown to produce a bacteriocin that was active against *Staphylococcus aureus* MTCC96 and *Pseudomonas aeruginosa* MTCC741 (Barman et al. 2018).

Thermophilic cultures used in fermentation of milk include *Lb. delbrueckii* subsp. *bulgaricus*, *Lb. helveticus, Lb. acidophilus*, *Lb. paracasei* subsp paracasei, Bifidobacterium species, and *Streptococcus thermophilus*. These cultures are mainly used for the production of yogurt, Bulgarian buttermilk, etc. Below we describe the beneficial health effects of some fermented dairy products having probiotic bacteria, beside the thermophilic starter cultures.

11.4.2 Yogurt

The popularity of yogurt as one of the most consumed fermented dairy products has increased in recent years due to its high nutritional-physiological values (Lourens-Hattingh and Viljoen 2001). The nutritional aspects of probiotic yogurt are mainly due to the presence of the beneficial bacteria with health-benefiting potentials. Some of the health benefits of thermophilic probiotic fermented milks are summarized as below.

Cancers are results of lethal cellular damage caused by free radicals, while antioxidant compounds prevent generation of these free radicals and consequently could prevent cancer generation (Urso and Clarkson 2003). The antioxidant property of probiotic yogurt that contained *Lb. bulgaricus, S. thermophiles, B. lactis* Bb12, and *Lb. acidophilus* La5 was evaluated in 30- to 60-year-old patients with type 2 diabetes (Ejtahed et al. 2012). These patients consumed 300 g/day probiotic yogurt for 6 weeks. The results showed that consumption of probiotic yogurt increased erythrocyte superoxide dismutase and glutathione peroxidase activities as well as total antioxidant status, significantly ($P < 0.05$), compared to the control group. Furthermore, the level of insulin was not significantly different in comparison with the healthy control group, while the content of serum malondialdehyde significantly decreased.

Some bioactive peptides have shown to possess antioxidant activity. Sah et al. (2014) showed that peptides separated from probiotic yogurt had antioxidant and antimutagenicity activity. These researchers showed that the IC50 of two peptides including 1,1-diphenyl-2-picrylhydrazyl and 2,20-azino-bis (3ethylbenzothiazoline-6-sulphonic acid) were 1.51 and 1.63 mg/mL, respectively.

The effect of metabolites produced by probiotic starter cultures on the texture and other sensory properties of the fermented milk products has been assessed. A number of bacterial species such as *S. thermophilus, Lb. kefiranofaciens, Lb. helveticus, Lb. sake, Lb. delbrueckii* subsp. *bulgaricus, Lc. lactis* subsp. *cremoris, B. longum,* and *B. infantis are* known to produce exopolysaccharide (EPS) (Surono and Hosono 2011). EPS plays an important role in the rheology, texture, and mouthfeel of fermented milks, and are found in yogurt, kefir, viili, and some other fermented dairy products. In a study, Hess and his co-investigators (1997), studied the effect of exopolysaccharide (EPS) produced by *Lb. delbrueckii* ssp. *bulgaricus* strain C1 and *S. thermophilus* strain B1 on the texture of the produced yogurt. They found that the susceptibility to syneresis was decreased, whereas the ropiness or extensibility was increased significantly. In addition, the shear stress of yogurt with EPS was increased from 0.1 to 0.3 s^{-1} and then a steady increase as shear rate was increased from 0.3 to 100 s^{-1}. Similar findings were reported by Guzel-Seydim et al. (2005), who stated that whey separation was decreased in probiotic yogurt with ropy polysaccharide-producing culture. There results also indicated increase in the contents of lactic acid, volatile fatty acids, and tyrosine whereas the acetaldehyde concentration and pH values were decreased. In another research, the effect of EPS on the texture of inulin-containing probiotic yogurt during 21 days of storage showed that the firmness in the texture of yogurt was not influenced by EPS significantly (Ramchandran and Shah 2010). Whereas the influence of EPS on yield stress (Pa), consistency index (Pa s), and thixotrophic behavior (Pa/s) was observable only after day 7. In addition, the use of the EPS from *Lb. fermentum* Lf2, as an additive, in yogurt augmented hardness and improved the water holding capacity of the product (Ale et al. 2016). The sensory properties of non-fat yogurts with 300 and 600 mg EPS/L showed that 600 mg/L of EPS extract had the highest values of consistency after 3 days of storage, while this property was not detected at the end of the shelf life. Similar to other results, the EPS333 produced by *S. thermophiles* strain AR333 in yogurts could increase the viscosity and water holding capacity which resulted in improved quality of yogurt (Zhang et al. 2018). These exopolysaccharides are composed of galactose, glucose, and galactosamine in a molar ratio of 3:2:1.

The immunostimulatory effects of EPS produced from *Lb. delbrueckii* ssp. *bulgaricus* OLL1073R-1 was described by Makino et al. 2016. They found that the oral administration of EPS or yogurt fermented with OLL1073R and *S. thermophilus* OLS305 to mice models, increased natural killer cell activity and also the production of IFN-γ production in spleen cells of mice after 3 weeks.

Another metabolite produced by probiotic bacteria includes folate which is B-group vitamin that is not synthesized in humans. Hence, humans require this vitamin which they recieve through exogenous source such as food. Crittenden et al. (2003) showed that *S. thermophiles, bifidobacteria,* and *E. faecium* can produce folate. Based on his results, *S. thermophiles* produced higher content of folate than the other tested strains. In another research study, combination of *B. lactis, L. acidophilus,* and *S. thermophilus* used in the production of probiotic yogurts was shown to demonstrate higher content of folate than yogurts produced by traditional starter cultures (*Lb. delbrueckii* subsp. *Bulgaricus* and *S. thermophiles*). Similar findings were reported by Laiño et al. (2012), who showed that *Lb. delbrueckii* subsp. *bulgaricus* CRL 863 isolated from artisanal Argentinean yogurts could produce folate. This strain produced both intracellular and extracellular folate after 6 h of growth in folate-free culture medium and then decreased after 10 h, while extracellular folates remained constant up to 24 h but intracellular forms reduced slightly. Besides, the content of folate in milk fermented by *Lb. delbrueckii* subsp. *bulgaricus* CRL 863 was shown to increase after 24 h of incubation at 37 °C.

The effect of fermented milk like yogurt on cholesterol has been investigated widely. Ataie-Jafari et al. (2009) had shown that consumption of probiotic yogurt was able to reduce total cholesterol levels in the serum of 14 healthy subjects. The subjects in study were given 300 g of ordinary yogurt or probiotic yogurt for 6 weeks, after a 4-week washout period, the study continued for another 6 weeks. In this study, probiotic yogurts were fermented by *S. thermophilus* and *Lb. delbrueckii* subsp. *Bulgaricus, Lb. acidophilus* and *B. lactis*. They concluded that the two probiotic strains *Lb. acidophilus or B. lactis* had important role on the observed hypocholesterolemia effects. Similarly, Baroutkoub et al. (2010) showed that total cholesterol and LDL (low-density lipoprotein) levels were decreased, while HDL (high-density lipoprotein) increased by consumption of probiotic yogurt. Furthermore, consumption of 300 g of probiotic yogurt containing *Lb. acidophilus* La5 and *B. lactis* Bb12 for 6 weeks in people with type 2 diabetes also showed that the total cholesterol and LDL-C were decreased 4.54% and 7.45%, respectively (Ejtahed et al. 2011). The total cholesterol:HDL-C ratio and LDL-C:HDL-C was also significantly decreased.

Some major antibacterial metabolites produced by probiotic bacteria include bacteriocin and other antimicrobial compounds like organic acid, hydrogen peroxide, and low-molecular-weight substances like Reuterin (Ammor et al. 2006). In a study, *S. thermophile was* shown to produce a bacteriocin which could decrease *L. monocytogenes* counts below detectable levels (Benkerroum et al. 2002). Zaeim et al. (2014) identified bacteriocin from *Lb. bulgaricus* isolated from local yogurt samples. This bacteriocin had activity against Gram-positive bacteria like

L. monocytogenes and *Bacillus cereus* and Gram-negative bacteria such as *E. coli* O157:H7 and *Proteus vulgaris*.

11.4.3 Dahi

Dahi is another fermented dairy product belonging to yogurt group, which is produced in India for long time. It is prepared by buffalo milk fermented by lactic acid bacteria. This product has been reported to be highly nutritious and possess therapeutic effects, and hence can be used as a functional food (Abbas and Jafri 1992; Sinha and Sinha 2000). The effect of probiotic dahi fermented by *Lb. acidophilus* and *Lb. casei* in rats with type 2 diabetes showed that HDL levels were reduced slightly and the thiobarbituric acid-reactive substances were lower than control group (Yadav et al. 2007a).

Kaushal and Kansal (2012) also evaluated the activity of antioxidant enzymes in dahi prepared from buffalo milk with different bacteria. Based on the type of bacteria used in the preparations of dahi, they divided this product into 2 groups: 1. *Lc. lactis* ssp. *cremoris* NCDC-86 and *Lc. lactis* ssp. *lactis biovar diacetylactis* NCDC60 along with selected strain of *Lb. acidophilus* LaVK2 (La-Dahi) and 2. *Lb. acidophilus* and *B. bifidum* BbVK3 (LaBb-Dahi). The activity of antioxidant enzymes was evaluated in mice fed with 5 g/day dahi for 4 months. The results showed the content of oxidation products, thiobarbituric acid-reactive substances (TBARS) and protein carbonyls, in red blood corpuscles (RBCs), heart tissues, liver and kidney was increased. Moreover, the level of superoxide dismutase (SOD) activity in RBCs and hepatic tissues during aging of mice was increased in both groups, while the CAT activity increased in RBCs and heart tissue of only LaBb-Dahi group.

In another study, the EPS-producing non-ropy strain of *Leuconostoc* sp. CFR 2181 isolated from dahi (Vijayendra et al. 2008) was evaluated. This EPS consisted mostly of glucose (91%), and hamnose and arabinose (1.8% each) and its molecular weight was in the range of 1.0×10^4 to 1.5×10^6 Da. The EPS produced from *Lb. fermentum* in low-fat dahi improved the rheological quality by lesser whey separation, higher viscosity, increased adhesiveness, and stickiness (Behare et al. 2013). Similar results were shown that demonstrated that some LAB like *Lb. fermentum* and *Lb. plantarum* isolated from traditional dahi could produce EPS (Patel et al. 2014) that could have beneficial effects when used in dairy products.

The production of free fatty acids (FFAs) and conjugated linoleic acid (CLA) in probiotic dahi containing *Lb. acidophilus* and *Lb. casei* during storage at 4°C for 10 days indicated an increase in butyric acids, linoleic acids, and also CLA content in probiotic dahi (Yadav et al. 2007b).

Immunomodulatory effects of bacteria isolated from dahi have also been reported. In a study conducted by Jain and his colleagues (2009), the ability of dahi containing probiotic *Lb. casei* to modulate immune response against *Salmonella enteritidis* infection in mice was evaluated. They found that the levels of the secretory immunoglobulin A (sIgA) and proliferation of spleen lymphocytes rate were significantly increased. In addition, interleukin (IL) IL-2, IL-6, and IFN-γ increased, but IL-4 reduced. Mitra et al. (2007) showed that the bacteriocin produced by *Lb. lactis* isolated from homemade dahi had activity against *B. cereus, L. monocytogenes, C. perfringens, E. faecalis,* and *St. aureus*. Similarly, the bacteriocin produced by *Lb. acidophilus* isolated from dahi was investigated by Mahmood et al. (2014). The bacteriocin showed activity against *E. coli, S. aureus, P. aeruginosa,* and *L. monocytogenes* and had antibacterial activity of 5369.13 AU/mg.

11.4.4 Zabadi

Zabadi is another traditional dairy product recognized as Egyptian yogurt. Zabadi is mostly made from cow milk and is produced at a local level by boiling, then cooling the milk and inoculation with a day-old previous batch of zabadi (serves as starter). Later, the milk is inoculated with starter cultures (*S. thermophilus* and *Lb. bulgaricus*), and incubated at temperature ranging ~30–38 °C) for 12–15 h. This product can be consumed as a fresh product or stored in a refrigerator. Zabadi (plain full-fat yogurt) has a smooth consistency and a thin body and is not solid like manufactured yogurt (Eissa et al. 2011). According to reports, increased production of polysaccharides at low-temperature fermentations could contribute to a smoother perceived texture of this traditional yogurt (Driessen 1984).

11.4.4.1 Therapeutic or Probiotic Fermented Milk Product

Food matrices are known to play significant role in the beneficial health effects of probiotic bacteria in the host (do Espírito Santo et al. 2011). Acidophilus milk is one of the well-known probiotic dairy foods that is recognized as a functional food and is made by the addition of *Lb. acidophilus*. Some potential benefits of acidophilus milk were reviewed by Gilliland (1989). These beneficial effects include antibacterial activity against pathogens, cholesterol lowering, and anticarcinogenic activity.

Yeast: Lactic Fermentations

Kefir and koumiss are categorized in this group. Kefir is a fermented drink traditionally made by adding kefir grains to cow or goat milk. These grains with cauliflower like appearance are a mixture of bacteria (*Lb. caucasius*) and yeasts (*Saccharomyces kefir and Torula kefir*). Kefir drink is thought to aid in digestion and calm upset stomachs (Hertzler and Clancy 2003). The role of LAB bacteria in kefir is to ferment lactose to lactic acid and provides the tangy flavor, while the yeasts ferment the available fermentable sugars in milk to yield small amounts of alcohol and CO_2, which gives kefir its fizz and effervescence.

Immunomodulatory properties of some yeasts of the kefir like *Kluyveromyces marxianus* B0399 have also been reported. This yeast has the ability to adhere to Caco-2cells and cause a reduction in the secretion of IL-10, IL-12, IL-8, and IFN-γ. In addition, *K. marxianus B0399* caused a reduction in the secretion of proinflammatory cytokines TNF-α, IL-6, and MIP1α in peripheral blood mononuclear cells stimulated with lipopolysaccharide (Maccaferri et al. 2012). Another yeast strain that is of probiotic potential is *S. boulardii.* This yeast has been known for its ability to improve diarrhea symptoms caused by *Clostridium difficile* (Bourrie et al. 2016). The antioxidant activity of 28 *Saccharomyces cerevisiae* isolates from kefir has been described by de Lima et al. (2017), and antioxidant activity was higher than 90. In addition, all strains showed intracellular β-galactosidase activity.

Functional properties of kumis were evaluated by several research (Ohashi et al. 2000; Osorio et al. 2011; Chaves-López et al. 2012), but research on probiotic metabolites has been very limited. Furthermore, Chaves-López et al. (2012) isolated ninety-three yeast strains from Colombian Kumis with Angiotensin I-converting enzyme (ACE) inhibitory activity. *Clavispora lusitaniae* KL4A, *Galactomyces geotrichum* KL20B and *Pichia kudriavzevii* KL52 showed the higher level of ACE peptide production while *Torulaspora delbrueckii* KL66A had the lowest.

Mold: Lactic Fermentations Such as Viili

Viili is a traditional fermented milk from Scandinavia, which is produced from LAB and the mold *Geotrichum candidum* (Kahala et al. 2008; Wang et al. 2008). Kahala et al. (2008) isolated *Lc. lactis* subsp. *lactis biovar diacetylactis* and *Leu. mesenteroides* subsp. *cremoris Lc. lactis* subsp. *cremoris* from Viili.

11.5 Probiotic Non-fermented Milks

There is limited data available regarding probiotic milk from different livestock and most available data is related to cow milk. The angiotensin-converting enzyme inhibitory (ACEI) activity of two peptides of milk containing *Enterococcus faecalis* CECT 5727 named β-casein *f* (133–138) (LHLPLP) and β-casein *f* (58–76) (LVYPFPGPIPNSLPQNIPP) demonstrated (IC50) values as low as 5 mM in rat (Quirós et al. 2007). Although T β -casein *f* (58–76) peptides showed lower antihypertensive activity in spontaneously hypertensive rats compared with LHLPLP. In another research, the impact of unfermented milk containing *Lb. fermentum* MTCC 5898 on immunity system, antioxidant capacity, and severity of pathogenic infection in aging mice was evaluated by Sharma et al. (2014). They found that the activity of some antioxidant enzymes like Catalase and glutathione peroxidase were increased significantly ($P < 0.05$); followed by activity of these enzymes, immune system improved due to enhanced free radical clearance system. While there was no significant difference in IgG2a compared with control group in mice. The pathogen colonization in the intestine, liver, and spleen was also decreased significantly ($P < 0.05$). Balakrishnan and Agrawal (2014) compared the antioxidant activity of fermented cow, goat, and camel milk with *Ped. pentosaceus.* They found that the activity was the highest in goat milk followed by camel and cow milk. In another study, the different function of postbiotics in camel milk and bovine milk probiotic strain *Lc.* lactis KX881782 in vitro was investigated by Ayyash et al. (2018). The fermented camel milk showed grater inhibitions of α-amylase than bovine milk. Whereas, the inhibition of α-glucosidase was not significantly different in both milk. This inhibition resulted in decrease carbohydrate hydrolysis, so reduces the possibility of sugars being absorbed by the human intestine. Furthermore, the proliferation of Caco-2, MCF-7, and HELA cells were more inhibited by fermented camel milk. In contrast the 1,1-diphenyl-2-picrylhydrazyl (DPPH) antioxidant capacity in camel milk was lower than bovine milk.

Fatty acids particularly conjugated linoleic acid (CLA) is another metabolite of probiotics. The content of CLA in fermented milk containing only *S. thermophilus* and *Lb. bulgaricus* was the highest among 10 commercial fermented cow milk products (Manzo et al. 2015). The fatty acid profiles of fermented cow, goat, and camel milk with *Ped. pentosaceus* indicated that oleic acid was higher in camel milk and also the content of linoleic and linolenic acids was low in all fermented milk (Balakrishnan and Agrawal 2014).

Some strains isolated from raw milk showed antimicrobial activity by producing bacteriocin. *Lb. plantarum* isolated from raw cow's milk samples produced bacteriocin with molecular weight approximately 9.5 kDa and could tolerate high temperature up to 121 °C (Sankar et al. 2012). Another strains of lactobacillus, Lb. sakei GM3 isolated from goat milk produced bacteriocin which had antimicrobial activity against *C. albicans*, *C. tropicalis*, *S. aureus*, *P. aeruginosa*, *S. enterica*, and *L. monocytogenes* (Avaiyarasi et al. 2016). The Molecular weight of this bacteriocin was 4.811 KDa and could withstand heat treating at 100 °C for 20 min. Furthermore, bacteriocin produced from various lactobacillus strains which isolated from raw cow, buffalo, and goat milk showed activity against some mastitis pathogens such as *S. aureus*, *E. coli*, *Y. enterocolitica*, *S. uberis*, and *S. xylosus* (Eid et al. 2016).

11.6 Probiotic Cheeses

Chesses can be divided into different groups like very hard and hard (≤38 g 100 g^{-1} moisture), semi-hard (averages ~40 g 100 g^{-1} moisture), Brined cheeses (50–55 g 100 g^{-1} moisture), soft cheeses, and other kinds of cheese (Tamime and Thomas 2018). However, we categorize different chesses based on the metabolite produced by the probiotic bacteria.

Cottage cheeses prepared with *Lb. casei*, *Lb. rhamnosus* GG and Himalayan cheese (Kalari) prepared by different probiotic strains *(Lb. casei, Lb. plantarum, and Lb. brevis*) showed higher antioxidant activity due to higher concentration of bioactive peptides (Abadía-García et al. 2013; Mushtaq et al. 2016). This may be related to the proteolysis occurred by these bacteria. Additionally, probiotic Minas Frescal cheese added with *Lb. casei* 01 showed higher ACEI (antioxidant and angiotensin I-converting enzyme inhibitory) activity compared to conventional cheese (Sperry et al. 2018).

ACEI peptides that are produced by a number of probiotic bacteria in many fermented dairy products have antihypertensive properties. Also adding probiotic bacteria like *Lb. casei* and *Lb. plantarum* to Cheddar cheese showed higher DPPH (2,2-diphenyl-picrylhydrazyl) and antioxidant activity during the ripening time at 16th weeks compared to the control samples (Chen et al. 2019). The DPPH reached its maximum at 16th weeks while other properties of probiotic cheese like texture and sensory analysis was not affected by probiotics.

The EPS-producing probiotic *Lb. plantarum* used in low-fat akawi cheese showed higher antioxidant activities and angiotensin-converting enzyme (ACE) inhibition in comparison with cheese made with non-EPS producers (Ayyash et al. 2012). In addition, the antioxidant activity increased when the storage time was prolonged. Donkor et al. (2012) also showed that inhibition of α-amylase in cheese with EPS-producing culture was higher, while the α-glucosidase inhibition was not significantly increased. The inhibition of α-amylase and α-glucosidase resulted in controlling diabetes and reducing carbohydrate hydrolase. This inhibition might be due to the presence of bioactive peptides.

In another study, it was shown that the content of free fatty acids (FFA) increased in cheeses inoculated either with *Lb. casei* and *B. lactis* at 60 days of ripening (Rodrigues et al. 2012). In addition, three conjugated linoleic acids (CLA) isomers (i.e., *cis*-9, *trans*11-C18:2, CLA1; *trans*-10, *cis*-12-C18:2, CLA2; *trans*-9, *trans*-12-C18:2, CLA3), α-linolenic acid (ALA) and γ-linolenic acid (GLA) were increased during ripening in 15 days that raised up to 60 days. The content of CLA (*cis*-9, *trans*-11-octadecadienoic acid) from different probiotic white cheese showed that *B. longum* cheese and *E. faecium* cheese had the highest and lowest CLA, respectively (Gursoy et al. 2012). Moreover, linoleic acid has been shown to increase in cheese samples prepared with *Lb. paracasei* and *Lb. acidophilus*. Besides, reports showed that addition of *Lb. casei* 01 to Minas Frescal cheeses showed higher level of medium- and long-chain fatty acids (Sperry et al. 2018). In addition, monounsaturated fatty acid such as oleic acid was higher than conventional cheese.

Nine strains of *E. faecium* isolated from Tafı Cheese (a homemade traditional cheese from Tucuman, Argentina) showed cholesterol reduction in vitro (Saavedra et al. 2003). All these strains had bile salts hydrolase activity (BSH), while some strains with negative BSH activity could not reduce cholesterol levels. Similar to these studies, it was shown that *Lb. plantarum* and *Lb. paracasei* isolated from Italian Castelmagno PDO cheese could reduce cholesterol levels in vitro (Belviso et al. 2009). While some research also showed in vivo cholesterol lowering effects of the cheese that harbored probiotic bacteria. The fresh Brazilian cheese containing *Lb. acidophilus* LA14 and *B. longum* BL05 fed for 2 weeks to rats, probiotic cheddar cheese with *L. plantarum* K25 in mice for 4 weeks could decrease total cholesterol and LDL cholesterol, while HDL cholesterol increased compared to control groups (Lollo et al. 2012; Zhang et al. 2013; Lollo et al. 2015).

The effect of probiotic Minas Frescal cheese on hypertension parameters in spontaneously hypertensive rats indicated that the systolic, diastolic, and mean blood pressure over 15 days decreased significantly ($P < 0.05$) compared to the control groups (Lollo et al. 2015). In addition, cheese containing LGG and *Lb. rhamnosus* LC 705 could reduce salivary microbial counts in young adults (Ahola et al. 2002). The subjects ate 5×15 g cheese per day for 3 weeks and the result showed that *S. mutans* and yeast counts decreased in 20% and 27%, respectively, in all the subjects.

The bacteriocinogenic *Lb. paraplantarum* FT259 isolated from Brazilian semi-hard Minas cheese was evaluated by Tulini et al. (2013). This bacterium produced bacteriocin against *L. monocytogenes*, *L. innocua*, and *L. sakei*, while no activity against Gram-negative bacteria was reported. Furthermore, *E. faecium* AQ71 isolated from Azerbaijani Motal cheese produced bacteriocin named enterocins P which had activity against *L. monocytogenes* and *B. cereus* (Ahmadova et al. 2013).

In another study, the inhibitory effects of the probiotics *Lb. acidophilus, Lb. casei* subsp. *paracasei,* and *B. lactis* in a Brazilian semi-hard goat cheese (coalho) was reported, during storage time (de Oliveira et al. 2014). The *Lb. casei* subsp. *paracasei* showed the highest inhibitory activity against *L.monocytogenes* and *S. aureus* on the 14th and 21st days of storage, respectively. While, *B.lactis* had activity against *S. aureus* on the 1st (16.32%), 14th (10.12%), and 21st (3.67%) days of storage, and only on 1st day of storage they had activity against *L. monocytogenes*. *Lb.plantarum* isolated from Traditional Iranian Cheese (Kouzeh) showed high activity against *S.aureus* and *S.epidermidis* (Jabbari et al. 2017). Furthermore, *Kluyveromyces marxianus* S-2-05 and *Kluyveromyces lactis* S-3-05 isolated from a traditional French cheese had inhibitory activity against *Salmonella typhimurium, Salmonella enteritidis,* and *Salmonella paratyphi* B (Ceugniez et al. 2017).

The availability of some minerals is significantly affected by probiotic bacteria in the different cheeses. As demonstrated by Aljewicz and Cichosz (2015), addition of *Lb. rhamnosus* increased in calcium availability in Dutch-type cheese. However, the availability of magnesium and phosphorus decreased in Swiss Dutch-type cheese after 6 weeks of ripening, respectively. In addition, the zinc availability Dutch-type cheese was increased and the availability of potassium lowered in Swiss-type cheese.

11.7 Probiotic Ice Cream

Probiotic ice cream containing 1×10^6 CFU of bacterial strains *B. lactis* Bb-12 and *Lb. acidophilus* La-5 per gram could reduce levels of salivary *S.mutans* in school children after 10 days (Singh et al. 2011). In addition, the goat's milk ice cream with *B.animalis* subsp. *Lactis* showed good sensory characteristics, while other physico-chemical properties such as overrun and melting behavior were not affected by adding probiotic (Da Silva et al. 2015).

11.8 Dried Probiotic Dairy Products

Drying milk is one of the ways to extend the shelf life and decrease the loss of milk in long chain between farmer and consumer. Dried products include milk powder, whey powder, whey protein concentrate, yogurt powder, and infant formula. The methods of drying can be divided into two groups, thermal and non-thermal (Aadinath et al. 2017). In dried probiotic dairy product, some factors play essential role for maintaining the viability of probiotic microorganisms, such as the methods of drying, the type and size of packaging used, condition of storage like temperature and humidity, quality of powder, the process of rehydration and handling of rehydration product, etc. (Gilliland 2001). One of the most widely used dried dairy products of this group is infant formula which is described below.

11.8.1 Infant Formula

The effect of probiotic supplemented infant formula on plasma lipid of infant was described by Kankaanpää et al. (2002). The *Bifidobacterium* Bb-12 supplemented formula could increase the α-linolenic acid in phospholipids, while the *Lb.* GG had no effect on this fatty acid. In addition, both probiotics were able to increase the percentage of the total monounsaturated fatty acids (MUFA).

Saavedra et al. (2004) showed that probiotic supplemented formula could lower the frequency of reported colic or irritability. According to their results, consumption of formulas containing *B. lactis* and *S. thermophiles* for long time resulted in reduction of colic, irritability and reduced the frequent use of antibiotics. Similar to these findings, it was reported that *B. lactis* (BB-12) and *Lb. reuteri* in infant formula could reduce infections in healthy 4- to 10-month-old infants after 4 weeks (Weizman et al. 2005). The control group had more diarrheas with longer duration, whereas *Lb. reuteri* group, compared to BB-12, had fewer days of diarrhea and lower visits to the clinic or child care unit. In addition, the effect of supplemented formula with *Lb.johnsonii* La1on fecal microbiota composition of infants was evaluated by Brunser et al. (2006). Ninety infants close to 4 months of age were divided into various groups and received probiotic supplemented infant formula for 13 weeks. The results showed that the fecal *Lactobacillus* count was higher than controls, while the count of *Clostridium*, *Bacteroides*, or *Enterococcus* were not significantly difference between the groups.

The safety of a prebiotic-containing starter formula supplemented with *Lb. paracasei* ssp. *paracasei* and *B. animalis* ssp. *lactis* for first 3 month in 126 newborns and then continued in 80 infants for 6 month was evaluated by Vlieger et al. (2009). The growth, clinical outcomes like crying and sleeping hours, number of gastrointestinal or upper respiratory tract infections, the amount of antibiotics used and visits to the general practitioner were not significantly difference compared to the control group.

11.8.2 Dairy-Based Dried Products

The traditional dairy-based dried products like Kashk, Tarhana, and Kurut are produced for many centuries. The most data about these kinds of products were related to isolation lactic acid bacteria, while the probiotic characteristics were not assessed. In addition, some nutrition value of these products was evaluated by researcher. The Kashk is produced under various names in different countries, Kishk (Lebanon, Syria), Zhum (Yemen), Kushuk (Iraq) (Tamime and O'connor 1995). These traditional dairy-based products are a good source of endogenous probiotics. The dried Kashk has been produced in a wide geographical region especially in rural parts of Iran for many years. Dried Kashk is produced from cow's and/or sheep's milk which is boiled and then cooled and inoculated with traditional yogurt made earlier as starter culture. The butter is isolated from sour yogurt and the remaining sour buttermilk is boiled followed by sieving by cloth bag. Finally, the thick whitish semi-solid part of buttermilk, which is sieved, is shaped in the form of conic or cubic balls and then sun-dried for 3–4 days (Iranmanesh et al. 2018). Ebrahimi et al. (2011) isolated *Lb. agilis* from Kashk which had the ability to assimilate cholesterol *in vitro* conditions. The Kishk samples containing *Lb. casei* (10^8–10^9 CFU mL^{-1}) showed antimicrobial activity against *E.coli* O157:H7 and reduced its count during storage at 4 °C after 20 days (Sadrizadeh et al. 2018).

Tarhana is another traditional fermented cereal dairy product produced in Turkey. It is made from cereal flours, yogurt, and different vegetables, and after fermentation it is sun dried and used as soap (Ozdemir et al. 2007). Sengun et al. (2009) isolated various LAB from Tarhana such as *S. thermophilus, L. fermentum, E. faecium, Ped. pentosaceus, Leu. pseudomesenteroides, Weissella cibaria, Lb. plantarum, Lb. delbrueckii* spp. *bulgaricus, Leu. citreum, Lb. paraplantarum*, and *Lb. casei.*

Kurut is traditional foods of Tibetan people. Kurut is prepared by natural fermentation of yak milk in a custom-made specially treated Tung made big jar, at ambient temperatures for 7–8 days. Kurut like kefir and koumiss have both alcohol and lactic acid. Kurut is almost known to all regions of Qinghai (Zhang et al. 2008). Sun et al. (2010) isolated *Lb. helveticus, Lb. suntoryeus, Lb. fermentum, Lb. plantarum*, and *Lb. delbrueckii* subsp. *bulgaricus* from Kurut. Furthermore, other cocci isolates were identified as *Lc. lactis* subsp. *lactis, Lc. lactis* subsp. *cremoris, Leu. lactis, Leu. mesenteroides* subsp. *mesenteroides*, and *S. thermophiles*. Luo et al. (2011) showed that some LAB strains isolated from Kurut had antimicrobial activity. Among the isolates, *Lb. casei, Lc. lactis* and *Leu lactis*, possessed bacteriocin-producing ability that could inhibit both *S. aureus* and *E. aerogenes*. The probiotic *Lb. helveticus* H9 isolated from Kurut was shown to produce antihypertensive peptides during milk fermentation (Chen et al. 2015).

11.9 Conclusions

It is evident that in last decades probiotic bacteria have been used in various dairy products due to their valuable health effects. Not only can probiotic be added to dairy products, but some traditional dairy products already are enriched with abundant of LAB species which show significant probiotic potential. These probiotic bacteria are able to add nutritious values to the milk and their products, as well as producing a number of metabolic products during fermentation they show therapeutic properties. Probiotic bacteria and their postbiotics (secretory metabolic compounds) can affect the quality of the dairy products by showing health benefits and might also improve the flavor, texture, and other sensory properties of the product. Furthermore, probiotic strains reveal other properties such as cholesterol lowering and immunomodulatory properties, anticancer and antioxidant activity. Besides, the consumption of fermented dairy products can simulate the immunity system. Further research studies are required that investigate the role of probiotic metabolite in some dairy products especially traditional dairy products that have intrinsic probiotic microbes. Consequently, these studies could lead to development of more nutritious and safe foods with proven therapeutic potential.

References

Aadinath T, Ghoshl PH, Amaladhas C (2017) Dried dairy products and their trends in the global market. In: Anandharamakrishnan C (ed) Handbook of drying for dairy products. Wiley-Blackwell, Hoboken, pp 15–18

Abadía-García L, Cardador A, del Campo STM, Arvízu SM, Castaño-Tostado E, Regalado-González C, Amaya-Llano SL (2013) Influence of probiotic strains added to cottage cheese on generation of potentially antioxidant peptides, anti-listerial activity, and survival of probiotic microorganisms in simulated gastrointestinal conditions. Int Dairy J 33(2):191–197

Abbas Z, Jafri W (1992) Yoghurt (dahi): a probiotic and therapeutic view. J Pak Med Assoc 42:221–224

Ahmadova A, Todorov SD, Choiset Y, Rabesona H, Zadi TM, Kuliyev A, Haertlé T (2013) Evaluation of antimicrobial activity, probiotic properties and safety of wild strain *Enterococcus faecium* AQ71 isolated from Azerbaijani Motal cheese. Food Control 30(2):631–641

Ahola AJ, Yli-Knuuttila H, Suomalainen T, Poussa T, Ahlström A, Meurman JH, Korpela R (2002) Short-term consumption of probiotic-containing cheese and its effect on dental caries risk factors. Arch Oral Biol 47(11):799–804

Alakomi HL, Skyttä E, Saarela M, Mattila-Sandholm T, Latva-Kala K, Helander IM (2000) Lactic acid permeabilizes gram-negative bacteria by disrupting the outer membrane. Appl Environ Microbiol 66:2001–2005

Ale EC, Perezlindo MJ, Pavón Y, Peralta GH, Costa S, Sabbag N, Binetti AG (2016) Technological, rheological and sensory characterizations of a yogurt containing an exopolysaccharide extract from *Lactobacillus fermentum* Lf2, a new food additive. Food Res Int 90:259–267

Aljewicz M, Cichosz G (2015) The effect of probiotic *Lactobacillus rhamnosus* HN001 on the in vitro availability of minerals from cheeses and cheese-like products. LWT-Food Sci Technol 60(2):841–847

Allgeyer LC, Miller MJ, Lee SY (2010) Sensory and microbiological quality of yogurt drinks with prebiotics and probiotics. J Dairy Sci 93(10):4471–4479

Ammor S, Tauveron G, Dufour E, Chevallier I (2006) Antibacterial activity of lactic acid bacteria against spoilage and pathogenic bacteria isolated from the same meat small-scale facility: 1—screening and characterization of the antibacterial compounds. Food Control 17(6):454–461

Ataie-Jafari A, Larijani B, Majd HA, Tahbaz F (2009) Cholesterol-lowering effect of probiotic yogurt in comparison with ordinary yogurt in mildly to moderately hypercholesterolemic subjects. Ann Nutr Metab 54(1):22–27

Avaiyarasi ND, Ravindran AD, Venkatesh P, Arul V (2016) In vitro selection, characterization and cytotoxic effect of bacteriocin of *Lactobacillus sakei* GM3 isolated from goat milk. Food Control 69:124–133

Ayyash MM, Sherkat F, Shah NP (2012) The effect of NaCl substitution with KCl on Akawi cheese: chemical composition, proteolysis, angiotensin-converting enzyme-inhibitory activity, probiotic survival, texture profile, and sensory properties. J Dairy Sci 95(9):4747–4759

Ayyash M, Al-Dhaheri AS, Al Mahadin S, Kizhakkayil J, Abushelaibi A (2018) In vitro investigation of anticancer, antihypertensive, antidiabetic, and antioxidant activities of camel milk fermented with camel milk probiotic: a comparative study with fermented bovine milk. J Dairy Sci 101(2):900–911

Baba AS, Najarian A, Shori AB, Lit KW, Keng GA (2014) Viability of lactic acid bacteria, antioxidant activity, and invitro inhibition of angiotensin-I-converting enzyme of *Lycium barbarum* yoghurt. Arab J Sci Eng 30(7):5355–5362

Balakrishnan G, Agrawal R (2014) Antioxidant activity and fatty acid profile of fermented milk prepared by *Pediococcus pentosaceus*. J Food Sci Technol 51(12):4138–4142

Barman S, Ghosh R, Mandal NC (2018) Production optimization of broad spectrum bacteriocin of three strains of *Lactococcus lactis* isolated from homemade buttermilk. Ann Agrarian Sci 16 (3):286–296

Baroutkoub A, Mehdi RZ, Beglarian R, Hassan J, Zahra S, Mohammad MS, Hadi EM (2010) Effects of probiotic yoghurt consumption on the serum cholesterol levels in hypercholestromic cases in Shiraz, Southern Iran. Sci Res Essays 5(16):2206–2209

Barros CP, Guimarães JT, Esmerino EA, Duarte MCK, Silva MC, Silva R, Cruz AG (2020) Paraprobiotics and postbiotics: concepts and potential applications in dairy products. Curr Opin Food Sci 32:1–8

Behare PV, Singh R, Nagpal R, Rao KH (2013) Exopolysaccharides producing *Lactobacillus fermentum* strain for enhancing rheological and sensory attributes of low-fat dahi. J Food Sci Technol 50(6):1228–1232

Belviso S, Giordano M, Dolci P, Zeppa G (2009) In vitro cholesterol-lowering activity of *Lactobacillus plantarum* and *Lactobacillus paracasei* strains isolated from the Italian Castelmagno PDO cheese. Dairy Sci Technol 89(2):169–176

Benkerroum N, Oubel H, Ben Mimoun LAMIAE (2002) Behavior of Listeria monocytogenes and *Staphylococcus aureus* in yogurt fermented with a bacteriocin-producing thermophilic starter. J Food Prot 65(5):799–805

Bourrie BC, Willing BP, Cotter PD (2016) The microbiota and health promoting characteristics of the fermented beverage kefir. Front Microbiol 7:647

Brunser O, Figueroa G, Gotteland M, Haschke-Becher E, Magliola C, Rochat F, Haschke F (2006) Effects of probiotic or prebiotic supplemented milk formulas on fecal microbiota composition of infants. Asia Pac J Clin Nutr 15(3)

Caballero-Franco C, Keller K, De Simone C, Chadee K (2007) The VSL#3 probiotic formula induces mucin gene expression and secretion in colonic epithelial cells. Am J Physiol Gastrointest Liver Physiol 292:G315–G322

Ceugniez A, Coucheney F, Jacques P, Daube G, Delcenserie V, Drider D (2017) Anti-Salmonella activity and probiotic trends of *Kluyveromyces marxianus* S-2-05 and *Kluyveromyces lactis* S-3-05 isolated from a French cheese, Tomme d'Orchies. Res Microbiol 168(6):575–582

Chaves-López C, Tofalo R, Serio A, Paparella A, Sacchetti G, Suzzi G (2012) Yeasts from Colombian Kumis as source of peptides with Angiotensin I converting enzyme (ACE) inhibitory activity in milk. Int J Food Microbiol 159(1):39–46

Chen Y, Zhang W, Sun Z, Meng B, Zhang H (2015) Complete genome sequence of *Lactobacillus helveticus* H9, a probiotic strain originated from kurut. J Biotechnol 194:37–38

Chen P, Liu L, Zhang X, Massounga Bora AF, Li X, Zhao M, Wang Y (2019) Antioxidant activity of Cheddar cheese during its ripening time and after simulated gastrointestinal digestion as affected by probiotic bacteria. Int J Food Prop 22(1):218–229

Crittenden RG, Martinez NR, Playne MJ (2003) Synthesis and utilisation of folate by yoghurt starter cultures and probiotic bacteria. Int J Food Microbiol 80(3):217–222

Da Silva PDL, de Fátima Bezerra M, dos Santos KMO, Correia RTP (2015) Potentially probiotic ice cream from goat's milk: characterization and cell viability during processing, storage and simulated gastrointestinal conditions. LWT-Food Sci Technol 62(1):452–457

de Almada CN, Almada CN, Martinez RC, Sant'Ana AS (2016) Paraprobiotics: evidences on their ability to modify biological responses, inactivation methods and perspectives on their application in foods. Trends Food Sci Technol 58:96–114

de Lima MDSF, de Souza KMS, Albuquerque WWC, Teixeira JAC, Cavalcanti MTH, Porto ALF (2017) *Saccharomyces cerevisiae* from Brazilian kefir-fermented milk: an in vitro evaluation of probiotic properties. Microb Pathog 110:670–677 de Oliveira MEG, Garcia EF, de Oliveira CEV, Gomes AMP, Pintado MME, Madureira ARMF, de Souza EL (2014) Addition of probiotic bacteria in a semi-hard goat cheese (coalho): survival to simulated gastrointestinal conditions and inhibitory effect against pathogenic bacteria. Food Res Int 64:241–247

do Espírito Santo AP, Perego P, Converti A, Oliveira MN (2011) Influence of food matrices on probiotic viability–a review focusing on the fruity bases. Trends Food Sci Technol 22 (7):377–385

Donkor ON, Stojanovska L, Ginn P, Ashton J, Vasiljevic T (2012) Germinated grains–sources of bioactive compounds. Food Chem 135(3):950–959

Driessen FM (1984) Modern trends in the manufacture of yoghurt. Int Dairy Found Bull 179:107–115

Ebrahimi MT, Ouweh AC, Hejazi MA, Jafari P (2011) Traditional Iranian dairy products: a source of potential probiotic lactobacilli. Afr J Microbiol Res 5(1):20–27

Ebringer L, Ferenčík M, Krajčovič J (2008) Beneficial health effects of milk and fermented dairy products. Folia Microbiol 53(5):378–394

Eid R, Jakee JE, Rashidy A, Asfour H, Omara S, Kandil MM, Seida AA (2016) Potential antimicrobial activities of probiotic *Lactobacillus* strains isolated from raw milk. J Probiotics Health 4:1e8

Eissa EA, Babiker EE, Yagoub AEA (2011) Physicochemical, microbiological and sensory properties of *Sudanese yoghurt* (zabadi) made from goat's milk. Anim Prod Sci 51(1):53–59

Ejtahed HS, Mohtadi-Nia J, Homayouni-Rad A, Niafar M, Asghari-Jafarabadi M, Mofid V, Akbarian-Moghari A (2011) Effect of probiotic yogurt containing *Lactobacillus acidophilus* and *Bifidobacterium lactis* on lipid profile in individuals with type 2 diabetes mellitus. J Dairy Sci 94(7):3288–3294

Ejtahed HS, Mohtadi-Nia J, Homayouni-Rad A, Niafar M, Asghari-Jafarabadi M, Mofid V (2012) Probiotic yogurt improves antioxidant status in type 2 diabetic patients. Nutrition 28 (5):539–543

FAO/WHO (2001) Evaluation of health and nutritional properties of probiotics in food including powder Milk with live lactic acid bacteria, report of a joint FAO/WHO expert consultation, food and agriculture Organization of the United. Nations, Rome

FAO/WHO (2002) Guidelines for the evaluation of probiotics in food, report of a joint FAO/WHO working group on drafting guidelines for the evaluation of probiotics in food, London Ontario in Canada. Food and Agriculture Organization of the United Nations, Rome

García-Burgos M, Moreno-Fernández J, Alférez MJ, Díaz-Castro J, López-Aliaga I (2020) New perspectives in fermented dairy products and their health relevance. J Funct Foods 72:104059

Gardiner GE, Bouchier P, O'Sullivan E, Kelly J, Collins JK, Fitzgerald G, Ross RP, Stanton C (2002) A spray dried culture for probiotic cheese manufacture. Int Dairy J 12:749–756

Gaudier E, Michel C, Segain JP, Cherbut C, Hoebler C (2005) The VSL# 3 probiotic mixture modifies microflora but does not heal chronic dextran-sodium sulfate-induced colitis or reinforce the mucus barrier in mice. J Nutr 135:2753–2761

Gilliland SE (1989) Acidophilus milk products: a review of potential benefits to consumers. J Dairy Sci 72(10):2483–2494

Gilliland SE (2001) Probiotics and prebiotics. In: Marth EH, Steele JL (eds) Applied dairy microbiology, 2nd edn. Marcel Dekker, New York, pp 327–344

Gogineni VK, Morrow LE, Malesker MA (2013) Probiotics: mechanisms of action and clinical applications. J Prob Health 1:101

Granato D, Branco GF, Cruz AG, Faria JDAF, Shah NP (2010) Probiotic dairy products as functional foods. Compr Rev Food Sci Food Saf 9(5):455–470

Granier A, Goulet O, Hoarau C (2013) Fermentation products: immunological effects on human and animal models. Pediatr Res 74(2):238–244

Guimarães JT, Balthazar CF, Scudino H, Pimentel TC, Esmerino EA, Ashokkumar M, Cruz AG (2019) High-intensity ultrasound: a novel technology for the development of probiotic and prebiotic dairy products. Ultrason Sonochem 57:12–21

Gursoy O, Seckin AK, Kinik O, Karaman AD (2012) The effect of using different probiotic cultures on conjugated linoleic acid (CLA) concentration and fatty acid composition of white pickle cheese. Int J Food Sci Nutr 63(5):610–615

Guzel-Seydim ZB, Sezgin E, Seydim AC (2005) Influences of exopolysaccharide producing cultures on the quality of plain set type yogurt. Food Control 16(3):205–209

Hekmat S, Reid G (2006) Sensory properties of probiotic yogurt is comparable to standard yogurt. Nutr Res 26(4):163–166

Hertzler SR, Clancy SM (2003) Kefir improves lactose digestion and tolerance in adults with lactose maldigestion. J Am Diet Assoc 103(5):582–587

Hess SJ, Roberts RF, Ziegler GR (1997) Rheological properties of nonfat yogurt stabilized using *Lactobacillus delbrueckii* ssp. *bulgaricus* producing exopolysaccharide or using commercial stabilizer systems. J Dairy Sci 80(2):252–263

Iranmanesh M, Ezzatpanah H, Mojgani N, Torshizi MAK, Aminafshar M, Maohamadi M (2012) Isolation of lactic acid bacteria from ewe milk, traditional yoghurt and sour buttermilk in Iran. Eur J Nutr Food Saf:79–92

Iranmanesh M, Ezzatpanah H, Mojgani N (2014) Antibacterial activity and cholesterol assimilation of lactic acid bacteria isolated from traditional Iranian dairy products. LWT-Food Sci Technol 58(2):355–359

Iranmanesh M, Ezzatpanah H, Mojgani N, Torshizi M (2015) Characterization and kinetics of growth of bacteriocin like substance produced by lactic acid bacteria isolated from ewe milk and traditional sour buttermilk in Iran. J Food Process Technol 6(12):1–9

Iranmanesh M, Ezzatpanah H, Akbari-Adergani B, Karimi Torshizi MA (2018) SPME/GC-MS characterization of volatile compounds of Iranian traditional dried Kashk. Int J Food Prop 21 (1):1067–1079

Jabbari V, Khiabani MS, Mokarram RR, Hassanzadeh AM, Ahmadi E, Gharenaghadeh S, Kafil HS (2017) *Lactobacillus plantarum* as a probiotic potential from kouzeh cheese (traditional Iranian cheese) and its antimicrobial activity. Probiotics Antimicrob Proteins 9(2):189–193

Jain S, Yadav H, Sinha PR (2009) Probiotic dahi containing *Lactobacillus casei* protects against *Salmonella enteritidis* infection and modulates immune response in mice. J Med Food 12 (3):576–583

Kahala M, Mäki M, Lehtovaara A, Tapanainen JM, Katiska R, Juuruskorpi M, Joutsjoki V (2008) Characterization of starter lactic acid bacteria from the Finnish fermented milk product viili. J Appl Microbiol 105(6):1929–1938

Kankaanpää PE, Yang B, Kallio HP, Isolauri E, Salminen SJ (2002) Influence of probiotic supplemented infant formula on composition of plasma lipids in atopic infants. J Nutr Biochem 13(6):364–369

Kaushal D, Kansal VK (2012) Probiotic dahi containing *Lactobacillus acidophilus* and *Bifidobacterium bifidum* alleviates age-inflicted oxidative stress and improves expression of biomarkers of ageing in mice. Mol Biol Rep 39(2):1791–1799

Kelsall BL (2008) Innate and adaptive mechanisms to control of pathological intestinal inflammation. J Pathol 214:242–259

Kollath W (1953) Nutrition and the tooth system; general review with special reference to vitamins. Dtsch Zahnärztl Z 8:7–16

Kostantinov SR, Kuipers EJ, Peppelenbosch MP (2013) Functional genomic analyses of the gut microbiota for CRC screening. Nat Rev Gastroenterol Hepatol 10:741–745

Kumar BV, Vijayendra SVN, Reddy OVS (2015) Trends in dairy and non-dairy probiotic products-a review. J Food Sci Technol 52(10):6112–6124

Kumura H, Tanoue Y, Tsukahara M, Tanaka T, Shimazaki K (2004) Screening of dairy yeast strains for probiotic applications. J Dairy Sci 87(12):4050–4056

Laiño JE, LeBlanc JG, Savoy de Giori G (2012) Production of natural folates by lactic acid bacteria starter cultures isolated from artisanal Argentinean yogurts. Can J Microbiol 58(5):581–588

Liévin-Le Moal V, Servin AL (2006) The front line of enteric host defense against unwelcome intrusion of harmful microorganisms: mucins, antimicrobial peptides, and microbiota. Clin Microbiol Rev 19:315–337

Lollo PCB, Cruz AG, Morato PN, Moura CS, Carvalho-Silva LB, Oliveira CAFD, Amaya-Farfan J (2012) Probiotic cheese attenuates exercise-induced immune suppression in Wistar rats. J Dairy Sci 95(7):3549–3558

Lollo PC, Morato PN, Moura CS, Almada CN, Felicio TL, Esmerino EA, Silva MC (2015) Hypertension parameters are attenuated by the continuous consumption of probiotic Minas cheese. Food Res Int 76:611–617

Lourens-Hattingh A, Viljoen BC (2001) Yogurt as probiotic carrier food. Int Dairy J 11(1-2):1–17

Luo F, Feng S, Sun Q, Xiang W, Zhao J, Zhang J, Yang Z (2011) Screening for bacteriocin-producing lactic acid bacteria from kurut, a traditional naturally-fermented yak milk from Qinghai–Tibet plateau. Food Control 22(1):50–53

Maccaferri S, Klinder A, Brigidi P, Cavina P, Costabile A (2012) Potential probiotic *Kluyveromyces marxianus* B0399 modulates the immune response in Caco-2 cells and peripheral blood mononuclear cells and impacts the human gut microbiota in an in vitro colonic model system. Appl Environ Microbiol 78(4):956–964

Mahmood T, Masud T, Sohail A (2014) Some probiotic and antibacterial properties of *Lactobacillus acidophilus* cultured from dahi a native milk product. Int J Food Sci Nutr 65(5):582–588

Makino S, Sato A, Goto A, Nakamura M, Ogawa M, Chiba Y, Asami Y (2016) Enhanced natural killer cell activation by exopolysaccharides derived from yogurt fermented with *Lactobacillus delbrueckii* ssp. *bulgaricus* OLL1073R-1. J Dairy Sci 99(2):915–923

Malcata FX, Gomes AM, Pintado ME (2005) Functional dairy foods – an overview. Egypt J Dairy Sci 33:1–12

Manzo N, Pizzolongo F, Montefusco I, Aponte M, Blaiotta G, Romano R (2015) The effects of probiotics and prebiotics on the fatty acid profile and conjugated linoleic acid content of fermented cow milk. Int J Food Sci Nutr 66(3):254–259

Marsili RT (1981) Monitoring bacterial metabolites in cultured buttermilk by high performance liquid chromatography and headspace gas chromatography. J Chromatogr Sci 19(9):451–456

Matsuguchi T, Takagi A, Matsuzaki T, Nagaoka M, Ishikawa K, Yokokura T, Yoshikai Y (2003) Lipoteichoic acids from *Lactobacillus* strains elicit strong tumor necrosis factor alphainducing activities in macrophages through toll-like receptor 2. Clin Diagn Lab Immunol 10(2):259–266

Mattar AF, Teitelbaum DH, Drongowski RA, Yongyi F, Harmon CM, Coran AG (2002) Probiotics up-regulate MUC-2 mucin gene expression in a Caco-2 cell-culture model. Pediatr Surg Int 18:586–590

Mitra S, Chakrabartty PK, Biswas SR (2007) Production of nisin Z by *Lactococcus lactis* isolated from Dahi. Appl Biochem Biotechnol 143(1):41–53

Moeller C, de Vrese M (2004) Review: probiotic effects of selected acid bacteria. Milchwissenschaft 59:11–12

Mollet B, Rowland I (2002) Functional foods: at the frontier between food and pharma. Curr Opin Biotechnol 5(13):483–485

Mondel M, Schroeder BO, Zimmermann K, Huber H, Nuding S, Beisner J, Fellermann K, Stange EF, Wehkamp J (2009) Probiotic *Escherichia coli* treatment mediates antimicrobial human betadefensin synthesis and fecal excretion in humans. Mucosal Immunol 2:166–172

Mushtaq M, Gani A, Masoodi FA, Ahmad M (2016) Himalayan cheese (Kalari/Kradi)–effect of different probiotic strains on oxidative stability, microbiological, sensory and nutraceutical properties during storage. LWT-Food Sci Technol 67:74–81

Nakamura F, Ishida Y, Sawada D, Ashida N, Sugawara T, Sakai M, Fujiwara S (2016) Fragmented lactic acid bacteria cells activate peroxisome proliferator-activated receptors and ameliorate dyslipidemia in obese mice. J Agric Food Chem 64:2549–2559

Oelschlaeger TA (2010) Mechanisms of probiotic actions–a review. Int J Med Microbiol 300 (1):57–62

Ohashi K, Bohgaki T, Shibuya H (2000) Antihypertensive substance in the leaves of kumis kucing (Orthosiphon aristatus) in Java Island. Yakugaku zasshi. J Pharmaceut Soc Jpn 120(5):474–482

Osorio JA, Ramirez C, Novoa CF, Gutiérrez LF (2011) Conjugated linoleic acid, fatty acid profile and process properties in kumis-fermented milk consumed in Colombia. Vitae 18(2):144–152

Ozdemir S, Gocmen D, Yildirim Kumral A (2007) A traditional Turkish fermented cereal food: Tarhana. Food Rev Intl 23(2):107–121

Parmjit S (2011) Fermented dairy products: starter cultures and potential nutritional benefits. Food and Nutrition Sciences

Patel A, Prajapati JB, Holst O, Ljungh A (2014) Determining probiotic potential of exopolysaccharide producing lactic acid bacteria isolated from vegetables and traditional Indian fermented food products. Food Biosci 5:27–33

Penner R, Fedorak RN, Madsen KL (2005) Probiotics and nutraceuticals: non-medicinal treatments of gastrointestinal diseases. Curr Opin Pharmacol 5:596–603

Quirós A, Ramos M, Muguerza B, Delgado MA, Miguel M, Aleixandre A, Recio I (2007) Identification of novel antihypertensive peptides in milk fermented with *Enterococcus faecalis*. Int Dairy J 17(1):33–41

Ramchandran L, Shah NP (2010) Characterization of functional, biochemical and textural properties of synbiotic low-fat yogurts during refrigerated storage. LWT-Food Sci Technol 43 (5):819–827

Robinson RK, Tamime AY (1990) Microbiology of fermented milks. In: Robinson RK (ed) Dairy microbiology, vol 2, 2nd edn. Elsevier Applied Science, London, pp 291–343

Rodrigues D, Rocha-Santos TA, Gomes AM, Goodfellow BJ, Freitas AC (2012) Lipolysis in probiotic and synbiotic cheese: the influence of probiotic bacteria, prebiotic compounds and ripening time on free fatty acid profiles. Food Chem 131(4):1414–1421

Saad N, Delattre C, Urdaci M, Schmitter JM, Bressollier P (2013) An overview of the last advances in probiotic and prebiotic field. LWT Food Sci Technol 50:1–16

Saavedra L, Taranto MP, Sesma F, de Valdez GF (2003) Homemade traditional cheeses for the isolation of probiotic *Enterococcus faecium* strains. Int J Food Microbiol 88(2-3):241–245

Saavedra JM, Abi-Hanna A, Moore N, Yolken RH (2004) Long-term consumption of infant formulas containing live probiotic bacteria: tolerance and safety. Am J Clin Nutr 79(2):261–267

Sadrizadeh N, Khezri S, Dehghan P, Mahmoudi R (2018) Antibacterial effect of *Teucrium polium* essential oil and *Lactobacillus casei* probiotic on *Escherichia coli* O157: H7 in Kishk. Appl Food Biotechnol 5(3):131–140

Sah BNP, Vasiljevic T, McKechnie S, Donkor ON (2014) Effect of probiotics on antioxidant and antimutagenic activities of crude peptide extract from yogurt. Food Chem 156:264–270

Sánchez B, Delgado S, Blanco-Míguez A, Lourenço A, Gueimonde M, Margolles A (2017) Probiotics, gut microbiota, and their influence on host health and disease. Mol Nutr Food Res 61(1):1600240

Sanders ME (2009) How do we know when something called "probiotic" is really a probiotic? A guideline for consumers and health care professionals. Funct Food Rev 1(1):3–12

Sankar NR, Priyanka VD, Reddy PS, Rajanikanth P, Kumar VK, Indira M (2012) Purification and characterization of bacteriocin produced by *Lactobacillus plantarum* isolated from cow milk. Int J Microbiol Res 3(2):133–137

Santiago Lopez L, Hernandez-Mendoza A, Garcia HS, Mata Haro V, Vallejo C, Gonzalez Cordova B (2015) The effects of consuming probiotic fermented milk on the immune system: a review of scientific evidence. Int J Dairy Technol 68:153–165

Sawada H, Furushiro M, Hirai K, Motoike M, Watanabe T, Yokokura T (1990) Purification and characterization of an antihypertensive compound from *Lactobacillus casei*. Agric Biol Chem 54(12):3211–3219

Schlee M, Harder J, Köten B, Stange EF, Wehkamp J, Fellermann K (2008) Probiotic lactobacilli and VSL#3 induce enterocyte betadefensin 2. Clin Exp Immunol 151:528–535

Sengun IY, Nielsen DS, Karapinar M, Jakobsen M (2009) Identification of lactic acid bacteria isolated from Tarhana, a traditional Turkish fermented food. Int J Food Microbiol 135 (2):105–111

Shah NP (2007) Functional cultures and health benefits. Int Dairy J 17:262–1277

Sharma M, Devi M (2014) Probiotics: a comprehensive approach toward health foods. Crit Rev Food Sci Nutr 54:537–552

Sharma G, Ghosh BC (2006) Probiotic dairy foods and prebiotics for health benefit. Indian Food Ind 25(1):68–73

Sharma R, Kapila R, Kapasiya M, Saliganti V, Dass G, Kapila S (2014) Dietary supplementation of milk fermented with probiotic *Lactobacillus fermentum* enhances systemic immune response and antioxidant capacity in aging mice. Nutr Res 34(11):968–981

Shiby VK, Mishra HN (2013) Fermented milks and milk products as functional foods - a review. Crit Rev Food Sci Nutr 53:482–496

Shin HS, Park SY, Lee DK, Kim SA, An HM, Kim JR, Kim JM, Cha MG, Lee SW, Kim KJ, Lee KO, Ha NJ (2010) Hypocholesterolemic effect of sonication killed Bifidobacterium longum isolated from healthy adult Koreans in high cholesterol fed rats. Arch Pharmacal Res 33:1425–1431

Singh R, Damle SG, Chawla A (2011) Salivary mutans streptococci and lactobacilli modulations in young children on consumption of probiotic ice-cream containing *Bifidobacterium lactis* Bb12 and *Lactobacillus acidophilus* La5. Acta Odontol Scand 69(6):389–394

Sinha PR, Sinha RN (2000) Importance of good quality dahi in food. Indian Dairyman 41:45–47

Sobrino-López A, Martín-Belloso O (2008) Use of nisin and other bacteriocins for preservation of dairy products. Int Dairy J 18(4):329–343

Sperry MF, Silva HL, Balthazar CF, Esmerino EA, Verruck S, Prudencio ES, Rocha RS (2018) Probiotic Minas Frescal cheese added with *L. casei* 01: physicochemical and bioactivity characterization and effects on hematological/biochemical parameters of hypertensive overweighted women–a randomized double-blind pilot trial. J Funct Foods 45:435–443

Stanton C, Ross RP, Fitzgerald GF, Van Sinderen D (2005) Fermented functional foods based on probiotics and their biogenic metabolites. Curr Opin Biotechnol 16(2):198–203

Sun Z, Liu W, Gao W, Yang M, Zhang J, Wu L, Zhang H (2010) Identification and characterization of the dominant lactic acid bacteria from kurut: the naturally fermented yak milk in Qinghai, China. J Gen Appl Microbiol 56(1):1–10

Surono IS, Hosono A (2011) Fermented milks types and standards of identity

Tamime AY (2002) Fermented milks: a historical food with modern applications–a review. Eur J Clin Nutr 56(4):S2–S15

Tamime AY, O'connor TP (1995) Kishk—a dried fermented milk/cereal mixture. Int Dairy J 5 (2):109–128

Tamime AY, Thomas LV (2018) Probiotic dairy products (Society of Dairy Technology). Wiley Blackwell, Hoboken, pp 99–110

Tamime AY, Marshall VM, Robinson RK (1995) Microbiological and technological aspects of milks fermented by bifidobacteria. J Dairy Res 62(1):151–187

Taverniti V, Guglielmetti S (2011) The immunomodulatory properties of probiotic microorganisms beyond their viability (ghost probiotics: proposal of paraprobiotic concept). Genes Nutr 6 (3):261

Tsilingiri K, Rescigno M (2013) Postbiotics: what else? Benefic Microbes 4(1):101–107

Tulini FL, Winkelströter LK, De Martinis EC (2013) Identification and evaluation of the probiotic potential of *Lactobacillus paraplantarum* FT259, a bacteriocinogenic strain isolated from Brazilian semi-hard artisanal cheese. Anaerobe 22:57–63

Tunick MH, van Hekken DL (2015) Dairy products and health: recent insights. J Agric Food Chem 63:9381–9388

Urso ML, Clarkson PM (2003) Oxidative stress, exercise, and antioxidant supplementation. Toxicology 189(1-2):41–54

Vijayendra SVN, Palanivel G, Mahadevamma S, Tharanathan RN (2008) Physico-chemical characterization of an exopolysaccharide produced by a non-ropy strain of Leuconostoc sp. CFR 2181 isolated from dahi, an Indian traditional lactic fermented milk product. Carbohydr Polym 72(2):300–307

Vlieger AM, Robroch A, van Buuren S, Kiers J, Rijkers G, Benninga MA, te Biesebeke R (2009) Tolerance and safety of *Lactobacillus paracasei* ssp. *paracasei* in combination with *Bifidobacterium animalis* ssp. *lactis* in a prebiotic-containing infant formula: a randomised controlled trial. Br J Nutr 102(6):869–875

Wang SY, Chen HC, Liu JR, Lin YC, Chen MJ (2008) Identification of yeasts and evaluation of their distribution in Taiwanese kefir and viili starters. J Dairy Sci 91(10):3798–3805

Weizman Z, Asli G, Alsheikh A (2005) Effect of a probiotic infant formula on infections in child care centers: comparison of two probiotic agents. Pediatrics 115(1):5–9

Yadav H, Jain S, Sinha PR (2007a) Antidiabetic effect of probiotic dahi containing *Lactobacillus acidophilus* and *Lactobacillus casei* in high fructose fed rats. Nutrition 23(1):62–68

Yadav H, Jain S, Sinha PR (2007b) Production of free fatty acids and conjugated linoleic acid in probiotic dahi containing *Lactobacillus acidophilus* and *Lactobacillus casei* during fermentation and storage. Int Dairy J 17(8):1006–1010

Yan F, Polk DB (2006) Probiotics as functional food in the treatment of diarrhea. Curr Opin Clin Nutr Metab Care 9:717–721

Young J (2000) Functional foods and the European consumer. Special Publ Roy Soc Chem 248:75–81

Zaeim D, Soleimanian-Zad S, Sheikh-Zeinoddin M (2014) Identification and partial characterization of a bacteriocin-like inhibitory substance (BLIS) from *Lb. bulgaricus* K41 isolated from indigenous yogurts. J Food Sci 79(1):M67–M73

Zhang H, Xu J, Wang J, Sun T, Li H, Guo M (2008) A survey on chemical and microbiological composition of kurut, naturally fermented yak milk from Qinghai in China. Food Control 19 (6):578–586

Zhang L, Zhang X, Liu C, Li C, Li S, Li T, Yang Z (2013) Manufacture of Cheddar cheese using probiotic *Lactobacillus plantarum* K25 and its cholesterol-lowering effects in a mice model. World J Microbiol Biotechnol 29(1):127–135

Zhang H, Ren W, Guo Q, Xiong Z, Wang G, Xia Y, Ai L (2018) Characterization of a yogurt-quality improving exopolysaccharide from *Streptococcus thermophilus* AR333. Food Hydrocoll 81:220–228

Encountering the Antibiotic Resistance by Bioactive Components and Therapies: Probiotics, Phytochemicals and Phages 12

Sheikh Ajaz Rasool, Muhammad Salman Rasool, and Munazza Ajaz

Abstract

Improper antimicrobial practices expedite the evolution of antibiotic-resistant and super-drug-resistant bacteria, and result in varied and elevated antibiotic resistance throughout the world. The dominant problem with antibiotic therapy is the antibiotic resistance acquisition in bacteria. Therefore, it is important to use non-antibiotic agents which could prevent the microbial propagation and control their virulence. The agents like probiotics, phytochemicals and bacteriophages have been found to tackle drug-resistant bacteria. The application of probiotics in medicines, agriculture and food industry is becoming influential to contain drug-resistant and virulent bacteria. The probiotic bacteria (Lactobacilli, Bifidobacteria, etc.) directly and indirectly by their products (bacteriocins, organic acids, short-chain fatty acids, polysaccharides, etc.) can counter or evade the entero-virulent, pathogenic and drug-resistant bacteria such as *Salmonella typhi*, *Clostridium difficile*, MRSA, Carbapenem-resistant representatives of enterobacteriaceae, *H. pylori*, *E. coli* O157:H7, etc. Intestinal probiotic bacteria (Lactobacilli) can also help to reduce diarrhoea, post-antibiotic therapy complications and inflammatory bowel disease. Phytochemicals offer strong antimicrobial action against virulent and resistant bacteria. These substances unaccompanied or accompanied with antibiotics can enhance antibacterial effect. Several plant procured compounds (Alkaloids, Phenolics, Cumarins, Terpenes

S. A. Rasool (✉)
Jinnah University for Women, Karachi, Pakistan

M. S. Rasool
D. J. Sindh Govt. Science College, Karachi, Pakistan

M. Ajaz
Federal Urdu University for Arts, Science and Technology, Karachi, Pakistan

N. Mojgani, M. Dadar (eds.), *Probiotic Bacteria and Postbiotic Metabolites: Role in Animal and Human Health*, Microorganisms for Sustainability 2,
https://doi.org/10.1007/978-981-16-0223-8_12

and Sulphur-compounds) have been observed to carry antimicrobial effects with a vast variety of mechanisms. Bacteriophages and the related enzymes (endolysins and virion-associated peptidoglycan hydrolases) having antimicrobial effect are also considered as therapeutic candidates to encounter drug-resistant bugs. Based on the literature review, probiotics, plant procured compounds and phages in solo or in combo with antibiotics can be the favourable and valuable therapeutic options.

Keywords

Antibiotics · Drug resistance · Probiotics · Bacteriophages · Phytochemicals

12.1 Introduction

Extraordinary expansion in the occurrence and predominance of antibiotic resistance has been witnessed after the administration and extensive application of antibiotics. This resistance is considered to have prevailed much before antibiotic consumption by humans (Broaders et al. 2013; Rasool et al. 2019a). Several ecological interaction-based studies have proved that microorganisms produce antimicrobials in microbial communities to ensure their survival (Samuels et al. 2013; Cawoy et al. 2014; Téllez et al. 2015). During natural selection events, the microbes which are able to accommodate and withstand antimicrobial stress can produce resistant microbial populations; therefore, antibiotic resistance is recognized as an innate aspect (Sherpa et al. 2015; Imperial and Ibana 2016; Rasool et al. 2019b). After knowing the antibiosis phenomena, antimicrobials were welcomed in chemotherapy of infectious diseases (Scanlon et al. 2014; Sherpa et al. 2015). Moreover, antibiotics are employed in agriculture and livestock (animals) for the enhancement of their yield and growth ratio, and treatment of animals' infections (Allen and Stanton 2014; Xiao et al. 2015). Nevertheless, unchecked and unrestrained antimicrobial practices accelerate the evolution of resistant and super-resistant bugs and lead to varied and heightened antibiotic resistance resulted in response to discriminative antibiotic stress on microbes (Rosander et al. 2008; Verraes et al. 2013; Hu et al. 2014; Card et al. 2015). The ratio of emerging resistant bugs is more rapid than the revelation of novel antibiotics (Sherpa et al. 2015). There are about 23,317 genes responsible for targeting 249 antibiotics (Hu et al. 2013). The microbial ecology considerably influences the existence of antibiotic resistance genes (Gibson et al. 2015). Further, vertical transmission of resistance genes through mobile genetic elements (MGEs) supports the prosperity of resistant microorganisms in the presence of antibiotics (Penders et al. 2013; Fouhy et al. 2014). Antibiotic resistance genes (with the help of plasmids) can travel between virulent and normal bacterial flora of humans (Broaders et al. 2013; Imperial and Ibana 2016).

The major issue of antibiotic therapy is the development and acquisition of antibiotic resistance in bacteria through various systems; drug inactivation, antibiotic target alteration, advancement of outer membrane permeability and efflux pumps,

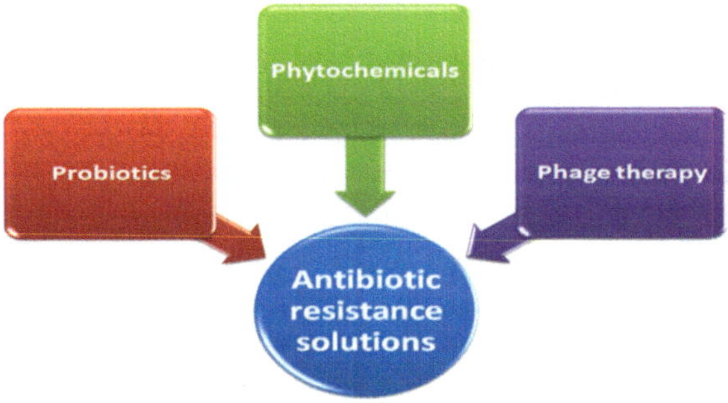

Fig. 12.1 Possible solutions for antibiotic resistance

circumvention of biochemical pathways, mutagenesis and horizontal gene transmission (Walsh 2000; Wright 2005; Naderi et al. 2014). Since, it is almost impossible to revert the resistance phenomena, the prevention of pathogenic microbial propagation, microbial killing, and subverting their virulence factors may be targeted by non-antibiotic agents; probiotics, phytochemicals and bacteriophages (Fig. 12.1). Due to growing antibiotic resistance dilemma, there is an urgent demand to recognize the performance of probiotics and normal microbiota or commensal microorganisms in subverting antimicrobial resistance and virulence (Liévin-Le Moal and Servin 2014).

12.2 The Probiotics: Significance in Agriculture and Human Health

Apart from antibiotic resistance issue, symbiotic tie up of microorganisms with gastrointestinal tract (GIT) of livestock animals facilitates fermentation. Fermentation of complicated polysaccharides by GIT normal microbial flora provides 70% energy required by farm animals and about 30% energy for monogastric herbivores (Téllez et al. 2015). Several researches have indicated that probiotics can be the alternatives of antibiotics in growth enhancement and restricting diseases in farm animals (Muñoz-Atienza et al. 2013; Téllez et al. 2015). The probiotics prevent turkeys and chickens from infections of *Salmonella* spp. by colonizing GIT as normal microflora. Moreover, probiotics can lower the severity of diarrhoea of unknown cause in turkeys. Extensive trials of suitable probiotics use in turkeys and chickens witnessed the decrease in complete production cost and boosted the performance. Livestock animals excrete *Escherichia coli* O157:H7 in their faeces which can be transmitted to humans (and produce bloody diarrhoea). The use of various bacterial probiotic blends can diminish the faecal discharge of *Escherichia coli* O157:H7 by sheep and cattle, and hence can reduce the chances of human infection (Téllez et al. 2015). Antimicrobial action of probiotics has also been determined against fish pathogenic Gram positive and Gram negative bacteria (Muñoz-Atienza et al. 2013).

Probiotics are prominent for the improvement of human health in general. Probiotic bacteria are being incorporated in several commercial foods to maintain GIT microbial flora (Songisepp et al. 2012). They are involved in various important functions like controlling lipid storage, inflammatory bowel disease (IBD), GIT epithelial growth, and inflammatory reactions (Collins and Gibson 1999; Ventura et al. 2009). Probiotics are medically verified to adjust baby's GIT normal flora upset after antibiotic therapy (Collado et al. 2012). Antibiotic treatment is harmful for gut normal flora of neonates resulting in killing the gut flora leading to diarrhoea (Varankovich et al. 2015). The application of probiotics before and after birth of baby has been found to avoid the onset of hypersensitivities (e.g. asthma) and GIT infections (Luoto et al. 2010). In adults, regular oral probiotics consumption has shown to cure intermittent diarrhoea inflicted by *Clostridium difficile* which is offered by post extended antibiotic therapy. Such antibiotic treatment approach kills GIT normal flora, which can enhance the *Clostridium difficile* growth (Ursell et al. 2013; Varankovich et al. 2015).

The application of probiotics in medicines and agriculture is becoming common since the evolution of drug-resistant microbial strains (Muñoz-Atienza et al. 2013; D'Orazio et al. 2015; Téllez et al. 2015; Varankovich et al. 2015). Reducing the antibiotics usage and employment of probiotics instead of antibiotics may assist in lowering the proportion of flourishing resistant microbial strains (Muñoz-Atienza et al. 2013; Varankovich et al. 2015). Probiotics application is the adoption of active advantageous microorganisms to get required consequences, for example, inhibiting ailments and promoting health in living beings (Collins and Gibson 1999). Prior to employment of probiotics, it is necessary to check probiotics for the presence of antibiotic resistance genes (Sanders et al. 2010). An important step in screening of antibiotic resistance is to differentiate between acquired or transmissible resistance and inherent resistance of probiotic bacteria (Chang et al. 2009; Hammad and Shimamoto 2010).

12.3 Probiotic Bacteria, Their Anti-Pathogen Products and Potentials

Probiotic bacteria are well established because they can control pathogens by producing organic acids and lowering the clustering and production of adhesins. Further, they are normal microbial flora of humans and animals and produce bacteriocins (Reid 1999; Ennahar et al. 2000; McAuliffe et al. 2001; Anas et al. 2008). They promote gut microbial habitat by attachment with mucous of intestine, and therefore prohibit the adherence of pathogenic microorganisms and also challenge pathogens for nutrient acquisition and provoke immunity in intestine. They are hostile to a wide variety of urinary and GI tract associated pathogenic bacteria (Gilliland and Walker 1990; Hutt et al. 2006). Interestingly, repression of virulence factor genes in *E. coli* O157:H7 is carried out by biologically active compounds released by probiotic bacteria. Probiotics may also reduce the attachment of *E. coli* O127:H6 and *E. coli* O157:H7 to the epithelial cells. Cell-free supernatants of

B. longum, *L. plantarum*, *L. rhamnosus* and *L. helveticus* were observed to possess anti-biofilm effect against MDR-*E. coli* (Abdelhamid et al. 2018).

The most familiar probiotics reside in the lactic acid bacteria (LAB) group, which involves various species of Lactobacilli and Enteroccci (Tompkins et al. 2008; Chang et al. 2009; Nueno-Palop and Narbad 2011; Songisepp et al. 2012; Gueimonde et al. 2013; Devi et al. 2015; Senan et al. 2015). In addition to other health benefits in animals and humans, lactobacilli can destroy *Helicobacter pylori* (responsible for peptic ulcer, gastric cancer and gastritis) (Téllez et al. 2015; Varankovich et al. 2015). Enterococci as probiotic quickly heal diarrhoea (Vankerckhoven et al. 2008; Varankovich et al. 2015). According to studies, a few lactic acid bacterial species have inherent resistance to beta-lactams, teicoplanin, bacitracin, vancomycin and kanamycin. Such resistance is advantageous when combine therapy (probiotic with antibiotic) is required (Hammad and Shimamoto 2010; Varankovich et al. 2015).

The dominant probiotic, bifidobacteria are crucial component of the human and animal GIT normal microbial flora. They are capable of preventing the attachment of *C. difficile*, enterotoxigenic and enteroptahogenic *E. coli* with the cells of gut epithelia. A combo of bifidobacteria and lactobacilli is responsible to reduce the adverse effects of anti-*H. pylori* therapy. Furthermore, some strains of bifidobacteria down-regulate the genes in human epithelia, inflicted by *H. pylori*, and mitigate inflammatory bowel disease (IBD) and diarrhoea (Varankovich et al. 2015). The other frequently employed probiotics in medicines, agriculture and food involve Streptococcus, Lactobacillus, Bacillus, Lactococcus, Enterococcus, Bifidobacterium, Pseudomonas, Pediococcus, Bacteroides, Trichoderma, yeast and Aspergillus, etc. Many food supplements have Lactobacillus (*L. rhamnosus, L. acidophilus, L. plantarum, L. farciminis, L. casei*), Bacillus (*B. subtilis, B. licheniformis, B. cereus var. toyoi*), *Escherichia coli* Nissle 1917, Enterococcus (*E. faecium*), Streptococcus (*S. infantarius*), Pedicoccus (*P. acidilactici*) and a few fungi like Kluyveromyces, *Saccharomyces cerevisiae* and *Saccharomyces boulardii* (Anadon et al. 2005; Cheng et al. 2014).

Important characteristics: avirulent, acid and bile stable, adherence and propagation in GIT, antimicrobial activity and production of valuable metabolites are needed to be a good probiotic (Fijan 2016). Many research trials have shown antibacterial potential of various probiotic strains against virulent and drug-resistant strains of *P. aeruginosa*, *K. pneumoniae*, *E. coli*, *Salmonella* spp., *E. fecalis* and *S. aureus* and also have antifungal action on *C. albicans* (Manzoor et al. 2016; Prabhurajeshwar and Chandrakanth 2017). Both in vivo and in vitro *L. acidophilus* (isolate of human stool) was found antagonistic against *H. pylori* (Coconnier et al. 1998). While human milk-isolated lactobacilli displayed antibacterial action over *S. typhi*, *S. flexneri*, *B. cereus* and *P. aeruginosa*, the lactobacilli of neonatal stool origin are antagonistic for enterotoxigenic *E. coli* (ETEC) (Tsai et al. 2008; Sharma et al. 2017). Moreover, antiparasitic action of *L. acidophilus* was also witnessed against *Trichomonas vaginalis* (Valadkhani et al. 2016). The vaginal lactobacilli are effective against anaerobic Gram negatives (*Mobiluncus* spp., *Prevotella bivia* and *Gardnerella vaginalis*), and Herpes Simplex Virus (HSV-2). Antimicrobial effect of probiotics is peculiar to the pathogenic strains (Matu et al. 2010; Ranjbar et al. 2015; Saud et al. 2020).

12.4 Bacteriocins

The production of bacteriocins by Gram positive probiotic bacteria (normal intestinal inhabitants) has killing and suppression effect on very similar relevant bacteria (Dobson et al. 2012; O'Shea et al. 2012; Cotter et al. 2013). Many bacteriocins target cell wall of the bacteria and some can alter gene expression (McAuliffe et al. 2001). Another conventional mode of action of bacteriocin is the formation of pores and channels in bacterial cell membrane which results in outflow of cellular contents. Additionally, normal *E. coli* (normal inhabitant of intestine) also yield plasmid-borne bacteriocins, specifically of low molecular mass (microcins) (Gordon and O'Brien 2006; Duquesne et al. 2007; O'Shea et al. 2012) and of greater size (colicins) (Cascales et al. 2007). Microcins upset various essential mechanisms in the focused bacterial cell like activity of DNA gyrase and ATP synthase. Colicins kill bacteria by forming pore, prohibiting the activity of nucleases and murein biosynthesis. In LAB, formation of bacteriocins is regulated by quorum sensing (QS) mechanism which is controlled by quorum sensing molecules or pheromones (bacteriocin like peptide) (Eijsink et al. 2002; Risoen et al. 2000; Sturme et al. 2007). These quorum sensing molecules get stimulated during infection (Moslehi-Jenabian et al. 2011). Numerous clinical studies express metabolites based antimicrobial activities of Lactobacillus that could serve as a substitute to antibiotics (Liévin-Le Moal and Servin 2014). Assorted bacteriocins are being applied as preservatives in foodstuff to prevent the growth of food-related pathogenic bacteria (Dobson et al. 2012; Cotter et al. 2013). Usually, it is considered that bacteriocins of lactobacilli origin are less effective against Gram negative bacteria, but it has been noticed that various bacteriocin-mimicking molecules and bacteriocins have hostile effect on entero-virulent Gram negatives (Campylobacter, EHEC, Salmonella, *H. pylori* and Shigella) (Zamfir et al. 1999; Kim et al. 2003; Han et al. 2007; Pascual et al. 2008; Messaoudi et al. 2012). Anti-*H. pylori* bacteriocins have been recognized in *L. casei* Shirota and *L. johnsonii* NCC 533 (Morency et al. 2001; Kim et al. 2003; Ryan et al. 2009; Simova et al. 2009).

12.5 Probiotics Potential Against Drug Resistance

Multidrug-resistant (MDR) bacteria are responsible for diverse global health dilemma. Oral introduction of lactobacilli can prevent many MDR-bacterial infections by producing hydrogen peroxide, lactic acid and other metabolites which discourage growth of pathogens (Jamalifar et al. 2011). Several prebiotics (products of probiotics) consist of polysols, polysaccharides, oligosaccharides [(mannanoligosaccharide (MOS), fructooligosaccharide (FOS)], hydrolysates of proteins, etc. Prebiotics can discriminatively propagate gut bacterial flora, improve immunity and offer antiviral activity. On the other hand, synbiotics, combination of pre- and probiotics, can improve immunity and decrease diarrhoea-related illness and fatality in piglets (Andersson et al. 2001; Gaggia et al. 2010).

Currently, carbapenem-resistant Enterobacteriaceae (CRE) is an important threat associated with illness and deaths because of the availability of narrow treatment options (Rodriguez-Bano et al. 2018; Chi-Chung Chen et al. 2019). Lactobacilli synthesize lactic acid, formic acid, acetic acid, etc. to lower the pH of intestine which results in antimicrobial effect. Various antimicrobials of lactobacilli like bacteriocins, fatty acids, ethanol and hydrogen peroxide employ antimicrobial action (Inglin et al. 2015). Such antimicrobials are accountable for the inhibition of pathogens, *P. aeruginosa*, *S. aureus*, *Streptococcus mutans*, *E. coli*, *Shigella* spp., and *C. difficile* (Jamalifar et al. 2011; McFarland 2015; Kumar et al. 2016; Kang et al. 2017; Ahn et al. 2018). Another notorious pathogen methicillin-resistant *Staphylococcus aureus* (MRSA) causes high mortality and morbidity and has few therapeutic options (Drew 2007; Raygada and Levine 2009). Specific probiotic therapy can solve this multidrug resistance problem (Tagg and Dierksen 2003; Roghmann and McGrail 2006). Many LAB on the basis of their antimicrobial metabolites can stop the growth of antibiotic-resistant bacteria (Petrova et al. 2009). Much pronounced antimicrobial activity was observed when cell lysate and entire broth of 3 LAB mix (1:1:1 ratio) was employed. This assay displayed 85% suppression of MDR-*S. aureus* (Bhola and Bhadekar 2019).

Propionibacterium is mostly recruited from farm cattle milk and the products obtained from milk (Rossi and Dellaglio 2007; Quigley et al. 2013). They are immobile, Gram positive probiotic bacteria which yield short-chain fatty acids and more metabolic products in gut (Huang and Adams 2004). *Propionibacterium freudenreichii* is popular as probiotic, in food and dairy industry and for the production of cheese and vitamins (Falentin et al. 2010; Thierry et al. 2011; Cousin et al. 2012; Ganan et al. 2013; Yuksekdag et al. 2014; Rabah et al. 2017). They are categorized in Qualified Presumption of Safety (QPS) and Generally Recognized as Safe (GRAS) for the application in food products (EFSA 2013; FDA 2014). Anti-virulence activity of *P. freudenreichii* subsp. shermanii (PS) and *P. freudenreichii* subsp. freudenreichii (PF) has been noted against multi-drug-resistant (MDR) Salmonella (Nair and Kollanoor-Johny 2017). Dairy recruited *L. helveticus* R0052 and *L. rhamnosus* R0011 have been proved to harbour activities against MDR bacterial infections clinically (Hagen et al. 2010; Foster et al. 2011; Tompkins et al. 2012; Nair and Kollanoor-Johny 2018).

12.6 The Antimicrobial Role of Gut-Related Microbiota or Probiotics

The probiotic bacteria in gut produce metabolites of low molecular mass which move towards systemic circulation and heal diseases. These metabolites establish metabolome which include polyamines and short-chain fatty acids (Matsumoto et al. 2012). It is important to note that antibiotic therapy leads to changes in the content of intestinal microbiota or probiotics which influence the intestinal metabolome and homeostasis (Antunes et al. 2011). The probiotics also cooperate in the anatomical and functional development of epithelial cell linings, hence improving immunity of the intestine (Kamada and Nunez 2013). Several gut probiotics or microflora prohibit the encroachment and virulence factors mediated damaging responses of pathogens by offering challenge for nutrient acquisition, enhancing gut immunity, release of organic acids and antimicrobial chemicals (Brown et al. 2013; Buffie and Pamer 2013). Moreover, they regulate and activate receptors on innate immune cells (Kinnebrew and Pamer 2012).

12.7 Human Intestinal Lactobacillus Strains

In many clinical analyses, lactobacilli as probiotics have been found effective against rotaviruses and gastroentero-virulent pathogens. The antibacterial action is based on metabolites: non-bacteriocin components (proteolytic enzymes), lactic acid and non-protein molecules. These antibacterial components directly kill bacteria, repress virulence genes and minimize detrimental response of virulence factors on intestinal cells (Kleerebezem et al. 2010; Lebeer et al. 2010; Bron et al. 2012; Dobson et al. 2012; Cotter et al. 2013; van Baarlen et al. 2013). Probiotic-stimulated immunomodulatory activities have been noticed in in vitro studies (Sanchez et al. 2008; Sanchez et al. 2010). Six strains of lactobacilli in randomized controlled trials (RCTs) proved to have antirotaviral and antimicrobial activities. These include *L. casei* strain Shirota YIT9029, *L. casei* DN-114 001, *L. acidophilus* strain LB (rearranged as *L. fermentum* LB-f), *L. jhonsonii* NCC 533, *L. rhamnosus* GG (ATCC 53103) and *L. reuteri* DSM17938 (Rosander et al. 2008; Liévin-Le Moal and Servin 2014).

Killing effect on entero-virulent bacteria (Gram negative and Gram positive) of lactobacilli cultures directly has been investigated. But in a few studies cell free spent culture supernatants (CFCSs) explored for bactericidal response. It is noteworthy that bactericidal response causes >3-log decline of living bacterial cell number after incubating the target bacterial culture for specified time and under controlled conditions. A fall in Shigella growth till 4-log CFU/mL was noticed that was caused by 4 h exposure of *L. reuteri* ATCC 55730, *L. jhonsonii* NCC 533, *L. acidophilus* LB and *L. rhamnosus* GG (Bernet-Camard et al. 1997; Hutt et al. 2006; Spinler et al. 2008; Zhang et al. 2011). Similar findings can be seen for other entero-virulent bacteria (Table 12.1).

L. reuteri ATCC 55730 behaves unfriendly with *Vibrio cholerae*. It is necessary to treat entero-virulent bacteria before their entrance in gut cells. Gut concerned antibiotic treatment often failed by entero-invasive bacteria because of their internalization by host gut cells in a vacuole. Interestingly, metabolites of *L. acidophilus* LB can efficiently kill *S. Typhimurium* nested in intracellular vesicle of enterocyte Caco-2/TC7 (Coconnier et al. 2000). Destructive effect of *L. acidophilus* LB, *L. johnsonii* NCC 533, *L. casei* DN-114 001, *L. casei* Shirota, and *L. rhamnosus* GG on entero-virulent bacteria is due to the presence of their metabolites or

Table 12.1 Decline in growth (log CFU/mL) of entero-virulent bacteria in 4 h after direct exposure to various lactobacilli strains

Targeted entero-pathogens	Direct exposure of lactobacilli strains	Decline in log CFU/mL	References
Entero-virulent *E. coli*	*L. acidophilus* LB *L. casei* Shirota *L. rhamnosus* GG *L. reuteri* ATCC 55730	3–4	Ogawa et al. (2001), Spinler et al. (2008), Zhang et al. (2011), Liévin-Le Moal and Servin (2014)
Listeria	*L. johnsonii* NCC 533, *L. acidophilus* LB	3–4	Bernet-Camard et al. (1997), Liévin-Le Moal and Servin (2014)
Shigella	*L. reuteri* ATCC 55730, *L. johnsonii* NCC 533 *L. acidophilus* LB *L. rhamnosus* GG	4	Bernet-Camard et al. (1997), Spinler et al. (2008), Liévin-Le Moal and Servin (2014)
S. Typhimurium	*L. rhamnosus* GG, *L. johnsonii* NCC 533, *L. casei* Shirota, *L. casei* DN-114 001 *L. reuteri* ATCC 55730, *L. acidophilus* LB	5	Bernet-Camard et al. 1997, Coconnier et al. (2000), Coconnier-Polter et al. (2005), Fayol-Messaoudi et al. (2005), Hutt et al. (2006), Makras et al. (2006), Vizoso Pinto et al. (2006), Fayol-Messaoudi et al. (2007), Pridmore et al. (2008), Spinler et al. (2008), Burkholder and Bhunia 2009, Atassi and Servin (2010), Marianelli et al. (2010), Asahara et al. 2011, Zhang et al. (2011), Liévin-Le Moal and Servin (2014)

compounds in CFCSs. These metabolites act either cooperatively or in solo (Fayol-Messaoudi et al. 2005; Makras et al. 2006). Bactericidal activity is also offered by low pH. Lactobacilli produce lactic acid through fermentation which lowers the intracellular pH of pathogens and makes cell membrane more pervious resulting in death. The killing effect of lactic acid over *S. Typhimurium* rises linearly as its concentration increases (Makras et al. 2006; Zhang et al. 2011). This effect is further supported by production of H_2O_2 (Pridmore et al. 2008; Atassi and Servin 2010). Several emitted compounds (non-proteinaceous) in CFCSs of *L. acidophilus* LB, *L. rhamnosus* GG and *L. johnsonii* NCC 533 screened for susceptibility of *S. Typhimurium* and found bactericidal. Few of these antibacterial compounds are thermo-stable and are effective at low pH (De Keersmaecker et al. 2006). Furthermore, some heat stable small peptides are active against *Salmonella typhi* and entero-aggregative *E. coli* (EAEC) (Lu et al. 2009). Lactobacilli have bactericidal potential over *H. pylori* straight forwardly and indirectly. CFCSs of *L. casei* Shirota have low pH-based killing effect on *H. pylori* (Sgouras et al. 2005). In the same manner, exposure of CFCSs of *L. acidophilus* LB and *L. johnsonii* NCC 533 leads to quick decline of *H. pylori* growth (6 log CFU/mL) (Michetti et al. 1999).

12.8 Effects of Probiotics on the Expression and Functionality of Virulence Factors

L. rhamnosus GG is capable of lowering the stx2A mRNA (shiga toxin mRNA) content in *E. coli* O157:H7 (Carey et al. 2008). Mobility of *S. Typhimurium* is adversely affected when it is treated with CFCSs of *L. casei* Shirota and *L. acidophilus* LB. This destruction of motility is due to the depolarization of cell membrane that disturbs the performance of flagella (Lievin-Le Moal et al. 2011; Lievin-Le Moal et al. 2013). Lactobacilli treatment can also adversely affect the morphology of *H. pylori*. Accordingly, a switch from spiral form to coccoid was observed (Sgouras et al. 2005).These forms are less virulent than the spiral form (Sisto et al. 2000).

Urease activity is crucial for the survival of *H. pylori* under acidic environment of stomach. Functionality of urease is diminished by *L. casei* Shirota and *L. acidophilus* LB (Coconnier et al. 1998; Sgouras et al. 2005). In addition to spiral form, motility of *H. pylori* is important for colonization. Non-proteinaceous substances (1000 Da) released by *L. johnsonii* NCC 533 are capable to hinder the *H. pylori* motility (Lertsethtakarn et al. 2011; Isobe et al. 2012). Similarly, motility and spiral form are irreversibly affected by CFCS of *L. casei* Shirota (Lievin-Le Moal et al. 2013).

12.9 Findings of Probiotics Antimicrobial Potential by Various Methods

Various researchers proved the antimicrobial potential of numerous probiotic strains. Choi and Chang planned the strategy to check the antimicrobial effect of *Lactobacillus plantarum* EM by applying spot-on-lawn procedure against serious pathogens including *S. aureus* ATCC 29123, *E. coli* O157:H7 ATCC 43895, *M. luteus* ATCC 1530, *P. aeruginosa* ATCC 27853, *B. cereus* KCTC 3624, *S. enterica serovar* Typhi ATCC 19430 and *V. parahaemolyticus* ATCC 17802. Most powerful bacterial-foe effect was noted against *V. parahaemolyticus* ATCC 17802 and the very low effect was found against *S. aureus* ATCC 29123. The researcher concluded that *Lactobacillus plantarum* EM bears vast antibacterial spectrum, and hence executed the advantageous need of probiotics (Choi and Chang 2015; Chan et al. 2018). In another study, *Lactobacillus plantarum* KL-1 was observed to generate bacteriocins of broad antibacterial potential. The examination of bacteriocins activity was carried out by spot-on lawn procedure. These bacteriocins were able to produce zones of inhibition against various Gram positives but, highest activity was noted against closely relevant *L. sakei* (Pilasombut et al. 2015).Various strains of lactobacilli were recorded antibacterial against *Pseudomonas aeruginosa*, *Shigella* spp., *S. aureus, S. typhi, E. faecalis, E. coli, and K. pneumonia by altered agar well method. The most efficient probiotic strains displayed the area of inhibition against aforementioned pathogens varies from 19 to 33 mm* (Prabhurajeshwar and Chandrakanth 2019).

Cell-less supernatants of lactobacilli and bifidobacteria species were investigated for probiotic potential against *E. coli* and *S. aureus* by applying agar well diffusion protocol. Effectiveness of supernatants was measured by the expansion of one of inhibition. Interestingly, supernatants were less active against *E. coli* comparatively (Ali et al. 2013). By the same technique, Naderi and colleagues evaluated the antibacterial activity of *L. rhamnosus*, *L. casei* and *L. acidophilus* against uropathogens (*Enterobacter* spp., *Enterococcus* spp., *Klebsiella pneumoniae* and *E. coli*). Accordingly, *L. casei* was the most competent probiotic against MDR-*E. coli* (Naderi et al. 2014). A spot-on-lawn and agar diffusion-based study was conducted in Iran in which child faecal isolate *Lactobacillus acidophilus* presented powerful effect against MDR-*P. aeruginosa*. But the commercial *Lactobacillus acidophilus* and *Lactobacillus reuteri* indicated comparatively weak activity (Jamalifar et al. 2011). Co-culturing of multi-species probiotics (species of Lactobacilli, Bifidobacterium and Enterococcus) with pathogenic *P. aeruginosa* has resulted in decline in *P. aeruginosa* growth from 9.2 log/mL (without probiotics) to 5.2 log/mL (with probiotics) (Koning et al. 2010; Fijan 2016). PCR and gene sequencing-based detection of bacteriocins genes (*pln*A to G, *pln*I to K, and *pln*N) was reported in yogurt-isolated Lactobacilli strains by Qian et al. (2020).

12.10 Plant Procured Chemicals (Phytochemicals)

Due to the incidences of drug resistance evolution the currently used antibacterial drugs have been unsuccessful (WHO 2014; Baym et al. 2016). One of the approved approaches is to include the combo of additional molecules (non-antibiotic or phytochemicals) with unsuccessful antibiotic in order to bring back its antibacterial potential and generate the chances for advanced treatment options (Brown 2015; Vandevelde et al. 2016; Rana et al. 2018). Hereof, phytochemicals present strong antimicrobial action while numerous scientists have employed plant-oriented natural substances against bacterial drug resistance (Cowan 1999; Khameneh et al. 2016; Fazly Bazzaz et al. 2018; Shakeri et al. 2018). These substances or compounds, unaccompanied or accompanied with antibiotics, can boost antibacterial effect up to extended spectrum (Fazly Bazzaz et al. 2010; Betts and Wareham 2014; Fazly Bazzaz et al. 2016; Fazly Bazzaz et al. 2018). The bacterial-foe effect of phytochemicals and other antibacterial compounds is chiefly connected with two processes: chemically obstruction of production and activity of integral constituents of bacteria and/or bypassing the typical antibacterial resistance systems. Many targets have been observed to be encountered by antibacterial compounds; (1) demolition of plasma membrane (2) bio-formation of cell wall (3) polypeptide synthesis (4) DNA restoration and bio-formation and (5) metabolism pathway hindrance. In response bacteria may offer resistance to antibacterial compounds by an array of processes (Khameneh et al. 2019).

Despite that microbes-foe artificial compounds have been authorized in various countries, the application of natural agents obtained from animals or plants and microbes captivates the concentration of numerous scientists (Gyawali and Ibrahim 2014; Moloney 2016). These agents displayed encouraging results in defeating bacterial antibiotic resistance (Rossiter et al. 2017). A wide variety of antibacterial,

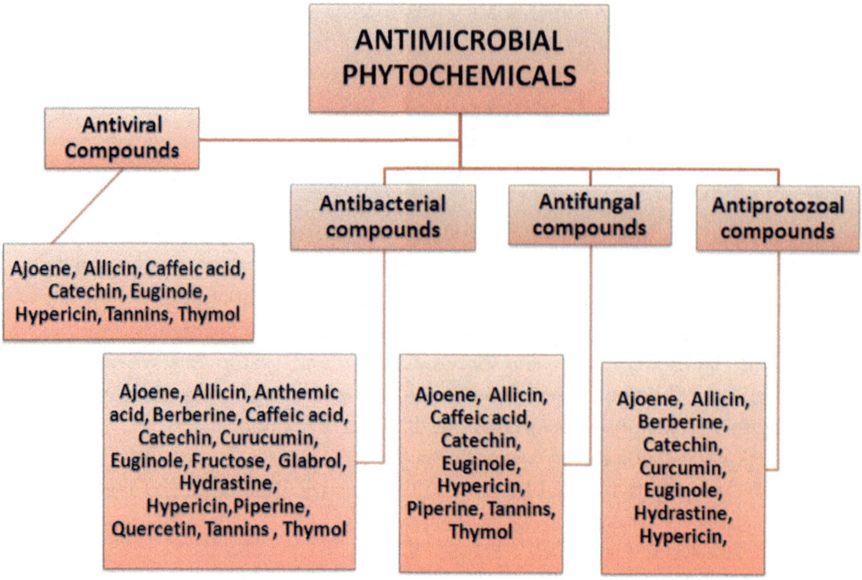

Fig. 12.2 Plant procured potentially antimicrobial compounds

antioxidant, anti-protozoan, antiviral and antifungal chemical compounds are derived from plants (Fig. 12.2). They can improve clinical utilization of earlier antibiotics by enhancing their potential of evading the resistance emergence (Barbieri et al. 2017).

The antimicrobially active plant chemicals can be categorized into various large groups depending upon their chemical configuration consisting of terpenoids, alkaloids, polyphenols and sulphur-bearing chemicals. These substances bear various potent strategies to encounter pathogenic microbes (Fig. 12.3) (Khameneh et al. 2019).

12.11 Terpenes or Terpenoids

Terpenes (isoprenoids) are acknowledged as divergent group of natural products and found in plants and in cell structures (Paduch et al. 2007; Oldfield and Lin 2012). Comparatively, Gram positives are more vulnerable to terpenes than Gram positives. Monoterpenes are inclined to decrease the density and raise the permeability of plasma membrane by changing the shape of its proteins and disrupt the respiratory pathway (Paduch et al. 2007). Alcoholic compounds of terpene (phytol, linalool, geranylgeraniol, geraniol, farnesol, plaunotol and nerolidol) suppress the propagation of *S. aureus*. Only nerolidol and farnesol could exhibit the bactericidal action relied on the damage to cell membrane (Togashi et al. 2010). Dehydroabietic acid (resin acid) and its derivatives (especially Carvone) were recorded efficient

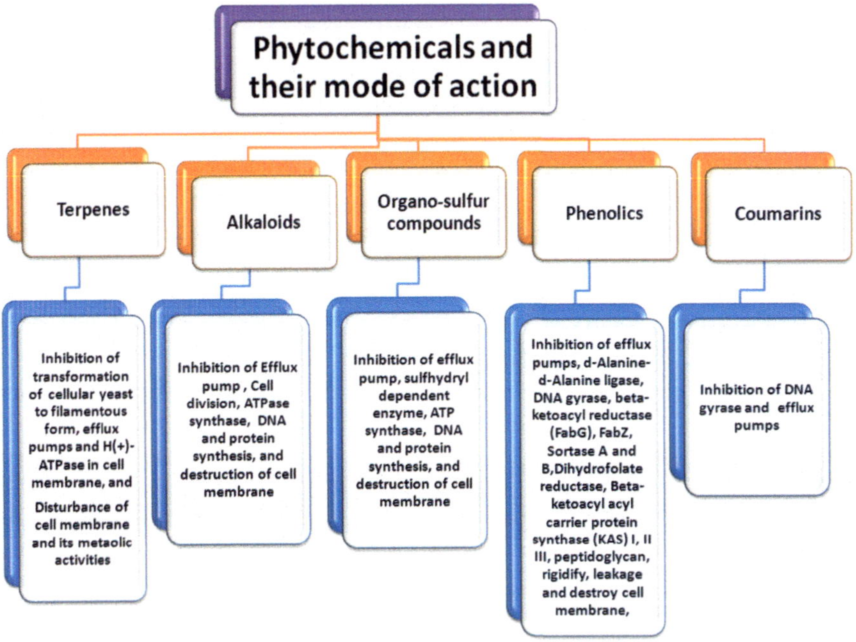

Fig. 12.3 Groups of plant-oriented compounds and their modes of antimicrobial action

antibacterial to *Campylobacter jejuni*, *S. aureus*, *E. coli*, *L. monocytogenes*, *E. faecium* and *C. albicans* (De Carvalho and Da Fonseca 2006; Savluchinske-Feio et al. 2006; Paduch et al. 2007). Furthermore, an antifungal, antibacterial and antiviral agent, Thymol has expressed dynamic effect over *Candida krusei*, *Candida glabrata* and *C. albicans* in solo and together with fluconazole (Sharifzadeh et al. 2018). Because of their extensive antifungal effect, thymol, menthol, eugenol and carvacrol are applied in food industries against food decomposing fungi (*Aspergillus* spp., *Penicillium* spp., *Rhizopus oryzae*, *Botrytis cinerea*, *Fusarium oxysporum* and *Alternaria alternata*) (Abbaszadeh et al. 2014). In addition to antifungal potential, thymol and carvacrol have efflux pump (EP) inactivation ability, anti-biofilm effect, and disintegration of plasma membrane and are bactericidal to *P. aeruginosa*, *E. coli*, *S. aureus* and *Enterobacter aerogenes* (Chauhan and Kang 2014; Amaral et al. 2015; Miladi et al. 2016). Eugenol presents anti-biofilm action against *S. aureus*, disruption of plasma membrane, and down-regulation of enterotoxin and biofilm genes. Finally, several terpenes have anti-mycobacterial effect (Yadav et al. 2015).

12.12 Alkaloids

They belong to heterocyclic compounds of nitrogen and have been verified for antibacterial effectiveness and infectious disease therapy (Cushnie et al. 2014). Many alkaloids work as efflux pumps inhibitor (EPI) that is an important process of most antibacterials. The extracts of *Piper longum* and *Piper nigrum* (Piperine), i.e. piperidine-like alkaloid, when used with Ciprofloxacin prevents the propagation of resistant *S. aureus* and lowers MICs for *S. aureus* (Khan et al. 2006). Similarly, co-application of gentamicin and Piperine was potent against methicillin-resistant *S. aureus* (MRSA) ailments (Khameneh et al. 2015). Additionally, its anti-NorA efflux pump activity has been noted in MRSA and non-MRSA (Kumar et al. 2008; Khameneh et al. 2015). Berberine (applied as traditional medicine) is recognized as alkaloid of isoquinoline and is present in bark and roots of Berberis species and other plants. This agent is anti-protozoan and antibacterial by intercalating within DNA, inhibiting topoisomerase IV, RNA polymerase and gyrase, and restriction of cell division (Iwasa et al. 2001; Yi et al. 2007; Domadia et al. 2008). Cell division restriction is linked with the inactivation of protein FtsZ needed for cell division (Boberek et al. 2010). Berberine has emerged as a vigorous bacteria-foe compound aimed to replace antibiotics and to defeat resistance.

Ungeremine is the alkaloid of iso-quinoline acquired from *Pancratium illyricum* which also has bacterial-foe activity. It is responsible for the DNA breaking as it inactivates topoisomerase IA (Casu et al. 2011; Schrader et al. 2013). Alkaloids of quinolone (maculine, dictamine, and kokusagine) have shown antibacterial activity and are separated from *Teclea afzelii*. They hinder the DNA biosynthesis by targeting topoisomerase II (Heeb et al. 2011). Reduction in oxygen utilization in bacteria has been observed when they are treated with Alkyl methyl quinolones (respiratory inhibitors) (Tominaga et al. 2002). Alkaloid of indole (Reserpine) acquired from *Rauwolfia serpentina* and has an excellent EPI effect (Abdelfatah and Efferth 2015). It improves the antibiotic effect against *Micrococcus* spp., *Staphylococcus* spp. and *Streptococcus* spp. when co-employed with antibiotics (Sridevi et al. 2017). Furthermore, it boosts the sensitivity of MDR-*A. baumannii* to antibiotics. Reduction in EP-based resistance to flouroquinolones in *Stenotrophomonas maltophilia* is noticed with the use of reserpine (Jia et al. 2015). It acts as EPI against Gram negative and positive bacteria (Sun et al. 2017). Extracts of *Macleaya cordata*, *Chelidonium majus* and *Sanguinaria canadensis* consist of Sanguinarine which is found as anti-MRSA. It is responsible for the discharge of autolytic enzyme and causes the disruption of MRSA. It also modifies altogether the process of septa development during cell division of MRSA (Obiang-Obounou et al. 2011). In addition to anti-MRSA effect, it has been reported as anti-mycobacterial against few species of Mycobacteria (Newton et al. 2002). Berberine and Sanguinarine both are vigorous transcriptional and DNA biosynthesis inhibitors (Al-Ani et al. 2015). Steroidal alkaloid (Tomatidine) are produced in eggplant, tomato and potato plant. In combo with aminoglycoside or in solo, it has demonstrated the killing activity against *S. aureus* (Mitchell et al. 2012; Jiang et al. 2016). It is an antibiotic enhancer which can synergistically improve the effects of Cefepime, Ampicillin, Gentamicin and Ciprofloxacin against *P. aeruginosa*, *S. aureus* and *E. fecalis* (Soltani et al. 2017). Another steroidal alkaloid (conessine) is obtained from *Holarrhena antidysenterica* and has broad spectrum bacterial-foe potential, EPI against AdeIJK efflux pump of *A. baumannii* and synergistic effect with antibiotics (Kumar et al. 2007; Damier-Piolle et al. 2008; Siddiqui et al. 2012; Siriyong et al. 2016; Zhou et al. 2017). Tricyclic ergot alkaloid (Chanoclavine) procured from *Ipomoea muricata* was noted effective against MDR-*E. coli* when delivered together with Tetracycline. Its antibacterial act is EPI (Dwivedi et al. 2019).

12.13 Sulphur-Containing Compounds or Organosulphur Compounds

The vast spectrum bacterial-foe effects of plant procured sulphur-containing compounds (isothiocyanates, allicin, *S*-allay-mercaptocystein, ajoene, dialkyl sulphides, *S*-allay cysteine and diakyl) have been observed (Sobolewska et al. 2015; Barbieri et al. 2017). Allicin (diallylthiosulphinate) is derived from garlic and is potentially effective against *P. aeruginosa*, *S. epidermidis*, MRSA and *Streptococcus aglactiae* (Reiter et al. 2017). Its co-acting anti-*P. aeruginosa* effect has been realized concomitantly with Ciprofloxacin, Cefoperazone and Tobramycin (Cai et al. 2008). The antimicrobial mode of action of allicin is because of blocking sulfhydryl-reliant enzymes (RNA polymerase, alcohol dehydrogenase and thioredoxinreductase), protein and DNA biosynthesis (Lanzotti et al. 2014). Ajoene is also present in garlic mainly as *E*- and *Z*-stereoisomers which strive extended spectrum antimicrobial effect (Gram negative and Gram positive bacteria, protozoa, fungi and viruses). It has similar bacterial-foe mode of action as that of allicin (Rehman and Mairaj 2013).

The volatile sulphur-containing compounds like isothiocyanates (ITCs) are derived through plant glucosinolates after interaction with myrosinase. These are considerable antibacterial contenders against *H. pylori* by lowering the urease activity (Fahey et al. 2013; Park et al. 2013). These agents probably bind with proteins or enzymes, bearing sulfhydryl groups and block various biochemical systems (Dufour et al. 2015). They intrude in ATP linking sites of *E. coli* ATPase by targeting cysteine. A novel Sulforaphane belongs to ITCs and has bacterial-foe behaviour for *H. pylori, Listeria monocytogenes* and *S. aureus* (Benzekri et al. 2016). Allyl ITCs (AITCs) are present in the *Eutrema japonicum* and *Armoracia rusticana* and are effective against *S. aureus* and *E. coli*. They also lower synergistically Erythromycin MICs against *Streptococcus pyogenes* and Streptomycin MICs against *P. aeruginosa* and *E. coli* (Palaniappan and Holley 2010; Saavedra et al. 2010; Lu et al. 2016). Their antibacterial effect is mediated by cell wall disruption, pore formation in cell membrane and oxidative breakage of disulphide linkage (Luciano and Holley 2009; Nedorostova et al. 2009). Phenethyl isothiocyanate (PEITC) is an efficient antimicrobial (anti-Gram positives and antifungal) chemical (Aires et al. 2009). Its antifungal effect is due to the accretion of reactive oxygen species (ROS) resulting in damage to mitochondria (Calmes et al. 2015). Benzyl ITCs (BITCs) have bactericidal response towards MRSA and act on plasma membrane as cationic protein (Sofrata et al. 2011; Dias et al. 2014).

12.14 Phenol-Containing Plant Compounds

These chemicals involve an extensive spectrum of natural substances which are broadly employed for medical objectives and intensify the antibiotic effect against resistant microorganisms by many mechanisms. They have indicated varied modes of action and synergistic effect through targeting cell membrane, EPs inactivation, and several vital enzymes like dihydrofolate reductase, urease and sortase A inactivation. The following noticed activities of phenol-containing compounds are impressive which make them marvellous candidates for clinical application (Farhadi et al. 2019a, b; Górniak et al. 2019).

They are expert in inhibiting efflux pumps, e.g. CmeABC of *C. jejuni* (Lechner et al. 2008; Klancnik et al. 2017). In addition to EPI effect, resveratrol causes the accretion of Ethidium bromide in *Arcobacter cryaerophilus* and *Arcobacter butzleri* (Ferreira et al. 2014). Baicalein can extraordinarily bring back the performance of Ciprofloxacin, Tetracycline and Beta-lactams against MRSA through inactivation of NorA efflux pumps (Chan et al. 2011). It also acts synergistically with Tetracycline in order to inactivate *E. coli* efflux pump (Fujita et al. 2005). Similar to Baicalein, Biochanin A (isoflavon) can also subvert the MRSA efflux pump (NorA) (Zou et al. 2014). It has growth-preventive response towards *Chlamydia* spp. and is a dominant EPI of Mycobacterium (Lechner et al. 2008; Cannalire et al. 2017). Isoflavonoid, flavonolignans (synergistic with Norfloxacin and Berberin) and many otherflavonoids (flavones) like Chrysoplenetin and Chrysosplenol-D (synergistic with Berberin) also restrict NorA activity (Stermitz et al. 2001; Morel et al. 2003; Stermitz et al. 2003). For the decline of EP efficiency, concentration of scientists has been captivated towards flavonoid-antibiotic hybridization. Studies have proved the intracellular aggregation of antibiotics and extraordinarily heightened effect of co-molecule (flavonoid-antibiotic hybrid) that validate required double action (Xiao et al. 2014). Kaempferol (flavonoid) is an impressive antimicrobial against *C. albicans* (Fluconazole resisted) and MRSA (Randhawa et al. 2016; Shao et al. 2016). Its anti-MRSA effect is due to the NorA EPI activity (Holler et al. 2012). Chalcones can impede NorA EP activity and lower erythromycin MICs (0.4–0.1 μg/mL) (Belofsky et al. 2004). Phenol-containing Catechin gallates like Epigallocatechin gallate (EGCG) can efficiently eradicate MRSA. They can attach themselves at ATP adhesive site on DNA gyrase, leading to the blockage of DNA gyrase activity (Gibbons et al. 2004).

Antimicrobial potential of plant-oriented phenols is not restricted to EPI effect however; other mechanisms are also known (Farhadi et al. 2019a, b). Green tea procured Tannins/polyphenols (anthraquinones and chebulinic acid) have been found to have anti-DNA gyrase effect (Patel et al. 2015). Semisynthetic Haloemodins (of natural anthraquinone) are the vital inhibitors of DNA gyrase in

Enterococcus faecium (vancomycin resistant) and MRSA (Duan et al. 2014). A novel plasma membrane destructive, 3-*p*-*trans*-coumaroyl-2-hydroxyquinic acid (CHQA) has been found effective against food-related pathogens (Wu et al. 2016). Compounds of phenol bind with essential enzyme β-Ketoacyl acyl synthase (KAS), responsible for the linking process of FabZ, FabL and FabG (fatty acid) biosynthesis. These fatty acids are the precursors of the cell membrane formation in bacteria. Curcumin (Turmeric derivative) can destroy *E. coli* and *S. aureus* by disrupting their plasma membrane (Tyagi et al. 2015). Cell wall formation can be interrupted by Apigenin and Quercetin (flavonoids) as they inactivate *d*-alanine:*d*-alanine ligase, an essential enzyme for cell wall formation (Wu et al. 2008). However, there are few other compounds like Sophoraflavanone B that can also target peptidoglycan elongation in MRSA (Mun et al. 2014). Other mechanisms for microbial enzymes inhibition by phenolic compounds have been studied such as inactivation of sortase, dihydrofolate reductase and urease (Navarro-Martínez et al. 2005; Maresso and Schneewind 2008; Xiao et al. 2013). EGCG have indicated anti-dihydrofolate reductase activity in *Stenotrophomonas maltophilia* (Navarro-Martínez et al. 2005). Sortase A enzyme of *S. aureus* can be targeted by Curcumin and Morin (Park et al. 2005). The functionality of urease of *H. pylori* is hindered by acetohydroxamic acid 4′,7,8-trihydroxyl-2-isoflavene (Xiao et al. 2013).

12.15 Coumarins

These compounds are not only obtained from plants but also from microbes (Smyth et al. 2009). They can inhibit DNA gyrase, inactivate EP and abolish quorum sensing resulting in reduction of biofilm and virulence components synthesis (Gutiérrez-Barranquero et al. 2015; D'Almeida et al. 2017; Reen et al. 2018; Zhang et al. 2018). The various derivatives Pyranocoumarins and Coumarins (Aegelinol benzoate, Agasyllin and Grandivittin) exert antioxidant and widened spectrum antibacterial effects. Mainly, agasyllin and aegelinol have more bactericidal effects against *S. aureus*, *Salmonella typhi* and *Enterobacter* spp., and *H. pylori* (Melliou et al. 2005; Basile et al. 2009). However, other compounds of Coumarins (osthole and 4′-senecioiloxyosthol) have exhibited bactericidal activity against *Klebsiella pneumoniae* and *S. aureus* (Tan et al. 2017). Reduction in norfloxacin MIC against MRSA has been witnessed when Coumarins are used concomitantly with Norfloxacin (Roy et al. 2013).

12.16 Bacteriophages

The earliest analysis on the application of phage (phage therapy) was carried out by Bruynoghe and Maisin in Belgium, 1921. They cutaneously infused phage suspension peculiar to *S. aureus* in order to cure carbuncle and fruncle. In 1940s, phages were first launched commercially by the companies of France and United States (O'Flaherty et al., 2009). The dilemma of antibiotic resistance and the demand of

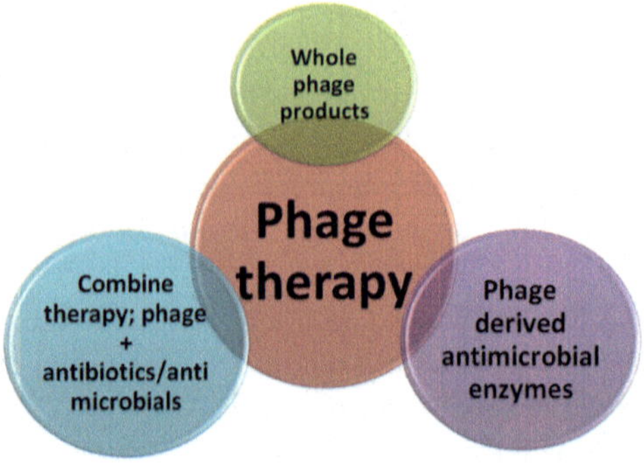

Fig. 12.4 Targeting drug-resistant and virulent bacteria by the application of phages and their enzymes

novel antimicrobials have provoked the attention of the researchers towards non-clinical and clinical utilization of bacteriophages (phages) and their procured enzymes having antimicrobial effect (Van Boeckel et al. 2014; Van Boeckel et al. 2015; Webber et al. 2015; Soumet et al. 2016). Phages in comparison with antibiotics also occupy therapeutic benefits. The undistinguishing antimicrobial effects of antibiotic treatment lead to the loss of microflora. On the other hand, therapeutic employment of phages can diminish this loss (Meader et al. 2013; Faber et al. 2016; Zhao et al. 2017).

Various infections are being dealt in Europe with phage treatment, taking the benefit of whole phage having lytic style of replication (Abedon et al. 2011; Viertel et al. 2014).A few food products containing whole phage are employed in Western countries, while phage procured enzymes are introduced in man and animal medicines (Sulakvelidze 2013; Cooper 2016; Totté et al. 2017; Cooper et al. 2018). Phage-oriented proteins of therapeutic interest are usually endolysins, needed for the release of freshly formed phages (Briers and Lavigne 2015; Rodríguez-Rubio et al. 2016). Nevertheless, some phage-oriented enzymes (e.g. spanins and holins) bearing antimicrobial response have also been explored (Roach, 2015; Song et al. 2016).Various phage therapeutic strategies have been adopted: (1) whole-phage occupying treatments consisting of various preferred phages of desired properties and selective antimicrobial effect in a blend (Gill and Hyman 2010; Weber-Dabrowska et al. 2016), (2) mix therapy (Phages together with antibiotics) has been found to kill drug-resistant bacteria, efficiently(Ryan et al. 2012; Daikos et al. 2014) and (3) synergistic effect has been noted with various blends of antibiotics and endolysins and endolysins alone against Gram negatives and Gram positives (Fig. 12.4) (Becker et al. 2008; García et al. 2010; Schuch et al. 2014).

Phages have been found antibacterial to popular entero-virulent bacteria like Campylobacter, Salmonella and *E. coli* O157:H7 (Huff et al. 2005; Johnson et al. 2008). US-FDA accepted LMP-102TM phage cocktail having six different types of phages in 2006, for targeting Listeria in meat. United States Department of Agriculture (USDA) in 2007 endorsed a phage therapeutic product for the eradication of *E. coli* from cattle. However, various phage products are under investigation for therapeutic scope. Phages reproduced in targeted bacteria and give rise to numerous fresh lytic phages, and also produce mutations in pathogens (Cheng et al. 2014).

12.17 Phage-Originated Antibacterial Enzymes

Some phage-oriented enzymes (e.g. endolysins and hydrolases) bearing antimicrobial response have also been explored (Roach and Donovan 2015; Song et al. 2016). Phage-oriented proteins of therapeutic interest are usually endolysins, needed for the release of freshly build-up phages (Briers and Lavigne 2015; Rodríguez-Rubio et al. 2016). Some of these enzymes are discussed below.

12.18 Endolysins

During late stage of lytic cycle several endolysins (transglycosylase, glucosidase, endopeptidase and amidase) are produced and are capable of deteriorating mucopeptide/murein of bacterial cell wall, in order to promote the new phages discharge. During 1950s, these enzymes were first time detected and found antibacterial to *Clostridium butyricum*, Staphylococcus, *L. monocytogenes* and *Bacillus anthracis* (Low et al. 2005). They are found effective in sepsis and bacterial infections of Group B Streptococcus, *Enterococcus faecalis* and *C. perfringens* (Fenton et al. 2010). Endolysin PAL has the capability to destroy Group-A Streptococci. Together with Endopeptidase Cpl-1 and Amidase PAL can lower the occurrence of pneumonia (both systemic and localized) (Fischetti 2005). Phage K originated Endolysin LysK can destroy MRSA (O'Flaherty et al. 2005). Similarly, PlyV12 Endolysin expresses excellent lytic action on Vancomycin-resistant-*E. faecium* and *E. faecalis* (Yoong et al. 2004). The phage phi3626 based endolysins can successfully cure Clostridium related ailments (Courchesne et al. 2009). Endolysins have exclusive Gram positive antibacterial spectrum and lead to very rapid bacterial lysis by targeting their murein of cell wall (Loeffler et al. 2003). So, that there is no chance for bacteria to evolve into resistant forms. But they are of little or no worth for targeting Gram negatives (Fischetti 2005).

12.19 Virion-Associated Peptidoglycan Hydrolases (VAPGHs)

They are a sort of lyases which break down murein or peptidoglycan to facilitate phage entrance in bacteria (Rodriguez-Rubio et al. 2013). Various viron-associated peptidoglycan hydrolases (VAPGHs) have been identified and their antimicrobial potential has been verified. For example, HydH5 (philPLA88 phage origin), protein gp61 (produced by phiMR11 phage) and Protein 17 (produced by P68 phage) were found remarkably antibacterial (at exponential growth phase) against MRSA and non-MRSA (Takac and Blasi 2005; Rashel et al. 2008; Rodriguez et al. 2011). Similarly, P5 protein produced by phage 6 bears antibacterial efficiency against

Proteus vulgaris, Pseudomonas, *S. Typhimurium* and *E. coli.* In addition to aforementioned Gram negatives, gp181 (produced by KZ phage) have shown lytic effect against Yersinia and *Ralstonia solanacearum.* Gp36 (produced by bacteriophage KMV) is heat stable and impressive against *E. coli* and *P. aeruginosa* (Lavigne et al. 2004). VAPGHs produced by phages targeting Gram negatives are of extensively antibacterial spectrum while, VAPGHs generated by phages of Gram positives are effective against only limited bacteria. They can also efficiently encounter drug-resistant pathogenic bacteria by lowering the expression of efflux system (Paul et al. 2001). Most of VAPGHs can tolerate and remain effective at high temperature, hence can be employed in food industry (Rodriguez-Rubio et al. 2013). We salute to the scientists (working in above referred domains) whose efforts are indeed laudable.

References

Abbaszadeh S, Sharifzadeh A, Shokri H, Khosravi AR, Abbaszadeh A (2014) Antifungal efficacy of thymol, carvacrol, eugenol and menthol as alternative agents to control the growth of food-relevant fungi. J Mycol Med 24(2):e51–e56

Abdelfatah SA, Efferth T (2015) Cytotoxicity of the indole alkaloid reserpine from *Rauwolfia serpentina* against drug-resistant tumor cells. Phytomedicine 22(2):308–318

Abdelhamid AG, Esaam A, Hazaa MM (2018) Cell free preparations of probiotics exerted antibacterial and antibiofilm activities against multidrug resistant *E. coli.* Saudi Pharm J 26:603–607

Abedon ST, Kuhl SJ, Blasdel BG, Kutter EM (2011) Phage treatment of human infections. Bacteriophage 1:66–85

Ahn KB, Baik JE, Park OJ, Yun CH, Han SH (2018) *Lactobacillus plantarum* lipoteichoic acid inhibits biofilm formation of *Streptococcus mutans.* PLoS One 13:e0192694

Aires A, Mota VR, Saavedra MJ, Rosa EA, Bennett RN (2009) The antimicrobial effects of glucosinolates and their respective enzymatic hydrolysis products on bacteria isolated from the human intestinal tract. J Appl Microbiol 106(6):2086–2095

Al-Ani I, Zimmermann S, Reichling J, Wink M (2015) Pharmacological synergism of bee venom and melittin with antibiotics and plant secondary metabolites against multi-drug resistant microbial pathogens. Phytomedicine 22(2):245–255

Ali FS, Saad OAO, Salwa AH (2013) Antimicrobial activity of probiotic bacteria. Egypt Acad J Biol Sci 5(2):21–34

Allen HK, Stanton TB (2014) Altered egos: antibiotic effects on food animal microbiomes. Annu Rev Microbiol 68:297–315

Amaral VCS, Santos PR, da Silva AF, dos Santos AR, Machinski M, Mikcha JMG (2015) Effect of carvacrol and thymol on *Salmonella* spp. biofilms on polypropylene. Int J Food Sci Technol 50 (12):2639–2643

Anadon A, Abroix Arzo M, Bories G, Brantom P, Brufau de Barbera J, Chesson A et al (2005) Opinion of the FEEDAP panel on the safety and efficacy of the product Farmatan for rabbits and piglets. EFSA J 222:1–20

Anas M, Jamal Eddine H, Mebrouk K (2008) Antimicrobial activity of *Lactobacillus* species isolated from Algerian raw goat's milk against *Staphylococcus aureus.* World J DairyFood Sci 23:39–49

Andersson H, Asp NG, Bruce A, Roos S, Wadstrom T, Wold A (2001) Health effects of probiotics and prebiotics: a literature review on human studies. Scand J Nutr 45:58–75

Antunes LC, Han J, Ferreira RB, Lolic P, Borchers CH, Finlay BB (2011) Effect of antibiotic treatment on the intestinal metabolome. Antimicrob AgentsChemother 55:1494–1503

Asahara T, Shimizu K, Takada T, Kado S, Yuki N, Morotomi M, Tanaka R, Nomoto K (2011) Protective effect of *Lactobacillus casei*s train Shirota against lethal infection with multi-drug resistant *Salmonella enterica* serovar Typhimurium DT104 in mice. J Appl Microbiol 110:163–173

Atassi F, Servin AL (2010) Individual and co-operative roles of lactic acid and hydrogen peroxide in the killing activity of enteric strain *Lactobacillus johnsonii* NCC933 and vaginal strain *Lactobacillus gasseri* KS120.1 against enteric, uropathogenic and vaginosis-associated pathogens. FEMS Microbiol Lett 304:29–38

Barbieri R, Coppo E, Marchese A, Daglia M, Sobarzo-Sanchez E, Nabavi SF et al (2017) Phytochemicals for human disease: an update on plant-derived compounds antibacterial activity. Microbiol Res 196:44–68

Basile A, Sorbo S, Spadaro V, Bruno M, Maggio A, Faraone N et al (2009) Antimicrobial and antioxidant activities of coumarins from the roots of *Ferulago campestris* (Apiaceae). Molecules 14(3):939–952

Baym M, Stone LK, Kishony R (2016) Multidrug evolutionary strategies to reverse antibiotic resistance. Science 351(6268):aad3292

Becker SC, Foster-Frey J, Donovan DM (2008) The phage K lytic enzyme LysK and lysostaphin act synergistically to kill MRSA. FEMS Microbiol Lett 287:185–191

Belofsky G, Percivill D, Lewis K, Tegos GP, Ekart J (2004) Phenolic metabolites of *Dalea versicolor* that enhance antibiotic activity against model pathogenic bacteria. J Nat Prod 67 (3):481–484

Benzekri R, Bouslama L, Papetti A, Snoussi M, Benslimene I, Hamami M et al (2016) Isolation and identification of an antibacterial compound from *Diplotaxis harra* (Forssk). Boiss Ind Crop Prod 80:228–234

Bernet-Camard MF, Lievin V, Brassart D, Neeser JR, Servin AL, Hudault S (1997) The human *Lactobacillus acidophilus* strain LA1 secretes a nonbacteriocin antibacterial substance(s) active in vitro and in vivo. Appl Environ Microbiol 63:2747–2753

Betts JW, Wareham DW (2014) In vitro activity of curcumin in combination with epigallocatechin gallate (EGCG) versus multidrug-resistant *Acinetobacter baumannii*. BMC Microbiol 14:172

Bhola J, Bhadekar R (2019) Invitro synergistic activity of lactic acid bacteria against multi-drug resistant staphylococci. BMC Complement Altern Med 19:70

Boberek JM, Stach J, Good L (2010) Genetic evidence for inhibition of bacterial division protein FtsZ by berberine. PLoS One 5(10):e13745

Briers Y, Lavigne R (2015) Breaking barriers: expansion of the use of endolysins as novel antibacterials against Gram-negative bacteria. FutureMicrobiol 10:377–390

Broaders E, Gahan CG, Marchesi JR (2013) Mobile genetic elements of the human gastrointestinal tract: potential for spread of antibiotic resistance genes. Gut Microbes 4:271–280

Bron PA, van Baarlen P, Kleerebezem M (2012) Emerging molecular insights into the interaction between probiotics and the host intestinal mucosa. Nat Rev Microbiol 10:66–78

Brown D (2015) Antibiotic resistance breakers: can repurposed drugs fill the antibiotic discovery void? Nat Rev Drug Discov 14(12):821–832

Brown EM, Sadarangani M, Finlay BB (2013) The role of the immune system in governing host-microbe interactions in the intestine. Nat Immunol 14:660–667

Buffie CG, Pamer EG (2013) Microbiota-mediated colonization resistance against intestinal pathogens. Nat Rev Immunol 13:790–801

Burkholder KM, Bhunia AK (2009) *Salmonella enterica* serovar Typhimurium adhesion and cytotoxicity during epithelial cell stress is reduced by *Lactobacillus rhamnosus* GG. Gut Pathog 1:14

Cai Y, Wang R, An MM, Liang BB, Fang Y (2008) In vitro bactericidal activity of allicin combined with cefoperazone, tobramycin and ciprofloxacin. Int J Antimicrob Agents 31(2):179–180

Calmes B, N'Guyen G, Dumur J, Brisach CA, Campion C, Iacomi B et al (2015) Glucosinolate-derived isothiocyanates impact mitochondrial function in fungal cells and elicit an oxidative stress response necessary for growth recovery. Front Plant Sci 6:414

Cannalire R, Machado D, Felicetti T, Santos Costa S, Massari S, Manfroni G et al (2017) Natural isoflavone biochanin a as a template for the design of new and potent 3-phenylquinolone efflux inhibitors against *Mycobacterium avium*. Eur J Med Chem 140:321–330

Card RM, Mafura M, Hunt T, Kirchner M, Weile J, Rashid MU et al (2015) Impact of ciprofloxacin and clindamycin administration on Gram negative bacteria isolated from healthy volunteers and characterization of the resistance genes they harbor. Antimicrob Agents Chemother 59:4410–4416

Carey CM, Kostrzynska M, Ojha S, Thompson S (2008) The effect of probiotics and organic acids on Shiga-toxin 2 gene expression in enterohemorrhagic *Escherichia coli* O157:H7. J Microbiol Methods 73:125–132

Cascales E, Buchanan SK, Duche D, Kleanthous C, Lloubes R, Postle K, Riley M, Slatin S, Cavard D (2007) Colicin biology. Microbiol Mol Biol Rev 71:158–229

Casu L, Cottiglia F, Leonti M, De Logu A, Agus E, Tse-Dinh YC et al (2011) Ungeremine effectively targets mammalian as well as bacterial type I and type II topoisomerases. Bioorg Med Chem Lett 21(23):7041–7044

Cawoy H, Mariutto M, Henry G, Fisher C, Vasilyeva N, Thonart P et al (2014) Plant defense stimulation by natural isolates of Bacillus depends on efficient surfactin production. Mol Plant-Microbe Interact 27:87–100

Chan BCL, Ip M, Lau CBS, Lui SL, Jolivalt C, Ganem-Elbaz C et al (2011) Synergistic effects of baicalein with ciprofloxacin against NorA over-expressed methicillin-resistant *Staphylococcus aureus* (MRSA) and inhibition of MRSA pyruvate kinase. J Ethnopharmacol 137(1):767–773

Chan AP, Choil Y, Brinkac LM, Krishnakumar R, De Pew J, Kim M et al (2018) Multidrug resistant pathogens respond differently to the presence of co-pathogen, commensal, probiotic and host cells. Sci Rep 8:8656

Chang L, Zhang ZY, Ke D, Jian-Ping Y, Xiao-Kui G (2009) Antibiotic resistance of probiotic strains of lactic acid bacteria isolated from marketed foods and drugs. Biomed Environ Sci 22:401–412

Chauhan AK, Kang SC (2014) Thymol disrupts the membrane integrity of *Salmonella ser. Typhimurium* invitro and recovers infected macrophages from oxidative stress in an ex vivo model. Res Microbiol 165(7):559–565

Chen C-C, Lai C-C, Huang H-L, Huang W-Y, Toh H-S, Weng T-C et al (2019) Antimicrobial activity of *Lactobacillus* species against carbapenem-resistant *enterobacteriaceae*. Front Microbiol 10:789

Cheng G, HaoH XS, Wang X, Dail M, Huang L, Yuanl Z (2014) Antibiotic alternatives: the substitution of antibiotics in animal husbandry. Front Microbiol 5:217

Choi EA, Chang HC (2015) Cholesterol-lowering effects of a putative probiotic strain *Lactobacillus plantarum* EM isolated from kimchi. LWT Food Sci Technol 62(1):210–217

Coconnier MH, Lievin V, Hemery E et al (1998) Antagonistic activity against helicobacter infection in vitro and in vivo by the human *Lactobacillus acidophilus* strain LB. Appl Environ Microbiol 64:4573–4580

Coconnier MH, Lievin V, Lorrot M, Servin AL (2000) Antagonistic activity of *Lactobacillus acidophilus* LB against intracellular *Salmonella enterica serovar Typhimurium* infecting human enterocyte-like Caco-2/ TC-7 cells. Appl Environ Microbiol 66:1152–1157

Coconnier-Polter MH, Lievin-Le Moal V, Servin AL (2005) A *Lactobacillus acidophilus* strain of human gastrointestinal microbiota origin elicits killing of enterovirulent *Salmonella enterica* serovar Typhimurium by triggering lethal bacterial membrane damage. Appl Environ Microbiol 71:6115–6120

Collado MC, Cernada M, Baüerl C, Vento M, Pérez-Martínez G (2012) Microbial ecology and host-microbiota interactions during early life stages. Gut Microbes 3:352–365

Collins MD, Gibson GR (1999) Probiotics, prebiotics, and synbiotics: approaches for modulating the microbial ecology of the gut. Am J Clin Nutr 69:1052s–1057s

Cooper IR (2016) A review of current methods using bacteriophages in live animals, food and animal products intended for human consumption. J Microbiol Methods 130:38–47

Cooper CJ, Koonjan S, Nilsson AS (2018) Enhancing whole phage therapy and their derived antimicrobial enzymes through complex formulation. Pharmaceuticals (Basel) 11(2):34

Cotter PD, Ross RP, Hill C (2013) Bacteriocins—a viable alternative to antibiotics? Nat Rev Microbiol 11:95–105

Courchesne NM, Parisien A, Lan CQ (2009) Production and application of bacteriophage and bacteriophage-encoded lysins. Recent Pat Biotechnol 3:37–45

Cousin FJ, Foligné B, Deutsch SM, Massart S, Parayre S, Le Loir Y et al (2012) Assessment of the probiotic potential of a dairy product fermented by *Propionibacterium freudenreichii* in piglets. J Agric Food Chem 60:7917–7927

Cowan MM (1999) Plant products as antimicrobial agents. Clin Microbiol Rev 12(4):564–582

Cushnie TP, Cushnie B, Lamb AJ (2014) Alkaloids: an overview of their antibacterial, antibiotic-enhancing and antivirulence activities. Int J Antimicrob Agents 44(5):377–386

D'Orazio G, Di Gennaro P, Boccarusso M, Presti I, Bizzaro G, Giardina S et al (2015) Microencapsulation of new probiotic formulations for gastrointestinal delivery: in vitro study to assess viability and biological properties. Appl Microbiol Biotechnol 99:9779–9789

Daikos GL, Tsaousi S, Tzouvelekis LS, Anyfantis I, Psichogiou M, Argyropoulou A et al (2014) Carbapenemase-producing *Klebsiella pneumoniae* blood stream infections: lowering mortality by antibiotic combination schemes and the role of carbapenems. Antimicrob Agents Chemother 58:2322–2328

D'Almeida R, Molina R, Viola C, Luciardi M, Peñalver CN, Bardon A et al (2017) Comparison of seven structurally related coumarins on the inhibition of quorum sensing of *Pseudomonas aeruginosa* and *Chromobacterium violaceum*. Bioorg Chem 73:37–42

Damier-Piolle L, Magnet S, Bremont S, Lambert T, Courvalin P (2008) Ade IJK, a resistance-nodulation-cell division pump effluxing multiple antibiotics in *Acinetobacter baumannii*. Antimicrob Agents Chemother 52(2):557–562

De Carvalho CC, Da Fonseca MMR (2006) Carvone: why and how should one bother to produce this terpene. Food Chem 95(3):413–422

De Keersmaecker SC, Verhoeven TL, Desair J, Marchal K, Vanderleyden J, Nagy I (2006) Strong antimicrobial activity of *Lactobacillus rhamnosus* GG against *Salmonella typhimurium* is due to accumulation of lactic acid. FEMS Microbiol Lett 259:89–96

Devi SM, Archer AC, Halami PM (2015) Screening, characterization and in vitro evaluation of probiotic properties among lactic acid bacteria through comparative analysis. Probiotics Antimicrob Proteins 7:181–192

Dias C, Aires A, Saavedra MJ (2014) Antimicrobial activity of isothiocyanates from cruciferous plants against methicillin-resistant *Staphylococcus aureus* (MRSA). Int J Mol Sci 15 (11):19552–19561

Dobson A, Cotter PD, Ross RP, Hill C (2012) Bacteriocin production: a probiotic trait? Appl Environ Microbiol 78:1–6

Domadia PN, Bhunia A, Sivaraman J, Swarup S, Dasgupta D (2008) Berberine targets assembly of *Escherichia coli* cell division protein FtsZ. Biochemistry 47(10):3225–3234

Drew RH (2007) Emerging options for treatment of invasive, multidrug-resistant *Staphylococcus aureus* infections. Pharmacotherapy 27:227–249

Duan F, Li X, Cai S, Xin G, Wang Y, Du D et al (2014) Haloemodin as novel antibacterial agent inhibiting DNA gyrase and bacterial topoisomerase I. J Med Chem 57(9):3707–3714

Dufour V, Stahl M, Baysse C (2015) The antibacterial properties of isothiocyanates. Microbiology 161(Pt 2):229–243

Duquesne S, Destoumieux-Garzon D, Peduzzi J, Rebuffat S (2007) Microcins, gene-encoded antibacterial peptides from enterobacteria. Nat Prod Rep 24:708–734

Dwivedi GR, Maurya A, Yadav DK, Singh V, Khan F, Gupta MK et al (2019) Synergy of clavine alkaloid 'chanoclavine' with tetracycline against multi-drugresistant *E. coli*. J Biomol Struct Dyn 37(5):1307–1325

EFSA (2013) EFSA panel on biological hazards (BIOHAZ) scientific opinion on the maintenance of the list of QPS biological agents intentionally added to food and feed (2013 update): QPS 2013 update. EFSA J 11:3449

Eijsink VG, Axelsson L, Diep DB, Havarstein LS, Holo H, Nes IF (2002) Production of class II bacteriocins by lactic acid bacteria; anexample of biological warfare and communication. Antonie Van Leeuwenhoek 81:639–654

Ennahar S, Sashihara T, Sonomoto K, Ishizaki A (2000) Class IIa bacteriocins: biosynthesis, structure and activity. FEMS Microbiol Rev 24:85–106

Faber F, Tran L, Byndloss MX, Lopez CA, Velazquez EM, Kerrinnes T et al (2016) Host-mediated sugar oxidation promotes post-antibiotic pathogen expansion. Nature 534:697–699

Fahey JW, Stephenson KK, Wade KL, Talalay P (2013) Urease from helicobacter pylori is inactivated by sulforaphane and other isothiocyanates. Biochem Biophys Res Commun 435 (1):1–7

Falentin H, Deutsch SM, Jan G, Loux V, Thierry A, Parayre S et al (2010) The complete genome of *Propionibacterium freudenreichii* CIRM-BIA1T, a hardy Actinobacterium with food and probiotic applications. PLoS One 5:e11748

Farhadi F, Khameneh B, Iranshahi M, Iranshahy M (2019a) Antibacterial activity of flavonoids and their structure–activity relationship: An update review. Phytother Res 33(1):13–40

Farhadi F, Khameneh B, Iranshahi M, Iranshahy M (2019b) Antibacterial activity of flavonoids and their structure–activity relationship: An update review. Phytother Res 33(1):13–40

Fayol-Messaoudi D, Berger CN, Coconnier-Polter MH, Lievin-Le Moal V, Servin AL (2005) pH-, lactic acid-, and non-lactic aciddependent activities of probiotic lactobacilli against *Salmonella enterica* serovar Typhimurium. Appl Environ Microbiol 71:6008–6013

Fayol-Messaoudi D, Coconnier-Polter MH, Moal VL, Atassi F, Berger CN, Servin AL (2007) The *Lactobacillus plantarum* strain ACA-DC287 isolated from a Greek cheese demonstrates antagonistic activity in vitro and in vivo against *Salmonella enterica* serovar Typhimurium. J Appl Microbiol 103:657–665

Fazly Bazzaz BS, Iranshahi M, Naderinasab M, Hajian S, Sabeti Z, Masumi E (2010) Evaluation of the effects of galbanic acid from ferula szowitsiana and conferol from *F. badrakema*, as modulators of multi-drug resistance in clinical isolates of *Escherichia coli* and *Staphylococcus aureus*. Res Pharm Sci 5(1):21–28

Fazly Bazzaz BS, Sarabandi S, Khameneh B, Hosseinzadeh H (2016) Effect of Catechins, green tea extract and Methylxanthines in combination with gentamicin against *Staphylococcus aureus* and *Pseudomonas aeruginosa*: combinationtherapy againstresistantbacteria. J Pharmacopuncture 19(4):312–318

Fazly Bazzaz BS, Khameneh B, Zahedian Ostad MR, Hosseinzadeh H (2018) In vitro evaluation of antibacterial activity of verbascoside, lemon verbena extract and caffeine in combination with gentamicin against drug-resistant *Staphylococcus aureus* and *Escherichia coli* clinical isolates. Avicenna J Phytomed 8(3):246–253

FDA (2014) Agency response letter GRAS notice no. GRN 000415. Available at: https://wayback.archive-it.org/7993/20171031010359/https://www.fda.gov/Food/IngredientsPackagingLabeling/GRAS/NoticeInventory/ucm335742.htm

Fenton M, Ross P, McAuliffe O, O'Mahony J, Coffey A (2010) Recombinant bacteriophage lysins as antibacterials. Bioeng Bugs 1:9–16

Ferreira S, Silva F, Queiroz JA, Oleastro M, Domingues FC (2014) Resveratrol against *Arcobacter butzleri* and *Arcobacter cryaerophilus*: activity and effect on cellular functions. Int J Food Microbiol 180:62–68

Fijan S (2016) Antimicrobial effect of probiotics against common pathogens. In: Probiotics and prebiotics in human nutrition and health. IntechOpen pp 193–221. https://doi.org/10.5772/63141

Fischetti VA (2005) Bacteriophage lytic enzymes: novel anti-infectives. Trends Microbiol 13:491–496

Foster LM, Tompkins TA, Dahl WJ (2011) A comprehensive post market review of studies on a probiotic product containing *Lactobacillus helveticus* R0052 and *Lactobacillus rhamnosus* R0011. Benef Microbes 2:319–334

Fouhy F, Ross RP, Fitzgerald GF, Stanton C, Cotter PD (2014) A degenerate Pcr-based strategy as a means of identifying homologues of aminoglycoside and β-lactam resistance genes in the gut microbiota. BMC Microbiol 14:25

Fujita M, Shiota S, Kuroda T, Hatano T, Yoshida T, Mizushima T et al (2005) Remarkable synergies between baicalein and tetracycline, and baicalein and beta-lactams against methicillin-resistant *Staphylococcus aureus.* MicrobiolImmunol 49(4):391–396

Gaggia F, Mattarelli P, Biavati B (2010) Probiotics and prebiotics in animal feeding for safe food production. Int J Food Microbiol 141(Suppl. 1):S15–S28

Ganan M, Martinez-Rodriguez AJ, Carrascosa AV, Vesterlund S, Salminen S, Satokari R (2013) Interaction of *Campylobacter* spp. and human probiotics in chicken intestinal mucus. Zoonoses Public Health 60:141–148

García P, Martínez B, Rodríguez L, Rodríguez A (2010) Synergy between the phage endolysin lysH5 and nisin to kill *Staphylococcus aureus* in pasteurized milk. Int J Food Microbiol 141:151–155

Gibbons S, Moser E, Kaatz GW (2004) Catechin gallates inhibit multidrug resistance (MDR) in *Staphylococcus aureus.* Planta Med 70(12):1240–1242

Gibson MK, Forsberg KJ, Dantas G (2015) Improved annotation of antibiotic resistance determinants reveals microbial resistomes cluster by ecology. ISMEJ 9:207–216

Gill JJ, Hyman P (2010) Phage choice, isolation, and preparation for phage therapy. Curr Pharm Biotechnol 11:2–14

Gilliland SE, Walker DK (1990) Factors to consider when selecting a culture of *Lactobacillus acidophilus* as a dietary adjunct to produce a hypocholesterolemic effect in humans. J Dairy Sci 73:905–911

Gordon DM, O'Brien CL (2006) Bacteriocin diversity and the frequency of multiple bacteriocin production in *Escherichia coli.* Microbiology 152:3239–3244

Górniak I, Bartoszewski R, Króliczewski J (2019) Comprehensive review of antimicrobial activities of plant flavonoids. Phytochem Rev 18(1):241–272

Gueimonde M, Sánchez B, De Los Reyes-Gavilán CG, Margolles A (2013) Antibiotic resistance in probiotic bacteria. Front Microbiol 4:202

Gutiérrez-Barranquero JA, Reen FJ, McCarthy RR, O'Gara F (2015) Deciphering the role of coumarin as a novel quorum sensing inhibitor suppressing virulence phenol types in bacterial pathogens. Appl Microbiol Biotechnol 99(7):3303–3316

Gyawali R, Ibrahim SA (2014) Natural products as antimicrobial agents. FoodControl 46:412–429

Hagen KE, Tramp CA, Altermann E, Welker DL, Tompkins TA (2010) Sequence analysis of plasmid pIR52-1 from *Lactobacillus helveticus* R0052 and investigation of its origin of replication. Plasmid 63:108–117

Hammad A, Shimamoto T (2010) Towards a compatible probiotic–antibiotic combination therapy: assessment of antimicrobial resistance in the Japanese probiotics. J Appl Microbiol 109:1349–1360

Han KS, Kim Y, Kim SH, Oh S (2007) Characterization and purification of acidocin 1B, a bacteriocin produced by *Lactobacillus acidophilus* GP1B. J Microbiol Biotechnol 17:774–783

Heeb S, Fletcher MP, Chhabra SR, Diggle SP, Williams P, Camara M (2011) Quinolones: from antibiotics to autoinducers. FEMS Microbiol Rev 35(2):247–274

Holler JG, Christensen SB, Slotved HC, Rasmussen HB, Gúzman A, Olsen CE et al (2012) Novel inhibitory activity of the *Staphylococcus aureus* NorA efflux pump by a kaempferol rhamnoside isolated from *Persea lingue* Nees. J Antimicrob Chemother 67(5):1138–1144

Hu Y, Yang X, Qin J, Lu N, Cheng G, Wu N et al (2013) Metagenome-wide analysis of antibiotic resistance genes in a large cohort of human gut microbiota. Nat Commun 4:2151

Hu Y, Yang X, Lu N, Zhu B (2014) The abundance of antibiotic resistance genes in human guts has correlation to the consumption of antibiotics in animal. Gut Microbes 5:245–249

Huang Y, Adams MC (2004) In vitro assessment of the upper gastrointestinal tolerance of potential probiotic dairy Propionibacteria. Int J Food Microbiol 91:253–260

Huff WE, Huff GR, Rath NC, Balog JM, Donoghue AM (2005) Alternatives to antibiotics: utilization of bacteriophage to treat colibacillosis and prevent food borne pathogens. Poult Sci 84:655–659

Hutt P, Shchepetova J, Loivukene K, Kullisaar T, Mikelsaar M (2006) Antagonistic activity of probiotic lactobacilli and bifidobacteria against entero- and uropathogens. J Appl Microbiol 100:1324–1332

Imperial ICVJ, Ibana JA (2016) Addressing the antibiotic resistance problem with probiotics: reducing the risk of its double-edged sword effect. Front Microbiol 7:1983

Inglin RC, Stevens MJ, Meile L, Lacroix C, Meile L (2015) High throughput screening assays for antibacterial and antifungal activities of *Lactobacillus* species. J Microbiol Methods 114:26–29

Isobe H, Nishiyama A, Takano T, Higuchi W, Nakagawa S, Taneike I et al (2012) Reduction of overall *Helicobacter pylori* colonization levels in the stomach of Mongolian gerbil by *Lactobacillus johnsonii* La1 (LC1) and its in vitro activities against *H. pylori* motility and adherence. Biosci Biotechnol Biochem 76:850–852

Iwasa K, Moriyasu M, Yamori T, Turuo T, Lee DU, Wiegrebe W (2001) In vitro cytotoxicity of the protoberberine-type alkaloids. J Nat Prod 64(7):896–898

Jamalifar H, Rahimi HR, Samadi N, Shahverdi AR, Sharifian Z, Hosseini F, Eslahi H, Fazeli MR (2011) Antimicrobial activity of different *Lactobacillus* species against multidrug resistant clinical isolates of *Pseudomonas aeruginosa*. IJM 3(1):21–25

Jia W, Wang J, Xu H, Li G (2015) Resistance of *Stenotrophomonas maltophilia* to fluoroquinolones: prevalence in a university hospital and possible mechanisms. Int J Environ Res Public Health 12(5):5177–5195

Jiang QW, Chen MW, Cheng KJ, Yu PZ, Wei X, Shi Z (2016) Therapeutic potential of steroidal alkaloids in cancer and other diseases. Med Res Rev 36(1):119–143

Johnson RP, Gyles CL, Huff WE, Ojha S, Huff GR, Rath NC et al (2008) Bacteriophages for prophylaxis and therapy in cattle, poultry and pigs. Anim Health Res Rev 9:201–215

Kamada N, Nunez G (2013) Role of the gut microbiota in the development and function of lymphoid cells. J Immunol 190:1389–1395

Kang MS, Lim HS, Oh JS, Lim YJ, Wuertz-Kozak K, Harro JM et al (2017) Antimicrobial activity of *Lactobacillus salivarius* and *Lactobacillus fermentum* against *Staphylococcus aureus*. Pathog Dis 75:ftx009

Khameneh B, Iranshahy M, Ghandadi M, Ghoochi Atashbeyk D, Fazly Bazzaz BS, Iranshahi M (2015) Investigation of the antibacterial activity and efflux pump inhibitory effect of co-loaded piperine and gentamicin nanoliposomes in methicillin-resistant *Staphylococcus aureus*. Drug Dev Ind Pharm 41(6):989–994

Khameneh B, Diab R, Ghazvini K, Fazly Bazzaz BS (2016) Breakthroughs in bacterial resistance mechanisms and the potential ways to combat them. MicrobPathog 95:32–42

Khameneh B, Iranshahy M, Soheili V, Bazzaz BSF (2019) Review on plant antimicrobials: a mechanistic viewpoint. Antimicrob Resist Infect Control 8:118

Khan IA, Mirza ZM, Kumar A, Verma V, Qazi GN (2006) Piperine, a phytochemical potentiator of ciprofloxacin against *Staphylococcus aureus*. Antimicrob Agents Chemother 50(2):810–812

Kim TS, Hur JW, Yu MA, Cheigh CI, Kim KN, Hwang JK, Pyun YR (2003) Antagonism of *Helicobacter pylori* by bacteriocins of lactic acid bacteria. J Food Prot 66:3–12

Kinnebrew MA, Pamer EG (2012) Innate immune signaling in defense against intestinal microbes. Immunol Rev 245:113–131

Klancnik A, Sikic Pogacar M, Trost K, Tusek Znidaric M, Mozetic Vodopivec B, Smole MS (2017) Anti-campylobacter activity of resveratrol and an extract from waste pinot noir grape skins and seeds, and resistance of camp. Jejuni planktonic and biofilm cells, mediated via the CmeABC efflux pump. J Appl Microbiol 122(1):65–77

Kleerebezem M, Hols P, Bernard E, Rolain T, Zhou M, Siezen RJ, Bron PA (2010) The extracellular biology of the lactobacilli. FEMS Microbiol Rev 34:199–230

Koning CJ, Jonkers D, Smidt H, Rombouts F, Pennings HJ, Wouters E et al (2010) The effect of a multispecies probiotic on the composition of the faecal microbiota and bowel habits in chronic obstructive pulmonary disease patients treated with antibiotics. Br J Nutr 103:1452–1460

Kumar N, Singh B, Bhandari P, Gupta AP, Kaul VK (2007) Steroidal alkaloids from *Holarrhena antidysenterica* (L.) WALL. Chem Pharm Bull (Tokyo) 55(6):912–914

Kumar A, Khan IA, Koul S, Koul JL, Taneja SC, Ali I et al (2008) Novel structural analogues of piperine as inhibitors of the NorA efflux pump of *Staphylococcus aureus*. J Antimicrob Chemother 61(6):1270–1276

Kumar M, Dhaka P, Vijay D, Vergis J, Mohan V, Kumar A et al (2016) Antimicrobial effects of *Lactobacillus plantarum* and *Lactobacillus acidophilus* against multidrug-resistant enteroaggregative *Escherichia coli*. Int J Antimicrob Agents 48:265–270

Lanzotti V, Scala F, Bonanomi G (2014) Compounds from allium species with cytotoxic and antimicrobial activity. Phytochem Rev 13(4):769–791

Lavigne R, Briers Y, Hertveldt K, Robben J, Volckaert G (2004) Identification and characterization of a highly thermostable bacteriophage lysozyme. Cell Mol Life Sci 61:2753–2759

Lebeer S, Vanderleyden J, De Keersmaecker SC (2010) Host interactions of probiotic bacterial surface molecules: comparison with commensals and pathogens. Nat Rev Microbiol 8:171–184

Lechner D, Gibbons S, Bucar F (2008) Plant phenolic compounds as ethidium bromide efflux inhibitors in *Mycobacterium smegmatis*. J Antimicrob Chemother 62(2):345–348

Lertsethtakarn P, Ottemann KM, Hendrixson DR (2011) Motility and chemotaxis in campylobacter and helicobacter. Annu Rev Microbiol 65:389–410

Liévin-Le Moal V, Servin AL (2014) Anti-infective activities of *Lactobacillus* strains in the human intestinal microbiota: from probiotics to gastrointestinal anti infectious biotherapeutic agents. Clin Microbiol Rev 27(2):167–199

Lievin-Le Moal V, Amsellem R, Servin AL (2011) Impairment of swimming motility by antidiarrheic *Lactobacillus acidophilus* strain LB retards internalization of *Salmonella enterica serovar Typhimurium* within human enterocyte-like cells. Antimicrob Agents Chemother 55:4810–4820

Lievin-Le Moal V, Fayol-Messaoudi D, Servin AL (2013) Compound(s) secreted by *Lactobacillus casei* strain Shirota YIT9029 irreversibly and reversibly impair the swimming motility of *Helicobacter pylori* and *Salmonella enterica* serovar Typhimurium, respectively. Microbiology 159:1956–1971

Loeffler JM, Djurkovic S, Fischetti VA (2003) Phagelytic enzyme Cpl-1 as a novel antimicrobial for pneumococcal bacteremia. Infect Immun 71:6199–6204

Low LY, Yang C, Perego M, Osterman A, Liddington RC (2005) Structure and lytic activity of a *Bacillus anthracis* prophage endolysin. J Biol Chem 280:35433–35439

Lu R, Fasano S, Madayiputhiya N, Morin NP, Nataro J, Fasano A (2009) Isolation, identification, and characterization of small bioactive peptides from *Lactobacillus* GG conditional media that exert both antiGram-negative and Gram-positive bactericidal activity. J Pediatr Gastroenterol Nutr 49:23–30

Lu Z, Dockery CR, Crosby M, Chavarria K, Patterson B, Giedd M (2016) Antibacterial activities of wasabi against *Escherichia coli* O157:H7 and *Staphylococcus aureus*. Front Microbiol 7:1403

Luciano FB, Holley RA (2009) Enzymatic inhibition by allyl isothiocyanate and factors affecting its antimicrobial action against *Escherichia coli* O157:H7. Int J Food Microbiol 131(2-3):240–245

Luoto R, Kalliomäki M, Laitinen K, Isolauri E (2010) The impact of perinatal probiotic intervention on the development of overweight and obesity: follow-up study from birth to 10 years. Int J Obes 34:1531–1537

Makras L, Triantafyllou V, Fayol-Messaoudi D, Adriany T, Zoumpopoulou G, Tsakalidou E et al (2006) Kinetic analysis of the antibacterial activity of probiotic lactobacilli towards *Salmonella enterica serovar Typhimurium* reveals a role for lactic acid and other inhibitory compounds. Res Microbiol 157:241–247

Manzoor A, Ul-Haq I, Baig S, Qazi JI, Seratlic S (2016) Efficacy of locally isolated lactic acid bacteria against antibiotic-resistant uropathogens. Jundishapur J Microbiol 9(1):e18952

Maresso AW, Schneewind O (2008) Sortase as a target of anti-infective therapy. Pharmacol Rev 60 (1):128–141

Marianelli C, Cifani N, Pasquali P (2010) Evaluation of antimicrobial activity of probiotic bacteria against *Salmonella enterica subsp. enterica serovar typhimurium* 1344 in a common medium under different environmental conditions. Res Microbiol 161:673–680

Matsumoto M, Kibe R, Ooga T, Aiba Y, Kurihara S, Sawaki E, Koga Y, Benno Y (2012) Impact of intestinal microbiota on intestinal luminal metabolome. Sci Rep 2:233

Matu MN, Orinda GO, Njagi ENM, Cohen CR, Bukusi EA (2010) In vitro inhibitory activity of human vaginal lactobacilli against pathogenic bacteria associated with bacterial vaginosis in Kenyan women. Anaerobe 16(3):210–215

McAuliffe O, Ross RP, Hill C (2001) Lantibiotics: structure, biosynthesis and mode of action. FEMS Microbiol Rev 25:285–308

McFarland LV (2015) Probiotics for the primary and secondary prevention of *C. difficile* infections: a meta-analysis and systematic review. Antibiotics 4:160–178

Meader E, Mayer MJ, Steverding D, Carding SR, Narbad A (2013) Evaluation of bacteriophage therapy to control *Clostridium difficile* and toxin production in an in vitro human colon model system. Anaerobe 22:25–30

Melliou E, Magiatis P, Mitaku S, Skaltsounis A-L, Chinou E, Chinou I (2005) Natural and synthetic 2, 2-dimethylpyranocoumarins with antibacterial activity. J Nat Prod 68(1):78–82

Messaoudi S, Kergourlay G, Dalgalarrondo M, Choiset Y, Ferchichi M, Prevost H et al (2012) Purification and characterization of a new bacteriocin active against campylobacter produced by *Lactobacillus salivarius* SMXD51. Food Microbiol 32:129–134

Michetti P, Dorta G, Wiesel PH, Brassart D, Verdu E, Herranz M et al (1999) Effect of whey-based culture supernatant of *Lactobacillus acidophilus* (johnsonii) La1 on *Helicobacter pylori* infection in humans. Digestion 60:203–209

Miladi H, Zmantar T, Chaabouni Y, Fedhila K, Bakhrouf A, Mahdouani K et al (2016) Antibacterial and efflux pump inhibitors of thymol and carvacrol against food-borne pathogens. Microb Pathog 99:95–100

Mitchell G, Lafrance M, Boulanger S, Seguin DL, Guay I, Gattuso M et al (2012) Tomatidine acts in synergy with aminoglycoside antibiotics against multiresistant *Staphylococcus aureus* and prevents virulence gene expression. J Antimicrob Chemother 67(3):559–568

Moloney MG (2016) Natural products as a source for novel antibiotics. Trends Pharmacol Sci 37 (8):689–701

Morel C, Stermitz FR, Tegos G, Lewis K (2003) Isoflavones as potentiators of antibacterial activity. J Agric Food Chem 51(19):5677–5679

Morency H, Mota-Meira M, LaPointe G, Lacroix C, Lavoie MC (2001) Comparison of the activity spectra against pathogens of bacterial strains producing a mutacin oral antibiotic. Can J Microbiol 47:322–331

Moslehi-Jenabian S, Vogensen FK, Jespersen L (2011) The quorum sensing luxS gene is induced in *Lactobacillus acidophilus* NCFM in response to *Listeria monocytogenes*. Int J Food Microbiol 149:269–273

Mun SH, Joung DK, Kim SB, Park SJ, Seo YS, Gong R et al (2014) The mechanism of antimicrobial activity of sophoraflavanone B against methicillin-resistant *Staphylococcus aureus*. Foodborne Pathog Dis 11(3):234–239

Muñoz-Atienza E, Gómez-Sala B, Araújo C, Campanero C, Del Campo R, Hernández PE et al (2013) Antimicrobial activity, antibiotic susceptibility and virulence factors of lactic acid bacteria of aquatic origin intended for use as probiotics in aquaculture. BMC Microbiol 13:15

Naderi A, Kasra-Kermanshahi A, Gharavi S, Fooladi AAI, Alitappeh MA, Saffarian P (2014) Study of antagonistic effects of *Lactobacillus* strains as probiotics on multi drug resistant (MDR) bacteria isolated from urinary tract infections (UTIs). Iran J Basic Med Sci 17(3):201–208

Nair DVT, Kollanoor-Johny A (2017) Effect of *Propionibacterium freudenreichii* on salmonella multiplication, motility, and association with avian epithelial cells. Poult Sci 96:1376–1386

Nair DVT, Kollanoor-Johny A (2018) Characterizing the antimicrobial function of a dairy-originated *Propionibacterium freudenreichii*, against multidrug-resistant *Salmonella enterica* serovar Heidelberg in Turkey poults. Front Microbiol 9:1475

Navarro-Martínez MD, Navarro-Perán E, Cabezas-Herrera J, Ruiz-Gómez J, García Cánovas F, Rodríguez-López JN (2005) Antifolate activity of epigallocatechin gallate against *Stenotrophomonas maltophilia*. Antimicrob Agents Chemother 49(7):2914–2920

Nedorostova L, Kloucek P, Kokoska L, Stolcova M, Pulkrabek J (2009) Antimicrobial properties of selected essential oils in vapour phase against foodborne bacteria. Food Control 20(2):157–160

Newton SM, Lau C, Gurcha SS, Besra GS, Wright CW (2002) The evaluation of fortythree plant species for in vitro antimycobacterial activities; isolation of active constituents from *Psoralea corylifolia* and *Sanguinaria canadensis*. J Ethnopharmacol 79(1):57–67

Nueno-Palop C, Narbad A (2011) Probiotic assessment of *Enterococcus faecalis* CP58 isolated from human gut. Int J Food Microbiol 145:390–394

O'Flaherty S, Coffey A, Meaney W, Fitzgerald GF, Ross RP (2005) The recombinant phage lysin LysK has a broad spectrum of lytic activity against clinically relevant staphylococci, including methicillin-resistant *Staphylococcus aureus*. J Bacteriol 187:7161–7164

O'Shea EF, Cotter PD, Stanton C, Ross RP, Hill C (2012) Production of bioactive substances by intestinal bacteria as a basis for explaining probiotic mechanisms: bacteriocins and conjugated linoleic acid. Int J Food Microbiol 152:189–205

Obiang-Obounou BW, Kang OH, Choi JG, Keum JH, Kim SB, Mun SH et al (2011) The mechanism of action of sanguinarine against methicillin-resistant *Staphylococcus aureus*. J Toxicol Sci 36(3):277–283

Ogawa M, Shimizu K, Nomoto K, Tanaka R, Hamabata T, Yamasaki S, Takeda T, Takeda Y (2001) Inhibition of in vitro growth of Shiga toxin-producing *Escherichia coli* O157:H7 by probiotic *Lactobacillus* strains due to production of lactic acid. Int J Food Microbiol 68:135–140

Oldfield E, Lin FY (2012) Terpene biosynthesis: modularity rules. Angew Chem Int Ed 51 (5):1124–1137

Paduch R, Kandefer-Szerszen M, Trytek M, Fiedurek J (2007) Terpenes: substances useful in human healthcare. Arch Immunol Ther Exp 55(5):315–327

Palaniappan K, Holley RA (2010) Use of natural antimicrobials to increase antibiotic susceptibility of drug resistant bacteria. Int J Food Microbiol 140(2-3):164–168

Park BS, Kim JG, Kim MR, Lee SE, Takeoka GR, Oh KB et al (2005) *Curcuma longa* L. constituents inhibit sortase A and *Staphylococcus aureus* cell adhesion to fibronectin. J Agric Food Chem 53(23):9005–9009

Park HW, Choi KD, Shin IS (2013) Antimicrobial activity of isothiocyanates (ITCs) extracted from horseradish (*Armoracia rusticana*) root against oral microorganisms. Biocontrol Sci 18 (3):163–168

Pascual LM, Daniele MB, Giordano W, Pajaro MC, Barberis IL (2008) Purification and partial characterization of novel bacteriocin L23 produced by *Lactobacillus fermentum* L23. Curr Microbiol 56:397–402

Patel K, Tyagi C, Goyal S, Jamal S, Wahi D, Jain R et al (2015) Identification of chebulinic acid as potent natural inhibitor of *M. tuberculosis* DNA gyrase and molecular insights into its binding mode of action. Comput Biol Chem 59:37–47

Paul VD, Rajagopalan SS, Sundarrajan S, George SE, Asrani JY, Pillai R et al (2001) A novel bacteriophage tail-associated muralytic enzyme (TAME) from phage K and its development into a potent anti staphylococcal protein. BMC Microbiol 11:226

Penders J, Stobberingh EE, Savelkoul PH, Wolffs P (2013) The human microbiome as a reservoir of antimicrobial resistance. Front Microbiol 4:87

Petrova M, Georgieva R, Dojchinovska L, Kirilov N, Iliev I, Antonova S, Hadjieva N, Ivanova I, Danova S (2009) Lactic acid bacteria against pathogenic microbes. Trakia J Sci 7(2):33–39

Pilasombut K, Rumjuankiat K, Ngamyeesoon N, Duyle ND (2015) In vitro characterization of bacteriocin produced by lactic acid bacteria isolated from Nem Chua, a traditional Vietnamese fermented pork. Korean J Food Sci Anim 35(4):473–478

Prabhurajeshwar C, Chandrakanth RK (2017) Probiotic potential of *Lactobacilli* with antagonistic activity against pathogenic strains: An in vitro validation for the production of inhibitory substances. Biom J 40(5):270–283

Prabhurajeshwar C, Chandrakanth K (2019) Evaluation of antimicrobial properties and their substances against pathogenic bacteria in-vitro by probiotic Lactobacilli strains isolated from commercial yoghurt. Clin Nut Exp 23:97–115

Pridmore RD, Pittet AC, Praplan F, Cavadini C (2008) Hydrogen peroxide production by *Lactobacillus johnsonii* NCC 533 and its role in anti-*Salmonella* activity. FEMS Microbiol Lett 283:210–215

Qian Z, Zhao D, Yin Y, Zhu H, Chen D (2020) Antibacterial activity of *Lactobacillus* strains isolated from Mongolian yogurt against *Gardnerella vaginalis*. Bio Med Res Intern:3548618. https://doi.org/10.1155/2020/3548618

Quigley L, O'sullivan O, Stanton C, Beresford TP, Ross RP, Fitzgerald GF et al (2013) The complex microbiota of raw milk. FEMS Microbiol Rev 37:664–698

Rabah H, Rosa do Carmo FL, Jan G (2017) Dairy Propionibacteria: versatile probiotics. Microorganisms 5:24

Rana R, Sharma R, Kumar A (2018) Repurposing of existing statin drugs for treatment of microbial infections: how much promising? Infect Disord Drug Targets. https://doi.org/10.2174/1871526518666180806123230

Randhawa HK, Hundal KK, Ahirrao PN, Jachak SM, Nandanwar HS (2016) Efflux pump inhibitory activity of flavonoids isolated from *Alpinia calcarata* against methicillin-resistant *Staphylococcus aureus*. Biologia 71(5):484–493

Ranjbar R, Goudarzi MM, Jounaidi N (2015) *Lactobacillus acidophilus* and assessment for its antiviral effect against herpes simplex virus type I. Biosci Biotech Res Asia 12:1351–1356

Rashel M, Uchiyama J, Takemura I, Hoshiba H, Ujihara T, Takatsuji H et al (2008) Tail-associated structural protein gp61 of *Staphylococcus aureus* phage phi MR11 has bifunctional lytic activity. FEMS Microbiol Lett 284:9–16

Rasool SA, Rasool MS, Ajaz M (2019a) Evolution of super-drug resistant microbial strains: mechanisms and strategies for containment. Biologia (Pak) 65(II):1–17

Rasool MS, Siddiqui F, Ajaz M, Rasool SA (2019b) Prevalence and antibiotic resistance profiles of gram negative bacilli associated with urinary tract infections (UTIs) in Karachi. Pakistan Pak J Pharm Sci 32(6):2617–2623

Raygada JL, Levine DP (2009) Methicillin-resistant *Staphylococcus aureus*: a growing risk in the hospital and in the community. Am Health Drug Benefits 2(2):86–95

Reen FJ, Gutiérrez-Barranquero JA, Parages ML (2018) Coumarin: a novel player in microbial quorum sensing and biofilm formation inhibition. Appl Microbiol Biotechnol 102 (5):2063–2073

Rehman F, Mairaj S (2013) Antimicrobial studies of allicin and ajoene. Int J Pharm Bio Sci 4 (2):1095–1105

Reid G (1999) The scientific basis for probiotic strains of *Lactobacillus*. Appl Environ Microbiol 65:3763–3766

Reiter J, Levina N, van der Linden M, Gruhlke M, Martin C, Slusarenko AJ (2017) Diallylthiosulfinate (Allicin), a volatile antimicrobial from garlic (*Allium sativum*), kills human lung pathogenic bacteria, including MDR strains, as a vapor. Molecules 22(10):1711

Risoen PA, Brurberg MB, Eijsink VG, Nes IF (2000) Functional analysis of promoters involved in quorum sensing-based regulation of bacteriocin production in *Lactobacillus*. Mol Microbiol 37:619–628

Roach DR, Donovan DM (2015) Antimicrobial bacteriophage-derived proteins and therapeutic applications. Bacteriophage 5:e1062590

Rodriguez L, Martinez B, ZhouY RA, Donovan DM, Garcia P (2011) Lytic activity of the virion-associated peptidoglycan hydrolase HydH5 of *Staphylococcus aureus* bacteriophage vB_SauS-philPLA88. BMC Microbiol 11:138

Rodriguez-Bano J, Gutierrez-Gutierrez B, Machuca I, Pascual A (2018) Treatment of infections caused by extended-spectrum-beta-lactamase-, ampC-, and carbapenemase-producing Enterobacteriaceae. Clin Microbiol Rev 31:e00079–e00017

Rodriguez-Rubio L, Martinez B, Donovan DM, Rodriguez A, Garcia P (2013) Bacteriophage virion-associated peptidoglycan hydrolases: potential new enzybiotics. Crit Rev Microbiol 39:427–434

Rodríguez-Rubio L, Gutiérrez D, Donovan DM, MartínezB RA, García P (2016) Phage lytic proteins: biotechnologicalapplicationsbeyondclinicalantimicrobials. Crit Rev Biotechnol 36:542–552

Roghmann MC, McGrail L (2006) Novel ways of preventing antibiotic-resistant infections: what might the future hold? Am J InfectControl 34:469–475

Rosander A, Connolly E, Roos S (2008) Removal of antibiotic resistance gene-carrying plasmids from *Lactobacillus reuteri* ATCC 55730 and characterization of the resulting daughterstrain, *L. reuteri* DSM 17938. Appl Environ Microbiol 74:6032–6040

Rossi F, Dellaglio F (2007) Quality of silages from Italian farms as attested by number and identity of microbial indicators. J Appl Microbiol 103:1707–1715

Rossiter SE, Fletcher MH, Wuest WM (2017) Natural products as platforms to overcome antibiotic resistance. Chem Rev 117(19):12415–12474

Roy SK, Kumari N, Pahwa S, Agrahari UC, Bhutani KK, Jachak SM et al (2013) NorA efflux pump inhibitory activity of coumarins from *Mesua ferrea*. Fitoterapia 90:140–150

Ryan KA, O'Hara AM, van Pijkeren JP, Douillard FP, O'Toole PW (2009) *Lactobacillus salivarius* modulates cytokine induction and virulence factor gene expression in *Helicobacter pylori*. J Med Microbiol 58:996–1005

Ryan EM, Alkawareek MY, Donnelly RF, Gilmore BF (2012) Synergistic phage-antibiotic combinations for the control of *Escherichia coli* biofilms in vitro. FEMS Immunol Med Microbiol 65:395–398

Saavedra MJ, Borges A, Dias C, Aires A, Bennett RN, Rosa ES et al (2010) Antimicrobial activity of phenolics and glucosinolate hydrolysis products and their synergy with streptomycin against pathogenic bacteria. Med Chem 6(3):174–183

Samuels RI, Mattoso TC, Moreira DD (2013) Chemical warfare: leafcutting ants defend themselves and their gardens against parasite attack by deploying antibiotic secreting bacteria. Commun Integr Biol 6:e23095

Sanchez B, Bressollier P, Urdaci MC (2008) Exported proteins in probiotic bacteria: adhesion to intestinal surfaces, host immunomodulation and molecular cross-talking with the host. FEMS Immunol Med Microbiol 54:1–17

Sanchez B, Urdaci MC, Margolles A (2010) Extracellular proteins secreted by probiotic bacteria as mediators of effects that promote mucosa bacteria interactions. Microbiology 156:3232–3242

Sanders ME, Akkermans LM, Haller D, Hammerman C, Heimbach JT, Hörmannsperger G et al (2010) Safety assessment of probiotics for human use. Gut Microbes 1:164–185

Saud B, Pandey P, Paudel G, Dhungana G, Shrestha V (2020) In-vitro antibacterial activity of probiotic against human multidrug resistant pathogens. Arch Vet Sci Med 3(1):31–39

Savluchinske-Feio S, Curto MJM, Gigante B, Roseiro JC (2006) Antimicrobial activity of resin acid derivatives. Appl Microbiol Biotechnol 72(3):430–436

Scanlon TC, Dostal SM, Griswold KE (2014) A high-through put screen for antibiotic drug discovery. Biotechnol Bioeng 111:232–243

Schrader KK, Avolio F, Andolfi A, Cimmino A, Evidente A (2013) Ungeremine and its hemisynthesized analogues as bactericides against *Flavobacterium columnare*. J Agric Food Chem 61(6):1179–1183

Schuch R, Lee HM, Schneider BC, Sauve KL, Law C, Khan BK et al (2014) Combination therapy with lysin CF-301 and antibiotic is superior to antibiotic alone for treating methicillin-resistant *Staphylococcus aureus*-induced murine bacteremia. J Infect Dis 209(9):1469–1478

Senan S, Prajapati J, Joshi C (2015) Feasibility of genome-wide screening for biosafety assessment of probiotics: a case study of *Lactobacillus helveticus* MTCC 5463. Probiotics Antimicrob Proteins 7:249–258

Sgouras DN, Panayotopoulou EG, Martinez-Gonzalez B, Petraki K, Michopoulos S, Mentis A (2005) *Lactobacillus johnsonii* La1 attenuates *Helicobacter pylori*-associated gastritis and reduces levels of proinflammatory chemokines in C57BL/6 mice. Clin Diagn Lab Immunol 12:1378–1386

Shakeri A, Sharifi MJ, Fazly Bazzaz BS, Emami A, Soheili V, Sahebkar A et al (2018) Bioautography detection of antimicrobial compounds from the essential oil of salvia Pachystachys. Curr Bioact Compd 14(1):80–85

Shao J, Zhang M, Wang T, Li Y, Wang C (2016) The roles of CDR1, CDR2, and MDR1 in kaempferol-induced suppression with fluconazole-resistant *Candida albicans*. Pharm Biol 54 (6):984–992

Sharifzadeh A, Khosravi AR, Shokri H, Shirzadi H (2018) Potential effect of 2isopropyl-5-methylphenol (thymol) alone and in combination with fluconazole against clinical isolates of *Candida albicans, C. glabrata* and *C. krusei*. J Mycol Med 28(2):294–299

Sharma C, Singh BP, Thakur N, Gulati S, Gupta S, Mishra SK, Panwar H (2017) Antibacterial effects of *Lactobacillus* isolates of curd and human milk origin against foodborne and human pathogens. 3 Biotech 7:31

Sherpa RT, Reese CJ, Montazeri Aliabadi H (2015) Application of iChip to grow "uncultivable" microorganisms and its impact on antibiotic discovery. J Pharm Pharm Sci 18:303–315

Siddiqui BS, Ali ST, Rizwani GH, Begum S, Tauseef S, Ahmad A (2012) Antimicrobial activity of the methanolic bark extract of *Holarrhena pubescens* (Buch.Ham), its fractions and the pure compound conessine. Nat Prod Res 26(11):987–992

Simova ED, Beshkova DB, Dimitrov ZPB (2009) Characterization and antimicrobial spectrum of bacteriocins produced by lactic acid bacteria isolated from traditional Bulgarian dairy products. J Appl Microbiol 106:692–701

Siriyong T, Chusri S, Srimanote P, Tipmanee V, Voravuthikunchai SP (2016) *Holarrhena antidysenterica* extract and its steroidal alkaloid, Conessine, as resistance-modifying agents against extensively drug-resistant *Acinetobacter baumannii*. Microb Drug Resist 22(4):273–282

Sisto F, Brenciaglia MI, Scaltrito MM, Dubini F (2000) *Helicobacter pylori*: ureA, cagA and vacA expression during conversion to the coccoid form. Int J Antimicrob Agents 15:277–282

Smyth T, Ramachandran V, Smyth W (2009) A study of the antimicrobial activity of selected naturally occurring and synthetic coumarins. Int J Antimicrob Agents 33(5):421–426

Sobolewska D, Podolak I, Makowska-Was J (2015) Allium ursinum: botanical, phytochemical and pharmacological overview. Phytochem Rev 14(1):81–97

Sofrata A, Santangelo EM, Azeem M, Borg-Karlson AK, Gustafsson A, Putsep K (2011) Benzyl isothiocyanate, a major component from the roots of *Salvadora persica* is highly active against Gram-negative bacteria. PLoS One 6(8):e23045

Soltani R, Fazeli H, Bahri Najafi R, Jelokhanian A (2017) Evaluation of the synergistic effect of Tomatidine with several antibiotics against standard and clinical isolates of *Staphylococcus aureus*, *Enterococcus faecalis*, *Pseudomonas aeruginosa* and *Escherichia coli*. Iran J Pharm Res 16(1):290–296

Song J, Xia F, Jiang H, Li X, Hu L, Gong P et al (2016) Identification and characterization of HolGH15: the holin of *Staphylococcus aureus* bacteriophage GH15. J Gen Virol 97:1272–1281

Songisepp E, Hütt P, Rätsep M, Shkut E, Kõljalg S, Truusalu K et al (2012) Safety of a probiotic cheese containing *Lactobacillus plantarum* Tensia according to a variety of health indices in different age groups. J Dairy Sci 95:5495–5509

Soumet C, Méheust D, Pissavin C, Le Grandois P, Frémaux B, Feurer C et al (2016) Reduced susceptibilities to biocides and resistance to antibiotics in food-associated bacteria following exposure to quaternary ammonium compounds. J Appl Microbiol 121:1275–1281

Spinler JK, Taweechotipatr M, Rognerud CL, Ou CN, Tumwasorn S, Versalovic J (2008) Human-derived probiotic *Lactobacillus reuteri* demonstrate antimicrobial activities targeting diverse enteric bacterial pathogens. Anaerobe 14:166–171

Sridevi D, Shankar C, Prakash P, Park JH, Thamaraiselvi K (2017) Inhibitory effects of reserpine against efflux pump activity of antibiotic resistance bacteria. Chem Biol Lett 4(2):69–72

Stermitz FR, Beeson TD, Mueller PJ, Hsiang J, Lewis K (2001) *Staphylococcus aureus* MDR efflux pump inhibitors from a Berberis and a Mahonia (sensu strictu) species. Biochem Syst Ecol 29 (8):793–798

Stermitz FR, Cashman KK, Halligan KM, Morel C, Tegos GP, Lewis K (2003) Polyacylated neohesperidosides from *Geranium caespitosum*: bacterial multidrug resistance pump inhibitors. Bioorg Med Chem Lett 13(11):1915–1918

Sturme MH, Francke C, Siezen RJ, de Vos WM, Kleerebezem M (2007) Making sense of quorum sensing in lactobacilli: aspecial focus on *Lactobacillus plantarum* WCFS1. Microbiology 153:3939–3947

Sulakvelidze A (2013) Using lytic bacteriophages to eliminate or significantly reduce contamination of food by foodborne bacterial pathogens. J Sci Food Agric 93:3137–3146

Sun N, Du RL, Zheng YY, Huang BH, Guo Q, Zhang RF et al (2017) Antibacterial activity of N-methylbenzofuro[3,2-b]quinoline and Nmethylbenzoindolo[3,2-b]-quinoline derivatives and study of their mode of action. Eur J Med Chem 135:1–11

Tagg JR, Dierksen KP (2003) Bacterial replacement therapy: adapting "germ warfare" to infection prevention. Trends Biotechnol 21:217–223

Takac M, Blasi U (2005) Phage P68 virion-associated protein 17 displays activity against clinical isolates of *Staphylococcus aureus*. Antimicrob Agents Chemother 49:2934–2940

Tan N, Bilgin M, Tan E, Miski M (2017) Antibacterial activities of pyrenylated coumarins from the roots of Prangos hulusii. Molecules 22(7):1098

Téllez G, Lauková A, Latorre JD, Hernandez-Velasco X, Hargis BM, Callaway T (2015) Food-producing animals and their health in relation to human health. Microb Ecol Health Dis 26:25876

Thierry A, Deutsch SM, Falentin H, Dalmasso M, Cousin FJ, Jan G (2011) New insights into physiology and metabolism of *Propionibacterium freudenreichii*. Int J Food Microbiol 149:19–27

Togashi N, Hamashima H, Shiraishi A, Inoue Y, Takano A (2010) Antibacterial activities against *Staphylococcus aureus* of terpene alcohols with aliphatic carbon chains. J Essent Oil Res 22 (3):263–269

Tominaga K, Higuchi K, Hamasaki N, Hamaguchi M, Takashima T, Tanigawa T et al (2002) In vivo action of novel alkyl methyl quinolone alkaloids against *Helicobacter pylori*. J AntimicrobChemother 50(4):547–552

Tompkins TA, Hagen KE, Wallace TD, Fillion-Forte V (2008) Safety evaluation of two bacterial strains used in Asian probiotic products. Can J Microbiol 54:391–400

Tompkins TA, Barreau G, Broadbent JR (2012) Complete genome sequence of *Lactobacillus helveticus* R0052, a commercial probiotic strain. J Bacteriol 194:6349

Totté JEE, van Doorn MB, Pasmans SGMA (2017) Successful treatmentof chronic*Staphylococcusaureus*-related dermatoses with the topical endolysin staphefekt SA.100: a report of 3 cases. Case Rep Dermatol 9:19–25

Tsai CC, Lin PP, Hsieh YM (2008) Three *Lactobacillus* strains from healthy infant stool inhibit enterotoxigenic *Escherichia coli* grown in vitro. Anaerobe 14:61–67

Tyagi P, Singh M, Kumari H, Kumari A and Mukhopadhyay K (2015). Bactericidal activity of curcumin I is associated with damaging of bacterial membrane. PLoS one., 10(3):e0121313-e.

Ursell LK, Treuren WV, Metcalf JL, Pirrung M, Gewirtz A, Knight R (2013) Replenishing our defensive microbes. BioEssays 35:810–817

Valadkhani Z, Hassan N, Aghighi Z et al (2016) Protective role of *Lactobacillus acidophilus* against vaginal infection with *Trichomonas vaginalis*. Mediterranean. J Biosci 1(2):50–54

Van Boeckel TP, Gandra S, Ashok A, Caudron Q, Grenfell BT, Levin SA, Laxminarayan R (2014) Global antibiotic consumption 2000 to 2010: An analysis of national pharmaceutical sales data. Lancet Infect Dis 14:742–750

Van Boeckel TP, Brower C, Gilbert M, Grenfell BT, Levin SA, Robinson TP et al (2015) Global trends in antimicrobial use in food animals. Proc Natl Acad SciUSA 112:5649–5654

van Baarlen P, Wells JM, Kleerebezem M (2013) Regulation of intestinal homeostasis and immunity with probiotic lactobacilli. Trends Immunol 34:208–215

Vandevelde NM, Tulkens PM, Van Bambeke F (2016) Modulating antibiotic activity towards respiratory bacterial pathogens by co-medications: a multi-target approach. Drug DiscovToday 21(7):1114–1129

Vankerckhoven V, Huys G, Vancanneyt M, Snauwaert C, Swings J, Klare I et al (2008) Genotypic diversity, antimicrobial resistance, and virulence factors of human isolates and probiotic cultures constituting two intraspecific groups of *Enterococcus faecium* isolates. Appl Environ Microbiol 74:4247–4255

Varankovich NV, Nickerson MT, Korber DR (2015) Probiotic-based strategies for therapeutic and prophylactic use against multiple gastrointestinal diseases. Front Microbiol 6:685

Ventura M, Turroni F, Canchaya C, Vaughan EE, O'toole PW, Sinderen DV (2009) Microbial diversity in the human intestine and novel insights from metagenomics. Front Biosci (Landmark Ed.) 14:3214–3221

Verraes C, Van Boxstael S, Van Meervenne E, Van Coillie E, Butaye P, Catry B et al (2013) Antimicrobial resistance in the food chain: a review. Int J Environ Res Public Health 10:2643–2669

Viertel TM, Ritter K, Horz HP (2014) Viruses versus bacteria—novel approaches to phage therapy as a tool against multidrug-resistant pathogens. J Antimicrob Chemother 69:2326–2336

Vizoso Pinto MG, Franz CM, Schillinger U, Holzapfel WH (2006) *Lactobacillus* spp. with in vitro probiotic properties from human faeces and traditional fermented products. Int J Food Microbiol 109:205–214

Walsh C (2000) Molecular mechanisms that confer antibacterial drug resistance. Nature 17 (406):775–781

Webber MA, Whitehead RN, Mount M, Loman NJ, Pallen MJ, Piddock LJV (2015) Parallel evolutionary pathways to antibiotic resistance selected by biocide exposure. J Antimicrob Chemother 70:2241–2248

Weber-Dabrowska B, Jonczyk-Matysiak E, Zaczek M, Łobocka M, Łusiak-Szelachowska M, Górski A (2016) Bacteriophage procurement for therapeutic purposes. Front Microbiol 7:1177

WHO. Antimicrobial resistance: global report on surveillance, vol. 2014. WHO, Geneva, 2014

Wright GD (2005) Bacterial resistance to antibiotics: enzymatic degradation and modification. Adv Drug Deliv Rev 57:1451–1470

Wu D, Kong Y, Han C, Chen J, Hu L, Jiang H et al (2008) d-Alanine:d-alanine ligase as a new target for the flavonoids quercetin and apigenin. Int J Antimicrob Agents 32(5):421–426

Wu Y, Bai J, Zhong K, Huang Y, Qi H, Jiang Y et al (2016) Antibacterial activity and membrane-disruptive mechanism of 3-p-trans-coumaroyl-2-hydroxyquinic acid, a novel phenolic compound from pine needles of *Cedrus deodara*, against *Staphylococcus aureus*. Molecules 21 (8):1084

Xiao ZP, Pengn ZY, Dong JJ, He J, Ouyang H, Feng YT et al (2013) Synthesis, structure activity relationship analysis and kinetics study of reductive derivatives of flavonoids as *Helicobacter pylori* urease inhibitors. Eur J Med Chem 63:685–695

Xiao ZP, Wang XD, Wang PF, Zhou Y, Zhang JW, Zhang L et al (2014) Design, synthesis, and evaluation of novel fluoroquinolone-flavonoid hybrids as potent antibiotics against drug-resistant microorganisms. Eur J Med Chem 80:92–100

Xiao H, Shao F, Wu M, Ren W, Xiong X, Tan B et al (2015) The application of antimicrobial peptides as growth and health promoters for swine. J Anim Sci Biotechnol 6:19

Yadav MK, Chae SW, Im GJ, Chung JW, Song JJ (2015) Eugenol: a phyto-compound effective against methicillin-resistant and methicillin-sensitive *Staphylococcus aureus* clinical strain biofilms. PLoS One 10(3):e0119564

Yi ZB, Yan Y, Liang YZ, Bao Z (2007) Evaluation of the antimicrobial mode of berberine by LC/ESI-MS combined with principal component analysis. J Pharm Biomed Anal 44(1):301–304

Yoong P, Schuch R, Nelson D, Fischetti VA (2004) Identification of a broadly active phage lytic enzyme with lethal activity against antibiotic-resistant *Enterococcus faecalis* and *Enterococcus faecium*. J Bacteriol 186:4808–4812

Yuksekdag ZN, Darilmaz DO, Beyatli Y (2014) Dairy Propionibacterium strains with potential as biopreservatives against food borne pathogens and their tolerance–resistance properties. Eur Food Res Technol 238:17–26

Zamfir M, Callewaert R, Cornea PC, Savu L, Vatafu I, De Vuyst L (1999) Purification and characterization of a bacteriocin produced by *Lactobacillus acidophilus* IBB 801. J Appl Microbiol 87:923–931

Zhang Y, Zhang L, Du M, Yi H, Guo C, Tuo Y, Han X, Li J, Yang L (2011) Antimicrobial activity against *Shigella sonnei* and probiotic properties of wild lactobacilli from fermented food. Microbiol Res 167:27–31

Zhang Y, Sass A, Van Acker H, Wille J, Verhasselt B, Van Nieuwerburgh F et al (2018) Coumarin reduces virulence and biofilm formation in *Pseudomonas aeruginosa* by affecting quorum sensing, type III secretion and c-di-GMP levels. Front Microbiol 9:1952

Zhao J, Liu Y, Xiao C, He S, Yao H, Bao G (2017) Efficacyof phage therapyin controlling rabbit colibacillosis and changes in cecal microbiota. Front Microbiol 8:957

Zhou LN, Ge XL, Dong TT, Gao HY, Sun BH (2017) Antibacterial steroidal alkaloids from *Holarrhena antidysenteriaca*. Chin J Nat Med 15(7):540–545

Zou D, Xie K, Wang H, Chen Y, Xie M (2014) Inhibitory effects of biochanin a on the efflux pump of methicillin-resistant *Staphylococcus aureus* (MRSA). Wei Sheng Wu Xue Bao 54 (10):1204–1211

Probiotic Bacteria as a Functional Delivery Vehicle for the Development of Live Oral Vaccines

13

Maryam Dadar, Youcef Shahali, and Naheed Mojgani

Abstract

Probiotics improve the immune system and help to fight with different infection. The LAB ability in eliciting the immune response against foreign antigens has led to their use as candidate vectors for mucosal vaccines. Here, the use of LAB as oral vaccine carriers and various expression systems intended to the production of heterologous proteins are reviewed and discussed. *Lactococcus lactis, Lactobacillus* strains, and *Streptococcus gordonii* are lactic acid bacteria (LAB) currently being advocated for use as live antigen delivery vehicles to mucosal sites. Since these vehicles differ in their life span and mode of antigen delivery within the small intestine, in this chapter we tried to determine the promising LAB candidates for the development of oral vaccines.

Keywords

Probiotics · *Lactococcus lactis* · *Streptococcus gordonii* · Lactobacillus strains · Vaccine

13.1 Introduction

The development of effective mucosal vaccines could have several advantages in modern vaccinology. Due to high cost, storage and delivery condition of vaccines, further improvements in the vaccination coverage are actively demanded in developing world. The development of a new generation of vaccines which could be

M. Dadar (✉) · Y. Shahali · N. Mojgani
Agriculture Research, Education and Extension Organization (AREEO), Razi Vaccine and Serum Research Institute (RVSRI), Karaj, Iran

N. Mojgani, M. Dadar (eds.), *Probiotic Bacteria and Postbiotic Metabolites: Role in Animal and Human Health*, Microorganisms for Sustainability 2,
https://doi.org/10.1007/978-981-16-0223-8_13

applied orally, with the ability to stimulate both mucosal and systemic immune responses, offers promising perspectives for the eradication of infection diseases in many regions worldwide. For this purpose, over the last decade probiotic lactic acid bacteria (LAB) have been extensively studied for their potential use in the development of new generation of oral vaccines (Mojgani et al. 2020). Probiotic bacteria are known as profitable bacteria that can be easily stored, delivered, and relatively inexpensive to produce (Dadar et al. 2017; Mojgani et al. 2020). Another important advantage of these bacterial vectors is their ability to deliver immune-protective antigens at the mucosal surfaces, leading to cell and/or humoral-mediated responses. The induction of targeted and specific local immune responses in the mucosa-associated lymphoid tissue (MALT) can be most effectively achieved by the direct application of the vaccine antigen to the mucosal surface by bacterial vectors. In this regard, a number of LAB that are commonly found in the intestine of animals and humans appeared to be promising candidates (Jandhyala et al. 2015; Boersma et al. 2000). Some strains of *Lactobacillus and Lactococcus* spp. are termed as biotherapeutic agents due to their immune regulatory responses. The main immune effects of probiotic LAB could be summarized in their immune modulatory features that have been reported in the treatment of many disorders such as autoimmune diseases and different cancers as well as their anti-inflammatory, immunogenic, and immuno-adjuvant properties favoring their use as live vaccine delivery system (Kassayova et al. 2014). Some randomized, placebo-controlled clinical trials (RCTs) highlighted the promising and effectiveness of probiotics as delivery vehicle able to promote both parenterally and mucosally administered vaccine-specific immune responses (Amdekar et al. 2010). Particularly, *Lactobacilli* were found to increase the performance of several candidate mucosal vaccines for infantile diarrhea, HIV, and malaria in preclinical studies involving experimental animals. One of the most important advantages of probiotics is the ease of its administration (Licciardi and Tang 2011), while stimulating the immune responses against the vaccine antigen. These outstanding properties could help to provide effective protection against infectious diseases particularly in regions with low vaccination coverage. In this chapter we will highlight different aspects of promising probiotics species, effective transformation systems, and putative live oral vaccines. The promising results achieved with a number of these beneficial constructs will be discussed.

13.2 Probiotics and Their Safety Status

The profitable live microorganisms colonizing our digestive tract have been named “probiotics”(Morelli and Capurso 2012). They preserve gut health by induction of immunomodulatory impacts to maintain the intestinal homeostasis and by modulation of the gastrointestinal microbiota (Adel et al. 2017; Azcárate-Peril et al. 2011; Kałużna-Czaplińska et al. 2017; Mojgani et al. 2020). Lilly and Stillwell introduced the term “probiotic” to explore components released by protozoa, which improved the growth of other organisms (Lilly and Stillwell 1965). Several investigation supported the effective role of probiotic strains as promising supplements to improve mucosal immunity while having limited side effects (Rolfe 2000; Van Doan et al. 2020). It is now well documented that probiotics help to preserve healthy digestion (Cremon et al. 2018). Very limited reports of any adverse effects of probiotics have been reported. However, the safety of probiotic strains is highly associated with specific health claims and their purported advantages (Pradhan et al. 2020). Furthermore, probiotic can provide benefits to multiple parts of the body, where other supplements commonly focus on specific organs (Reid et al. 2003; Varankovich et al. 2015; Bisanz et al. 2015). The beneficial effects of probiotics have been reported on improving the bioavailability of essential nutrients (Pandey et al. 2015), gastrointestinal health (Varankovich et al. 2015), ameliorating the adaptive and innate immune response (Galdeano et al. 2019), decreasing common symptoms of food intolerance (Oak and Jha 2019), preventing the onset of certain cancers (Sharma 2019), and preventing atopic sensitization among susceptible subjects (Allen et al. 2014). Probiotics also clearly benefit patients who have had broad-spectrum antibiotic therapy by restoring a healthy intestinal microbiota (Sartor 2004; Peterson et al. 2015; Sánchez et al. 2017). Probiotics also appeared to be effective in the stimulation of both mucosal and systemic antigen-specific immune responses throughout the nasal and oral route (Anand et al. 2019; Neto et al. 2018; Cervin 2018; Mercenier et al. 2000).

13.3 Probiotics as Delivery Vehicles for Vaccine

Probiotics are described as live microorganisms that live in the gastrointestinal (GI) tract and, when administered in acceptable amounts, may confer a health benefit to the host (Tang 2009). About 100 years ago, Elie Metchnikoff discovered the health properties of probiotics and attributed the long life of Bulgarian peasants to their consumption of Lactobacillus probiotic bacteria (Rizzardini et al. 2012; Youngster et al. 2011). More recently, many reports pointed out the beneficial effects of probiotics on adaptive and innate immune responses in vivo and in vitro (Adel et al. 2017; Mojgani et al. 2020; Safari et al. 2016; Van Doan et al. 2020). Probiotics can regulate immunological responses directly by their interaction with epithelial cells and intestinal immune cells and/or indirectly by regulation of the intestinal microbiota (Adel et al. 2017; Power et al. 2014). Moreover, the effective role of probiotics in health improvement appeared to be the consequence of their combined impacts on epithelial barrier integrity, immune modulation, and gut microbiota (Mojgani et al. 2020). It was shown that interactions between immune system and the microbiota are important for the safe and natural improvement of healthy immune responses. Several investigation have reported that intestinal dysbiosis can lead to chronic inflammatory conditions such as inflammatory bowel disease and allergic disease likely as a result of abnormal regulation of immune system (Johansson et al. 2011; Westerholm-Ormio et al. 2010). Immune effects of probiotics can vary according to the selected probiotic, indicating its effects are strain- and species-specific (Licciardi and Tang 2011). Nowadays, the most broadly evaluated probiotic bacteria in clinical trials and animal models are the Bifidobacteria and *Lactobacilli* species (Licciardi and Tang 2011). The immunomodulatory effects of probiotic bacteria also have been described by the beneficial role of some probiotic strains such as *Lactococcus lactis* as a protein expression system (Singh et al. 2017; Wang et al. 2020). Both commensal and pathogenic intestinal bacteria could interact with the gastrointestinal mucosal lymphoid system (GALT) by Pathogen Recognition Receptors (PRR) expressed on specialized epithelial DCs and M cells, and antigen-presenting cells (APC) (Amdekar et al. 2010). These signaling pathways are important for the homeostasis of the intestinal immune response, preventing immune dysregulation by promoting tolerogenic responses as well as by stimulating the host protection against intestinal pathogens at the same time (Licciardi and Tang 2011). Furthermore, lactobacilli can be delivered orally, thereby providing a convenient presentation of antigens that can be applied on a large scale as an inexpensive option in less industrialized countries (Pouwels et al. 1998). Molecular-biological features such as efficient adhesion and colonization to human tissue surfaces play a major role in specific and nonspecific immune responses to LAB (Bermúdez-Humarán 2009). The role of the human microbiome as a regulator of both systemic and mucosal immunity is now well documented, and some probiotics appeared to be promising vaccine candidates (Ferreira et al. 2005; Vitetta et al. 2017). The applicable LAB for use as delivery vectors of vaccine commonly comprise multiple *Lactobacillus* species, *Lactococcus lactis,* and *Streptococcus gordonii.* Several comprehensive reviews of recombinant *L. lactis* vaccines describing the immune response of these recombinant bacteria against viral and bacterial antigens have been published (Bahey-El-Din et al. 2010; Pontes et al. 2011; Bermúdez-Humarán et al. 2011). However, because of the large number of published articles detailing *L. lactis, S. gordonii,* and lactobacilli as vaccine vectors, this chapter will particularly focus on the results of *in vivo* studies.

13.4 Bacteria Antigen Expressed by Probiotics

Numerous studies have introduced engineered LAB strains as delivery vehicles for bacterial antigens. The efficacy of *L. lactis*-based vaccine is associated with the route of administration and is related to the nature and amount of antigen produced. For example, it has been shown that oral immunization with recombinant *L. lactis* expressing the pneumococcal protective protein A (PppA) induced sufficient protection against respiratory pneumococcal infection (Villena et al. 2008). However, another study suggested that a better immunostimulation is achieved by intranasal (IN) administration of recombinant *L. lactis* strains expressing *Yersinia pseudotuberculosis* low-calcium response V (LcrV) antigen (Daniel et al. 2009). Furthermore, IN-vaccinated mice had developed protection against both oral and systemic infections with *Y. pseudotuberculosis*, although oral vaccination failed to protect against an oral challenge with *Y. pseudotuberculosis* and did not induce any specific immune response (Daniel et al. 2009). *L. lactis*-secreting listeriolysin O (LLO) induces specific CD8+ T cells and sufficient protection against *Listeria monocytogenes* in a murine infection model (Bahey-El-Din et al. 2010). However, in vivo and in vitro characterization of DNA delivery by recombinant *L. lactis* secreting a mutated form of *L. monocytogenes* Internalin A (InlA) confirmed the production of the mutated InlA at the *L. lactis* surface and appeared to be a promising strategy for plasmid transfer (De Azevedo et al. 2012).

In another experiment, recombinant *L. lactis* MG1363 expressing urease subunit B (UreB) of *Helicobacter pylori* failed to produce a sufficient immune response through the different tested oral vaccination regimens against *H. pylori* challenge (Lee et al. 2001). In contrast, another study reported that oral vaccination with *L. lactis* MG1363 expressing the antigen CagL of *Helicobacter pylori* can be evaluated as a potentially live vaccine able to induce a significant immune response in mice (Aliramaei et al. 2020). Another study also confirmed the significant immunoreactivity of BALB/c mice to the *H. pylori* Lpp20 antigen expressed in an engineered *L. lactis* strain and administrated orally (Zhang et al. 2016). Orally and intraperitoneally immunized mice with recombinant *L. lactis* producing Omp31 antigen of *Brucella melitensis* also promoted the production of serum IgM and IgG antibodies as well as IFN-γ and IL-10 (Shirdast et al. 2020). The successful cloning and expression of *B. melitensis* bp26 gene in *L. lactis* also confirmed the efficacy of this vector for the production of an oral vaccine conferring protection against brucellosis (Maghvan et al. 2019). Similarly, oral immunization using the recombinant probiotic *Lactobacillus casei* expressing the outer membrane protein OMP19, which is specific to *Brucella* species, induced strong mucosal immune responses in mice and protection against *Brucella abortus* (Mohammadi and Golchin 2020). A recombinant *Lactobacillus casei* expressing a flagellar antigen from *Salmonella enterica* alsoinduced cell-mediated immune responses and offered an efficient mucosal protection (Kajikawa et al. 2007). Another important LAB vector that has been used in numerous studies is *S. gordonii*. The subcutaneous injection of *S. gordonii* cells expressing S1 subunit of pertussis from *Bordetella pertussis* on their surface efficiently immunized New Zealand white rabbits (Lee et al. 1999). Furthermore, nasal or oral immunization of dd-Y mice with recombinant *L. lactis* expressing surface protective antigen (SpaA) led to the production of antigen-specific fecal IgA and serum IgG protecting mice against a challenge with *Erysipelothrix rhusiopathia* (Cheun et al. 2004). However, there are several oral vaccinations with LAB-based vaccine vectors that are highly variable in immune responses and antibody production; therefore, further works are needed to improve the efficiency and delivery route of the live bacterial vector.

13.5 Viral Antigen Expressed by Probiotics

A number of viral antigens have been produced in LAB-based vectors. Remarkably Xin et al. reported the protective effect and immunogenicity of an orally administered recombinant *L. lactis* expressing surface-bound (Env) protein of human immunodeficiency virus type 1 (HIV-1), showing sufficient protection in mice challenged intraperitoneally with the virus (Xin et al. 2003). Another study also confirmed that the construction of paired cell surface display of *Salmonella enterica* Serovar Typhimurium *FliC* and HIV-1 *Gag* in *Lactobacillus acidophilus* as mucosal vaccine vectors could induce an increase of Gag-specific IgA-secreting cells. Interestingly, the expression of FliC in LAB conferred an adjuvant impact on the local production of IgA (Kajikawa et al. 2012). The oral administration of the recombinant *L. acidophilus* expressing the membrane proximal external region (MPER) of HIV-1 within the major S-layer protein (SlpA) led to a Th1 and Th17 dominance in the immune responses (Kajikawa et al. 2015). The use of lactobacilli expressing HIV antigens was found to be an effective way to improve the systemic (serum IgG) and mucosal (IgA) immune responses against HIV proteins (Gag, MPER) in orally immunized mice (LeCureux and Dean 2018). The subcutaneous injection of a genetically modified *Lactobacillus casei* producing VLPs of human papillomavirus (HPV) by the L1 protein led to an increased expression of serum-specific IgG in BALB/c mice (Aires et al. 2006). Other studies performed on C57BL/6 mice have effectively used a recombinant *L. casei*, displaying on its surface the minor capsid protein L2 as well as the early oncoproteins E6 and E7 proteins of HPV, to induce T cell-regulated cellular immunity with antitumor impacts (Lee et al. 2010; Poo et al. 2006). In Balb/c mice, the oral utilization of the L2 protein of HPV-16 expressed in *L. casei* was able to stimulate mucosal and systematic cross-neutralizing responses (Yoon et al. 2012). Similarly, the oral administration of the genetically modified *L. casei* displaying E7 protein of HPV-16 promoted the generation of antigen-specific cytotoxic T lymphocytes against HPV16 in C57BL/6 mice (Adachi et al. 2010). The same approach was used to immunize C57BL/6 mice with the severe acute respiratory syndrome coronavirus (SARS-CoV)-spike protein expressed in *L. casei*. The nasal and oral inoculations of the recombinant *L. casei* displaying this protein generated high level of neutralizing-antibodies and an effective protection against the SARS-CoV (Lee et al. 2006). In this study, the authors also reported an elevated content of mucosal IgA in bronchoalveolar and intestinal lavage fluids of immunized C57BL/6 mice following the intranasal or oral immunization. The N protein of SARS-CoV expressed into the cytoplasm or secreted in the medium of *L. lactis* modified by genetic engineering was also able to stimulate the production of N-specific IgG as a promising mucosal vaccine candidate (Pei et al. 2005). *L. lactis* was also used for the production of a recombinant flaviviral E protein of Dengue virus and effectively promoted the humoral immune response to dengue infection in immunized mice (Crill et al. 2009).

The oral immunization with recombinant porcine rotavirus VP4 and VP4-LTB expressed in *L. casei* stimulated neutralizing serum IgG and mucosal IgA antibody responses in female Balb/c mice (Qiao et al. 2009). An efficient protection was also achieved by using a recombinant *L. rhamnosus* GG expressing protein G of rotavirus in mouse model, decreasing the severity, prevalence, and duration of diarrhea-associated rotavirus infections (Günaydın et al. 2014). In a fish model, the oral administration of *L. casei* expressing protein antigens of the infectious pancreatic necrosis virus (IPNV)(Min et al. 2012; Li-Li et al. 2012), Koi herpesvirus (KHV) (Cui et al. 2015), spring viremia of carp virus (SVCV) (Cui et al. 2015; Dadar et al. 2018), and viral hemorrhagic septicemia (VHS) (Naderi-Samani et al. 2020) significantly stimulated the generation of serum IgM and was found to be a promising vaccine vector candidate against viral disease in fish.

Different study also confirm the protective role of the oral administration of an *L. plantarum* strain expressing hemagglutinin (HA) gene of H9N2 and H5N1 avian influenza virus by increasing the production of serum IgG, fecal IgA, and bronchiolar IgA in BALB/c mice (Shi et al. 2014; Wang et al. 2012).

13.6 Protozoal Antigens Expressed by Probiotics

L. lactis have also been applied as a promising candidate for the delivery and expression of protozoal antigens. Oral immunization of BALB/c and C57BL/6 mice against rodent malaria with recombinant *L. lactis* expressing the C terminal fragment of the merozoite surface protein 1 (MSP-119) of *Plasmodium yoelii* provided protection at the asexual erythrocytic stage of malaria (Zhang et al. 2005). The oral administration of the *Giardia lamblia* cyst wall protein 2 (CWP2) displayed at the surface of *L. lactis* led to increased levels of mucosal anti-CWP2 IgA in the intestine of mice along with a significant decrease in fecal cyst shedding following challenge experiments with *Giardia muris* (Lee and Faubert 2006).

The protective effect of vaccine antigens expressed in LAB against parasite infection could be significantly affected by inoculation doses, the vaccination regimen used, the route of administration, its cellular expression in LAB strain utilize and nature and amount of the expressed antigen.

13.7 Immunomodulatory Responses of Probiotic Bacteria

It is now well documented that immune response could be directly or indirectly affected by the gut microbiota, influencing the pathways involved in both innate and adaptive immune responses (Frei et al. 2015; Hoseinifar et al. 2017). Therefore, it was suggested that the condition of many diseases could be improved by improving the gut microbiota. This hypothesis was supported by the weak performance of oral vaccines in developing countries because of a higher proportion of children with impaired or dysbiotic gut microbiota (Rosshart et al. 2017; Sánchez et al. 2017). The protective mechanisms of probiotic are commonly related to the interaction of probiotics with host cells or to the cross-talk of probiotics with other microbiota or pathogenic microorganisms (Taverniti and Guglielmetti 2011). Direct interaction of probiotic bacteria with the host cell can be regulated by the bacterial cells, independent of their viability and by the multitude of specific components or products released by bacteria as well as through their effects on specific human cells, such as those of the mucosa-associated lymphoid tissue (MALT, playing an important immunoregulatory role on the specific immunity (Adams 2010). Furthermore, the immunomodulatory effects of genomic DNA of probiotic bacteria have been reported on the human peripheral blood mononuclear cells (PBMC) responses through elevated IL-1 and IL-10 production(Lammers et al. 2003). The significant differences in magnitude and kinetics of IL-10 and IL-1β release in response to genomic DNA of probiotic revealed the influence of gut bacterial components on the intestinal response of the mucosal immune system (Delcenserie et al. 2008; Lammers et al. 2003).

Interestingly, a study demonstrated that heat-inactivated probiotics such as *Lactobacillus casei* Shirota were marginally less effective in stimulating some proinflammatory cytokines, including interleukin tumor necrosis factor (TNF)-a and (IL)-12, when compared to viable cells, while similar induction of IL-10 was reported for viable and inactivated cells. Moreover, live Gram-negative probiotic bacteria of *Escherichia coli* Nissle 1917 stimulated the release of a higher amount of IL-10 and proinflammatory cytokines by the murine monocyte/macrophage cell line (J774A) when compared to heat-killed bacteria (Cross et al. 2004). The immunogenic effects reported by probiotics are directly associated with their immune-modulating responses (production of IL-6 and IL-10) and their tumor-reducing activities (Reid et al. 2003; Van Hoang et al. 2018; Amdekar et al. 2010). However, species- and strain-specific effects have been attributed to different LAB based on their ability to promote particular patterns of mucosal cytokine expression (Mojgani et al. 2020). Through the oral route, probiotic bacteria enter the gut and may promote the induction of cytokines through mucosal lymphoid cells. The up or down modulation of the immune response could be affected by a broad range of immune cells such as endothelial cells, fibroblasts, and stromal cells which are responsible for the production of specific cytokines. Furthermore, the interaction between probiotic bacteria and the gut epithelial cells leads to the stimulation of a cascade of signals regulating the immune response (Amdekar et al. 2010; Galdeano and Perdigon 2006).The nonspecific immune response induced by probiotics mainly occurs through inflammatory pathways resulting in the activation of macrophages and phagocytic cells [polymorphonuclear (PMN)](Mojgani et al. 2020; Perdigon et al. 1995). After arrival of probiotic bacteria to the colon, they are absorbed by the overlying M cells in the Peyer's patches or across the overlying normal epithelium in the lamina propria of the small intestine. The intact probiotics cells are then handled by phagocytizing cells including antigen-presenting cells (APC), the macrophages, B and T lymphocytes, and dendritic cells (Perdigon et al. 1995; Perdigón et al. 2001). The mucosal immune epithelial cells stimulated by probiotics could regulate the immune defense pathways by the release of IL-2, IL-6, and IL-10 that are able to modulate both nonspecific and specific immune mechanisms through their actions on immune cells. Probiotic bacteria could also affect B cells in axillary lymph nodes by pulses of IL-6 expression, which is known as an important B cell differentiation factor (Akira et al. 2001). Probiotics are also capable to stimulate the release of cytokines through the macrophages and T cells, leading to the regulation of the mucosal immune response (Kawashima et al. 2018; Galdeano et al. 2019). Furthermore, the release of luminal secretory IgA could be promoted by some probiotic bacteria, regulating both systemic and mucosal immunity (Perdigon et al. 1995, Perdigón et al. 2001). LAB also regulates the immune responses by stimulating the production of type 1 interferons (IFNs), which has a critical effect on the anti-viral immune response (Kawashima et al. 2018). Moreover, it was shown that the absorption of particular LAB strains by DCs promotes the fractional maturation of dendritic cells implicated in the immune response to foreign- and self-antigens (Foligne et al. 2007).

13.8 The Design and Construct of Probiotic Bacteria as Putative Vaccines

A promising application of LAB is its administration for the improvement of live mucosal vaccines. Some LAB species such as *L. lactis* have been reported as promising vaccine vector candidates for a multitude of bacterial and viral antigens (Pouwels et al. 1998; Wells et al. 1996). Several delivery systems such as cell wall, extracellular or cytoplasm medium have been designed to target different heterologous proteins to a particular location of cell (Le Loir et al. 2005). For the construction of recombinant protein in LAB vehicles, several parts, including genes coding for heterologous proteins, and multiple cloning site (MCS) promoters and terminator (T) should be accurately designed (Yeng et al. 2009). Promoters commonly are species specific and originate from the bacteria used for protein expression. Moreover, several efficient expression systems have been designed to produce various heterologous proteins in LAB (Boersma et al. 2000; Villatoro-Hernández et al. 2012). The regulation of the expression of recombinant immunogenic proteins in LAB may be performed by strong or inducible constitutive promoters. The nisin-inducible promotor Pnis, acting as one of the key components in the nisin-inducible controlled expression (NICE) system, is now widely applied (Villatoro-Hernández et al. 2012). The NICE system is largely used to express heterologous proteins in *L. lactis* because of several advantages, including efficiently induced and tightly controlled expression leading to high yields of protein (Roshan and Souza 2012), large-scale production process, and easy use (Mierau and Kleerebezem 2005). Fermentation parameters, nisin amounts, and growth conditions have been optimized to increase the yield of recombinant proteins. In addition, another constitutive promoter that could not be controlled by any regulator or growth conditions are thought to be constitutive under laboratory growth conditions. The most generally used constitutive promoters originating from *L. lactis* genomic library include P21, P23, and P59 (strong promotors) as well as P32 and P44 (weak promotor). These strong and weak promoters have been used to express different heterologous proteins in *L. lactis* (Morello et al. 2008). Furthermore, the expression of proteins bearing an N-terminal signal peptide in the growth medium could be performed by the Sec pathway (Mierau and Kleerebezem 2005). However, multistep process and different factors localized in all cell compartments could impact on the protein secretion in Gram-positive bacteria. For example, intracellular targeting factors such as bacterial *ffh* genes which are contributed in protein folding and secretion and encode the protein elements of signal recognition particle (Tjalsma et al. 2004). The translocation machinery in *L. lactis* comprised of the ATPase-dependent motor, SecA, partly prepare the energy demanded for preprotein translocation, and integral membrane proteins, i.e., SecG, SecE, and SecY, produce the conducting channel via the hydrophobic membrane environment. Moreover, signal peptide cleavage such as *SipL* in *L. lactis* strains is another factor that is removed by type I signal peptidase during or shortly after preprotein translocation across the membrane. Surface quality control proteins *PmpA* and *HtrA* in the *L. lactis* are other factors involved in the protein secretion of Gram-positive bacteria. These proteins are transported by the Sec machinery and involved several folding factors like folding catalysts and chaperones, and are directly involved in the degradation of misfolded/unfolded proteins through housekeeping proteases (Morello et al. 2008).

13.9 Advantage and Disadvantage of Probiotics as Delivery Vectors of Vaccine

To date, human vaccines are categorized under 4 main groups including (1) live attenuated vaccines, (2) recombinant, subunit, conjugate, and polysaccharide vaccines, (3) toxoid vaccines, and (4) whole inactivated vaccines (Tong 2019). Some live attenuated vaccines raised safety and quality concerns leading to the further development of subunit and or killed vaccines during the last decades (Mercenier et al. 2000; del Rio et al. 2018; Jiang et al. 2019). After several studies on the nonpathogenic and pathogenic microorganisms naturally found in food, some of them have been applied in the development of safer live bacterial vaccines that can induce efficient immune response to one or more expressed antigens has been proposed (Glenting et al. 2007; Detmer and Glenting 2006). In this regard, some probiotics like LAB were proposed as potential mucosal delivery vehicles for vaccine development. The use of LAB as vaccine vectors showed several attractive benefits such as the maintenance and acceptance of genetic modifications, noninvasive administration (usually intranasal or oral), simple, high safety levels, and low cost. LAB also tends to induce high levels of mucosal and systemic antibodies against the expressed foreign antigen and minimal immune responses directed against LAB cells themselves after uptake by the mucosal immune system.

The main advantages of LAB such as *L. lactis* comprised of their well-recognized status as safe (GRAS) microorganisms for a long time in fermented foods, the lack of endotoxin lipopolysaccharides (LPS) commonly found in Gram-negative bacteria, safety administration for human, the accessibility to full genome sequencing of LAB, fewer native exoproteins, and smaller genome size in comparison with Gram-negative bacteria. The production of recombinant antigens in LAB is exempted from the use of complex and multistep purification and refolding procedures which complicate the protein production and is ideal for the development of mucosal vaccines (Bahey-El-Din et al. 2010; Bermúdez-Humarán 2009). Also, the secretion of heterologous proteins produced in LAB bacterial hosts into the medium can be performed easily, thereby facilitating their purification (Morello et al. 2008). However, a major disadvantage of the mucosal route for the use of recombinant L. lactis as oral vaccine vectors is the huge amount of protein which is required to compensate the natural protein degradation occurring at the mucosal surfaces of the gastrointestinal tract (Bermúdez-Humarán 2009). Moreover, the low efficiency of LAB transformation is a critical obstacle in the development of LAB vaccines and therefore the organization of effective transformation protocols is hotly demanded (Glenting et al. 2007; Tarahomjoo 2012). Thus, the development of novel vectors capable to minimize these limitations and optimize the delivery of immunogenic molecules to target tissues is a real challenge for future researches in the field.

13.10 Conclusions

LAB have been safely ingested by humans for several millenia. *L. lactis* appeared to be one of the most promising LAB in the production of oral vaccines and has been genetically engineered for the expression of a large variety of antigens and therapeutic proteins. This property has been used by scientists for the development of novel delivery vectors for immunogenic proteins into the mucosal tissues. New window for the administration of genetically engineered LAB as delivery vehicles has been opened by the successful Phase I clinical trial of a L. lactis strain with secretion of interleukin-10 for Crohn's disease. However, an improved potency of LAB-based delivery vehicles is required to elicit specific immune responses at lower doses. The capability of LAB vehicle carriers to stimulate effective protection against infective agents related to the antigen presentation mode (secreted, cell surface exposed or cytoplasmic), the sufficient delivery of antigen in vivo, immunization regimen (timing, dose and route), and development of LAB carrier's requirements for the most suitable LAB strains. Moreover, important steps for improvement of the vaccines efficiency is the coexpression of antigens with antigen-presenting cells targeting peptides and adjuvants in LAB vehicles along with the coadministration of immunoregulatory cytokines. The characterization of regulatory anchoring, and secretion signals from genome sequences could improve these features (Wells and Mercenier 2008). *L. lactis* provides a powerful genetic platform for the expression of different antigens which can be expressed and delivered mostly to the gut environment, often with very hopeful outcomes. The number of investigations involving lactobacilli and L. lactis has regularly increased over the last decade, smoothing the way for the development of a new generation of oral mucosal vaccines. In this regard, more works should be done to investigate the associated immune mechanisms within the mucosa, particularly the pathways of bacterial uptake into the immune inductive sites (DCs, M cells). The oral route for LAB administration also can have an impact on the immune response stimulated due to the diversity of mucosal inductive sites. The basic differences between the various location of antigen expression, including intracellular, surface display and secreted, as well as the properties specific to different LAB considerably influence the immune response, and therefore the selected strains should be properly assessed and used for specific antigens. However, the implementation of standardized model systems with emphasis on the most common mouse models (C57BL/6 and BALB/c), follow-up, experimental setup, and methodologies should be a valuable step forward the production and release of LAB-base vaccines.

References

Adachi K, Kawana K, Yokoyama T, Fujii T, Tomio A, Miura S, Tomio K, Kojima S, Oda K, Sewaki T (2010) Oral immunization with a Lactobacillus casei vaccine expressing human papillomavirus (HPV) type 16 E7 is an effective strategy to induce mucosal cytotoxic lymphocytes against HPV16 E7. Vaccine 28:2810–2817

Adams CA (2010) The probiotic paradox: live and dead cells are biological response modifiers. Nutr Res Rev 23:37–46

Adel M, El-Sayed A-FM, Yeganeh S, Dadar M, Giri SS (2017) Effect of potential probiotic Lactococcus lactis subsp. lactis on growth performance, intestinal microbiota, digestive enzyme activities, and disease resistance of Litopenaeus vannamei. Probiotics Antimicrob Proteins 9:150–156

Aires KA, Cianciarullo AM, Carneiro SM, Villa LL, Boccardo E, Pérez-Martinez G, Perez-Arellano I, Oliveira MLS, Ho PL (2006) Production of human papillomavirus type 16 L1 virus-like particles by recombinant Lactobacillus casei cells. Appl Environ Microbiol 72:745–752

Akira S, Takeda K, Kaisho T (2001) Toll-like receptors: critical proteins linking innate and acquired immunity. Nat Immunol 2:675

Aliramaei MR, Khorasgani MR, Rahmani MR, Esfahani SHZ, Emamzadeh R (2020) Expression of Helicobacter pylori CagL gene in Lactococcus lactis MG1363 and evaluation of its immunogenicity as an oral vaccine in mice. Microb Pathog 142:103926

Allen SJ, Jordan S, Storey M, Thornton CA, Gravenor MB, Garaiova I, Plummer SF, Wang D, Morgan G (2014) Probiotics in the prevention of eczema: a randomised controlled trial. Arch Dis Child 99:1014–1019

Amdekar S, Dwivedi D, Roy P, Kushwah S, Singh V (2010) Probiotics: multifarious oral vaccine against infectious traumas. FEMS Immunol Med Microbiol 58:299–306

Anand A, Sato M, Aoyagi H (2019) Screening of phosphate-accumulating probiotics for potential use in chronic kidney disorder. Food Sci Technol Res 25:89–96

Azcárate-Peril MA, Sikes M, Bruno-Bárcena JM (2011) The intestinal microbiota, gastrointestinal environment and colorectal cancer: a putative role for probiotics in prevention of colorectal cancer? Am J Physiol Gastrointest Liver Physiol 301:G401–G424

Bahey-El-Din M, Gahan CG, Griffin BT (2010) Lactococcus lactis as a cell factory for delivery of therapeutic proteins. Curr Gene Ther 10:34–45

Bermúdez-Humarán LG (2009) Lactococcus lactis as a live vector for mucosal delivery of therapeutic proteins. Hum Vaccin 5:264–267

Bermúdez-Humarán LG, Kharrat P, Chatel J-M, Langella P (2011) Lactococci and lactobacilli as mucosal delivery vectors for therapeutic proteins and DNA vaccines, microbial cell factories. BioMed Central:1–10

Bisanz JE, Enos MK, PrayGod G, Seney S, Macklaim JM, Chilton S, Willner D, Knight R, Fusch C, Fusch G (2015) Microbiota at multiple body sites during pregnancy in a rural Tanzanian population and effects of Moringa-supplemented probiotic yogurt. Appl Environ Microbiol 81:4965–4975

Boersma W, Shaw M, Claassen E (2000) Probiotic bacteria as live oral vaccines Lactobacillus as the versatile delivery vehicle. In: Probiotics 3. Springer, New York, pp 234–270

Cervin AU (2018) The potential for topical probiotic treatment of chronic rhinosinusitis, a personal perspective. Front Cell Infect Microbiol 7:530

Cheun H, Kawamoto K, Hiramatsu M, Tamaoki H, Shirahata T, Igimi S, Makino SI (2004) Protective immunity of SpaA-antigen producing Lactococcus lactis against Erysipelothrix rhusiopathiae infection. J Appl Microbiol 96:1347–1353

Cremon C, Barbaro MR, Ventura M, Barbara G (2018) Pre-and probiotic overview. Curr Opin Pharmacol 43:87–92

Crill WD, Hughes HR, Delorey MJ, Chang G-JJ (2009) Humoral immune responses of dengue fever patients using epitope-specific serotype-2 virus-like particle antigens. PLoS One 4:e4991

Cross ML, Ganner A, Teilab D, Fray LM (2004) Patterns of cytokine induction by gram-positive and gram-negative probiotic bacteria. FEMS Immunol Med Microbiol 42:173–180
Cui L-C, Guan X-T, Liu Z-M, Tian C-Y, Xu Y-G (2015) Recombinant lactobacillus expressing G protein of spring viremia of carp virus (SVCV) combined with ORF81 protein of koi herpesvirus (KHV): a promising way to induce protective immunity against SVCV and KHV infection in cyprinid fish via oral vaccination. Vaccine 33:3092–3099
Dadar M, Dhama K, Vakharia VN, Hoseinifar SH, Karthik K, Tiwari R, Khandia R, Munjal A, Salgado-Miranda C, Joshi SK (2017) Advances in aquaculture vaccines against fish pathogens: global status and current trends. Rev Fisher Sci Aquac 25:184–217
Dadar M, Chakraborty S, Dhama K, Prasad M, Khandia R, Hassan S, Munjal A, Tiwari R, Karthik K, Kumar D (2018) Advances in designing and developing vaccines, drugs and therapeutic approaches to counter human papilloma virus. Front Immunol 9:2478
Daniel C, Sebbane F, Poiret S, Goudercourt D, Dewulf J, Mullet C, Simonet M, Pot B (2009) Protection against Yersinia pseudotuberculosis infection conferred by a Lactococcus lactis mucosal delivery vector secreting LcrV. Vaccine 27:1141–1144
De Azevedo M, Karczewski J, Lefévre F, Azevedo V, Miyoshi A, Wells JM, Langella P, Chatel J-M (2012) In vitro and in vivo characterization of DNA delivery using recombinant Lactococcus lactis expressing a mutated form of L. monocytogenes Internalin A. BMC Microbiol 12:1–9
del Rio B, Redruello B, Fernandez M, Martin MC, Ladero V, Alvarez MA (2018) Lactic acid bacteria as a live delivery system for the in situ production of nanobodies in the human gastrointestinal tract. Front Microbiol 9:3179
Delcenserie V, Martel D, Lamoureux M, Amiot J, Boutin Y, Roy D (2008) Immunomodulatory effects of probiotics in the intestinal tract. Curr Issues Mol Biol 10:37
Detmer A, Glenting J (2006) Live bacterial vaccines–a review and identification of potential hazards. Microb Cell Factories 5:23
Ferreira L, Ferreira RC, Schumann W (2005) Bacillus subtilis as a tool for vaccine development: from antigen factories to delivery vectors. An Acad Bras Cienc 77:113–124
Foligne B, Zoumpopoulou G, Dewulf J, Younes AB, Chareyre F, Sirard J-C, Pot B, Grangette C (2007) A key role of dendritic cells in probiotic functionality. PLoS One 2:e313
Frei R, Akdis M, O'Mahony L (2015) Prebiotics, probiotics, synbiotics, and the immune system: experimental data and clinical evidence. Curr Opin Gastroenterol 31:153–158
Galdeano CM, Perdigon G (2006) The probiotic bacterium Lactobacillus casei induces activation of the gut mucosal immune system through innate immunity. Clin Vaccine Immunol 13:219–226
Galdeano CM, Cazorla SI, Dumit JML, Vélez E, Perdigón G (2019) Beneficial effects of probiotic consumption on the immune system. Ann Nutr Metab 74:115–124
Glenting J, Poulsen LK, Kato K, Madsen SM, Frøkiær H, Wendt C, Sørensen HW (2007) Production of recombinant peanut allergen Ara h 2 using Lactococcus lactis. Microb Cell Factories 6:28
Günaydın G, Zhang R, Hammarström L, Marcotte H (2014) Engineered Lactobacillus rhamnosus GG expressing IgG-binding domains of protein G: capture of hyperimmune bovine colostrum antibodies and protection against diarrhea in a mouse pup rotavirus infection model. Vaccine 32:470–477
Hoseinifar SH, Dadar M, Ringø E (2017) Modulation of nutrient digestibility and digestive enzyme activities in aquatic animals: the functional feed additives scenario. Aquac Res 48:3987–4000
Jandhyala SM, Talukdar R, Subramanyam C, Vuyyuru H, Sasikala M, Reddy DN (2015) Role of the normal gut microbiota. World J Gastroenterol: WJG 21:8787
Jiang B, Li Z, Ou B, Duan Q, Zhu G (2019) Targeting ideal oral vaccine vectors based on probiotics: a systematical view. Appl Microbiol Biotechnol 1-13
Johansson MA, Sjögren YM, Persson J-O, Nilsson C, Sverremark-Ekström E (2011) Early colonization with a group of Lactobacilli decreases the risk for allergy at five years of age despite allergic heredity. PLoS One 6:e23031

Kajikawa A, Satoh E, Leer RJ, Yamamoto S, Igimi S (2007) Intragastric immunization with recombinant Lactobacillus casei expressing flagellar antigen confers antibody-independent protective immunity against Salmonella enterica serovar Enteritidis. Vaccine 25:3599–3605

Kajikawa A, Zhang L, Long J, Nordone S, Stoeker L, LaVoy A, Bumgardner S, Klaenhammer T, Dean G (2012) Construction and immunological evaluation of dual cell surface display of HIV-1 gag and Salmonella enterica serovar Typhimurium FliC in Lactobacillus acidophilus for vaccine delivery. Clin Vaccine Immunol 19:1374–1381

Kajikawa A, Zhang L, LaVoy A, Bumgardner S, Klaenhammer TR, Dean GA (2015) Mucosal immunogenicity of genetically modified Lactobacillus acidophilus expressing an HIV-1 epitope within the surface layer protein. PLoS One 10:e0141713

Kałużna-Czaplińska J, Gątarek P, Chartrand MS, Dadar M, Bjørklund G (2017) Is there a relationship between intestinal microbiota, dietary compounds, and obesity? Trends Food Sci Technol 70:105–113

Kassayova M, Bobrov N, Strojný L, Kiskova T, Mikeš J, Demečková V, Orendáš P, Bojkova B, Péč M, Kubatka P (2014) Preventive effects of probiotic bacteria Lactobacillus plantarum and dietary fiber in chemically-induced mammary carcinogenesis. Anticancer Res 34:4969–4975

Kawashima T, Ikari N, Kouchi T, Kowatari Y, Kubota Y, Shimojo N, Tsuji NM (2018) The molecular mechanism for activating IgA production by pediococcus acidilactici K15 and the clinical impact in a randomized trial. Sci Rep 8:1–9

Lammers KM, Brigidi P, Vitali B, Gionchetti P, Rizzello F, Caramelli E, Matteuzzi D, Campieri M (2003) Immunomodulatory effects of probiotic bacteria DNA: IL-1 and IL-10 response in human peripheral blood mononuclear cells. FEMS Immunol Med Microbiol 38:165–172

Le Loir Y, Azevedo V, Oliveira SC, Freitas DA, Miyoshi A, Bermúdez-Humarán LG, Nouaille S, Ribeiro LA, Leclercq S, Gabriel JE (2005) Protein secretion in Lactococcus lactis: an efficient way to increase the overall heterologous protein production. Microb Cell Factories 4:2

LeCureux JS, Dean GA (2018) Lactobacillus mucosal vaccine vectors: immune responses against bacterial and viral antigens. Msphere 3

Lee P, Faubert GM (2006) Expression of the Giardia lamblia cyst wall protein 2 in Lactococcus lactis. Microbiology 152:1981–1990

Lee SF, March RJ, Halperin SA, Faulkner G, Gao L (1999) Surface expression of a protective recombinant pertussis toxin S1 subunit fragment in Streptococcus gordonii. Infect Immun 67:1511–1516

Lee MH, Roussel Y, Wilks M, Tabaqchali S (2001) Expression of Helicobacter pylori urease subunit B gene in Lactococcus lactis MG1363 and its use as a vaccine delivery system against H. pylori infection in mice. Vaccine 19:3927–3935

Lee J-S, Poo H, Han DP, Hong S-P, Kim K, Cho MW, Kim E, Sung M-H, Kim C-J (2006) Mucosal immunization with surface-displayed severe acute respiratory syndrome coronavirus spike protein on Lactobacillus casei induces neutralizing antibodies in mice. J Virol 80:4079–4087

Lee T-Y, Kim Y-H, Lee K-S, Kim J-K, Lee I-H, Yang J-M, Sung M-H, Park J-S, Poo H (2010) Human papillomavirus type 16 E6-specific antitumor immunity is induced by oral administration of HPV16 E6-expressing Lactobacillus casei in C57BL/6 mice. Cancer Immunol Immunother 59:1727–1737

Licciardi PV, Tang ML (2011) Vaccine adjuvant properties of probiotic bacteria. Discov Med 12:525–533

Li-Li Z, Min L, Jun-Wei G, Xin-Yuan Q, Yi-Jing L, Di-Qiu L (2012) Expression of infectious pancreatic necrosis virus (IPNV) VP2–VP3 fusion protein in Lactobacillus casei and immunogenicity in rainbow trouts. Vaccine 30:1823–1829

Lilly DM, Stillwell RH (1965) Probiotics: growth-promoting factors produced by microorganisms. Science 147:747–748

Maghvan MA, Jafari P, Hoseini SD, Behrozikhah AM (2019) Cloning and expression of B. mellitensis bp26 Gene in Lactococcus lactis as a food grade vaccine. Avicenna J Med Biotechnol 11:264

Mercenier A, Muller-Alouf H, Grangette C (2000) Lactic acid bacteria as live vaccines. Curr Issues Mol Biol 2:17–26
Mierau I, Kleerebezem M (2005) 10 years of the nisin-controlled gene expression system (NICE) in Lactococcus lactis. Appl Microbiol Biotechnol 68:705–717
Min L, Li-Li Z, Jun-Wei G, Xin-Yuan Q, Yi-Jing L, Di-Qiu L (2012) Immunogenicity of Lactobacillus-expressing VP2 and VP3 of the infectious pancreatic necrosis virus (IPNV) in rainbow trout. Fish Shellfish Immunol 32:196–203
Mohammadi E, Golchin M (2020) High protection of mice against Brucella abortus by oral immunization with recombinant probiotic Lactobacillus casei vector vaccine, expressing the outer membrane protein OMP19 of Brucella species. Comp Immunol Microbiol Infect Dis 101470
Mojgani N, Shahali Y, Dadar M (2020) Immune modulatory capacity of probiotic lactic acid bacteria and applications in vaccine development. Benefic Microbes 11:213–226
Morelli L, Capurso L (2012) FAO/WHO guidelines on probiotics: 10 years later. J Clin Gastroenterol 46:S1–S2
Morello E, Bermudez-Humaran L, Llull D, Solé V, Miraglio N, Langella P, Poquet I (2008) Lactococcus lactis, an efficient cell factory for recombinant protein production and secretion. J Mol Microbiol Biotechnol 14:48–58
Naderi-Samani M, Soltani M, Dadar M, Taheri-Mirghaed A, Zargar A, Ahmadivand S, Hassanzadeh R, Goudarzi LM (2020) Oral immunization of trout fry with recombinant Lactococcus lactis NZ3900 expressing G gene of viral hemorrhagic septicaemia virus (VHSV). Fish Shellfish Immunol 105:62–70
Neto MPC, de Souza Aquino J, da Silva LDFR, de Oliveira Silva R, de Lima Guimaraes KS, de Oliveira Y, de Souza EL, Magnani M, Vidal H, de Brito Alves JL (2018) Gut microbiota and probiotics intervention: a potential therapeutic target for management of cardiometabolic disorders and chronic kidney disease? Pharmacol Res 130:152–163
Oak SJ, Jha R (2019) The effects of probiotics in lactose intolerance: a systematic review. Crit Rev Food Sci Nutr 59:1675–1683
Pandey KR, Naik SR, Vakil BV (2015) Probiotics, prebiotics and synbiotics-a review. J Food Sci Technol 52:7577–7587
Pei H, Liu J, Cheng Y, Sun C, Wang C, Lu Y, Ding J, Zhou J, Xiang H (2005) Expression of SARS-coronavirus nucleocapsid protein in Escherichia coli and Lactococcus lactis for serodiagnosis and mucosal vaccination. Appl Microbiol Biotechnol 68:220–227
Perdigon G, Alvarez S, Rachid M, Agüero G, Gobbato N (1995) Immune system stimulation by probiotics. J Dairy Sci 78:1597–1606
Perdigón G, Fuller R, Raya R (2001) Lactic acid bacteria and their effect on the immune system. Curr Issues Intest Microbiol 2:27–42
Peterson CT, Sharma V, Elmén L, Peterson SN (2015) Immune homeostasis, dysbiosis and therapeutic modulation of the gut microbiota. Clin Exp Immunol 179:363–377
Pontes DS, De Azevedo MSP, Chatel J-M, Langella P, Azevedo V, Miyoshi A (2011) Lactococcus lactis as a live vector: heterologous protein production and DNA delivery systems. Protein Expr Purif 79:165–175
Poo H, Pyo HM, Lee TY, Yoon SW, Lee JS, Kim CJ, Sung MH, Lee SH (2006) Oral administration of human papillomavirus type 16 E7 displayed on Lactobacillus casei induces E7-specific antitumor effects in C57/BL6 mice. Int J Cancer 119:1702–1709
Pouwels PH, Leer RJ, Shaw M, den Bak-Glashouwer M-JH, Tielen FD, Smit E, Martinez B, Jore J, Conway PL (1998) Lactic acid bacteria as antigen delivery vehicles for oral immunization purposes. Int J Food Microbiol 41:155–167
Power SE, O'Toole PW, Stanton C, Ross RP, Fitzgerald GF (2014) Intestinal microbiota, diet and health. Br J Nutr 111:387–402
Pradhan D, Mallappa RH, Grover S (2020) Comprehensive approaches for assessing the safety of probiotic bacteria. Food Control 108:106872

Qiao X, Li G, Wang X, Li X, Liu M, Li Y (2009) Recombinant porcine rotavirus VP4 and VP4-LTB expressed in Lactobacillus casei induced mucosal and systemic antibody responses in mice. BMC Microbiol 9:1–11

Reid G, Jass J, Sebulsky MT, McCormick JK (2003) Potential uses of probiotics in clinical practice. Clin Microbiol Rev 16:658–672

Rizzardini G, Eskesen D, Calder PC, Capetti A, Jespersen L, Clerici M (2012) Evaluation of the immune benefits of two probiotic strains Bifidobacterium animalis ssp. lactis, BB-12® and Lactobacillus paracasei ssp. paracasei, L. casei 431® in an influenza vaccination model: a randomised, double-blind, placebo-controlled study. Br J Nutr 107:876–884

Rolfe RD (2000) The role of probiotic cultures in the control of gastrointestinal health. J Nutr 130:396S–402S

Roshan D, Souza DRP (2012) Lactococcus lactis: an efficient Gram positive cell factory for the production and secretion of recombinant protein

Rosshart SP, Vassallo BG, Angeletti D, Hutchinson DS, Morgan AP, Takeda K, Hickman HD, McCulloch JA, Badger JH, Ajami NJ (2017) Wild mouse gut microbiota promotes host fitness and improves disease resistance. Cell 171:1015–1028. e1013

Safari R, Adel M, Lazado CC, Caipang CMA, Dadar M (2016) Host-derived probiotics Enterococcus casseliflavus improves resistance against Streptococcus iniae infection in rainbow trout (Oncorhynchus mykiss) via immunomodulation. Fish Shellfish Immunol 52:198–205

Sánchez B, Delgado S, Blanco-Míguez A, Lourenço A, Gueimonde M, Margolles A (2017) Probiotics, gut microbiota, and their influence on host health and disease. Mol Nutr Food Res 61:1600240

Sartor RB (2004) Therapeutic manipulation of the enteric microflora in inflammatory bowel diseases: antibiotics, probiotics, and prebiotics. Gastroenterology 126:1620–1633

Sharma A (2019) Importance of probiotics in cancer prevention and treatment. In: Recent developments in applied microbiology and biochemistry. Elsevier, San Diego, CA, pp 33–45

Shi S-H, Yang W-T, Yang G-L, Cong Y-L, Huang H-B, Wang Q, Cai R-P, Ye L-P, Hu J-T, Zhou J-Y (2014) Immunoprotection against influenza virus H9N2 by the oral administration of recombinant Lactobacillus plantarumNC8 expressing hemagglutinin in BALB/c mice. Virology 464:166–176

Shirdast H, Ebrahimzadeh F, Taromchi AH, Mortazavi Y, Esmaeilzadeh A, Sekhavati MH, Nedaei K, Mirabzadeh E (2020) Recombinant Lactococcus Lactis displaying Omp31 antigen of brucella melitensis can induce an immunogenic response in BALB/c mice. Probiotics Antimicrob Proteins:1–10

Singh SK, Roeffen W, Mistarz UH, Chourasia BK, Yang F, Rand KD, Sauerwein RW, Theisen M (2017) Construct design, production, and characterization of Plasmodium falciparum 48/45 R0. 6C subunit protein produced in Lactococcus lactis as candidate vaccine. Microb Cell Factories 16:1–11

Tang ML (2009) Probiotics and prebiotics: immunological and clinical effects in allergic disease. In: Microbial host-interaction: tolerance versus allergy. Karger Publishers, Basel, pp 219–238

Tarahomjoo S (2012) Development of vaccine delivery vehicles based on lactic acid bacteria. Mol Biotechnol 51:183–199

Taverniti V, Guglielmetti S (2011) The immunomodulatory properties of probiotic microorganisms beyond their viability (ghost probiotics: proposal of paraprobiotic concept). Genes Nutr 6:261–274

Tjalsma H, Antelmann H, Jongbloed JD, Braun PG, Darmon E, Dorenbos R, Dubois J-YF, Westers H, Zanen G, Quax WJ (2004) Proteomics of protein secretion by Bacillus subtilis: separating the "secrets" of the secretome. Microbiol Mol Biol Rev 68:207–233

Tong CYW (2019) Tutorial topics in infection for the combined infection training programme. Oxford University Press, USA

Van Doan H, Hoseinifar SH, Ringø E, Ángeles Esteban M, Dadar M, Dawood MA, Faggio C (2020) Host-associated probiotics: a key factor in sustainable aquaculture. Rev Fisher Sci Aquac 28:16–42

Van Hoang V, Ochi T, Kurata K, Arita Y, Ogasahara Y, Enomoto K (2018) Nisin-induced expression of recombinant T cell epitopes of major Japanese cedar pollen allergens in Lactococcus lactis. Appl Microbiol Biotechnol 102:261–268

Varankovich NV, Nickerson MT, Korber DR (2015) Probiotic-based strategies for therapeutic and prophylactic use against multiple gastrointestinal diseases. Front Microbiol 6:685

Villatoro-Hernández J, Kuipers OP, Saucedo-Cárdenas O, Montes-de-Oca-Luna R (2012) Heterologous protein expression by Lactococcus lactis. In: Recombinant gene expression. Springer, New York, pp 155–165

Villena J, Medina M, Raya R, Alvarez S (2008) Oral immunization with recombinant Lactococcus lactis confers protection against respiratory pneumococcal infection. Can J Microbiol 54:845–853

Vitetta L, Saltzman ET, Thomsen M, Nikov T, Hall S (2017) Adjuvant probiotics and the intestinal microbiome: enhancing vaccines and immunotherapy outcomes. Vaccine 5:50

Wang Z, Yu Q, Gao J, Yang Q (2012) Mucosal and systemic immune responses induced by recombinant Lactobacillus spp. expressing the hemagglutinin of the avian influenza virus H5N1. Clin Vaccine Immunol 19:174–179

Wang L, Zhao D, Sun B, Yu M, Wang Y, Ru Y, Jiang Y, Qiao X, Cui W, Zhou H (2020) Oral vaccination with the porcine circovirus type 2 (PCV-2) capsid protein expressed by Lactococcus lactis induces a specific immune response against PCV-2 in mice. J Appl Microbiol 128:74–87

Wells JM, Mercenier A (2008) Mucosal delivery of therapeutic and prophylactic molecules using lactic acid bacteria. Nat Rev Microbiol 6:349–362

Wells J, Robinson K, Chamberlain L, Schofield K, Le Page R (1996) Lactic acid bacteria as vaccine delivery vehicles. Antonie Van Leeuwenhoek 70:317–330

Westerholm-Ormio M, Vaarala O, Tiittanen M, Savilahti E (2010) Infiltration of Foxp3-and Toll-like receptor-4–positive cells in the intestines of children with food allergy. J Pediatr Gastroenterol Nutr 50:367–376

Xin K-Q, Hoshino Y, Toda Y, Igimi S, Kojima Y, Jounai N, Ohba K, Kushiro A, Kiwaki M, Hamajima K (2003) Immunogenicity and protective efficacy of orally administered recombinant Lactococcus lactis expressing surface-bound HIV Env. Blood 102:223–228

Yeng HW, Shamsudin MN, Rahim RA (2009) Construction of an expression vector for Lactococcus lactis based on an indigenous cryptic plasmid. Afr J Biotechnol 8

Yoon S-W, Lee T-Y, Kim S-J, Lee I-H, Sung M-H, Park J-S, Poo H (2012) Oral administration of HPV-16 L2 displayed on Lactobacillus casei induces systematic and mucosal cross-neutralizing effects in Balb/c mice. Vaccine 30:3286–3294

Youngster I, Kozer E, Lazarovitch Z, Broide E, Goldman M (2011) Probiotics and the immunological response to infant vaccinations: a prospective, placebo controlled pilot study. Arch Dis Child 96:345–349

Zhang Z-H, Jiang P-H, Li N-J, Shi M, Huang W (2005) Oral vaccination of mice against rodent malaria with recombinant Lactococcus lactis expressing MSP-119. World J Gastroenterol 11:6975

Zhang R, Peng X, Duan G, Shi Q, Chen S, Wang C, Fan Q, Xi Y (2016) An engineered Lactococcus lactis strain exerts significant immune responses through efficient expression and delivery of Helicobacter pylori Lpp20 antigen. Biotechnol Lett 38:2169–2175

Promising Prospects of Probiotics and Postbiotics Derived from Lactic Acid Bacteria as Pharma Foods

14

Hooi Ling Foo, Laiella Shaahierra Jann Hishamuddin, and Teck Chwen Loh

Abstract

The focus of food consumption has shifted from satisfaction to health-promoting impact. Consumers are becoming more aware of their health. Food and health care industries procure myriad functional foods that contain additional nutritional components. Lactic acid bacteria (LAB) play an essential role in various industries owing to their health-promoting effects. Cancer originates at the sites of chronic inflammation. Cancer is a severe public health problem which is considered as among the common cause of death globally. The incidence and mortality rate of cancer have been steadily growing worldwide. Many chemotherapy regimens are effectively used to treat cancer; however, cancer cells often acquire drug resistance that generally leads to relapse and worsening of prognosis. Therefore, continuing endless effort in finding a safer alternative or add-on treatment with lower or no side effects through the healthy dietary constituents as well as practical and appropriate supplements is prior necessary. The promising

H. L. Foo (✉)
Department of Bioprocess Technology, Faculty of Biotechnology and Biomolecular Sciences, Universiti Putra Malaysia, Serdang, Selangor, Malaysia

Institute of Bioscience, Universiti Putra Malaysia, Serdang, Selangor, Malaysia
e-mail: hlfoo@upm.edu.my

L. S. J. Hishamuddin
Institute of Bioscience, Universiti Putra Malaysia, Serdang, Selangor, Malaysia

T. C. Loh
Department of Animal Science, Faculty of Agriculture, Universiti Putra Malaysia, Serdang, Selangor, Malaysia

Institute of Tropical Agriculture and Food Security, Universiti Putra Malaysia, Serdang, Selangor, Malaysia

N. Mojgani, M. Dadar (eds.), *Probiotic Bacteria and Postbiotic Metabolites: Role in Animal and Human Health*, Microorganisms for Sustainability 2,
https://doi.org/10.1007/978-981-16-0223-8_14

prospects of LAB probiotics and fermented products mediated by LAB have been extensively reported. Recently, there are numerous studies on LAB metabolites, postbiotics, a new preparation of non-viable supplement, exhibiting various health-promoting and beneficial physiological effects to the animal host by improving the mucosal gut barrier integrity and reducing the pathogen-induced inflammation. Therefore, a comparative review of the promising potential of probiotics and postbiotics as a functional food, ingredient or supplements and adjunctive therapeutic aids are revealed in this chapter.

Keywords

Postbiotic · Probiotic · Health impact · Anti-cancer · Anti-inflammation

14.1 Introduction

Natural herbs and foods have a long-standing tradition in many cultures to treat ailments. However, in recent years, "superfoods" have received tremendous attention. Interest is now rapidly expanding to foods with clinically enhanced properties. Currently, the health and life science sectors are undergoing significant change across all their industries. Both sectors are converged frequently to formulate medical nutrition products, having specific nutritional compositions for intervention in disease progression and symptom alleviation (Weenen et al. 2013). Hence, new markets and industries have emerged from the convergence of both health and life science sectors.

The core technology domain of medical nutrition industry is food. Nevertheless, technological development is mainly driven by pharmaceutical/pharmacological technologies. Hence, boundary-crossing developments are occurring between the food and pharmaceutical industries, particularly (Weenen et al. 2013). Pharma food products resulting from this convergence are known as Nutritional Supplements, Functional Foods and Medical Nutrition, which are food substances that are considered to improve health and exist between conventional foods and pharmaceuticals (Eussen et al. 2011; Henry 2010; Verhagen et al. 2010). Generally, pharma foods promote either in general or by specifically targeting a bodily function, such as improving digestion, bone density and so on. In the next decade, we can expect to see a shift from traditional "farmer foods" to more sophisticated pharma foods.

Cancer is a severe public health problem and it is considered as among the common cause of death globally. Throughout the past years, the incidence and mortality rate of cancer have been steadily growing worldwide. Based on the World Health Organization (WHO 2018), there were an estimated 17 million new cancer cases (excluding non-melanoma skin cancer) and 9.6 million deaths due to cancer in 2018 worldwide. In both sexes, lung cancer is the most diagnosed cancer incidence (11.6% of the total cases), followed by female breast cancer (11.6%) and colorectal cancer (10.2%). By mortality, the lung is also a leading cause of cancer death accounting for 18.4% of the total cancer deaths, followed by colorectal cancer

(9.2%) and stomach cancer (8.2%). However, women are mostly diagnosed with breast cancer and it is the leading cause of cancer death in women.

Approximately 70% of cancer deaths happened in low- and middle-income countries, mostly due to late detection of cancer and limited access to treatment. In Malaysia, cancer is the third leading cause of death. As of 2018, 43,837 cases were diagnosed and 26,395 death were reported due to cancer (GLOBOCAN 2019). Despite the advancement in cancer therapy, these treatments cause various side effects due to unspecific toxicity to normal cells (Zakuan et al. 2019). Furthermore, cancer cells often acquire drug resistance which leads to relapse and worsening of prognosis (Kovalchuk et al. 2008). The treatments of chemotherapy, radiotherapy and immunotherapy could cause weakness, fatigue, nausea, hair loss, vomiting, tissue damage and autoimmune diseases (Aslam et al. 2014; Kroschinsky et al. 2017). Exposure to chemotherapeutic drugs could result in the acquisition of multi-drug resistance (MDR).

MDR leads to significantly worse response to treatment. Conventional treatment with nanoparticles like titanium dioxide, silica, and gold complexes could also increase the risk of metastasis of cancer cells (Peng et al. 2019). Although the severity of some of these side effects are generally mild, life-threatening complications may also occur. Therefore, finding an alternative or adjunctive treatment which can reduce the side effects is necessary for the advancement of cancer treatment. Lactic acid bacteria (LAB) are Gram-positive, non-motile, non-sporulating rods and cocci (Bernardeau et al. 2008) bacteria. Probiotics are highly selected LAB, such as *Bifidobacterium* spp., *Lactobacillus* spp. and *Streptococcus* spp. (Rafter 2002) with emerging evidence as potential preventative and therapeutic agents for cancers (Zhong et al. 2014). Bioactive metabolites produced by probiotics, so-called postbiotic, confer various beneficial effects including anti-inflammatory, anti-cancer and anti-microbial. This chapter reveals the comparative beneficial impacts of different preparations of LAB.

14.2 Lactic Acid Bacteria

LAB are commonly classified as a group of facultative anaerobes that produce lactic acid as the main product during sugar metabolism. It consists of *Lactobacillus, Streptococcus, Enterococcus, Lactococcus, Leuconostoc, Pediococcus, Tetragenococcus, Weissella* and *Bifidobacterium* (Hutkins 2006; Masood et al. 2011). They are subdivided into rod and cocci based on their morphology. Homofermentative LAB produce mostly lactic acid and heterofermentative LAB produce acetic acid or alcohol in addition to lactic acid (Halász 2009). Phylogenetically, LAB are divided into two lines of descent. Gram-positive LAB with a DNA base composition of less than 53 mol % guanine and cytosine (G + C) are in the group of Clostridium branch. In contrast, DNA with higher base composition than 53 mol % G + C belongs to the Actinomycetes branch (Savadogo et al. 2007). The typical LAB, such as *Carnobacterium, Lactobacillus, Lactococcus,*

Leuconostoc, Pediococcus and *Streptococcus*, have a G + C content of less than 50 mol % and belong to the Clostridium branch (Schleifer and Ludwig 1995).

LAB are typically found as natural microbiota or microflora in animals and the human intestine. To date, LAB have been applied widely as a starter culture or co-culture of fermentation processes in food and alcoholic beverages industries in favour of producing natural and healthy products (Leroy and De Vuyst 2004). One of the core genera of LAB is *Lactobacillus.* It has been recently reclassified into 25 new genera based on the polyphasic approaches, which includes host-adapted LAB of *Lactobacillus delbrueckii* group, *Paralactobacillus* and 23 novel genera of *Acetilactobacillus, Agrilactobacillus, Amylolactobacillus, Apilactobacillus, Bombilactobacillus, Companilactobacillus, Dellaglioa, Fructilactobacillus, Furfurilactobacillus, Holzapfelia, Lacticaseibacillus, Lactiplantibacillus, Latilactobacillus, Lapidilactobacillus, Lentilactobacillus, Levilactobacillus, Ligilactobacillus, Limosilactobacillus, Liquorilactobacillus, Loigolactobacilus, Paucilactobacillus, Schleiferilactobacillus* and *Secundilactobacillus (*Zheng et al. 2020*).* The name of *Lactiplantibacillus plantarum* was suggested for the plantarum-group lactobacilli.

Some LAB have been proven to be a vital probiotic bacteria, which are generally recognised as safe (GRAS) microorganisms by the Food and Drug Administration (FDA) of USA and they are also given "quantified presumption of safety" (QPS) status by the European Union. Hence, probiotic LAB play an essential role in various industries, particularly in the food industry and livestock industry (EFSA Panel on Biological Hazards (BIOHAZ) 2013). Furthermore, LAB can produce an array of compounds which contribute to the improvement of nutritional value, organoleptic, technological and shelf life of the end product (Ayad et al. 2004). The prolong of shelf life by LAB fermentation is mainly due to the rapid acidification of food attributed to the production of organic acids, primarily lactic acid. However, the production of other metabolites, such as acetic acid, ethanol, aroma compounds, bacteriocins, exopolysaccharides and several enzymes, is of importance (De Vuyst and Leroy 2007).

14.3 Health Impacts of Bacterial Probiotics

Probiotic was initially defined as a live microbial feed supplement which beneficially affects the host animal by improving its intestinal microbial balance (Fuller 1989). However, this definition is unsatisfactory and imprecise as more probiotics were demonstrated to confer other beneficial health impacts. Food and Agriculture Organization and World Health Organization define probiotic bacteria as "live microorganisms which, when administered in adequate amounts confer a health benefit on the host" (FAO/WHO Expert Consultation 2001). To be classified as a probiotic strain, several aspects of functionality must acquire: 1) tolerance to acid and human gastric juice, 2) bile tolerance (for survival in the small bowel), 3) adherence to epithelial surfaces and persistence in the human gastrointestinal tract (GIT), 4) immuno-stimulation, but no pro-inflammatory effect, 5) antagonistic activity against pathogens and 6) antimutagenic and anticarcinogenic properties (Lee and Salminen 1995).

Probiotics are mainly associated with LAB. *Lactobacillus* and *Bifidobacterium* are the 2 genera of LAB that are mostly employed for various industrial applications due to their long history of safe applications (Linares et al. 2016). In addition, *Propionibacterium* and *Streptococcus* were also observed to possess beneficial properties and hence they are also vogue as probiotic microorganisms. Dairy propionibacteria promote bifidobacteria growth in the gut, enhance the use of nutrients, hypocholesterolemic and immunomodulation effects (Zárate 2012), while *Streptococcus thermophilus* produces a high amount of β-galactosidase in the GIT, which is essential for lactose hydrolysis, making it beneficial to improve lactose intolerance (Rul et al. 2011). However, other bacterial and yeast species have been suggested to be a potential probiotic strain. Recently, a few strains of short-chain fatty acids (SCFA) producing *Escherichia coli* isolated from adult human microbiota were proven to possess promising probiotic properties (Nakkarach et al. 2020).

In the past decade, more health impacts of LAB were demonstrated, such as enhanced immune response, colonic microbiota balance, vaccine adjuvant effect, reduction of faecal enzymes implicated in cancer initiation, treatment of diarrheal diseases, antibiotic therapy, control of rotavirus and *Clostridium difficile*, control of gastric ulcers, reduction of serum cholesterol, antagonism against food-borne pathogens and tooth decay organisms, and lactose intolerance and malabsorption (Masood et al. 2011). Disruptions of the balance of gastrointestinal microflora will disturb the homeostasis, leading to intestinal microbial dysbiosis and other diseases associated with an unhealthy gut. Goldenberg et al. (2017) reported that probiotics could reduce the risk of *Clostridium difficile*-Associated Diarrhoea (CDAD) incidence rate by 60% on average. Patients treated with *Lactobacillus acidophilus* and *Lactobacillus casei* demonstrated a lower incidence rate of CDAD and antibiotic-induced diarrhoea (Gao et al. 2010). Supplementation of probiotics in patients with gestational diabetes mellitus improved glycaemic control and decreased triglycerides and VLDL cholesterol concentrations (Karamali et al. 2016).

As for immunity responses, administration of probiotic could modulate both innate and adaptive immunity (Vitetta et al. 2017). Probiotics exert a positive effect on human immunological defence by stimulating macrophages, NK cells, antigen-specific cytotoxic T-lymphocytes, and the release of different cytokines (Ashraf and Shah 2014). Ferreira dos Santos et al. (2016) reported that *Lactobacillus plantarum* Lp62 decreased the IL-8 secretion by *Salmonella Typhi*-stimulated HT-29 cells and prevented the adhesion of pathogens to the epithelial cells. Moreover, *L. plantarum* Lp62 inhibited the inflammatory stimulation in epithelial cells and macrophages by secreting TNF-α, IL1-β, and IL-17, while increased IL-10 secretion by mononuclear cells. Supplementation of probiotic *Lactobacillus rhamnosus* GG during breastfeeding stimulates the maturation of humoral immune response by increasing the total number of immunoglobulin secreting cells, particularly, IgG, IgA and IgM (Rinne et al. 2005).

Beneficial bacteria may also stimulate wound healing. Recently, Han et al. (Han et al. 2019) studied the effect of probiotics on oral mesenchymal stem cells and wound healing. They revealed that the probiotic *Lactobacillus reuteri* could activate the potentials of gingival mesenchymal stem cells and enhanced the wound healing process by regulating the pathway of PI3K/AKT/β-catenin/TGFβ1. As for the wound healing effect, Mohseni et al. (2018) reported that 12 weeks supplementation of probiotic decreased ulcer length, depth and width of a wound in patients with a diabetic foot ulcer. Orally administrated yoghurt containing probiotic *Lactobacillus gasseri* has significantly accelerated the wound healing of acetic acid-induced gastric ulcer in rats (Uchida et al. 2010). Probiotic of *Lactobacilli*, *Bifidobacteria* and *Streptococcus* species accelerated the acetic acid-induced gastric ulcer in rats by expressing and production of angiogenesis promoting vascular endothelial growth factor (Dharmani et al. 2013).

Currently, emerging evidence that relates to gut microbiota and the function of the central nervous system (Tillisch 2014) are extensively reported. Daily administration of probiotic formulation consisting of *Lactobacillus helveticus* R0052 and *Bifidobacterium longum* R0175 could reduce anxiety-like behaviour in rats and alleviate psychological distress in human volunteers (Messaoudi et al. 2011). Folate is a vitamin B that plays a vital role in the health and physiological functions of human and animals. Some LAB have the capability of producing folate extracellularly. Interestingly, *L. plantarum* I-UL4 was suggested to be employed for the enhancement of folate level in milk and dairy products since it produced the highest folate extracellularly in comparison to other LAB species (Nor et al. 2010). Kobyliak et al. (2018) demonstrated that a multi-strain probiotic containing *Bifidobacterium*, *Lactobacillus*, *Lactococcus* and *Propionibacterium* reduced liver fat, aminotransferase activity, TNF-α and IL-6 levels in non-alcoholic fatty liver disease patients.

The focus of food consumption today has shifted from satisfaction to health-promoting impact. Consumers are becoming more aware of their health and the food industry is going hand in hand with the production of so-called functional foods, containing additional nutritional components. LAB play an important role in food and beverages industries owing to its GRAS and QPS status, which provide health-promoting effects when consumed in addition to several other reasons. The growth of LAB increases carbohydrate content of the foods they ferment with reduced pH condition resulting from the production of lactic acid from carbohydrate fermentation (Solioz et al. 2011). Certain LAB may secrete bacteriocin extracellularly. Bacteriocins have been suggested as an alternative to replacing chemical preservatives and heat treatment, to preserve food more naturally and rich in organoleptic and nutritional properties (Gálvez et al. 2007). Application of bacteriocins as bio-preservatives in food preparation is safe for consumers as they are inactivated by pancreatic or gastric enzymes (Liu et al. 2011). The combined action of both low pH and bacteriocins secretion is essential in the food industry to extend the shelf life by inhibiting the growth of food spoilage and pathogenic microorganisms (Gálvez et al. 2007). Bacteriocins are inhibitory peptide molecules that act against food-borne pathogens such as *Clostridium botulinum*, *Staphylococcus aureus* and *Listeria monocytogenes* (Nettles and Barefoot 1993).

L. plantarum IUL-4 is the first *L. plantarum* strain reported to harbour both class I *plw* and class II *plnEF* bacteriocin genes simultaneously (Tai et al. 2015). Both classes of plantaricin genes contributed to the broad anti-microbial activity against various pathogens such as *L. monocytogenes, Salmonella enterica, E. coli* and Vancomycin resistant *enterococci* (VRE) (Kareem et al. 2014). Several other LAB (*Lactococcus lactis* subsp. *lactis, Pediococcus pentosaceus, Leuconostoc mesenteroides Lactobacillus curvatus, Lactobacillus sakei* and *L. plantarum*) have also been documented for the secretion of bacteriocin (Todorov et al. 2006; Shin et al. 2008; Mataragas et al. 2003; Aasen et al. 2000; Kormin et al. 2001). Pediocin secretion from three *Pediococcus* species (*P. acidilactici* NCIM 2292, *P. pentosaceous* NCIM 2296 and *P. cervisiae* NCIM 2171) showed significant inhibition against *Bacillus cereus, L. monocytogenes*, and *S. aureus* but moderate against *E. coli* and *Pseudomonas* and less against *Clostridium perfringens (*Jamuna and Jeevaratnam 2004*)*. The food acidification caused by LAB changes the texture due to the precipitation of some proteins (Solioz et al. 2011). Exopolysaccharides forming LAB such as *Streptococcus thermophilus* and *Lactobacillus delbrueckii* subsp. *bulgaricus* are used in the manufacturing of yoghurt to improve texture, avoid syneresis and increase the viscosity of products (Ruas-Madiedo et al. 2002). LAB produce aromas and flavours and accelerate the cheese's maturation through its proteolytic and lipolytic activities (Kongo 2013).

Applications of probiotics have been extended from human to agricultural applications, including animals and plants. In response to consumers' demands of natural product, probiotics have been widely used in agriculture as an alternative growth promoter to replace conventional antibiotic treatments and synthetic chemical feed (Fuller 1989), owing to the broad range of anti-microbial activity against pathogens. Several reports have demonstrated the beneficial effects of probiotic supplementations in animal feed. Corn-soybean meal supplemented with probiotic (*Pediococcus acidlactici*) increased villus height in duodenum and ileum while lowering the number of coliforms of the ileum and serum cholesterol level (Taheri et al. 2010). Probiotic supplementation showed lower oxidative spoilage in broiler breast meat over a 7-day post-mortem ageing, therefore increasing the shelf life of chicken meat (Abdulla et al. 2018).

Recently, probiotics isolated from Malaysian foods have been reported to produce various versatile extracellular hydrolytic enzymes (Mohamad Zabidi et al. 2020). *L. plantarum* that grow on palm kernel cake secreted a cocktail of multi extracellular hydrolytic enzymes (Lee et al. 2019). These enzymes degrade fibrous and crystalline cellulosic materials, thereby improving the nutritional value of fermented palm kernel cake substantially to be used as alternative feed ingredients to reduce the production cost of animals (Mohamad Zabidi et al. 2020; Lee et al. 2019). Alshelmani et al. (2016) suggested that inclusion of 15% fermented palm kernel cake could replace up to 30% of yellow maize in broiler diet, which can be reflected in the cost savings of feed for the poultry industry.

Furthermore, *Lactobacillus* and *Bifidobacterium* strains have also been found useful in the rapid removal of toxic metals such as cadmium and lead from the water as reported by Halttunen et al. (Halttunen et al. 2007). A similar study has also been reported by Bhakta et al. (2012), whereby mud and sludge-isolated *Lactobaciullus reuteri* Cd70–13 and Pb71–1 were used as a heavy metal sorbent to eliminate heavy metals in the ambience. Mechanisms such as complex formation, ion exchange, adsorption, chelation and microprecipitation have been proposed to be involved in metal biosorption (Ahalya et al. 2003).

14.4 Health Impacts of Postbiotic

Postbiotics are defined as non-viable soluble bioactive metabolites produced by probiotic LAB, which exerts a myriad beneficial effect on the host, directly or indirectly. Several terms have been proposed for postbiotic preparation, such as metabiotics, biogenics, metabolites or simply cell-free supernatants (CFS), soluble factors secreted by live probiotic, which when administered in adequate amount will confer beneficial effects (Tsilingiri et al. 2012). The composition of soluble bioactive components that present in postbiotics could vary substantially amongst the producer cells, attributing to the fermentation condition, such as growth medium and physical parameters of the fermentation process. The bioactive compounds of postbiotics include organic acid, bacteriocin, hydrogen peroxide, ethanol, fatty acids, diacetyl, acetaldehyde, acetone, reuterin, reutericyclin, SCFA, hydrolytic enzymes, peptides, teichoic acids, peptidoglycan-derived muropeptides, endo- and exo-polysaccharides, cell surface proteins, vitamins, plasmalogens, bacteriocin-like compounds and other low molecular mass compounds with anti-microbial activities (Aguilar-Toalá et al. 2018; Konstantinov et al. 2013; Paul et al. 2018).

The bioactive compounds of postbiotic target the host-microbe-pathogen interface rescuing biotic and immune unbalances, as well as inflammation, thus providing new therapeutic opportunities (Puccetti et al. 2020). Postbiotics mimic the beneficial health effects of probiotics while avoiding the risk of administering live microorganisms. The supplementation of viable probiotics has been associated with systemic infections such as bacteraemia and fungemia, as well as transferring of antibiotic resistance gene and virulence factor, and risk of sepsis. The incidents of systemic infections by probiotics have been reported for premature infants, immunocompromised and impairment of epithelial barrier patients (Paul et al. 2018). Treatment with a combination of probiotics resulted in the death of 16% of acute pancreatitis patients, as compared to 6% in the placebo group (Besselink et al. 2018). Furthermore, consumption of live bacteria could also induce interaction with the compound of food matrix or ingredient which may cause bloating.

The growing evidence demonstrates that probiotic metabolites exert various beneficial health impacts, but not limited to, anti-microbial, antioxidant, and immunomodulatory in the same way as the parent-live probiotics over the last 10 years. Hypertension has been associated with gut microbiome dysbiosis. The manipulation of the gut microbiota can lead to the development of new antihypertensive therapies (Robles-Vera et al. 2017). The first evidence of SCFA that present in postbiotic metabolite could decrease blood pressure was reported in 1983, whereby hypotension was noted when acetate was added to haemodialysis lysate (Muralitharan et al.

2020). In addition, various *Lactobacillus* sp. have been identified to produce bioactive peptides, which are useful in promoting human health, including reducing the risk of hypertension (Raveschot et al. 2018).

Certain postbiotics may be a rich bacteriocin source for the inhibition of a wide range of pathogenic bacteria that are likely to reduce the occurrence of infection (Cicenia et al. 2014). Reuterin that produced by *Lactobacillus reuteri* is the first molecule identified with potent anti-pathogenic activity against a broad spectrum of microorganisms including *Escherichia, Salmonella, Shigella, Proteus, Pseudomonas, Clostridium, Staphylococcus*, fungi, and protozoa, many of which are pathogenic to humans (Talarico and Dobrogosz 1989). Six strains of *Lactiplantibacillus plantarum* (formerly was known as *Lactobacillus plantarum*) isolated from Malaysian foods exhibited antagonistic activity against *S. typhimurium, E. coli*, followed by *L. monocytogenes* and Vancomycin-resistant *enterococci* (Thanh et al. 2010).

Exopolysaccharides (EPS) produced by *Lactococcus lactis* F-mou showed high anti-microbial activity against *Staphylococcus aureus, Pseudomonas aeruginosa, E. coli, L. monocytogenes, B. cereus, Proteus mirabilis, Acinetobacter baumannii, Enterobacter cloacae* and *Candida albicans* (Nehal et al. 2019). In addition, *L. lactis* F-mou also displayed potent beta-carotene bleaching inhibition and high radical scavenging activity (> 90%). EPS produced by *Lactobacillus reuteri* SHA101 and *Lactobacillus vaginalis* SHA110 isolated from the gut cecum of healthy hen showed potent radical scavenging activity of hydroxyl DPPH (2,2-diphenyl-1-picrylhydrazyl) and superoxide radicals and reducing power (Rajoka et al. 2019).

Antioxidant properties of postbiotics derived from *L. plantarum* were demonstrated in the study conducted by Izuddin et al. (2020), whereby an increase of hepatic antioxidant enzyme glutathione peroxidase (GPx) concentration in serum and ruminal fluid were observed. He et al. (2017) demonstrated a preventative effect against neonatal gut-derived *E. coli* K1 infection through promoting the maturation of neonatal intestinal defence. The addition of *L. casei* subsp. *rhamnosus* reduced the adherence of pathogenic bacteria, enteropathogenic and enterotoxigenic *E. coli* and *Klebsiella pneumonia* to the tissue receptor of Caco-2 intestinal cells (Forestier et al. 2001). The CFS metabolites of four isolates of LAB isolated from the breast milk of healthy women demonstrated anti-viral activity (32% inhibition) against HIV-1 infection (Martín et al. 2010).

Postbiotics have also been reported to modulate inflammatory responses. CFS of *Lactobacillus fermentum* inhibited the pro-inflammatory response of HeLa 229 cells to *Yersinia enterocolitica* by inhibiting the production of IL8 (Frick et al. 2007). *Lactobacillus rhamnosus* and its CFS were evaluated for their immunomodulatory effects on human dendritic cells (DC) challenged with *E. coli* (Bermudez-Brito et al. 2014). The results showed that CFS was more effective than live probiotics in reducing the secretion of pro-inflammatory cytokines when DC was challenged with *E. coli*. Furthermore, in the presence of *E. coli*, both treatments induced the production of TGF-1β, an inhibitor for the synthesis of pro-inflammatory cytokines, as well as an activator for the toll-death receptor signalling molecule to enhance the innate immunity. Therefore, cultured CFS is a safer alternative to live bacteria to modulate immune responses of human DC. In another study, *L. casei* and its postbiotic reduced the mRNA level of (IL)-1α, IL-6, IL-8 and increased the secretion of IL10 in ileal and colonic mucosa in post-infectious irritable bowel syndrome (PI-IBS) (Compare et al. 2017).

Recently, *E. coli* KUB-36 metabolite has been reported to reduce the production of pro-inflammatory cytokines IL-1β, IL-6, IL-8 and TNF-α, but induced the secretion of IL-10 in lipopolysaccharide-induced THP-1 macrophage cells (Nakkarach et al. 2021). Postbiotics has also been shown to be a novel therapeutic agent for the prevention and treatment of severe acute respiratory syndrome coronavirus 2 (SARS-CoV-2), due to their immunomodulatory properties (Mantziari et al. 2020). Gou et al. (Gou et al. 2020) suggested that intestinal microbiome and their metabolites can serve as a potential preventive/treatment target for intervention, particularly among those who are susceptible to the SARS-CoV-2 infection. However, more clinical trials should be performed to verify the suggestion.

References

Aasen IM, Møretrø T, Katla T, Axelsson L, Storrø I (2000) Influence of complex nutrients, temperature and pH on bacteriocin production by *Lactobacillus sakei* CCUG 42687. Appl Microbiol Biotechnol 53:159–166

Abdulla NR, Sabow AB, Foo HL, Loh TC, Zamri AM (2018) Growth performance, fatty acid profile and lipid oxidative stability of breast muscle in chickens fed probiotics and antibiotics or their mixture. South Afr J Anim Sci 48

Aguilar-Toalá JE, Garcia-Varela R, Garcia HS, Mata-Haro V, González-Córdova AF, Vallejo-Cordoba B, Hernández-Mendoza A (2018) Postbiotics: an evolving term within the functional foods field. Trends Food Sci Technol 75:105–114

Ahalya N, Ramachandra TV, Kanamadi RD (2003) Biosorption of heavy metals. Res J Chem Environ 7:71–79

Alshelmani MI, Loh TC, Foo HL, Sazili AQ, Lau WH (2016) Effect of feeding different levels of palm kernel cake fermented by *Paenibacillus polymyxa* ATCC 842 on nutrient digestibility, intestinal morphology, and gut microflora in broiler chickens. Anim Feed Sci Technol 216:216–224

Ashraf R, Shah NP (2014) Immune system stimulation by probiotic microorganisms. Crit Rev Food Sci Nutr 54:938–956

Aslam MS, Naveed S, Ahmed A, Abbas Z, Gull I, Athar MA (2014) Side effects of chemotherapy in cancer patients and evaluation of patients opinion about starvation based differential chemotherapy. J Cancer Ther 5:817–822

Ayad EH, Nashat S, El-Sadek N, Metwaly H, El-Soda M (2004) Selection of wild lactic acid bacteria isolated from traditional Egyptian dairy products according to production and technological criteria. Food Microbiol 21:715–725

Bermudez-Brito M, Munoz-Quezada S, Gomez-Llorente C, Romero F, Gil A (2014) *Lactobacillus rhamnosus* and its cell-free culture supernatant differentially modulate inflammatory biomarkers in *Escherichia coli*-challenged human dendritic cells. Br J Nutr 111:1727–1737

Bernardeau M, Vernoux JP, Henri-Dubernet S, Gueguen M (2008) Safety assessment of dairy microorganisms: the *Lactobacillus* genus. Int J Food Microbiol 126:278–285

Besselink MG, van Santvoort HC, Buskens E, Boermeester MA, van Goor H, Timmerman HM, Rosman C (2018) Probiotic prophylaxis in predicted severe acute pancreatitis: a randomised, double-blind, placebo-controlled trial. Lancet 371:651–659

Bhakta JN, Ohnishi K, Munekage Y, Iwasaki K, Wei MQ (2012) Characterisation of lactic acid bacteria-based probiotics as potential heavy metal sorbents. J Appl Microbiol 112:1193–1206
Cicenia A, Scirocco A, Carabotti M, Pallotta L, Marignani M, Severi C (2014) Postbiotic activities of *lactobacilli*-derived factors. J Clin Gastroenterol 48:18–22
Compare D, Rocco A, Coccoli P, Angrisani D, Sgamato C, Iovine B, Salvatore U, Nardone G (2017) *Lactobacillus casei* DG and its postbiotic reduce the inflammatory mucosal response: an ex-vivo organ culture model of post-infectious irritable bowel syndrome. BMC Gastroenterol 17:53
De Vuyst L, Leroy F (2007) Bacteriocins from lactic acid bacteria: production, purification, and food applications. J Mol Microbiol Biotechnol 13:194–199
Dharmani P, De Simone C, Chadee K (2013) The probiotic mixture VSL# 3 accelerates gastric ulcer healing by stimulating vascular endothelial growth factor. PLoS One 8:e58671
EFSA Panel on Biological Hazards (BIOHAZ) (2013) Scientific opinion on the maintenance of the list of QPS biological agents intentionally added to food and feed (2013 update). EFSA J 11:3449
Eussen SR, Verhagen H, Klungel OH, Garssen J, van Loveren H, van Kranen HJ, Rompelberg CJ (2011) Functional foods and dietary supplements: products at the interface between pharma and nutrition. Eur J Pharmacol 668:S2–S9
FAO/WHO Expert Consultation (2001) Health and nutritional properties of probiotics in food including powder milk with live lactic acid bacteria
Ferreira dos Santos T, Alves Melo T, Almeida ME, Passos Rezende R, Romano CC (2016) Immunomodulatory effects of *Lactobacillus plantarum* Lp62 on intestinal epithelial and mononuclear cells. BioMed Res Int:1–8
Forestier C, De Champs C, Vatoux C, Joly B (2001) Probiotic activities of *Lactobacillus casei rhamnosus*: in vitro adherence to intestinal cells and anti-microbial properties. Res Microbiol 152:167–173
Frick JS, Schenk K, Quitadamo M, Kahl F, Köberle M, Bohn E, Aepfelbacher M, Autenrieth IB (2007) *Lactobacillus fermentum* attenuates the pro-inflammatory effect of *Yersinia enterocolitica* on human epithelial cells. Inflamm Bowel Dis 13:83–90
Fuller R (1989) Probiotics in man and animals. J Appl Bacteriol 66:365–378
Gálvez A, Abriouel H, López RL, Omar NB (2007) Bacteriocin-based strategies for food biopreservation. Int J Food Microbiol 120:51–70
Gao XW, Mubasher M, Fang CY, Reifer C, Miller LE (2010) Dose–response efficacy of a proprietary probiotic formula of *Lactobacillus acidophilus* CL1285 and *Lactobacillus casei* LBC80R for antibiotic-associated diarrhea and *Clostridium difficile*-associated diarrhea prophylaxis in adult patients. Am J Gastroenterol 105:1636–1641
GLOBOCAN (2019) Statistics of Cancer in Malaysia. https://gco.iarc.fr/today/data/factsheets/populations/458-malaysia-fact-sheets.pdf. Accessed January 2020
Goldenberg JZ, Yap C, Lytvyn L, Lo CK, Beardsley J, Mertz D, Johnston BC (2017) Probiotics for the prevention of *Clostridium difficile*-associated diarrhea in adults and children. Cochrane Database Syst Rev 12
Gou W, Fu Y, Yue L, Chen GD, Cai X, Shuai M, Xu F, Yi X, Chen H, Zhu YJ, Xiao ML (2020) Gut microbiota may underlie the predisposition of healthy individuals to COVID-19. MedRxiv
Halász A (2009) Lactic acid bacteria. Food Qual Stand 3:70–82
Halttunen T, Salminen S, Tahvonen R (2007) Rapid removal of lead and cadmium from water by specific lactic acid bacteria. Int J Food Microbiol 114:30–35
Han N, Jia L, Su Y, Du J, Guo L, Luo Z, Liu Y (2019) *Lactobacillus reuteri* extracts promoted wound healing via PI3K/AKT/β-catenin/TGFβ1 pathway. Stem Cell Res Ther 10:243
He X, Zeng Q, Puthiyakunnon S, Zeng Z, Yang W, Qiu J, Du L, Boddu S, Wu T, Cai D, Huang SH (2017) *Lactobacillus rhamnosus* GG supernatant enhance neonatal resistance to systemic *Escherichia coli* K1 infection by accelerating development of intestinal defense. Sci Rep 7:43305
Henry CJ (2010) Functional foods. Eur J Clin Nutr 64:657–659

Hutkins RW (2006) Fermented vegetables. In: Microbiology and technology of fermented foods. Wiley, Hoboken, NJ, pp 223–259

Izuddin WI, Humam AM, Loh TC, Foo HL, Samsudin AA (2020) Dietary postbiotic *Lactobacillus plantarum* improves serum and ruminal antioxidant activity and upregulates hepatic antioxidant enzymes and ruminal barrier function in post-weaning lambs. Antioxidants 9:250

Jamuna M, Jeevaratnam K (2004) Isolation and partial characterisation of bacteriocins from *Pediococcus* species. Appl Microbiol Biotechnol 65:433–439

Karamali MA, Dadkhah F, Sadrkhanlou M, Jamilian M, Ahmadi S, Tajabadi-Ebrahimi M, Jafari P, Asemi Z (2016) Effects of probiotic supplementation on glycaemic control and lipid profiles in gestational diabetes: a randomised, double-blind, placebo-controlled trial. Diabetes Metab 42:234–241

Kareem KY, Ling FH, Chwen LT, Foong OM, Asmara SA (2014) Inhibitory activity of postbiotic produced by strains of *Lactobacillus plantarum* using reconstituted media supplemented with inulin. Gut Pathogens 6:23

Kobyliak N, Abenavoli L, Mykhalchyshyn G, Kononenko L, Boccuto L, Kyriienko D, Dynnyk O (2018) A multi-strain probiotic reduces the fatty liver index, cytokines and aminotransferase levels in NAFLD patients: evidence from a randomised clinical trial. J Gastrointestin Liver Dis 27:41–49. https://pubmed.ncbi.nlm.nih.gov/29557414/

Kongo JM (2013) Lactic acid bacteria as starter-cultures for cheese processing: past, present and future developments. Lactic Acid Bacteria-R & D for Food, Health and Livestock Purposes. 1–22

Konstantinov SR, Kuipers EJ, Peppelenbosch MP (2013) Functional genomic analyses of the gut microbiota for CRC screening. Nat Rev Gastroenterol Hepatol 10:741–745

Kormin S, Rusul G, Radu S, Ling FH (2001) Bacteriocin-producing lactic acid bacteria isolated from traditional fermented food. Mal J Med Sci MJMS 8:63

Kovalchuk O, Filkowski J, Meservy J, Ilnytskyy Y, Tryndyak VP, Vasyl'F C, Pogribny IP (2008) Involvement of microRNA-451 in resistance of the MCF-7 breast cancer cells to chemotherapeutic drug doxorubicin. Mol Cancer Ther 7:2152–2159

Kroschinsky F, Stölzel F, von Bonin S, Beutel G, Kochanek M, Kiehl M, Schellongowski P (2017) New drugs, new toxicities: severe side effects of modern targeted and immunotherapy of cancer and their management. Crit Care 21:1–11

Lee YK, Salminen S (1995) The coming of age of probiotics. Trends Food Sci Technol 6:241–245

Lee FH, Wan SY, Foo HL, Loh TC, Mohamad R, Abdul Rahim R, Idrus Z (2019) Comparative study of extracellular proteolytic, cellulolytic, and hemicellulolytic enzyme activities and biotransformation of palm kernel cake biomass by lactic acid bacteria isolated from malaysian foods. Int J Mol Sci 20:4979

Leroy F, De Vuyst L (2004) Lactic acid bacteria as functional starter cultures for the food fermentation industry. Trends Food Sci Technol 15:67–78

Linares DM, Ross P, Stanton C (2016) Beneficial microbes: the pharmacy in the gut. Bioengineered 7:11–20

Liu G, Wang H, Griffiths MW, Li P (2011) Heterologous extracellular production of enterocin P in *Lactococcus lactis* by a food-grade expression system. Eur Food Res Technol 233:123–129

Mantziari A, Salminen S, Szajewska H, Malagón-Rojas JN (2020) Postbiotics against pathogens commonly involved in pediatric infectious diseases. Microorganisms 8:1510

Martín V, Maldonado A, Fernández L, Rodríguez JM, Connor RI (2010) Inhibition of human immunodeficiency virus type 1 by lactic acid bacteria from human breastmilk. Breastfeed Med 5:153–158

Masood MI, Qadir MI, Shirazi JH, Khan IU (2011) Beneficial effects of lactic acid bacteria on human beings. Crit Rev Microbiol 37:91–98

Mataragas M, Metaxopoulos J, Galiotou M, Drosinos EH (2003) Influence of pH and temperature on growth and bacteriocin production by *Leuconostoc mesenteroides* L124 and *Lactobacillus curvatus* L442. Meat Sci 64:265–271

Messaoudi M, Lalonde R, Violle N, Javelot H, Desor D, Nejdi A, Bisson JF, Rougeot C, Pichelin M, Cazaubiel M, Cazaubiel JM (2011) Assessment of psychotropic-like properties of a probiotic formulation (*Lactobacillus helveticus* R0052 and *Bifidobacterium longum* R0175) in rats and human subjects. Br J Nutr 105:755–764

Mohamad Zabidi NA, Foo HL, Loh TC, Mohamad R, Abdul Rahim R (2020) Enhancement of versatile extracellular cellulolytic and hemicellulolytic enzyme productions by *Lactobacillus plantarum* RI 11 isolated from Malaysian food using renewable natural polymers. Molecules 25:2607. https://doi.org/10.3390/molecules25112607

Mohseni S, Bayani M, Bahmani F, Tajabadi-Ebrahimi M, Bayani MA, Jafari P, Asemi Z (2018) The beneficial effects of probiotic administration on wound healing and metabolic status in patients with diabetic foot ulcer: a randomised, double-blind, placebo-controlled trial. Diabetes Metab Res Rev 34:e2970

Muralitharan RR, Jama HA, Xie L, Peh A, Snelson M, Marques FZ (2020) Microbial peer pressure: the role of the gut microbiota in hypertension and its complications. Hypertension 76:1674–1687

Nakkarach A, Foo HL, Song AA, Nitisinprasert S, Withayagiat U (2020) Promising discovery of beneficial *Escherichia coli* in the human gut. 3Biotech 10:296

Nakkarach A, Withayagiat U, Song AA, Abdul Mutalib NE, Nitisinprasert S, Foo HL (2021) Anti-cancer and anti-inflammatory effects elicited by short chain fatty acids produced by *Escherichia coli* isolated from healthy human gut microbiota. Microb Cell Factories 20:36

Nehal F, Sahnoun M, Smaoui S, Jaouadi B, Bejar S, Mohammed S (2019) Characterization, high production and anti-microbial activity of exopolysaccharides from *Lactococcus lactis* F-mou. Microb Pathog 132:10–19

Nettles CG, Barefoot SF (1993) Biochemical and genetic characteristics of bacteriocins of food-associated lactic acid bacteria. J Food Prot 56:338–356

Nor NM, Mohamad R, Foo HL, Rahim RA (2010) Improvement of folate biosynthesis by lactic acid bacteria using response surface methodology. Food Technol Biotechnol 48:243–250

Paul D, Manna S, Mandal SM (2018) Antibiotics associated disorders and post-biotics induced rescue in gut health. Curr Pharm Des 24:821–829

Peng F, Setyawati MI, Tee JK, Ding X, Wang J, Nga ME, Ho HK, Leong DT (2019) Nanoparticles promote in vivo breast cancer cell intravasation and extravasation by inducing endothelial leakiness. Nat Nanotechnol 14:279–286

Puccetti M, Xiroudaki S, Ricci M, Giovagnoli S (2020) Postbiotic-enabled targeting of the host-microbiota-pathogen interface: hints of antibiotic decline? Pharmaceutics 12:624

Rafter J (2002) Lactic acid bacteria and cancer: mechanistic perspective. Br J Nutr 88:89–94

Rajoka MS, Mehwish HM, Hayat HF, Hussain N, Sarwar S, Aslam H, Nadeem A, Shi J (2019) Characterization, the antioxidant and anti-microbial activity of exopolysaccharide isolated from poultry origin *Lactobacilli*. Probiotics Anti Microb Proteins 11:1132–1142

Raveschot C, Cudennec B, Coutte F, Flahaut C, Fremont M, Drider D, Dhulster P (2018) Production of bioactive peptides by *Lactobacillus* species: from gene to application. Front Microbiol 9:2354

Rinne M, Kalliomaki M, Arvilommi H, Salminen S, Isolauri E (2005) Effect of probiotics and breastfeeding on the *Bifidobacterium* and *Lactobacillus/Enterococcus* microbiota and humoral immune responses. J Pediatr 147:186–191

Robles-Vera I, Toral M, Romero M, Jiménez R, Sánchez M, Pérez-Vizcaíno F, Duarte J (2017) Antihypertensive effects of probiotics. Curr Hypertens Rep 19:26

Ruas-Madiedo P, Hugenholtz J, Zoon P (2002) An overview of the functionality of exopolysaccharides produced by lactic acid bacteria. Int Dairy J 12:163–171

Rul F, Ben-Yahia L, Chegdani F, Wrzosek L, Thomas S, Noordine ML, Gitton C, Cherbuy C, Langella P, Thomas M (2011) Impact of the metabolic activity of *Streptococcus thermophilus* on the colon epithelium of gnotobiotic rats. J Biol Chem 286:10288–10296

Savadogo A, Ouattara CA, Traore AS (2007) Potential of lactic acid bacteria in human nutrition. Food 1:79–84

Schleifer KH, Ludwig W (1995) Phylogenetic relationships of lactic acid bacteria. In: The genera of lactic acid bacteria. Springer, Boston, MA, pp 7–18

Shin MS, Han SK, Ryu JS, Kim KS, Lee WK (2008) Isolation and partial characterisation of a bacteriocin produced by *Pediococcus pentosaceus* K23-2 isolated from Kimchi. J Appl Microbiol 105:331–339

Solioz M, Mermod M, Abicht HK, Mancini S (2011) Responses of lactic acid bacteria to heavy metal stress. Stress responses of lactic acid bacteria. Springer, Boston, MA, pp 163–195

Taheri HR, Moravej H, Malakzadegan A, Tabandeh F, Zaghari M, Shivazad M, Adibmoradi M (2010) Efficacy of *Pediococcus acidlactici*-based probiotic on intestinal Coliforms and villus height, serum cholesterol level and performance of broiler chickens. Afr J Biotechnol 9:7564–7567

Tai HF, Foo HL, Rahim RA, Loh TC, Abdullah MP, Yoshinobu K (2015) Molecular characterisation of new organisation of plnEF and plw loci of bacteriocin genes harbour concomitantly in *Lactobacillus plantarum* I-UL4. Microb Cell Factories 14:89

Talarico TL, Dobrogosz WJ (1989) Chemical characterisation of an anti-microbial substance produced by *Lactobacillus reuteri*. Antimicrob Agents Chemother 33:674–679

Thanh NT, Chwen LT, Foo HL, Hair-Bejo M, Kasim AB (2010) Inhibitory activity of metabolites produced by strains of *Lactobacillus plantarum* isolated from Malaysian fermented food. Int J Probiotics Prebiotics 5:37

Tillisch K (2014) The effects of gut microbiota on CNS function in humans. Gut Microbes 5:404–410

Todorov SD, Danova ST, Van Reenen CA, Meincken M, Dinkova G, Ivanova IV, Dicks LM (2006) Characterisation of bacteriocin HV219, produced by *Lactococcus lactis* subsp. *lactis* HV219 isolated from human vaginal secretions. J Basic Microbiol 46:226–238

Tsilingiri K, Barbosa T, Penna G, Caprioli F, Sonzogni A, Viale G, Rescigno M (2012) Probiotic and postbiotic activity in health and disease: comparison on a novel polarised ex-vivo organ culture model. Gut 61:1007–1015

Uchida M, Shimizu K, Kurakazu K (2010) Yogurt containing *Lactobacillus gasseri* OLL 2716 (LG21 yogurt) accelerated the healing of acetic acid-induced gastric ulcer in rats. Biosci Biotechnol Biochem 74:1891–1894

Verhagen H, Vos E, Francl S, Heinonen M, van Loveren H (2010) Status of nutrition and health claims in Europe. Arch Biochem Biophys 501:6–15

Vitetta L, Saltzman ET, Thomsen M, Nikov T, Hall S (2017) Adjuvant probiotics and the intestinal microbiome: enhancing vaccines and immunotherapy outcomes. Vaccine 5:50

Weenen TC, Ramezanpour B, Pronker ES, Commandeur H, Claassen E (2013) Food-pharma convergence in medical nutrition–best of both worlds? PLoS One 8:e82609

WHO (2018) Latest global cancer data. https://www.who.int/cancer/PRGlobocanFinal.pdf. Accessed January 2020

Zakuan NM, Ling FH, Yazan LS (2019) Anti-microbial, anti-cancer and immunomodulatory properties of proteinaceous postbiotic metabolite produced by *Lactobacillus plantarum* I-UL4. Mal J Med Health Sci 15:81–84

Zárate G (2012) Dairy propionibacteria: less conventional probiotics to improve the human and animal health. Probiotic Anim:153–202

Zheng J, Wittouck S, Salvetti E, Franz CM, Harris HM, Mattarelli P, O'Toole PW, Pot B, Vandamme P, Walter J, Watanabe K (2020) A taxonomic note on the genus *Lactobacillus*: description of 23 novel genera, emended description of the genus *Lactobacillus* Beijerinck 1901, and union of *Lactobacillaceae* and *Leuconostocaceae*. Int J Syst Evol Microbiol 70:2782–2858

Zhong L, Zhang X, Covasa M (2014) Emerging roles of lactic acid bacteria in protection against colorectal cancer. World J Gastroenterol 20:7878

Nondairy Foods as Potential Carriers of Probiotic Bacteria and Postbiotics

15

Fereshteh Ansari and Hadi Pourjafar

Abstract

Dairy-based foods are suitable substrates as a carrier for probiotic microorganisms, nevertheless the great number of lactose intolerant people, their high fat content, and also by reason of the increasing vegetarianism the consumers are looking for substitutes. Consequently, studies have been extensively done on the possibility of probiotic microorganisms in nondairy-based carriers, for instance, vegetables, fruits, cereals, and meat products. This chapter reviews the utilization of probiotics in nondairy-based foodstuffs and some of the technical issues. These issues comprise the efficiency and viability of probiotic microorganisms in nondairy foods; sensory and acceptability of nondairy probiotic products, technological challenges and advancements of nondairy probiotic foods, and postbiotics and food applications of postbiotics.

Keywords

Nondairy foods · Probiotics · Carrier · Postbiotics

F. Ansari
Razi Vaccine and Serum Research Institute, Agricultural Research, Education and Extension Organization (AREEO), Tehran, Iran

Research Center for Evidence-Based Medicine, Health Management and Safety Promotion Research Institute, Tabriz University of Medical Sciences, Tabriz, Iran

H. Pourjafar (✉)
Alborz University of Medical Sciences, Dietary Supplements and Probiotic Research Center, Karaj, Iran

Department of Food Sciences, Maragheh University of Medical Sciences, Maragheh, Iran

N. Mojgani, M. Dadar (eds.), *Probiotic Bacteria and Postbiotic Metabolites: Role in Animal and Human Health*, Microorganisms for Sustainability 2,
https://doi.org/10.1007/978-981-16-0223-8_15

15.1 Introduction

Functional foods are nutrients that have possible positive impacts on healthiness beyond basic nourishment (Ansari et al. 2020b; Granato et al. 2020). The probiotic foodstuffs market is growing very quickly because of enlarged customer awareness about the influence of these functional foods on health and fitness, and currently probiotic products include 60–70% of the entire functional food market (Tripathi and Giri 2014). The universal market for probiotic products (foods and drinks) was about 24.8 billion € in 2011, over 31.1 billion € in 2015 and is predicted to reach about 43 billion € until 2020 (Aspri et al. 2020).

Dairy food products have been conventionally considered as the greatest carriers for probiotic microorganisms. Nevertheless, recently, alternative nondairy foodstuffs have been applied for the separation of possible probiotic isolates for the manufacture of innovative nondairy probiotic foods, and also on the other hand it is determined that nondairy probiotic products have been more appealing by reason of consumer demands (Aspri et al. 2020; Min et al. 2019). This chapter reviews the current knowledge concerning several nondairy probiotic products existing worldwide, for instance, vegetables, fruits, meat products, cereals, and confectionary products with the purpose of give a vision to the issue and to display a way advancing for the future.

15.2 Important Nondairy Probiotic Foods

Various probiotic products have been produced and marketed in different parts of the world, some of which are related to nondairy probiotic products. In the following, the most important nondairy probiotic foods are reviewed (see Table 15.1).

15.3 Fruits and Vegetables

The nutritional and biological potential of fruits and vegetables has led to the conversion of these foods into products with multiple properties in maintaining the balance of microorganisms. These beneficial effects cause the water of these products to be used to treat various diseases. Studies have shown that the presence of potassium salts, bioflavonoids, vitamins, and alkalis in fruit and vegetable and their lack of fat can have beneficial effects in the prevention and treatment of cardiovascular disease (DiRienzo 2014; Fernandez and Marette 2017). It has been shown that the beneficial effects of fruits and vegetables can be improved by a biological process such as lactic fermentation, so that today researchers are studying lactic fermentation of vegetables as a natural preservation method. In addition, some fruits and vegetables contain prebiotics that stimulate the growth of certain probiotics (Szutowska 2020).

Many studies have used probiotics in dairy products, but lactose intolerance and cholesterol levels of these products are the two main disadvantages that have led to

Table 15.1 Some selected publications related to nondairy probiotic products during the last 5 years

Probiotic product	Probiotic strains	Probiotic numbers (log/g or mL)	Reference
Carrot and orange juice	*L. plantarum*	~8 and 9	Valero-Cases et al. (2017)
Mango and guava juice	*L. casei, L. bulgaricus, Streptococcus thermophilus*	7–8	Maldonado et al. (2017)
Apricot juice	*B. lactis* Bb-12, *B. longum* Bb-46, *L. casei* 01, *L. acidophilus* La-5	~7	Bujna et al. (2018)
Tomato juice	*L. plantarum, L. casei*	–	Liu et al. (2018)
Sohiong juice	*L. plantarum* MCC 2974	10	Vivek et al. (2019)
Pomegranate juice	*L. plantarum* ATCC 14917	8.8	Mantzourani et al. (2019)
Banana, strawberry and juçara	*B. animalis* subsp. Lactis, *L. acidophilus*, *L. casei, L. plantarum*	~5 and 7	de Oliveira Ribeiro et al. (2020)
Cornelian cherry (*Cornus mas L.*) drink	*Saccharomyces cerevisiae* DDNd10, *Pichia kudriavzevii* DCNa, *Wickerhamomyces subpelliculosus* DFNb6	~8	Di Cagno et al. (2020)
Fermented beverage from maize and rice	*L. plantarum, L. acidophilus, Torulaspora delbrueckii*	7	Freire et al. (2017)
Fermented oat flour drink	*L. plantarum*	14	Gupta and Bajaj (2017)
maize-based substrate	*L. paracasei* LBC-81, *Saccharomyces cerevisiae* CCMA 0731, *Saccharomyces cerevisiae* CCMA 0732, *Pichia kluyveri* CCMA 0615	6	Menezes et al. (2018)
Legume sprouts	*L. plantarum* 299 V	9	Świeca et al. (2018)
Breadfruit flour drink	*L. plantarum* DPC 206, *L. acidophilus, L. casei* Shirota	7–8	Gao et al. (2019)
Wheat/rice cereal infant products	*B. animalis* subsp. lactis BB-12®	6	Leboš-Pavunc et al. (2019)
Dry-fermented pork neck and sausage	*L. acidophilus Bauer, B. animalis* BB-12, *L. rhamnosus* LOCK900	~6–8	Wójciak et al. (2017)
Beef sausage	*L. plantarum* TN8, *Pediococcus acidilactici* MA18/5 M	~8	Slima et al. (2018)
Bovine Salami	*L. plantarum* 299v	7	Blaiotta et al. (2018)
Fermented sausage	*B. longum* KACC 91563	~3–6	Song et al. (2018)

(continued)

Table 15.1 (continued)

Probiotic product	Probiotic strains	Probiotic numbers (log/g or mL)	Reference
Dry fermented sausage	*L. paracasei* LPC02	~7–8	Coelho et al. (2019)
Fermented sausage	*L. paracasei, L. rhamnosus* GG	–	Bis-Souza et al. (2019b)
Dry-cured meat sausages	*L. plantarum*	~8	Sirini et al. (2020b)
Spanish Salchichón	*L. paracasei, L. rhamnosus* GG	~8	Bis-Souza et al. (2020)

their nonconsumption by some people. Therefore, the use of probiotics in fruit and vegetable foods can be a good alternative for a group of people with special needs (i.e., vegetarians and people with allergic reactions to milk proteins) (Aspri et al. 2020; Min et al. 2019).

Another important point to consider is the viability of probiotics in fruit and vegetable products. Due to the fact that the pH of fruits and vegetables is low (pH 2.5–3.7), and also bacteria are sensitive to acidic conditions, it is necessary to use strains that can survive in these conditions and make the product healthier and increase the shelf life. Therefore, the addition of probiotics to fruit and vegetable-based foods and beverages is more complex than dairy products because of the need to protect them from acidic environmental conditions (Min et al. 2019). It has been shown that in fruit juices (pH 3.7–4.3) Lactobacilli can resist and survive better than Bifidobacterial (Patel 2017). Nevertheless, some fruit juices may contain components that sustain the survivability of probiotic microorganisms like ascorbic acid, that declines O/R potential, organic acids or saccharides that may be used as a carbon source or cellulose that can guard these microorganisms throughout processing and storing (Martins et al. 2013).

Beetroot, cabbage, carrot, olive, oranges, pineapple, mango, strawberry, blueberry, cranberry, sweet lime, cashew apple, and grapes are some instances of vegetable and fruit juices employed as food substrates for the delivery of probiotic microorganisms. Various types of probiotic vegetable/fruit products have been technologically advanced and commercialized, including fermented, juices, deserts, and dried forms. Extensive variety of probiotic strains, mostly species of Bifidobacteria and Lactobacillus, for instance, *Bifidobacterium bifidum*, *B. longum*, *B. breve*, *Lactobacillus acidophilus*, *L. rhamnosus*, *L. casei*, *L. paracasei*, *L. plantarum*, and *L. fermentum* have been broadly used in the advance of several vegetable/fruit probiotic products (Aspri et al. 2020; Min et al. 2019).

Probiotic vegetable/fruit can be produced either with straight adding of the probiotic strain, for example, into the juice of them or via the fermentation with the probiotic microorganism. The fermentation is more valuable because probiotic strain grows into the vegetable/fruit texture or juice to a more adapted probiotic strain and a low-sugar product, which may perhaps develop its survival rates (Pereira and Rodrigues 2018). Also, throughout the fermentation process, the metabolites of probiotics such as exopolysaccharides and bacteriocins can aid to improve the quality of the probiotic product and rise their shelf life through storing time (Aspri et al. 2020).

Newly, various studies focused on the manufacture of fermented probiotic and synbiotic vegetable/fruit, for instance pomegranate extracts, and Cornelian cherry drinks via delignified wheat bran (Kazakos et al. 2020; Liu et al. 2018; Valero-Cases et al. 2020), carrot-orange extracts and nectars with diverse inulin concentrations (Alizadeh et al. 2019; Lu et al. 2018), synbiotic apple juice or orange extract with oligofructose (Miranda et al. 2019; Pimentel et al. 2015; Zhu et al. 2020), blended drink of orange extract, hibiscus tea, and oligofructose (Miranda et al. 2019), and mixed red fruit drinks of papaya, blackberry, and strawberry added with three distinct prebiotics including inulin, galactooligosaccharides (GOS), and fructooligosaccharides (FOS) (Bernal-Castro et al. 2019).

Pereira et al. (2011) studied the production of probiotic cashew apple juice and they showed the cell counts of *L. casei* in the product after 6 weeks storage was about 8 log cfu/mL (Pereira et al. 2011). Similar results have been shown in melon juice (Fonteles et al. 2013) and pineapple juice (Sheehan et al. 2007).

Sheehan et al. (2007) described wide alterations relating to the acid resistance characteristic of Bifidobacterium and Lactobacillus in pineapple, orange, and cranberry juices. The survival rate of probiotics in pineapple and orange juices was higher than cranberry juice. The number of *L. rhamnosus*, *L. casei*, and *L. paracasei* was above 6 log cfu/mL in pineapple juice and over 7 log cfu/mL in orange juice for 12 weeks. In other study, a probiotic drink using coconut water was produced via fermenting it by means of *L. plantarum* (Prado et al. 2015).

It has been studied the suitability of beet juice (Yoon et al. 2005), cabbage juice (Ningrum et al. 2019; Yoon et al. 2006), and tomato (Dzandu 2019) by *L. casei*, *L. acidophilus*, *L. plantarum*, and *L. delbrueckii*, and the number of the viable cells the four probiotics in all fermented products raged from 5–8 log cfu/mL after 4 weeks storing at 4 °C.

Kun et al. (2008) showed that carrot juice can promote the growth of *B. bifidum* B 3.2, *B. bifidum* B7, and *B. lactis* Bb-12. All probiotic strains displayed high primary cell counts of 10 log cfu/mL (Kun et al. 2008).

Mantzourani et al. (2018) applied *L. plantarum* ATCC 14917 in producing probiotic Cornelian cherry juice. Consequences of their study disclosed that the number of viable cells of *L. plantarum* ATCC 14917 was acceptable during cold storage time and no significant organoleptic changes were observed in both fermented and non-fermented samples (Mantzourani et al. 2018). Also, in another study, they produced and surveyed fermented pomegranate juice by use of the same strain (*L. plantarum* ATCC 14917) (Mantzourani et al. 2019).

Bujna et al. (2018) investigated mono- and mixed probiotic cultures for production of apricot juice as a new nondairy probiotic beverage. Fermentation process developed using probiotic bacteria individually disclosed cell numbers of, 7.06, 7.16, 7.2, and 7.25, log (cfu/mL h) *L. casei* 01, *L. acidophilus* La5, *B. lactis*

Bb-12, and *B. longum* Bb-46 strains, respectively, even though sample fermented via a mixed culture a higher cell numbers was detected (Bujna et al. 2018).

Li et al. (2019) demonstrated that using *L. plantarum* ATCC 14917 in apple juice can improve the phenolic composition of apple juice and promote its total antioxidant capacity. Also, Peng et al. (2020) assessed the fermentation performance of a combination of Lactobacillus spp. in cloudy apple juices from nine cultivars. The consequences disclosed that cultivar impacted most the characteristics of the fermented cloudy apple juice. The highest probiotic number (6.37×10^8 CFU/mL) and acetic acid contents (2.67 mg/mL) achieved from the fermented cloudy apple juices made from Changfu (Peng et al. 2020).

In another study by Zhu et al. (2020), survival rate of *L. sanfranciscensis* into three diverse nondairy carrier (apple, tomato, and orange) were investigated throughout 4 weeks storage at 4 °C. Results showed that the survivability of probiotics in all samples met the recommended level of >6–7 log cfu/mL at the end of storage time.

15.4 Cereals

Cereals are one of the most important sources of protein, carbohydrates, vitamins, minerals, and fiber for humans. They can be used as sources of indigestible carbohydrates and, with their water-soluble fiber such as beta-glucan, arabinoxylan, and oligosaccharides such as galacto and fructo oligosaccharides and resistant starch, as a prebiotic, can selectively stimulate the growth of Lactobacilli and Bifidobacteria in the colon. Whole grains are sources of phytochemicals such as phytosterols, phenolic compounds, antioxidants, phytic acid, and sterols (Lamsal and Faubion 2009; Ogunremi et al. 2020).

The nutritional quality of grains is sometimes lower than that of milk due to their lower protein content, deficiency of certain amino acids such as lysine, inability to digest starch, hard nature of grains, and the presence of anti-nutritional compounds such as phytic acid, tannins, and polyphenols. These compounds vary widely in chemical structure and function, so fermentation can reduce the levels of indigestible carbohydrates, poly and oligosaccharides, improve protein quality, and increase lysine levels. Some amino acids may also be synthesized and the availability of B vitamins may increase. Fermentation also provides the optimum pH for the enzymatic degradation of phytate and the release of minerals such as manganese (an important factor for the growth of lactic acid bacteria), iron, zinc, and calcium. Various strains of Lactobacilli and Bifidobacteria need fermentable carbohydrates, amino acids, B vitamins, nitrogen, and minerals to grow, and different types of cereals (like wheat, barley, millet, maize, oats, rye, and sorghum) are an inexpensive culture and good carrier for these probiotics (Charalampopoulos et al. 2002; Chavan et al. 2018; Kocková et al. 2013). Some instances of traditional cereal-based fermented beverages (like Mahewa, Bushera, Boza, Togwa, and Pozol) are described here.

Mahewu is a sour cereal-based probiotic beverage and it is made by means of a multi-grain combination which can contains millet, maize, malt, sorghum, and wheat flour (Panghal et al. 2018). The natural fermentation process is made via the malt's microflora at about 25 °C. The chief bacteria isolated from Mahewu is *L. lactis* subsp. lactis (Blandino et al. 2003). Bushera is an old beverage made with millet or sorghum flour, and chiefly *L. brevis*, and another LAB like Enterococcus, Lactococcus, Leuconostoc, and Streptococcus were isolated from Bushera (Muyanja et al. 2003).

Boza is a traditional beverage from the natural fermentation of cereals like maize, wheat, millet, rye, and others that are mixed with sugar (Todorov et al. 2008). Boza displays a great variety of LAB and yeasts which contain *L. acidophilus*, *L. coprophilus*, *Lactobacillus brevis*, *L. plantarum*, *L. fermentum*, *Leuconostoc mesenteroides*, *Leuconostoc reffinolactis*, *Saccharomyces cerevisiae*, *Candida tropicalis*, *Candida glabrata*, *Geotrichum candidum*, and *Geotrichum penicillatum* (Heperkan et al. 2014). Togwa is another cereal-based fermented probiotic beverage from China and Japan. This traditional probiotic drink is produced via fermenting multi-grains like sorghum, finger millet, and maize flour with some probiotics, for instance Streptococcus and mainly *L. plantarum* (Mugula et al. 2003).

Various investigations have been done to promote cereal-based probiotic products and to assess the suitability of diverse cereal grains to improve probiotic microorganisms' growth and uphold their survivability into products during manufacturing and storage time and also throughout gastrointestinal conditions (in vitro and in vivo). Świeca et al. (2018) surveyed legume sprouts as a nondairy carrier for *L. plantarum* 299 V. The sprouts that have been supplemented with the probiotic, a lower mesophilic bacteria flora, particularly LAB, was detected in comparison with the control groups (without probiotic). The *L. plantarum* number was also steady throughout the cold storage period (Świeca et al. 2018).

In a study by Menezes et al. (2018), *L. paracasei* LBC-81 was employed lonely and in mix with *S. cerevisiae* CCMA0731, *S. cerevisiae* CCMA0732, and *Pichia kluyveri* CCMA0615 into maize-based substrate as a different functional food. Three out of the four strains displayed acceptable survivability with counts more than 6 log cfu/mL, which is the suggested for probiotic foodstuffs, excluding the *Pichia kluyveri* which reduced throughout fermentation and storage period(Menezes et al. 2018). Leboš-Pavunc et al. (2019) studied the effect of dehydrated wheat/rice media on probiotic activity of *B. animalis* ssp. lactis BB-12. The probiotic strain (*B. animalis* ssp. lactis BB-12) displayed the high survivability throughout the storage time of 106 weeks (Leboš-Pavunc et al. 2019). Gao et al. (2019) investigated the development of a probiotic beverage by means of breadfruit flour as a substrate, and *L. plantarum* DPC 206 and *L. acidophilus* as probiotic strains. The produced probiotic beverage was found to have adequate cell viability and also satisfactory sensory characteristic.

Soy milk is the main and well-known food in Asian countries that is now consumed all over the world. Soybean milk is a stable emulsion of oil, water, and protein which is prepared by soaking dried soybeans and grinding them. Soy milk is a rich source of high-quality plant protein, isoflavones and B vitamins that are free of milk sugar or lactose and is a good choice for people with lactose intolerance. Laboratory studies on probiotic microorganisms have shown that soy milk is a good substrate for some probiotic bacteria, such as *L. casei, L. helveticus, L. fermentum, L. reuteri,* and *L. acidophilus* (Niamah et al. 2017; Shilpa et al. 2011; Shimakawa et al. 2003; Taghizadeh et al. 2018).

In a related study, Homayouni Rad et al. (2020) investigated soy ice cream (produced via the powder of soy milk) as a carrier for effective delivering of *L. casei.* The viability of mentioned probiotic bacteria was assessed over storage time (180-day, at −25 °C). The results disclosed considerable changes in the count of *L. casei* in this product subsequently freezing and during storage time ($p < 0.05$). The most significant drop was perceived through the first 60 days approximately 1.83 logs after that the tendency of survival of *L. casei* leveled off over the next 120 days (Homayouni et al. 2020b). In similar study, Norouzi et al. (2019) surveyed the survival rate of probiotic *L. paracasei* ssp. paracasei into fermented and non-fermented frozen soy dessert during 180 days storage at −24 °C. The results showed a considerable rise ($p < 0.05$) in overrun (42.57 ± 8.5) values in fermented probiotic frozen soy dessert compared to other samples. In contrast to non-fermented samples, there was no considerable ($p < 0.05$) reduction in cell numbers of *L. paracasei* throughout storage time. Both probiotic samples have capable potential for application as functional foods. Nevertheless, fermentation could rise the stability of *L. paracasei* in frozen soy dessert. Likewise, the organoleptic and physicochemical properties of frozen soy dessert were enhanced via fermentation (Norouzi et al. 2019).

de Carvalho Marchesin et al. (2018) studied the impact of a soy-based probiotic drink as a carrier for *B. longum* ATCC 15707 and *Enterococcus faecium* CRL 183 on the fecal microbiota configuration, body weight and inflammatory parameters in diet-induced obese mice (de Carvalho Marchesin et al. 2018). In another study, Devanthi et al. (2018) investigated the impact of concurrent and consecutive inoculation of cultures (*Tetragenococcus halophilus* and *Zygosaccharomyces rouxii* as starter cultures) in moromi fermentation models, regarding survivability, physicochemical variations, and volatiles formation (Devanthi et al. 2018). Lima Moraes Filho et al. (2019) studied creamy soy sauce as carrier for *L. plantarum BG 112* (Moraes Filho et al. 2019).

Recently, Setta et al. (2020) reviewed potential of probiotics from fermented cereal-based beverages in enhancing healthiness of poor people in Africa. Also, several researches have exposed that traditional African fermented cereal-based drinks are possible probiotic carriers due to the probiotic bacteria (specially Lactobacillus and Bifidobacterium spp.) and yeasts which are involved in the fermentation of such foodstuffs. These probiotic products propose an occasion for the African cereal beverages to be used to deliver probiotic health advantages to the majority of populations. There are also other similar products in different countries that can replace the shortage of fermented dairy products and expensive probiotic foods (Setta et al. 2020).

15.5 Meat Products

The function of probiotics in dairy products is fully acceptable, while their function in meat products is still being studied. Fermented meat products are suitable for carrying probiotics because they are not heated at all during processing or receive very little heat and may improve the survival of probiotic bacteria in the gastrointestinal tract. In contrast, bacterial viability may be reduced due to high salt content and low water activity and low pH. Therefore, the results depend on the strain used. However, the combination of these microorganisms with fermented meat products can create some technological challenges. On the other hand, the acceptability of the new meat product for the consumer and the survival of sufficient probiotic microorganisms during the process is also among the issues under discussion (Kołożyn-Krajewska and Dolatowski 2012). Probiotic bacterial strains have been used successfully in the production of dairy products and some fruit juices, but their use in the production of raw fermented meat products is not very suitable. It seems that it is possible to add strains of meat-derived probiotic bacteria in the process of producing fermented raw meat products, but research is needed to find out which probiotic species can grow in which meat products (De Vuyst et al. 2008; Kołożyn-Krajewska and Dolatowski 2009).

Fermented sausages (as raw meat products) are auspicious target meat products with probiotic microorganisms, as such foodstuffs are treated without heat treatment and probiotic microorganisms can continue to be live in the final product (Aspri et al. 2020; Kumar et al. 2015). However, probiotics must maintain their stability under adverse conditions for the production of fermented sausages such as low pH (<4), nitrite (120 ppm), a_w (less than 0.85), and salt (1–3%) (Ordóñez et al. 1999; Vignolo et al. 2010). Probiotic cultures should be able to grow rapidly during fermentation and grow easily on industrial scales, be resistant to the freezing process, provide a longer shelf life, and also improve the sensory quality of the final product. In such fermented products, the addition of 3% sodium chloride and at least 120 ppm nitrite is mandatory to maintain the microbial safety of the product. Therefore, the use of salt-resistant medium is the first condition for the production of sausages with probiotic properties (Aspri et al. 2020; Papamanoli et al. 2003). Despite all the difficult conditions for the growth and survival of probiotics, fermented sausages are regarded suitable carriers for them due to the guard of the probiotic cells to bile salts and low pH which are applied from the fat molecules in the passageway over the GI zone and the motivation of probiotic growing by the existence of the prebiotic fibers (Bis-Souza et al. 2019a).

Important employed species of probiotic bacteria in fermented meat foods are *L. plantarum*, *L. casei*, *L. paracasei*, *L. sakei*, *L. acidophilus*, *L. rhamnosus*, *Pediococcus pentosaceus,* and *Pediococcus acidilactici*. The amalgamation of the probiotic microorganisms can be attained via substituting the traditional starter culture or via applying the traditional starter in association with the probiotic strain (Bis-Souza et al. 2019a). Various investigations reveal the effective utilization of probiotic strains into diverse fermented meat foods, for instance, different fermented sausages, salami, dry cured pork loins, sturgeon fermented sausage, mutton fermented sausage, and Longaniza de Pascua (Aspri et al. 2020).

Rubio et al. (2014) studied nutritionally improved fermented sausages as a carrier for delivery of lactobacilli probiotics. They used *L. rhamnosus* CTC1679 as a probiotic for the making of Fuet (low acid fermented sausage). According to the results, *L. rhamnosus* CTC1679 was able to grow and reach numbers of 8 log cfu/g without disturbing the organoleptic characteristics of the product (Rubio et al. 2014). In another study, Wójciak et al. (2017) investigated technical aspect of *L. rhamnosus* LOCK900, *L. acidophilus* Bauer, and *B. animalis* BB-12 usage in dry fermented pork neck and sausage. They showed that *L. acidophilus* retained the quality of the product better than the *L. rhamnosus* and *B. animalis* (Wójciak et al. 2017). Slima et al. (2018) demonstrated that probiotic strains of *Pediococcus acidilactici* MA 18/5 M and *L. plantarum* TN8 could be employed to improve sensory properties and cooking yield and also extending sausage shelf life. In another study, Pavli et al. (2020) investigated the potential of dry-fermented pork sausages as a carrier for *L. plantarum* L125 strain. The results of their study disclosed that the viability rate of *L. plantarum* was suitable (>6 log cfu/g) during the storage time without considerable impacting on the technological and the organoleptic properties of the final product.

In similar study, de L Agüero et al. (2020) studied the technological characteristics of LAB as starter cultures for dry fermented sausages. Eight strains were evaluated properties counting the capability to grow, gas formation, lactic acid production, hydrogen peroxide production, salt tolerance, nitrate reductase activity, catalase activity, lipolytic activity, proteolytic activity, decarboxylation of amino acids, performance at low temperatures, and antimicrobial activity against pathogen microorganisms related to the product. According to the results, *L. rhamnosus* Lr-32, *L. rhamnosus* R0011, *L. casei* Shirota, *L. paracasei* Lpc-37, and *Enterococcus faecium* MXVK29 were suitable candidates for use as fermented sausages starters. *L. rhamnosus* Lr-32 was the best enduring the low pH, salt, and nitrate throughout the simulated phases of fermentation and maturing of sausage (de L Agüero et al. 2020). Sirini et al. (2020b) studied the effect of chestnut flour and probiotic microorganism on the functionality of dry-cured meat sausages. Adding the chestnut flour diminished pH and remaining nitrite in Longaniza de Pascua. The results of the study showed that Longaniza de Pascua is a good carrier for *L. plantarum* (Sirini et al. 2020b). The same authors reviewed the use of probiotic microorganisms in the formulation of healthy meat products (Sirini et al. 2020a).

15.6 Other Traditional Products

There are many traditional fermented products with different names in different parts of the world that can be the origin of different types of probiotic microorganisms and may not have been researched yet. These products can be suitable carriers of native probiotics and cause the transfer of probiotics into the consumer's GI tract. However, there are some products that have been extensively researched and their health effects have been proven to be effective carriers of probiotic microorganisms, postbiotics, and other functional compounds. There are several non-dairy products that can be discussed in this regard, which in this section, we have explained two of these popular products; Kombucha and Chocolate.

One of the most famous of these products is Kombucha as a fermented beverage, with a history of some thousand years in the East. Kombucha drink has been claimed to be a nutritional supplement that consuming it supports the immune system and averts some diseases. Kombucha is a symbiotic evolution of acetic acid bacteria and osmophilic yeast species (SCOBY) which have to be cultivated in sweetened tea with glucose or sucrose. SCOBY includes some bacteria like *Acetobacter aceti* spp. xylinum, *Acetobacter xylinum*, *Corynebacterium glutamicum*, *Acetobacter pasteurianus*, and *Acetobacter xylinoides* and also some yeasts like *Saccharomyces bisporus*, *Saccharomyces cerevisiae*, *Saccharomyces ludwigii*, *Schizosaccharomyces pombe*, *Zygosaccharomyces bailii*, *Candida kefyer*, *Candia krusei*, *Pichia* sp., *Brettanomyces* sp., *Torulopsis* sp., and *Issatchenkia orientalis occidentalis*. Several complexes and postbiotics have been isolated from Kombucha including carbonic acid, glucuronic acid, gluconic acid, acetic acid, folic acid, butyric acid, oxalic acid, lactic acid, malic acid, nucleic acid, citric acid, carbon dioxide, ethanol, antibiotics, vitamins B including B1, B2, B6, and B12, and vitamin C (Ansari et al. 2017; Ansari et al. 2019; Chakravorty et al. 2016; Coelho et al. 2020; Villarreal-Soto et al. 2018).

Another product, as a functional food, not only does not unpleasantly impact healthiness, but also hinders some disorders such as cancer, osteoporosis, diabetes, and cardiovascular diseases. Cocoa is rich in proteins, minerals, carbohydrates, flavonoids, and polyphenolic antioxidants (Aspri et al. 2020). Several researchers have recommended that chocolate is a suitable substrate for probiotic microorganisms providing guard to probiotics throughout storing time and passage into GI lumen (dos Santos Filho et al. 2019; Kemsawasd et al. 2016; Konar et al. 2016). Klindt-Toldam et al. (2016) displayed that *B. lactis* HN019 and *L. acidophilus* NCFM combined into dark chocolate and milk chocolate remained viable throughout storage period and also during GI lumen (Klindt-Toldam et al. 2016). Zarić et al. (2016) showed that after 6 months of storage, the viability of *L. acidophilus* NCFM, *L. rhamnosus* HN001, and *B. lactis* HN01 was above 90%, with cell number of approximately 8 log cfu/g(Zarić et al. 2016). In another study, Mirković et al. (2018) studied the organoleptic quality and volatile profile of dark chocolate supplemented with microencapsulated probiotic *L. plantarum 564* and *L. plantarum* 299v. The consequences disclosed suitable survival of both probiotic strains after manufacture and throughout storage period (8 log cfu/g in the first 60 days and over 6 log cfu/g up to 180 days) (Mirković et al. 2018). Cielecka-Piontek et al. (2020) studied survival of commercial probiotic strains and their effect on dark chocolate synbiotic snack with raspberry content during the storage and after simulated digestion. The results showed the cell count of probiotics was steady (8 log cfu/g) and moderately high through 6 months of storage time (Cielecka-Piontek et al. 2020).

15.7 Commercially Available Nondairy-Based Probiotic Products

Various nondairy probiotic foods are previously extant for consumers. The primary nondairy probiotic food was produced via a Swedish corporation Skane Dairy (1994-ProViva) (Aspri et al. 2020; Bansal et al. 2016). Oatmeal gruel being fermented via *L. plantarum* 299v was the base substrate and malted barley was supplemented to expand liquefaction of the product, and finally it has been mixed with diverse fruit extracts like strawberry, blueberry, or tropical fruits. This probiotic product comprised 5×10^{10} CFU/L of *L. plantarum* (Molin 2001). A comparable product GoodBelly was the primary nondairy probiotic beverage in the United States in 2006 (Aspri et al. 2020; Panghal et al. 2018). Furthermore, some cereal probiotic products have been available in some markets, for example, probiotic flakes (Muesli®, Portugal), cereal bars (CornyActiv®, Germany), whole wheat breakfast cereals (Weetaflakes®, France), whole grain probiotic liquidR (Grainfields, Australia), whole grain oatmeal (United Kingdom), and snack bar (Goodness®, United Kingdom) (Aspri et al. 2020; Dornblaser 2007).

15.8 Viability of Probiotics in Nondairy Foods

The survival rate of probiotics is one of the most important factors in the study of these microorganisms. Dairy/nondairy Foods containing probiotics fall within the “functional foods” class and these nutrients should comprise as a minimum 7 log cfu/g probiotics and consumed at levels higher than 100 g/day to have supportive impacts on healthiness (Abdolhosseinzadeh et al. 2018). However, there are still some complications related to the low survival of probiotic microorganisms in foods along with GI environments (Mirzaei et al. 2011; Suvarna et al. 2018). Various investigations have obviously showed that the kind of carrier foods could impact not only the survivability of probiotic microorganisms throughout production and storing period, but also on their functional characteristics, for instance vulnerability to harsh situations in the GI tract (low pH, bile salt, and several enzymes), ability to stick to enterocytes, and immunomodulation (Kedia et al. 2009; Marco and Tachon 2013; Ranadheera et al. 2012; Ranadheera et al. 2014).

Dairy foodstuffs (rich in milk fat) can enhance the viability of probiotics during manufacturing and GI tract. Nevertheless, the physical structure of nondairy foods, for instance fruits and vegetables, may offer protecting milieu for probiotic microorganisms and decrease their contact to strict GI circumstances too (Ansari et al. 2020a; Homayouni et al. 2020a; Homayouni et al. 2018). Fermented meats such as sausage structure have also revealed a possible in retaining the survivability of probiotic cells over GI transportation (Klingberg and Budde 2006; Rouhi et al. 2013). Various published articles are available on how to advance the viability of probiotic microorganisms in nondairy foods. The most striking and effective approaches are fortification via prebiotic ingredients (for instance dietary fiber, resistant starch, inulin, cellulose), adding antioxidants, refrigerated storing in atmosphere improved carbon dioxide, and microencapsulation (Khosravi et al. 2021; Shah 2000; Tamime et al. 2005). In summary, the important point that has already been mentioned is to keep the number of probiotics in an acceptable range so that it can maintain its beneficial effects on the host, and this is very much affected by the type of food that carries these microorganisms.

15.9 Organoleptic/Sensory and Acceptability of Nondairy Probiotics

The organoleptic/sensory properties and acceptability of probiotic products are especially important in the case of nondairy products in terms of industrial and mass production. The organoleptic characteristics of nondairy probiotic foods can be impacted via interactions between diverse probiotics strains and food substrates, where taste, aroma, flavor, color, and textures may be improved via the creation of diverse metabolic ingredients, for instance, organic acids, exopolysaccharides, and other metabolites through processing and storage (Aspri et al. 2020; Panghal et al. 2018). Consequently, it is significant to assess not only the suitable probiotic viability, but also the organoleptic acceptance throughout manufacturing and storage time of probiotic nondairy foods. For example, in fruit products, depending on the type of fruit, processing and storing temperature, type of probiotic, and the addition of protectants and prebiotics, it can influence the organoleptic characteristics of the final product (Lebaka et al. 2018).

15.10 Summary of Technological Challenges and Advancements of Nondairy Probiotic Foods

The most investigated technologies which were involved in fermentation, rehydration, drying, microencapsulation, and storage have been advanced and effectively used to protect some probiotic microorganisms from environmental tensions related to several nondairy food media, nonetheless there are still various technological challenges in manufacturing and protecting probiotic foods. Certainly, the maintenance of adequate viable cells of probiotics is an important factor of quality. Merely applying any probiotic species such as Lactobacillus and Bifidobacterium does not assurance high viable content in fermented foodstuffs after fermentation and throughout the storage period (Min et al. 2019).

Technological challenges such as manufacturing, handing out, temperature of production and storage, pH rate, oxygen content, O/R potential, a_w rate, relative humidity (RH), antimicrobial agents, and external stresses can affect the use of probiotic microorganisms in food products (Min et al. 2019; Vasudha and Mishra 2013). For example, several food components such as sugar and salts can bind water and cause low a_w and improve the viability of probiotics (Holck et al. 2011). Conversely, surplus a_w, for example, in fruit juice, can decrease the survival of probiotics throughout storage period (Vasudha and Mishra 2013).

Microencapsulation of the probiotic microorganisms is one of the modern and most effectual methods. Microencapsulation by means of emulsion and extrusion methods has been used for the protection of probiotics against harsh circumstances (Pourjafar et al. 2018, 2020). Some studies have shown that the addition of protectants (e.g., lactose, trehahose cellobiose, and sucrose) to the probiotic media can advance survivability of probiotics in the nondairy foods (Min et al. 2019).

15.11 Postbiotics and Food Applications of Postbiotics

15.11.1 Classes of Postbiotics

Postbiotics can be secreted by live microbiota during its life cycle or may be released after bacterial lysis. These compounds are essential in regulating self-growth, development, reproduction and modulating the growth of other microorganisms and can also have an effect on the physiological responses of the host by modifying cellular processes and metabolic pathways (Aguilar-Toalá et al. 2018). In general, these postbiotics are classified either by their elemental composition or by their physiological function.

According the first classification method there are seven main classes of postbiotics; (1) Cell-Free Supernatants including biologically active metabolites secreted by bacteria and yeast into the surrounding liquid. This kind of postbiotics is obtained directly from cell cultures. For this purpose, the microbial cells are separated from overnight grown whole culture broth by centrifugations, and the supernatant containing postbiotics is then filtered to ensure sterility. (2) Exopolysaccharides which are biopolymers with different chemical properties released outside the bacterial cell wall. These biopolymers form heterogeneous molecules called exopolysaccharides (EPSs). EPSs has recently gained a lot of attraction and are widely studied for their potential positive biological properties. (3) Enzymes that are a part of the defense mechanisms of the organism against harmful effects of reactive oxygen species (ROS) damaging lipids, proteins, carbohydrates, and nucleic acids. Antioxidant enzymes, such as glutathione peroxidase (GPx), peroxide dismutase (SOD), catalase, and NADH-oxidase are classified within this category. (4) Cell Wall Fragments such as bacterial lipoteichoic acid (LTA). These components are immunogenic and can elicit immune responses. They are proven to have anti-infectious, anti-inflammatory and anti-cancer effects. (5) Short-Chain Fatty Acids (SCFAs) produced by fermentation of plant polysaccharides by intestinal microbiota. Acetic, propionic, and butyric acids are from the most well-known SCFAs. Helping to renew intestinal epithelium, modulating gene expression, immunosuppressive effects, and metabolic and anti-inflammatory function are from the properties of SCFAs. (6) Bacterial Lysates (BLs) which are results of the chemical or mechanical degradation of Gram-positive and Gram-negative bacteria commonly found in the environment. They have specific immunological activities like activating T and B lymphocytes. The beneficial effects of BLs in the case of infections and allergic diseases have been observed in several studies. (7) Metabolites Produced by Gut Microbiota including vitamins, phenolic-derived metabolites, and aromatic amino acids. These products have high bioavailability, antioxidative features, and signaling properties and are very important in host-microbiome crosstalk (Żółkiewicz et al. 2020).

According the physiological function of the postbiotics they are grouped in categories such as immunomodulator, anti-inflammatory, hypocholesterolemic, anti-obesogenic, anti-hypertensive, anti-proliferative, and antioxidant (Aguilar-Toalá et al. 2018). These classifications lead to better understanding of the function of postbiotics and is used for the application of them for clinical and industrial purposes.

15.11.2 Manipulation of Postbiotic Composition for Food Application

To apply postbiotics in nondairy food products several issues should be considered. First of all, the susceptibility of postbiotics to manipulation processes such as pH alterations, heat treatment and exposure to NaCl and proteolytic enzymes which can severely change the postbiotic properties. For instance, it has been shown that antimicrobial activity of bacteriocinogenic *Lactococcus lactis* subsp. lactis CWBI-B1410 and *L. curvatus* CWBI-B28 was completely lost after 8 days storage at 10 °C in the neutralized pH solution. The second issue is the interaction of postbiotics with other food compounds. Food ingredients may contain several postbiotic inhibitory elements like enzymes, proteins, and carbohydrates which may limit the postbiotic properties. And the last but not the least is the effects of postbiotics on consumer's overall acceptance of the food (Moradi et al. 2020).

In the case of application postbiotics for each kind of food products all the above issues should be assessed. There are also some hurdle technologies to assist overcoming the mentioned concerns. For instance, postbiotics may be protected and released in their target sites thorough encapsulation processes. Moreover, application of nanocarriers for hydrophobic postbiotics like fat-soluble vitamins improves their resistance to the light, heat, and oxygen during storage and augments bioavailability of the postbiotics. They also increase transparency of the product, which is highly important in the case of postbiotic beverages (Homayouni Rad et al. 2020). However, we need more studies in this field to develop effective and applicable methods to protect postbiotics in the food matrix (Moradi et al. 2020).

15.11.3 Interaction Between Postbiotics and Food Ingredients

As it has been discussed previously, postbiotics have several kinds of properties, such as anti-cancer, anti-oxidant, and anti-microbial effects. All the ingredients of a food matrix have the potential to increase or decrease the specific effects of the added postbiotics. For instance, it has been shown that the postbiotic Nisin is rapidly inactivated in raw beef because of potential interaction with raw components like proteases and glutathione. In this case heating the meat at 71 °C or application of liposome-encapsulated nisin was proposed for maintenance of the activity of nisin during processing and storage (Younes et al. 2017). Similarly, anti-microbial postbiotics may be applied to enhance the microbial safety of the food. Although the results of a research indicate that bacteriocins with strong inhibition on agar plates had only a reduced and transient inhibitory effect when applied in ground beef (Hartmann et al. 2011). Some metabolites like EPS can protect the pathogens from harsh environmental conditions, so their application in food products should be under specific cautions. Otherwise they may increase the proliferation of the bacteria and introduce a safety hazard for the consumers (Moradi et al. 2020).

On the other hand, some of the food ingredients lead to the production of valuable substances as postbiotics which are not normally produced by the bacterium. For example, some strains of *L. plantarum* produce 10-hydroxy-cis-12-octadecenoic acid (HYA) in the presence of linoleic acid. This substance is very beneficial for controlling colitis and diabetes (Moradi et al. 2020).

15.12 Conclusion

Nondairy food products (legumes, cereals, pseudocereals, fruits, and vegetables) provide a valuable opportunity for special groups of people such as people intolerant or allergic to milk proteins, those who are hypercholesterolemic, or those who are vegetarian to use the benefits of functional foods and specifically postbiotics. These products can be readily consumed by other groups, so they will have a wide market target. We need more in vitro and in vivo studies to address the growing international requirements for nondairy foods containing postbiotics.

The following research fields have been proposed for future studies:

- Developing products for consumers with certain dietary restrictions such as milk allergies, low cholesterol or fat content, diabetes, phenylketonuria, and lactose intolerance.
- Proposing novel products for specialized market segments such as children and chronically ill patients.
- Working on novel technologies for production of foods containing postbiotics.
- Evaluating the viability of postbiotics in different environmental conditions, in final product, and in gastrointestinal tube.
- Assessing Organoleptic and functional characteristics of the final product.
- Using encapsulation techniques for targeted delivery of postbiotics.
- Assurance of safety and effectivity of the final products.
- Developing international standards for products containing postbiotics.

The general plan for essential fields of future studies has been demonstrated in Fig. 15.1.

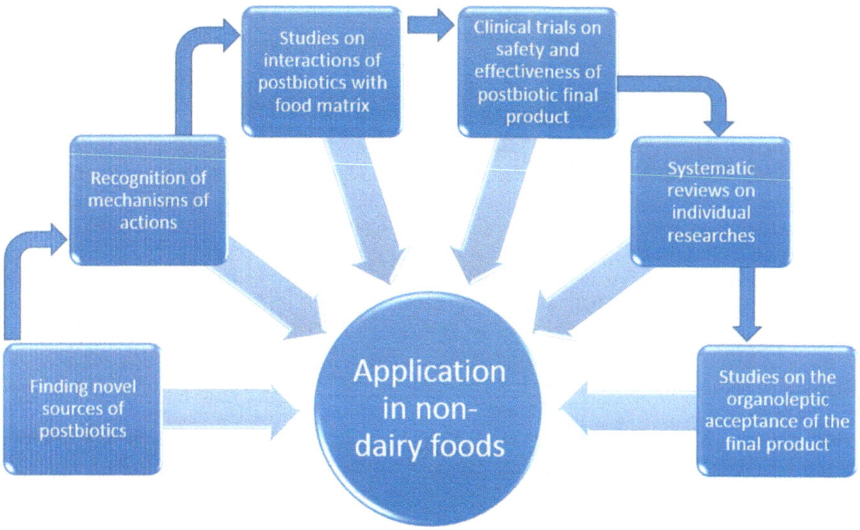

Fig. 15.1 Future perspectives of researches on the postbiotics in none dairy foods

References

Abdolhosseinzadeh E, Dehnad AR, Pourjafar H, Homayouni A, Ansari F (2018) The production of probiotic scallion yogurt: viability of lactobacillus acidoplilus freely and microencapsulated in the product. Carpathian J Food Sci Technol 10

Aguilar-Toalá J, Garcia-Varela R, Garcia H, Mata-Haro V, González-Córdova A, Vallejo-Cordoba-B, Hernández-Mendoza A (2018) Postbiotics: An evolving term within the functional foods field. Trends Food Sci Technol 75:105–114

Alizadeh A, Oskuyi AS, Amjadi S (2019) The optimization of prebiotic sucrose-free mango nectar by response surface methodology: The effect of stevia and inulin on physicochemical and rheological properties. Food Sci Technol Int 25:243–251

Ansari F, Pourjafar H, Esmailpour S (2017) Study on citric acid production and antibacterial activity of kombucha green tea beverage during production and storage. Annual Res Rev Biol 16:1–8

Ansari F, Pourjafar H, Kangari A, Homayouni A (2019) Evaluation of the Glucuronic Acid Production and Antibacterial Properties of Kombucha Black Tea. Curr Pharm Biotechnol 20:985–990

Ansari F, Pourjafar H, Bahadori MB, Pimentel TC (2020a) Effect of microencapsulation on the development of antioxidant activity and viability of lactobacillus acidophilus la5 in whey drink during fermentation. Biointerf Res Appl Chem 11:9762–9771

Ansari F, Pourjafar H, Tabrizi A, Homayouni A (2020b) The effects of probiotics and prebiotics on mental disorders: a review on depression, anxiety, Alzheimer, and autism spectrum disorders. Curr Pharm Biotechnol

Aspri M, Papademas P, Tsaltas D (2020) Review on non-dairy probiotics and their use in non-dairy based products. Fermentation 6:30

Bansal S, Mangal M, Sharma SK, Gupta RK (2016) Non-dairy based probiotics: a healthy treat for intestine. Crit Rev Food Sci Nutr 56:1856–1867

Bernal-Castro CA, Díaz-Moreno C, Gutiérrez-Cortés C (2019) Inclusion of prebiotics on the viability of a commercial Lactobacillus casei subsp. rhamnosus culture in a tropical fruit beverage. J Food Sci Technol 56:987–994

Bis-Souza C, Barba F, Lorenzo J, Penna AB, Barretto A (2019a) New strategies for the development of innovative fermented meat products: a review regarding the incorporation of probiotics and dietary fibers. Food Rev Intl 35:467–484

Bis-Souza CV, Pateiro M, Domínguez R, Lorenzo JM, Penna ALB, da Silva Barretto AC (2019b) Volatile profile of fermented sausages with commercial probiotic strains and fructooligosaccharides. J Food Sci Technol 56:5465–5473

Bis-Souza CV, Pateiro M, Domínguez R, Penna AL, Lorenzo JM, Barretto ACS (2020) Impact of fructooligosaccharides and probiotic strains on the quality parameters of low-fat Spanish Salchichón. Meat Sci 159:107936

Blaiotta G, Murru N, Di Cerbo A, Romano R, Aponte M (2018) Production of probiotic bovine salami using Lactobacillus plantarum 299v as adjunct. J Sci Food Agric 98:2285–2294

Blandino A, Al-Aseeri M, Pandiella S, Cantero D, Webb C (2003) Cereal-based fermented foods and beverages. Food Res Int 36:527–543

Bujna E, Farkas NA, Tran AM, Sao Dam M, Nguyen QD (2018) Lactic acid fermentation of apricot juice by mono-and mixed cultures of probiotic Lactobacillus and Bifidobacterium strains. Food Sci Biotechnol 27:547–554

Chakravorty S, Bhattacharya S, Chatzinotas A, Chakraborty W, Bhattacharya D, Gachhui R (2016) Kombucha tea fermentation: microbial and biochemical dynamics. Int J Food Microbiol 220:63–72

Charalampopoulos D, Pandiella S, Webb C (2002) Growth studies of potentially probiotic lactic acid bacteria in cereal-based substrates. J Appl Microbiol 92:851–859

Chavan M, Gat Y, Harmalkar M, Waghmare R (2018) Development of non-dairy fermented probiotic drink based on germinated and ungerminated cereals and legume. LWT 91:339–344

Cielecka-Piontek J, Dziedziński M, Szczepaniak O, Kobus-Cisowska J, Telichowska A, Szymanowska D (2020) Survival of commercial probiotic strains and their effect on dark chocolate synbiotic snack with raspberry content during the storage and after simulated digestion. Electron J Biotechnol 48:62–71

Coelho SR, Lima ÍA, Martins ML, Júnior AAB, de Almeida Torres Filho R, Ramos ADLS, Ramos EM (2019) Application of Lactobacillus paracasei LPC02 and lactulose as a potential symbiotic system in the manufacture of dry-fermented sausage. LWT 102:254–259

Coelho RMD, Almeida A, do Amaral RQG, da Mota RN, de Sousa PHM (2020) Kombucha. Int J Gastronomy Food Sci:100272

de Carvalho Marchesin J, Celiberto LS, Orlando AB, de Medeiros AI, Pinto RA, Zuanon JAS, Spolidorio LC, dos Santos A, Taranto MP, Cavallini DCU (2018) A soy-based probiotic drink modulates the microbiota and reduces body weight gain in diet-induced obese mice. J Funct Foods 48:302–313

de L Agüero N, Frizzo LS, Ouwehand AC, Aleu G, Rosmini MR (2020) Technological characterisation of probiotic lactic acid bacteria as starter cultures for dry fermented sausages. Foods:9–596

de Oliveira Ribeiro AP, dos Santos Gomes F, dos Santos KMO, da Matta VM, de Araujo Santiago MCP, Conte C, de Oliveira Costa SD, de Oliveira Ribeiro L, de Oliveira Godoy RL, Walter EHM (2020) Development of a probiotic non-fermented blend beverage with juçara fruit: effect of the matrix on probiotic viability and survival to the gastrointestinal tract. LWT 118:108756

De Vuyst L, Falony G, Leroy F (2008) Probiotics in fermented sausages. Meat Sci 80:75–78

Devanthi PVP, Linforth R, Onyeaka H, Gkatzionis K (2018) Effects of co-inoculation and sequential inoculation of Tetragenococcus halophilus and Zygosaccharomyces rouxii on soy sauce fermentation. Food Chem 240:1–8

Di Cagno R, Filannino P, Cantatore V, Polo A, Celano G, Martinovic A, Cavoski I, Gobbetti M (2020) Design of potential probiotic yeast starters tailored for making a cornelian cherry (Cornus mas L.) functional beverage. Int J Food Microbiol:108591

DiRienzo DB (2014) Effect of probiotics on biomarkers of cardiovascular disease: implications for heart-healthy diets. Nutr Rev 72:18–29

Dornblaser L (2007) Probiotics and prebiotics: What in the world is going on? Cereal Foods World 52:20

dos Santos Filho AL, Freitas HV, Rodrigues S, Abreu VKG, de Oliveira Lemos T, Gomes WF, Narain N, Pereira ALF (2019) Production and stability of probiotic cocoa juice with sucralose as sugar substitute during refrigerated storage. LWT 99:371–378

Dzandu BA (2019) In vitro and in vivo (mouse model) investigation of the efficacy of probiotic bacteria Lactobacillus Casei and Lactobacillus Rhamnosus incorporated in tomato juice

Fernandez MA, Marette A (2017) Potential health benefits of combining yogurt and fruits based on their probiotic and prebiotic properties. Adv Nutr 8:155S–164S

Fonteles TV, Costa MGM, de Jesus ALT, Fontes CPML, Fernandes FAN, Rodrigues S (2013) Stability and quality parameters of probiotic cantaloupe melon juice produced with sonicated juice. Food Bioproc Tech 6:2860–2869

Freire AL, Ramos CL, Schwan RF (2017) Effect of symbiotic interaction between a fructooligosaccharide and probiotic on the kinetic fermentation and chemical profile of maize blended rice beverages. Food Res Int 100:698–707

Gao Y, Hamid N, Gutierrez-Maddox N, Kantono K, Kitundu E (2019) Development of a probiotic beverage using breadfruit flour as a substrate. Foods 8:214

Granato D, Barba FJ, Bursać Kovačević D, Lorenzo JM, Cruz AG, Putnik P (2020) Functional foods: product development, technological trends, efficacy testing, and safety. Annu Rev Food Sci Technol 11:93–118

Gupta M, Bajaj BK (2017) Development of fermented oat flour beverage as a potential probiotic vehicle. Food Biosci 20:104–109

Hartmann HA, Wilke T, Erdmann R (2011) Efficacy of bacteriocin-containing cell-free culture supernatants from lactic acid bacteria to control Listeria monocytogenes in food. Int J Food Microbiol 146:192–199

Heperkan D, Daskaya-Dikmen C, Bayram B (2014) Evaluation of lactic acid bacterial strains of boza for their exopolysaccharide and enzyme production as a potential adjunct culture. Process Biochem 49:1587–1594

Holck AL, Axelsson L, Rode TM, Høy M, Måge I, Alvseike O, Trine M, Omer MK, Granum PE, Heir E (2011) Reduction of verotoxigenic Escherichia coli in production of fermented sausages. Meat Sci 89:286–295

Homayouni Rad A, Samadi Kafil H, Fathi Zavoshti H, Shahbazi N, Abbasi A (2020) Therapeutically effects of functional postbiotic foods. Clin Excell 10:33–52

Homayouni A, Javadi M, Ansari F, Pourjafar H, Jafarzadeh M, Barzegar A (2018) Advanced methods in ice cream analysis: a review. Food Anal Methods 11:3224–3234

Homayouni A, Ansari F, Azizi A, Pourjafar H, Madadi M (2020a) Cheese as a potential food carrier to deliver probiotic microorganisms into the human gut: a review. Curr Nutr Food Sci 16:15–28

Homayouni A, Mokarram RR, Norouzi S, Dehnad A, Barkhordari A, Homayouni H, Pourjafar H (2020b) Soy ice cream as a carrier for efficient delivering of Lactobacillus casei. Nutr Food Sci

Kazakos S, Mantzourani I, Plessas S (2020) Assessment of pomegranate juice as an alternative "substrate" for probiotic delivery. Recent advances and prospects. Fermentation 6:24

Kedia G, Vázquez JA, Charalampopoulos D, Pandiella SS (2009) In vitro fermentation of oat bran obtained by debranning with a mixed culture of human fecal bacteria. Curr Microbiol 58:338–342

Kemsawasd V, Chaikham P, Rattanasena P (2016) Survival of immobilized probiotics in chocolate during storage and with an in vitro gastrointestinal model. Food Biosci 16:37–43

Khosravi AR, Ghorbani-Choboghlo H, Nikaein D, Pourjafar H (2021) Probiotic efficacy of microencapsulated Saccharomyces cerevisiae on gastrointestinal tract integrity in rats. Biointerf Res Appl Chem 11:9456–9466

Klindt-Toldam S, Larsen SK, Saaby L, Olsen LR, Svenstrup G, Müllertz A, Knøchel S, Heimdal H, Nielsen DS, Zielińska D (2016) Survival of Lactobacillus acidophilus NCFM® and

Bifidobacterium lactis HN019 encapsulated in chocolate during in vitro simulated passage of the upper gastrointestinal tract. LWT 74:404–410

Klingberg TD, Budde BB (2006) The survival and persistence in the human gastrointestinal tract of five potential probiotic lactobacilli consumed as freeze-dried cultures or as probiotic sausage. Int J Food Microbiol 109:157–159

Kocková M, Dilongová M, Hybenovás E (2013) Evaluation of cereals and pseudocereals suitability for the development of new probiotic foods. J Chem 2013:414303

Kołożyn-Krajewska D, Dolatowski ZJ (2009) Probiotics in fermented meat products. Acta Sci Pol Technol Aliment 8:61–74

Kołożyn-Krajewska D, Dolatowski ZJ (2012) Probiotic meat products and human nutrition. Process Biochem 47:1761–1772

Konar N, Toker OS, Oba S, Sagdic O (2016) Improving functionality of chocolate: a review on probiotic, prebiotic, and/or synbiotic characteristics. Trends Food Sci Technol 49:35–44

Kumar BV, Vijayendra SVN, Reddy OVS (2015) Trends in dairy and non-dairy probiotic products-a review. J Food Sci Technol 52:6112–6124

Kun S, Rezessy-Szabó JM, Nguyen QD, Hoschke Á (2008) Changes of microbial population and some components in carrot juice during fermentation with selected Bifidobacterium strains. Process Biochem 43:816–821

Lamsal B, Faubion J (2009) The beneficial use of cereal and cereal components in probiotic foods. Food Rev Intl 25:103–114

Lebaka VR, Wee YJ, Narala VR, Joshi VK (2018) Development of new probiotic foods—A case study on probiotic juices. In: Therapeutic, probiotic, and unconventional foods. Elsevier, Amsterdam, pp 55–78

Leboš-Pavunc A, Penava L, Ranilović J, Novak J, Banić M, Butorac K, Petrović E, Mihaljević-Herman V, Bendelja K, Savić-Mlakar A (2019) Influence of dehydrated wheat/rice cereal matrices on probiotic activity of bifidobacterium animalis ssp. lactis BB-12®. Food Technol Biotechnol 57:147

Li Z, Teng J, Lyu Y, Hu X, Zhao Y, Wang M (2019) Enhanced antioxidant activity for apple juice fermented with Lactobacillus plantarum ATCC14917. Molecules 24:51

Liu Y, Chen H, Chen W, Zhong Q, Zhang G, Chen W (2018) Beneficial effects of tomato juice fermented by Lactobacillus plantarum and Lactobacillus casei: antioxidation, antimicrobial effect, and volatile profiles. Molecules 23:2366

Lu Y, Tan CW, Chen D, Liu SQ (2018) Potential of three probiotic lactobacilli in transforming star fruit juice into functional beverages. Food Sci Nutr 6:2141–2150

Maldonado RR, da Costa Araújo L, da Silva Dariva LC, Rebac KN, de Souza Pinto IA, Prado JPR, Saeki JK, Silva TS, Takematsu EK, Tiene NV (2017) Potential application of four types of tropical fruits in lactic fermentation. LWT 86:254–260

Mantzourani I, Nouska C, Terpou A, Alexopoulos A, Bezirtzoglou E, Panayiotidis MI, Galanis A, Plessas S (2018) Production of a novel functional fruit beverage consisting of cornelian cherry juice and probiotic bacteria. Antioxidants 7:163

Mantzourani I, Kazakos S, Terpou A, Alexopoulos A, Bezirtzoglou E, Bekatorou A, Plessas S (2019) Potential of the probiotic lactobacillus plantarum ATCC 14917 strain to produce functional fermented pomegranate juice. Foods 8:4

Marco ML, Tachon S (2013) Environmental factors influencing the efficacy of probiotic bacteria. Curr Opin Biotechnol 24:207–213

Martins EMF, Ramos AM, Vanzela ESL, Stringheta PC, de Oliveira Pinto CL, Martins JM (2013) Products of vegetable origin: a new alternative for the consumption of probiotic bacteria. Food Res Int 51:764–770

Menezes AGT, Ramos CL, Dias DR, Schwan RF (2018) Combination of probiotic yeast and lactic acid bacteria as starter culture to produce maize-based beverages. Food Res Int 111:187–197

Min M, Bunt CR, Mason SL, Hussain MA (2019) Non-dairy probiotic food products: an emerging group of functional foods. Crit Rev Food Sci Nutr 59:2626–2641

Miranda RF, da Silva JP, Machado ARF, da Silva EC, de Souza RC, Marcolino VA, Klososki SJ, Pimentel TC, Barão CE (2019) Impact of the addition of Lactobacillus casei and oligofructose on the quality parameters of orange juice and hibiscus tea mixed beverage. J Food Process Preserv 43:e14249

Mirković M, Seratlić S, Kilcawley K, Mannion D, Mirković N, Radulović Z (2018) The sensory quality and volatile profile of dark chocolate enriched with encapsulated probiotic Lactobacillus plantarum bacteria. Sensors 18:2570

Mirzaei H, Pourjafar H, Homayouni Rad A (2011) The effect of microencapsulation with calcium alginate and resistant starch on the Lactobacillus acidophilus (La5) survival rate in simulated gastrointestinal juice conditions. J Vet Res 66:337–342

Molin G (2001) Probiotics in foods not containing milk or milk constituents, with special reference to Lactobacillus plantarum 299v. Am J Clin Nutr 73:380s–385s

Moradi M, Kousheh SA, Almasi H, Alizadeh A, Guimarães JT, Yılmaz N, Lotfi A (2020) Postbiotics produced by lactic acid bacteria: the next frontier in food safety. Compr Rev Food Sci Food Saf 19:3390–3415

Moraes Filho ML, Busanello M, Garcia S (2019) Probiotic creamy soy sauce with Lactobacillus plantarum BG 112. Br Food J

Mugula J, Nnko S, Narvhus J, Sørhaug T (2003) Microbiological and fermentation characteristics of togwa, a Tanzanian fermented food. Int J Food Microbiol 80:187–199

Muyanja C, Narvhus JA, Treimo J, Langsrud T (2003) Isolation, characterisation and identification of lactic acid bacteria from bushera: a Ugandan traditional fermented beverage. Int J Food Microbiol 80:201–210

Niamah AK, Sahi AA, Al-Sharifi AS (2017) Effect of feeding soy milk fermented by probiotic bacteria on some blood criteria and weight of experimental animals. Probiotics Antimicrob Proteins 9:284–291

Ningrum DR, Budiwati TA, Nissa RC, Mawarda PC (2019) The production of freeze-dried probiotic from nappa cabbage juice by lactic acid bacteria. Teknologi Indonesia 41:33–41

Norouzi S, Pourjafar H, Ansari F, Homayouni A (2019) A survey on the survival of Lactobacillus paracasei in fermented and non-fermented frozen soy dessert. Biocatal Agric Biotechnol 21:101297

Ogunremi OR, Agrawal R, Sanni A (2020) Production and characterization of volatile compounds and phytase from potentially probiotic yeasts isolated from traditional fermented cereal foods in Nigeria. J Genet Eng Biotechnol 18:1–8

Ordóñez JA, Hierro EM, Bruna JM, Hoz LDL (1999) Changes in the components of dry-fermented sausages during ripening. Crit Rev Food Sci Nutr 39:329–367

Panghal A, Janghu S, Virkar K, Gat Y, Kumar V, Chhikara N (2018) Potential non-dairy probiotic products–A healthy approach. Food Biosci 21:80–89

Papamanoli E, Tzanetakis N, Litopoulou-Tzanetaki E, Kotzekidou P (2003) Characterization of lactic acid bacteria isolated from a Greek dry-fermented sausage in respect of their technological and probiotic properties. Meat Sci 65:859–867

Patel A (2017) Probiotic fruit and vegetable juices-recent advances and future perspective. Int Food Res J 24:1850–1857

Peng W, Meng D, Yue T, Wang Z, Gao Z (2020) Effect of the apple cultivar on cloudy apple juice fermented by a mixture of Lactobacillus acidophilus, Lactobacillus plantarum, and Lactobacillus fermentum. Food Chem 340:127922

Pereira ALF, Rodrigues S (2018) Turning fruit juice into probiotic beverages, fruit juices. Elsevier, Amsterdam, pp 279–287

Pereira ALF, Maciel TC, Rodrigues S (2011) Probiotic beverage from cashew apple juice fermented with Lactobacillus casei. Food Res Int 44:1276–1283

Pimentel TC, Madrona GS, Prudencio SH (2015) Probiotic clarified apple juice with oligofructose or sucralose as sugar substitutes: sensory profile and acceptability. LWT Food Sci Technol 62:838–846

Pourjafar H, Noori N, Gandomi H, Basti AA, Ansari F (2018) Stability and efficiency of double-coated beads containing lactobacillus acidophilus obtained from the calcium alginate-chitosan and eudragit S100 nanoparticles microencapsulation. Int J Probiotics Prebiotics 13

Pourjafar H, Noori N, Gandomi H, Basti AA, Ansari F (2020) Viability of microencapsulated and non-microencapsulated Lactobacilli in a commercial beverage. Biotechnolo Rep 25:e00432

Prado FC, Lindner JDD, Inaba J, Thomaz-Soccol V, Brar SK, Soccol CR (2015) Development and evaluation of a fermented coconut water beverage with potential health benefits. J Funct Foods 12:489–497

Ranadheera CS, Evans C, Adams M, Baines S (2012) In vitro analysis of gastrointestinal tolerance and intestinal cell adhesion of probiotics in goat's milk ice cream and yogurt. Food Res Int 49:619–625

Ranadheera CS, Evans CA, Adams MC, Baines SK (2014) Effect of dairy probiotic combinations on in vitro gastrointestinal tolerance, intestinal epithelial cell adhesion and cytokine secretion. J Funct Foods 8:18–25

Rouhi M, Sohrabvandi S, Mortazavian A (2013) Probiotic fermented sausage: viability of probiotic microorganisms and sensory characteristics. Crit Rev Food Sci Nutr 53:331–348

Rubio R, Jofré A, Aymerich T, Guàrdia MD, Garriga M (2014) Nutritionally enhanced fermented sausages as a vehicle for potential probiotic lactobacilli delivery. Meat Sci 96:937–942

Setta MC, Matemu A, Mbega ER (2020) Potential of probiotics from fermented cereal-based beverages in improving health of poor people in Africa. J Food Sci Technol:1–12

Shah N (2000) Probiotic bacteria: selective enumeration and survival in dairy foods. J Dairy Sci 83:894–907

Sheehan VM, Ross P, Fitzgerald GF (2007) Assessing the acid tolerance and the technological robustness of probiotic cultures for fortification in fruit juices. Innov Food Sci Emerg Technol 8:279–284

Shilpa V, Subrota H, Deepika Y (2011) Biofunctionality of probiotic soy yoghurt. Food Nutr Sci 2:502–509

Shimakawa Y, Matsubara S, Yuki N, Ikeda M, Ishikawa F (2003) Evaluation of Bifidobacterium breve strain Yakult-fermented soymilk as a probiotic food. Int J Food Microbiol 81:131–136

Sirini N, Frizzo L, Aleu G, Soto L, Rosmini M (2020a) Use of probiotic microorganisms in the formulation of healthy meat products. Curr Opin Food Sci 38:141–146

Sirini N, Roldán A, Lucas-González R, Fernández-López J, Viuda-Martos M, Pérez-Álvarez J, Frizzo L, Rosmini M (2020b) Effect of chestnut flour and probiotic microorganism on the functionality of dry-cured meat sausages. LWT 134:110197

Slima SB, Ktari N, Triki M, Trabelsi I, Abdeslam A, Moussa H, Makni I, Herrero AM, Jiménez-Colmenero F, Ruiz-Capillas C (2018) Effects of probiotic strains, Lactobacillus plantarum TN8 and Pediococcus acidilactici, on microbiological and physico-chemical characteristics of beef sausages. LWT 92:195–203

Song M-Y, Van-Ba H, Park W-S, Yoo J-Y, Kang H-B, Kim J-H, Kang S-M, Kim B-M, Oh M-H, Ham J-S (2018) Quality characteristics of functional fermented sausages added with encapsulated probiotic Bifidobacterium longum KACC 91563. Korean J Food Sci Anim Resour 38:981

Suvarna S, Dsouza J, Ragavan ML, Das N (2018) Potential probiotic characterization and effect of encapsulation of probiotic yeast strains on survival in simulated gastrointestinal tract condition. Food Sci Biotechnol 27:745–753

Świeca M, Kordowska-Wiater M, Pytka M, Gawlik-Dziki U, Bochnak J, Złotek U, Baraniak B (2018) Lactobacillus plantarum 299V improves the microbiological quality of legume sprouts and effectively survives in these carriers during cold storage and in vitro digestion. PLoS One 13:e0207793

Szutowska J (2020) Functional properties of lactic acid bacteria in fermented fruit and vegetable juices: a systematic literature review. Eur Food Res Technol:1–16

Taghizadeh G, Jahadi M, Abbasi H (2018) Physicochemical properties of probiotic soy milk chocolate mousse during refrigerated storage. Appl Food Biotechnol 5:79–86

Tamime A, Saarela M, Sondergaard AK, Mistry V, Shah N (2005) Production and maintenance of viability of probiotic microorganisms in dairy products. Probiotic Dairy Products 3:39–63
Todorov S, Botes M, Guigas C, Schillinger U, Wiid I, Wachsman M, Holzapfel W, Dicks L (2008) Boza, a natural source of probiotic lactic acid bacteria. J Appl Microbiol 104:465–477
Tripathi MK, Giri SK (2014) Probiotic functional foods: survival of probiotics during processing and storage. J Funct Foods 9:225–241
Valero-Cases E, Roy NC, Frutos MJ, Anderson RC (2017) Influence of the fruit juice carriers on the ability of Lactobacillus plantarum DSM20205 to improve in vitro intestinal barrier integrity and its probiotic properties. J Agric Food Chem 65:5632–5638
Valero-Cases E, Cerdá-Bernad D, Pastor J-J, Frutos M-J (2020) Non-dairy fermented beverages as potential carriers to ensure probiotics, prebiotics, and bioactive compounds arrival to the gut and their health benefits. Nutrients 12:1666
Vasudha S, Mishra H (2013) Non dairy probiotic beverages. Int Food Res J 20
Vignolo G, Fontana C, Fadda S (2010) Semidry and dry fermented sausages. In: Handbook of meat processing. Wiley, Ames, IA, pp 379–398. https://onlinelibrary.wiley.com/doi/10.1002/9780813820897.ch22
Villarreal-Soto SA, Beaufort S, Bouajila J, Souchard JP, Taillandier P (2018) Understanding kombucha tea fermentation: a review. J Food Sci 83:580–588
Vivek K, Mishra S, Pradhan RC, Jayabalan R (2019) Effect of probiotification with Lactobacillus plantarum MCC 2974 on quality of Sohiong juice. LWT 108:55–60
Wójciak KM, Libera J, Stasiak DM, Kołożyn-Krajewska D (2017) Technological aspect of Lactobacillus acidophilus Bauer, Bifidobacterium animalis BB-12 and Lactobacillus rhamnosus LOCK900 use in dry-fermented pork neck and sausage. J Food Process Preserv 41:e12965
Yoon KY, Woodams EE, Hang YD (2005) Fermentation of beet juice by beneficial lactic acid bacteria. LWT Food Sci Technol 38:73–75
Yoon KY, Woodams EE, Hang YD (2006) Production of probiotic cabbage juice by lactic acid bacteria. Bioresour Technol 97:1427–1430
Younes M, Aggett P, Aguilar F, Crebelli R, Dusemund B, Filipič M, Frutos MJ, Galtier P, Gundert-Remy U, Kuhnle GG, Lambré C, Leblanc J-C, Lillegaard IT, Moldeus P, Mortensen A, Oskarsson A, Stankovic I, Waalkens-Berendsen I, Woutersen RA, Wright M, Herman L, Tobback P, Pizzo F, Smeraldi C, Tard A, Papaioannou A, Gott D (2017) Safety of nisin (E 234) as a food additive in the light of new toxicological data and the proposed extension of use. EFSA J 15:e05063
Zarić DB, Bulatović ML, Rakin MB, Krunić TŽ, Lončarević IS, Pajin BS (2016) Functional, rheological and sensory properties of probiotic milk chocolate produced in a ball mill. RSC Adv 6:13934–13941
Zhu W, Lyu F, Naumovski N, Ajlouni S, Ranadheera CS (2020) Functional efficacy of probiotic lactobacillus sanfranciscensis in apple, orange and tomato juices with special reference to storage stability and in vitro gastrointestinal survival. Beverages 6:13
Żółkiewicz J, Marzec A, Ruszczyński M, Feleszko W (2020) Postbiotics—A step beyond pre-and probiotics. Nutrients 12:2189